Limit Theory for Mixing Dependent Random Variables

Mathematics and Its Application

Volume 378

Limit Theory for Mixing Dependent Random Variables

by

Lin Zhengyan

and

Lu Chuanrong

Department of Mathematics,
Hangzhou University,
Hangzhou, The People's Republic of China

Science Press
New York/Beijing

Kluwer Academic Publishers
Dordrecht/Boston/London

A C.I.P. Catalogue record for this book is available form the Library of Congress

ISBN 978-90-481-4748-9

Kluwer Academic Publishers incorporates
the publishing programmes of
D. Reidel, Martinas Nijhoff, Dr. W. Junk and MTP Press.

Sold and distributed in the U.S.A. and Canada
by Kluwer Academic Publishers,
101 Philip Drive, Norwell, MA 02061, U.S.A.

Sold and distributed in the People's Republic of China
by Science Press, Beijing.

In all other countries, sold and distributed
by Kluwer Academic Publishers Group,
P.O. Box 322, 3300 AH Dordrecht, The Netherlands.

Softcover reprint of the hardcover 1st edition 1996

SERIES EDITOR'S PREFACE

'Et moi, ..., si j'avait su comment en revenir, je n'y serais point allé.'

JulesVerne

The series is divergent; therefore we may be able to do something with it.

O.Heaviside

One service methematics has rendered the human race. It has put common sense back where it belongs, on the topmost shelf next to the dusty canister labelled 'discarded non-secse'.

Eric T. Bell

Mathematics is a tool for thought. A highly necessary tool in a world where both feedback and nonlinearities abound. Similarly, all kinds of parts of mathematics serve as tools for other parts and for other sciences.

Applying a simple rewriting rule to the quote on the right above one finds such statements as: 'One service topology has rendered mathematical physics ...'; 'One service logic has rendered computer science ...'; 'One service category theory has rendered mathematics ...'. All arguable true. And all statements obtainable this way form part of the raison d'être of this series.

This series, *Mathematics and Its Applications,* started in 1977. Now that over one hundred volumes have appeared it seems opportune to reexamine its scope. At the time I wrote

"Growing specialization and diversification have brought a host of monographs and textbooks on increasingly specialized topics. However, the 'tree' of knowledge of mathematics and related fields does not grow only by putting forth new branches. It also happens, quite often in fact, that branches which were thought to be completely disparate are suddenly seen to be related. Further, the kind and level of sophistication of mathematics applied in various sciences has changed drastically in recent years: measure theory is used (non-trivially) in regional and theoretical economics; algebraic geometry interacts with physics; the Minkowsky lemma, coding theory and the structure of water meet one another in packing and covering theory; quantume fields, crystal defects anf mathematical programming profit from homotopy theory; Lie algebras are relevant to filtering; and prediction and electrial engineering can use Stein spaces. And in addition to this there are such new emerging subdisciplines as "experimental methematics', 'CFD', 'completely integrable systems', 'chaos, synergetics and large-scale order', which are almost impossible to fit into the existing classification schemes. They draw upon widely different sections of mathematics."

By and large, all this this still applies today. It is still true that at first sight mathematics seems rather fragmented and that to find, see, and exploit the deeper underlying interrelations more effort is needed and so are books that can help mathematicians and scientists do so. Accordingly MIA will continue to try to make such book available.

If anything, the description I gave in 1977 is now an understatement. To the examples of interaction areas one should add string theory where Riemann surfaces, algebraic geometry, modular functions, knots, quantum field theory, Kac-Moody algebras, monstrous moonshine (and more) all come together. And to the examples of things which can be usefully applied let me add the topic 'finite geometry'; a combination of words which sounds like it might not even exist, let alone be applicable. And yet it is being applied: to statistics via designs, to radar/sonar detection arrays (via finite projective planes), and to bus connections of VLSI chips (via difference sets). There seems to be no part of (so-called pure) mathematics that is not in immediate danger of being applied. And, accordingly, the applied mathematician needs to be aware of much more. Besides analysis and numerics, the traditional workhorses, he may need all kinds of combinatorics, algebra, probability, and so on.

In addition, the applied scientist needs to cope increasingly with the nonlinear world and the extra mathematical sophistication that this requires. For that is where the rewards are. Linear models are honest and a bit sad and depressing: proportional efforts and results. It is in the nonlinear world that infinitesimal inputs may result in macroscopic outputs (or vice versa). To appreciate what I sm hinting at; if electronics were linear we would have no fun with transistors and computers; we would have no TV; in fact you would not be reading these lines.

There is also no safety in ignoring such outlandish things as nonstandard analysis, superspace and anticommuting integration, p-adic and ultrametric space. All three have applications in both electrical engineering and physics. Once, complex numbers were equally outlandish, but they frequently proved the shortest path between 'real' results. Similarly, the first two topics named have already provided a number of 'wormhole' paths. There is no telling where all this is leading-fortunately.

Thus the original scope of the series, which for various (sound) reasons now comprises five subseries: white (Japan), yellow (China), red (USSR), blue (Eastern Europe), and green (everything else), still applies. It has been enlarged a bit to include book treating of the tools from one subdiscipline which are used in others. Thus the series still aims at books dealing with:

a central concept which plays an improtant role in several different mathematical and/or scientific specialization areas;

New applications of the results and ideas from one area of scientific endeavour into another;

influences which the results, problems and concepts of one field of enquiry have, and have had, on the development of another.

The present volume, one of the first in the 'Chinese subseries' of MIA, also appropriately enough, one dealing with fundamental issues: interrelations between logic and computer science. The advent of computers has sparked off revived interest in a host of fundamental issues in science and mathematics such as computability, recursiveness, computational complexity and automated theorem proving to which latter topic ths author has made seminal contributions for which he was awarded the ATP prize in 1982.

It is a pleasure to welcome this volume in this series.

The shortest path between two truths in the real domain passes through the complex domain

J. Hadamard

La physique ne nous donne pas seulement ;'occasion de rèsoudr des problèmes ... elle nous fait presentir la solution.

H. Poincaré

Never lend books, for no one ever returns them; the only books I have in my library are books that other folk have lent me.

Anatole France

The function of an expert is not to be more right than other people, but to be wrong for more sophisticated reasons.

David Butler

Bussum, August 1989

Michiel Hazewinkel

Preface

The classical limit theorems of probability theory for independent random variables had been developed successfully in the thirties and forties. The basic results were summed up in Gnedenko and Kolmogorov's monograph "*Limit Distributions for Sums of Indenpendent Random Variables*" (1954) and Petrov's monograph "*Sums of Independent Random Variables*" (1975) . The modern limit theorems of probability theory, such as weak convergence of probability measures and strong approximations etc, have been studied by many authors since the fifties. The limit theory for weakly dependent random variables was also discussed deeply. In fact, the limit distributions of sums for non-independent random variables were studied early by some probabilitists and statisticians, such as Bernstein (1927), Hopf (1937), Hoeffding and Robbins (1948), etc. The dependence of random variables as a concept is developed not only in some branches of probability theory and mathematical statistics, such as Markov chains, random field theory and time series analysis, etc, but also appears in many practical problems. Although the assumption of independence is reasonable sometimes, it is difficult to check the independence of a sample. Moreover in many practical problems, the samples are not independent observations.

The definition of strong mixing (α-mixing) was first introduced by Rosenblatt (1956). Ibragimov (1959), Rozanov and Volconski (1959) also introduced this concept independently at the same time as they introduced the definition of φ-mixing. The definition of ρ-mixing was introduced by Kolmogorov and Rozanov (1960). All these concepts describe the asymptotic independence of random variables when the difference of their indices goes to infinity. The 1971 monograph by Ibragimov and Linnik, "*Independent and Stationary Sequence of Random Variables*" summed up main results of convergence in distribution for a mixing sequence up to the sixties. Since the theory of weak convergence of probability measures appears, particularly, following the monograph by Billingsley, "*Convergence of Probability Measures*" (1968), the weak convergence for a sequence of mixing random variables attracts the attentions of many authors and some ideal results have been obtained. The theory of strong approximations for a sequence of dependent random variables is discussed systematically in Philipp and Stout's monograph "*Almost Sure Invariance Principle for Partial Sums of Weakly Dependent Random Variables*" (1975). The results of this monograph have been improved comprehensively by us.

The modern limit theory for a sequence of mixing random variables has been studied deeply by many authors. This book will introduce them comprehensively, including Z. Y. Lin, C. R. Lu and Q. M. Shao's work for the weak convergence and strong approximations.

The book consists of four parts. The first part contains two chapters. We shall introduce the definitions of various mixing sequences and give a series of inequalities for mixing random variables, some of them are due to Shao. These inequalities are indispensable tools for the proofs of various limit theorems.

In the second part, which is separated into five chapters, the weak convergence, Berry-Esseen inequality and the rate of weak convergence are discussed. Some ideal results, such as weak convergence of ρ-mixing sequences, will be introduced.

In the third part, the almost sure convergence and strong approximations for the mixing random variables are studied. There are four chapters in this part. Some best results will be presented, such as strong approximations of partial sums for mixing sequences, which are done by Shao and Lu; the limiting behaviour of the increments of partial sums for a mixing sequence is obtained by Lin, et al.

In the fourth part, the weak convergence and strong approximations for some statistics with mixing dependent samples and some other kinds of dependent random variables are studied. Most results are profound.

Our best thanks are due to Dr. Q. M. Shao, whose results enrich greatly the book. We also want to thank all colleagues who help us to complete the book. We express our most gratitude to National Science Foundation of China and Zhejiang Province for their financial supports as well.

Lin, Z. Y.

Lu, C. R.

Hangzhou University, May 1996

Contents

Part I Introduction

In this part, we introduce some common and important definitions of weakly dependent random variables, establish some bounds of covariances for the various mixing sequences, and also discuss the relations between each other for different definitions. These will be given in Chapter 1.

In Chapter 2, we give the estimations of some kind of moments of partial sums of a mixing sequence, which play important roles in the limit theorems and will be used often.

Chapter 1 Definitions and Basic Inequalities

In this book, we always assume that $\{X_n, n \geq 1\}$ is a sequence of random variables defined on a probability space $(\Omega, \mathcal{F}, P)$. There are many ways to describe weak dependence or asymptotic independence of $\{X_n\}$. In Section 1.1, we give some common and important definitions of this kind. In Section 1.2, some basic inequalities on covariances of $\{X_n\}$ are established, which are useful for studying limit properties of $\{X_n\}$. In these sections, we also discuss the relations between each other for different definitions.

1.1 Definitions

Let $\mathcal{A}$ and $\mathcal{B}$ be sub-σ-fields of $\mathcal{F}, L_p(\mathcal{A})$ a set of all $\mathcal{A}$-measurable random variables with p-th moments. Define

$$\begin{aligned}
\alpha(\mathcal{A},\mathcal{B}) &= \sup_{A\in\mathcal{A},B\in\mathcal{B}} |P(AB)-P(A)P(B)|,\\
\rho(\mathcal{A},\mathcal{B}) &= \sup_{X\in L_2(\mathcal{A}),Y\in L_2(\mathcal{B})} \frac{|EXY-EXEY|}{\sqrt{\mathrm{Var}X\,\mathrm{Var}Y}},\\
\varphi(\mathcal{A},\mathcal{B}) &= \sup_{A\in\mathcal{A},B\in\mathcal{B},P(A)>0} |P(B|A)-P(B)|,\\
\psi(\mathcal{A},\mathcal{B}) &= \sup_{A\in\mathcal{A},B\in\mathcal{B},P(A)P(B)>0} \frac{|P(AB)-P(A)P(B)|}{P(A)P(B)},\\
\beta(\mathcal{A},\mathcal{B}) &= E(\mathrm{tvar}_{B\in\mathcal{B}}|P(B|\mathcal{A})-P(B)|),\\
\lambda(\mathcal{A},\mathcal{B}) &= \sup_{X\in L_{1/\alpha}(\mathcal{A}),Y\in L_{1/\beta}(\mathcal{B})} \frac{|EXY-EXEY|}{\|X\|_{1/\alpha}\|Y\|_{1/\beta}},
\end{aligned}$$

where tvar means total variation and $\|X\|_p = (E|X|^p)^{1/p}$. Let $\mathcal{F}_a^b = \sigma(X_i, a \leq i \leq b)$, $\mathbb{Z}$ a set of all integers, $\mathbb{Z}^+$ a set of all non-negative integers, $\mathbb{N}$ a set of all positive integers. Some commom and important definitions of mixing sequences are as follows:

Definition 1.1.1. A sequence $\{X_n, n \geq 1\}$ is said to be *α-mixing* or *strong mixing* if

$$\alpha(n) = \sup_{k \in \mathbb{N}} \alpha(\mathcal{F}_1^k, \mathcal{F}_{k+n}^\infty) \to 0 \quad \text{as } n \to \infty.$$

Definition 1.1.2. A sequence $\{X_n, n \geq 1\}$ is said to be *ρ-mixing* if

$$\rho(n) = \sup_{k \in \mathbb{N}} \rho(\mathcal{F}_1^k, \mathcal{F}_{k+n}^\infty) \to 0 \quad \text{as } n \to \infty.$$

Definition 1.1.3. A sequence $\{X_n, n \geq 1\}$ is said to be *φ-mixing* or *uniformly strong mixing* if

$$\varphi(n) = \sup_{k \in \mathbb{N}} \varphi(\mathcal{F}_1^k, \mathcal{F}_{k+n}^\infty) \to 0 \quad \text{as } n \to \infty.$$

Definition 1.1.4. A sequence $\{X_n, n \geq 1\}$ is said to be *ψ-mixing* or *$*$-mixing* if

$$\psi(n) = \sup_{k \in \mathbb{N}} \psi(\mathcal{F}_1^k, \mathcal{F}_{k+n}^\infty) \to 0 \quad \text{as } n \to \infty.$$

Definition 1.1.5. A sequence $\{X_n, n \geq 1\}$ is said to be *absolutely regular* if

$$\beta(n) = \sup_{k \in \mathbb{N}} \beta(\mathcal{F}_1^k, \mathcal{F}_{k+n}^\infty) \to 0 \quad \text{as } n \to \infty.$$

Definition 1.1.6. Let $0 \leq \alpha, \beta \leq 1, \alpha + \beta = 1$. A sequence $\{X_n, n \geq 1\}$ is said to be (α, β) *-mixing* if

$$\lambda(n) = \sup_{k \in \mathbb{N}} \lambda(\mathcal{F}_1^k, \mathcal{F}_{k+n}^\infty) \to 0 \quad \text{as } n \to \infty.$$

Remark 1.1.1. The versions of the above definitions for a sequence with time-parameter set R^+ or R or $\mathbb{Z}$ are trivial.

Remark 1.1.2. The concept of α-mixing was introduced by Rosenblatt (1956). The concept of ρ-mixing was introduced by Kolmogorov and Rozanov (1960). Dobrushin (1956) first introduced the definition of φ-mixing for a Markov process. This definition for a stationary process was presented by Ibragimov(1959) and Rozanov and Volconski (1959) respectively (one can also trace back to Hirschfeld 1935 and Gebelein 1941).

Absolute regularity was introduced by Kolmogorov (1959),(cf. Rosanov and Volconski 1959). Blum, Hanson and Koopmans (1963) presented the concept of ψ-mixing. (α, β)-mixing was introduced by Bradley (1985a) and Shao(1989a) independently.

Remark 1.1.3. Doob(1953) showed that a Döeblin irreducible Markov chain is φ-mixing with $\varphi(n) \le ab^n$ for some $a > 0$ and $0 \le b < 1$; Rosenblatt (1971) showed that a purely non-deterministic Markov chain is α-mixing; Davydov (1973) gave a class of Markov chains which are β-mixing.

Remark 1.1.4. For simplicity, we always assume that the mixing coefficients $\alpha(n), \rho(n), \cdots, \lambda(n)$ all are non-increasing.

It is clear from the definitions that

$$\rho(n) = \lambda_{1/2,1/2}(n), \quad \lambda_{1,0}(n) = \varphi(n) \le \psi(n),$$

and further

$$\alpha(n) \le \rho(n)$$

by taking $X = 1_A$ and $Y = 1_B$ in the definition of ρ-mixing.

Kolmogorov and Rozanov (1960) investigated the relation between α-mixing and ρ-mixing for a Gaussian sequence.

Theorem 1.1.1. *For a Gaussian sequence $\{X_n, n \ge 1\}$, we have*

$$\alpha(\mathcal{F}_1^k, \mathcal{F}_{k+n}^\infty) \le \rho(\mathcal{F}_1^k, \mathcal{F}_{k+n}^\infty) \le 2\pi\alpha(\mathcal{F}_1^k, \mathcal{F}_{k+n}^\infty).$$

Proof. The former inequality is obvious.

For any $\varepsilon > 0$, there exist two normal random variables $X \in L_2(\mathcal{F}_1^k), Y \in L_2(\mathcal{F}_{k+n}^\infty)$ such that $EX = EY = 0, \mathrm{Var}X = \mathrm{Var}Y = 1$ and

$$r := EXY \ge \rho(\mathcal{F}_1^k, \mathcal{F}_{k+n}^\infty) - \varepsilon.$$

Noting that $A := \{X > 0\} \in \mathcal{F}_1^k, \quad B := \{Y > 0\} \in \mathcal{F}_{k+n}^\infty$, we have

$$P(AB) = \frac{1}{4} + \frac{1}{2\pi}\arcsin r, \quad P(A)P(B) = \frac{1}{4} \tag{1.1.1}$$

by elementary calculations (see Cramér 1946, p.290). If $\alpha(\mathcal{F}_1^k, \mathcal{F}_{k+n}^\infty) > \frac{1}{4}$, it is clear that

$$2\pi\alpha(\mathcal{F}_1^k, \mathcal{F}_{k+n}^\infty) > \frac{\pi}{2} \ge \rho(\mathcal{F}_1^k, \mathcal{F}_{k+n}^\infty);$$

if $\alpha(\mathcal{F}_1^k, \mathcal{F}_{k+n}^\infty) \le \frac{1}{4}$, by (1.1.1) we obtain

$$\alpha(\mathcal{F}_1^k, \mathcal{F}_{k+n}^\infty) \ge P(AB) - P(A)P(B) = \frac{1}{2\pi}\arcsin r,$$

which implies

$$\rho(\mathcal{F}_1^k, \mathcal{F}_{k+n}^\infty) - \varepsilon \le r \le \sin 2\pi\alpha \le 2\pi\alpha.$$

The theorem is proved by arbitrariness of ε.

Kolmogorov and Rozanov (1960) also studied the relation between the spectral function of a (weakly) stationary sequence and ρ-mixing property. At first, we give some notations and concepts about a stationary sequence $\{X_n, n \in \mathbb{N}\}$. Let the covariance function of $\{X_n\}$

$$R(n) = EX_m X_{m+n}.$$

By the Herglotz theorem, there exists the spectral resolution for $R(n)$ as follows:

$$R(n) = \int_{-\pi}^{\pi} e^{in\lambda}\, dF(\lambda),$$

where $F(\lambda)$ is called the spectral function of the stationary sequence. When the spectral function is absolutely continuous, its derivative $f(\lambda) = F'(\lambda)$ is called the spectral density of the stationary sequence.

Theorem 1.1.2. *If the spectral function of a stationary sequence is not absolutely continuous, then $\rho(n) \equiv 1$, i.e. the sequence is not ρ-mixing. Conversely, if the spectral function is absolutely continuous, then*

$$\rho(n) = \inf_h \operatorname{ess\,sup}_\lambda \left| f(\lambda) - e^{i\lambda n} h(e^{-i\lambda}) \right| / f(\lambda),$$

where the inf is extended over all functions which is analytically continuable in unit circle; and further, if there exists an analytic function $h_0(z)$ in unit circle with the boundary value $h_0(e^{-i\lambda})$ such that $|f(\lambda)/h_0(e^{-i\lambda})| \ge \varepsilon > 0$ and $(f(\lambda)/h_0(e^{-i\lambda}))^{(k)}$ is bounded uniformly, then

$$\rho(n) \le cn^{-k}$$

for some $c > 0$. In particular, when $f(\lambda)$ is a rational function of $e^{i\lambda}$,

$$\rho(n) = e^{-cn}$$

for some $c > 0$.

The Proof of Theorem 1.1.2 is omitted(Kolmogorov, Rozanov 1960).

1.2 Basic inequalities

Let X be $\mathcal{F}_{-\infty}^{k}$ measurable and Y be $\mathcal{F}_{k+n}^{\infty}$ measurable.

In this section, we establish some bounds of the covariance $\mathrm{Cov}(X,Y) = EXY - EXEY$ for the various mixing sequences.

At first, we consider the α-mixing case.

Lemma 1.2.1. *Let $\{X_n, n \in \mathbb{Z}\}$ be an α-mixing sequence, $X \in \mathcal{F}_{-\infty}^{k}$ and $Y \in \mathcal{F}_{k+n}^{\infty}$ with $|X| \le C_1$ and $|Y| \le C_2$. Then*

$$|EXY - EXEY| \le 4C_1C_2\alpha(n). \tag{1.2.1}$$

Proof. By the property of conditional expectation, we have

$$\begin{aligned}|EXY - EXEY| &= \Big|E\{X(E(Y|\mathcal{F}_{-\infty}^{k}) - EY)\}\Big| \\ &\le C_1E|E(Y|\mathcal{F}_{-\infty}^{k}) - EY| \\ &= C_1|E\xi\{E(Y|\mathcal{F}_{-\infty}^{k}) - EY\}|,\end{aligned}$$

where $\xi = \mathrm{sgn}(E(Y|\mathcal{F}_{-\infty}^{k}) - EY) \in \mathcal{F}_{-\infty}^{k}$, i.e.

$$|EXY - EXEY| \le C_1|E\xi Y - E\xi EY|.$$

With the same argument procedure it follows that

$$|E\xi Y - E\xi EY| \le C_2|E\xi\eta - E\xi E\eta|,$$

where $\eta = \mathrm{sgn}\ (E(\xi|\mathcal{F}_{k+n}^{\infty}) - E\xi)$. Therefore

$$|EXY - EXEY| \le C_1C_2|E\xi\eta - E\xi E\eta|. \tag{1.2.2}$$

Put $A = \{\xi = 1\}, B = \{\eta = 1\}$. It is clear that $A \in \mathcal{F}_{-\infty}^{k}, B \in \mathcal{F}_{k+n}^{\infty}$. Using the definition of α-mixing, we obtain

$$\begin{aligned}&|E\xi\eta - E\xi E\eta| \\ &\quad = |P(AB) + P(A^cB^c) - P(A^cB) - P(AB^c) \\ &\qquad - (P(A) - P(A^c))(P(B) - P(B^c))| \\ &\quad \le 4\alpha(n).\end{aligned}$$

Inserting it into (1.2.2) yields (1.2.1).

Lemma 1.2.2. *Let $\{X_n, n \in \mathbb{Z}\}$ be an α-mixing sequence, $X \in \mathcal{F}_{-\infty}^k$ and $Y \in \mathcal{F}_{k+n}^{\infty}$ with $E|X|^p < \infty$ for some $p > 1$ and $|Y| \leq C$. Then*

$$|EXY - EXEY| \leq 6C\|X\|_p(\alpha(n))^{1/q}, \tag{1.2.3}$$

where $1/p + 1/q = 1$.

Proof. Let $X_N = XI(|X| \leq N)$, $X_N' = X - X_N$. Write

$$|EXY - EXEY| \leq |EX_NY - EX_NEY| + |EX_N'Y - EX_N'EY|.$$

By Lemma 1.2.1, $|EX_NY - EX_NEY| \leq 4CN\alpha(n)$. For the second term of the right hand side of the above inequality, we have

$$|EX_N'Y - EX_N'EY| \leq 2CE|X_N'| \leq 2CN^{-p+1}E|X|^p.$$

Taking $N = \|X\|_p(\alpha(n))^{-1/p}$ yields (1.2.3).

For a random variable X and a continuous non-decreasing function $f(x)$ on R^+ with $f(0) = 0$, which doesn't identically equal to zero, define

$$\|X\|_f = \inf\{t > 0, Ef(|X|/t) \leq 1\}.$$

From this definition, it is easy to know that

$$\|X\|_f = 0 \Longleftrightarrow X = 0 \quad \text{a.s.} \tag{1.2.4}$$

and if $0 < \|X\|_f < \infty$, then $Ef(|X|/\|X\|_f) \leq 1$. Moreover, if $|X_1| \leq |X_2|$ a.s., then $\|X_1\|_f \leq \|X_2\|_f$.

Lemma 1.2.3. *Let $\{X_n, n \in \mathbb{Z}\}$ be an α-mixing sequence, $X \in \mathcal{F}_{-\infty}^k$, $Y \in \mathcal{F}_{k+n}^{\infty}$, $f(x)$ and $g(x)$ be two continuous functions on R^+ with $f(0) = g(0) = 0$, $f(x)/x^{\frac{r+s}{r}} \nearrow \infty$ and $g(x)/x^{\frac{r+s}{s}} \nearrow \infty$ for some $r > 0, s > 0, \|X\|_f < \infty, \|Y\|_g < \infty$. Then*

$$|EXY - EXEY| \leq 10 \operatorname{inv} f\Big(\frac{1}{\alpha(n)}\Big) \operatorname{inv} g\Big(\frac{1}{\alpha(n)}\Big)\alpha(n)\|X\|_f\|Y\|_g. \tag{1.2.5}$$

Proof. It is easy to see that $E|X|^{1+s/r} < \infty$ and $E|Y|^{1+r/s} < \infty$ by the conditions of the lemma . If either $\|X\|_f = 0$ or $\|Y\|_g = 0$, (1.2.4) implies that (1.2.5) holds. If $\alpha(n) = 0$, (1.2.5) is trivial by independence of X and Y. Now we assume that $\|X\|_f > 0, \|Y\|_g > 0$ and $\alpha(n) > 0$. There are $M > 0$ and $N > 0$ such that

$$\alpha(n) = 1/f(M/\|X\|_f) = 1/g(N/\|Y\|_g).$$

Let

$$X_M = XI(|X| \le M), \quad X_M' = X - X_M,$$
$$Y_N = YI(|Y| \le N), \quad Y_N' = Y - Y_N.$$

We have

$$\begin{aligned} &|EXY - EXEY| \\ &\le |EX_MY_N - EX_MEY_N| + |EX_M'Y_N - EX_M'EY_N| \\ &\qquad + |EX_MY_N' - EX_MEY_N'| + |EX_M'Y_N' - EX_M'EY_N'| \\ &=: I_1 + I_2 + I_3 + I_4. \end{aligned} \tag{1.2.6}$$

By Lemma 1.2.1, $I_1 \le 4MN\alpha(n)$. Noting that $f(x)/x \nearrow \infty$ and $g(x)/x \nearrow \infty$, we have

$$\begin{aligned} E|X_M'| &= E(|X_M'|/\|X_M'\|_f)\|X_M'\|_f \\ &\le Ef(|X_M'|/\|X_M'\|_f)M/f(M/\|X_M'\|_f) \\ &\le M/f(M/\|X\|_f). \end{aligned}$$

Therefore

$$I_2 \le 2MN/f(M/\|X\|_f) = 2\,\mathrm{inv}\, f\Big(\frac{1}{\alpha(n)}\Big)\,\mathrm{inv}\, g\Big(\frac{1}{\alpha(n)}\Big)\alpha(n)\|X\|_f\|Y\|_g.$$

Similarly, we have the same estimation for I_3.

Furthermore, noting that $f(x)/x^{\frac{r+s}{r}} \nearrow \infty$ and $g(x)/x^{\frac{r+s}{s}} \nearrow \infty$, we have

$$\begin{aligned} EX_M'Y_N' &\le \Big(E(|X_M'|/\|X_M'\|_f)^{\frac{r+s}{r}}\Big)^{\frac{r}{r+s}} \\ &\quad \cdot \Big(E(|Y_N'|/\|Y_N'\|_g)^{\frac{r+s}{s}}\Big)^{\frac{s}{r+s}} \|X_M'\|_f\|Y_N'\|_g \\ &\le \Big(Ef(|X_M'|/\|X_M'\|_f)\Big)^{\frac{r}{r+s}}\Big(Eg(|Y_N'|/\|Y_N'\|_g)\Big)^{\frac{s}{r+s}} \\ &\quad \cdot MN/\Big(f(M/\|X_M'\|_f)\Big)^{\frac{r}{r+s}}\Big(g(N/\|Y_N'\|_g)\Big)^{\frac{s}{r+s}} \\ &\le MN/\Big(f(M/\|X\|_f)\Big)^{\frac{r}{r+s}}\Big(g(N/\|Y\|_g)\Big)^{\frac{s}{r+s}}. \end{aligned}$$

Hence

$$\begin{aligned} I_4 &\le 2MN/\Big(f(M/\|X\|_f)\Big)^{\frac{r}{r+s}}\Big(g(N/\|Y\|_g)\Big)^{\frac{s}{r+s}} \\ &\le 2\,\mathrm{inv}\, f\Big(\frac{1}{\alpha(n)}\Big)\,\mathrm{inv}\, g\Big(\frac{1}{\alpha(n)}\Big)\alpha(n)\|X\|_f\|Y\|_g. \end{aligned}$$

Now, inserting these estimations into (1.2.6) yields (1.2.5).

As some consequences of this lemma, we have

Lemma 1.2.4. *Let $\{X_n, n \in \mathbb{Z}\}$ be an α-mixing sequence, $X \in \mathcal{F}_{-\infty}^k$ and $Y \in \mathcal{F}_{k+n}^\infty$ with $E|X|^p < \infty$ and $E|Y|^q < \infty, \frac{1}{p} + \frac{1}{q} < 1$. Then*

$$|EXY - EXEY| \le 10\|X\|_p\|Y\|_q(\alpha(n))^{1-\frac{1}{p}-\frac{1}{q}}. \tag{1.2.7}$$

Lemma 1.2.5. *Let $\{X_n, n \in \mathbb{Z}\}$ be an α-mixing sequence, $X \in \mathcal{F}_{-\infty}^k$ and $Y \in \mathcal{F}_{k+n}^\infty$ with $E|X|^{2+\delta} \le C_1, E|Y|^{2+\delta} \le C_2$. Then*

$$|EXY - EXEY| \le 10(C_1C_2)^{\frac{1}{2+\delta}}(\alpha(n))^{\frac{\delta}{2+\delta}}. \tag{1.2.8}$$

For an (α, β)-mixing sequence and a ρ-mixing sequence, we have the following lemmas.

Lemma 1.2.6. *Let $\{X_n, n \in \mathbb{Z}\}$ be an (α, β)-mixing sequence, $X \in L_p(\mathcal{F}_{-\infty}^k)$ and $Y \in L_q(\mathcal{F}_{k+n}^\infty)$ with $p, q \ge 1$ and $1/p + 1/q = 1$. Then*

$$|EXY - EXEY| \le 4\lambda(n)^{\frac{1}{\alpha p} \wedge \frac{1}{\beta q}}\|X\|_p\|Y\|_q. \tag{1.2.9}$$

Proof. Without loss of generality, assume that $\alpha p \ge 1$, which implies that $\beta q \le 1$. Put

$$Y_1 = YI(|Y| \le C), \quad Y_2 = Y - Y_1,$$

where C is a positive constant specified later on. Write

$$|EXY - EXEY| \le |EXY_1 - EXEY_1| + |EXY_2 - EXEY_2|. \tag{1.2.10}$$

By the definition of (α, β)-mixing and the Hölder inequality

$$\begin{aligned}|EXY_1 - EXEY_1| &\le \lambda(n)\|X\|_{1/\alpha}\|Y_1\|_{1/\beta} \\ &\le \lambda(n)C^{1-\beta q}\|X\|_p\|Y\|_q^{\beta q},\end{aligned}$$

$$\begin{aligned}|EXY_2| &\le (E|Y_2|^q)^{1-\frac{1}{\alpha p}}(E|X|^{\alpha p}|Y_2|^{\beta q})^{\frac{1}{\alpha p}} \\ &\le (E|Y_2|^q)^{1-\frac{1}{\alpha p}}\left(E|X|^{\alpha p}E|Y_2|^{\beta q} + \lambda(n)(E|X|^p)^\alpha(|Y_2|^q)^\beta\right)^{\frac{1}{\alpha p}} \\ &\le (E|Y|^q)^{1-\frac{1}{\alpha p}}\left(E|X|^{\alpha p}E|Y|^q C^{-\alpha q} + \lambda(n)(E|X|^p)^\alpha(E|Y|^q)^\beta\right)^{\frac{1}{\alpha p}} \\ &\le \|X\|_p\|Y\|_q^q C^{-\frac{q}{p}} + \lambda^{\frac{1}{\alpha p}}(n)\|X\|_p\|Y\|_q\end{aligned}$$

and

$$|EXEY_2| \le \|X\|_p \|Y\|_q^q C^{-q/p}.$$

Inserting these estimations into (1.2.10) and taking $C = \|Y\|_q (\lambda(n))^{-1/\alpha q}$ we obtain (1.2.9).

Let $p = q = 2$ in (1.2.9). It is easy to see that

$$\rho(n) \le 4\lambda(n)^{\frac{1}{2\alpha} \wedge \frac{1}{2\beta}}. \tag{1.2.11}$$

As a consequence of Lemma 1.2.6, noting that $\rho(n) = \lambda_{1/2,1/2}(n)$, we have

Lemma 1.2.7. *Let $\{X_n, n \in \mathbb{Z}\}$ be a ρ-mixing sequence, $X \in L_p(\mathcal{F}_{-\infty}^k)$ and $Y \in L_q(\mathcal{F}_{k+n}^\infty)$ with $p, q \ge 1$ and $1/p + 1/q = 1$. Then*

$$|EXY - EXEY| \le 4\rho(n)^{\frac{2}{p} \wedge \frac{2}{q}} \|X\|_p \|Y\|_q.$$

For the φ-mixing case, we have the following three results.

Lemma 1.2.8. *Let $\{X_n, n \in \mathbb{Z}\}$ be a φ-mixing sequence, $X \in L_p(\mathcal{F}_{-\infty}^k)$ and $Y \in L_q(\mathcal{F}_{k+n}^\infty)$ with $p, q \ge 1$ and $1/p + 1/q = 1$. Then*

$$|EXY - EXEY| \le 2(\varphi(n))^{\frac{1}{p}} \|X\|_p \|Y\|_q. \tag{1.2.12}$$

Proof. At first, we assume that $X$and $Y$are simple functions, i.e.

$$X = \sum_i a_i I_{A_i}, \quad Y = \sum_j b_j I_{B_j},$$

where both $\sum_i$ and $\sum_j$ are finite sums and $A_i \cap A_k = \emptyset \quad (i \ne k)$, $B_j \cap B_l = \emptyset \quad (j \ne l)$, $A_i \in \mathcal{F}_{-\infty}^k$, $B_j \in \mathcal{F}_{k+n}^\infty$. So

$$EXY - EXEY = \sum_{i,j} a_i b_j P(A_i B_j) - \sum_{i,j} a_i b_j P(A_i) P(B_j).$$

By the Hölder inequality we have

$$
\begin{aligned}
&|EXY-EXEY| \\
&\quad = \Big|\sum_i a_i(P(A_i))^{1/p}\sum_j(P(B_j|A_i)-P(B_j))b_j(P(A_i))^{1/q}\Big| \\
&\quad \le \Big(\sum_i |a_i|^p P(A_i)\Big)^{1/p}\Big(\sum_i P(A_i)|\sum_j b_j(P(B_j|A_i)-P(B_j))|^q\Big)^{1/q} \\
&\quad \le \|X\|_p\Big|\sum_i P(A_i)\Big(\sum_j |b_j|^q(P(B_j|A_i) \\
&\qquad + P(B_j))\Big)\Big(\sum_j |P(B_j|A_i)-P(B_j)|\Big)^{\frac{q}{p}}\Big|^{\frac{1}{q}} \\
&\quad \le 2^{1/q}\|X\|_p\|Y\|_q \max_i\Big(\sum_j |P(B_j|A_i)-P(B_j)|\Big)^{1/p}. \qquad (1.2.13)
\end{aligned}
$$

Note that

$$
\begin{aligned}
\sum_j |P(B_j|A_i)-P(B_j)| &= (P(\cup_j^+ B_j|A_i)-P(\cup_j^+ B_j)) \\
&\quad - (P(\cup_j^- B_j|A_i)-P(\cup_j^- B_j)) \\
&\le 2\varphi(n), \qquad (1.2.14)
\end{aligned}
$$

where the union $\cup_j^+(\cup_j^-)$ is carried out over j such that $P(B_j|A_i)-P(B_j) > 0$ $(P(B_j|A_i)-P(B_j) < 0)$. Inserting (1.2.14) into (1.2.13) yields (1.2.12) for the simple function case.

In order to complete the proof of the lemma, let

$$
X_N = \begin{cases} 0 & \text{if } |X| > N. \\ k/N & \text{if } k/N < X \le (k+1)/N,\ |X| \le N; \end{cases}
$$

$$
Y_N = \begin{cases} 0 & \text{if } |Y| > N. \\ k/N & \text{if } k/N < Y \le (k+1)/N,\ |Y| \le N. \end{cases}
$$

We have showed that (1.2.12) is true for X_N and Y_N. Moreover, note

$$
E|X-X_N|^p \to 0,\ E|Y-Y_N|^q \to 0, \quad \text{as } N \to \infty.
$$

Letting $N \to \infty$, we obtain (1.2.12) for the general case.

Let $p = q = 2$ in (1.2.12). It is easy to see that

$$
\rho(n) \le 2\varphi^{1/2}(n). \qquad (1.2.15)
$$

From the proof of Lemma 1.2.8, we can see that

Lemma 1.2.9. *Let $\{X_n, n \in \mathbb{Z}\}$ be a φ-mixing sequence, $X \in \mathcal{F}_{-\infty}^k$ and $Y \in \mathcal{F}_{k+n}^{\infty}$ with $|X| \le C_1$ and $|Y| \le C_2$. Then*

$$|EXY - EXEY| \le 2C_1C_2\varphi(n). \tag{1.2.16}$$

Let $p = 1$ and $q = \infty$ in (1.2.12). From Lemma 1.2.8, we also have

Lemma 1.2.10. *Let $\{X_n, n \in \mathbb{Z}\}$ be a φ-mixing sequence, $X \in \mathcal{F}_{-\infty}^k$ and $Y \in \mathcal{F}_{k+n}^{\infty}$ with $E|X| < \infty$ and $|Y| \le C$. Then*

$$|EXY - EXEY| \le 2C\varphi(n)E|X|. \tag{1.2.17}$$

Finally, we consider the ψ-mixing case.

Lemma 1.2.11. *Let $\{X_n, n \in \mathbb{Z}\}$ be a ψ-mixing sequence, $X \in \mathcal{F}_{-\infty}^k$ and $Y \in \mathcal{F}_{k+n}^{\infty}$ with $E|X| < \infty$ and $E|Y| < \infty$. Then $E|XY| < \infty$ and*

$$|EXY - EXEY| \le \psi(n)E|X|E|Y|. \tag{1.2.18}$$

Proof. At first, we assume that X and Y are non-negative simple functions. We have

$$\begin{aligned}|EXY - EXEY| &= |\sum_{i,j} a_ib_j(P(A_iB_j) - P(A_i)P(B_j))| \\ &\le \sum_{i,j} a_ib_j\psi(n)P(A_i)P(B_j) \\ &= \psi(n)EXEY.\end{aligned}$$

From this, (1.2.18) holds for non-negative random variables X and Y.

For the general case, write $X = X^+ - X^-$, $Y = Y^+ - Y^-$. We have

$$\begin{aligned}&|EXY - EXEY| \\ &\quad \le |EX^+Y^+ - EX^+EY^+| + |EX^+Y^- - EX^+EY^-| \\ &\quad\ + |EX^-Y^+ - EX^-EY^+| + |EX^-Y^- - EX^-EY^-| \\ &\quad \le \psi(n)(EX^+ + EX^-)(EY^+ + EY^-) \\ &\quad \le \psi(n)E|X|E|Y|.\end{aligned}$$

Finally, we summarize the relations between one and another of variaous mixing properties. It is easy to verify that

$$2\alpha(n) \le \beta(n) \le \varphi(n). \tag{1.2.19}$$

With a necessary and sufficient condition for Markov processes to be ψ-mixing, one can show that a φ-mixing (Markov) sequence is not ψ-mixing (Blum, Hanson and Koopmans 1963). Ibragimov and Solev (1969) given an example of a stationary α-mixing Gaussian process which is not β-mixing; such a process is ρ-mixing but not β-mixing. Davydov (1973) constructed a stationary α-mixing Markov process with less than geometric rate of decay of the mixing coefficients, which is not ρ-mixing. It is possible that a geometrically ergodic Markov process which is not Doeblin recurrent is β-mixing and not φ-mixing (Andrews 1984). Combining these results and recalling Remark 1.1.4, (1.2.11) and (1.2.15) we have

$$\psi-\text{mixing}\left\{\begin{array}{c}\Longrightarrow\\ \not\Longleftarrow\end{array}\right\}\varphi-\text{mixing}\left\{\begin{array}{c}\Longrightarrow\\ \not\Longleftarrow\end{array}\right.\left\{\begin{array}{l}\beta-\text{mixing}\left\{\begin{array}{c}\Longrightarrow\\ \not\Longleftarrow\end{array}\right\}\alpha-\text{mixing}\\ \quad\not\Uparrow\quad\not\Downarrow\\ \rho-\text{mixing}\left\{\begin{array}{c}\Longrightarrow\\ \not\Longleftarrow\end{array}\right\}\alpha-\text{mixing}\end{array}\right.$$

$$\Uparrow$$

$$\lambda-\text{mixing}$$

Chapter 2 Moment Estimations of Partial Sums

The estimations of some kind of moments of partial sums of a mixing sequence play important roles in showing limit theorems. In Section 2.1, we give some forms of the variances of partial sums of mixing sequences of various kinds. Section 2.2 is devoted to deduce some inequalities for the moments of partial sums. In passing we also give some probability inequalities in this section.

2.1 Variances of partial sums

Let $\{X_n, n \in \mathbb{Z}\}$ be a (weakly) stationary sequence with $EX_1 = 0$, $EX_1^2 < \infty$. Put $S_n = \sum_{i=1}^n X_i$. We investigate its variance $\mathrm{Var}S_n$. Let $R(n)$ be the correlation function and $F(\lambda)$ be the spectral function of the sequence $\{X_n\}$.

At first, we give the representations of $\mathrm{Var}S_n$ by $R(n)$ or $F(\lambda)$.

Theorem 2.1.1.

$$\mathrm{Var}S_n = \sum_{|j|<n} (n - |j|)R(j), \tag{2.1.1}$$

$$\mathrm{Var}S_n = \int_{-\pi}^{\pi} \frac{\sin^2 \frac{n\lambda}{2}}{\sin^2 \frac{\lambda}{2}} dF(\lambda). \tag{2.1.2}$$

If the spectral function is absolutely continuous, i.e., there is a spectral density $f(\lambda)$, and further, if $f(\lambda)$ is continuous at $\lambda = 0$, then

$$\mathrm{Var}S_n = 2\pi f(0)n + o(n) \quad \text{as } n \to \infty. \tag{2.1.3}$$

The proof of the theorem can be found in the book of Ibragimov and Linnik (1971) and is not presented here.

When a stationary sequence satisfies a certain mixing condition, $\mathrm{Var}S_n$ possesses more evident form.

Theorem 2.1.2. *Suppose that a stationary sequence $\{X_n\}$ is φ-mixing and $\mathrm{Var}S_n \to \infty$ as $n \to \infty$. Then*

$$\mathrm{Var}S_n = nh(n), \tag{2.1.4}$$

where $h(n)$ is a slowly varying function of n and its domain of definition can be extended to R such that $h(x)$ is also slowly varying on R.

Proof. We first prove that $h(n)$ is slowly varying. Put $\sigma_n^2 = \mathrm{Var}S_n$. Equivalently, we show that for every positive integer k

$$\lim_{n\to\infty} \sigma_{kn}^2/\sigma_n^2 = k. \tag{2.1.5}$$

Let

$$\xi_j = \sum_{s=1}^{n} X_{(j-1)n+(j-1)r+s}, \quad j = 1, 2, \cdots, k,$$

$$\eta_j = \sum_{s=1}^{r} X_{jn+(j-1)r+s}, \quad j = 1, 2, \cdots, k-1,$$

$$\eta_k = -\sum_{s=1}^{(k-1)r} X_{nk+s},$$

where $r = [\log \sigma_n^2]$. By Theorem 1.1.2, $\{X_n\}$ possesses a spectral density $f(\lambda)$. Using (2.1.2) we obtain

$$\sigma_n^2 = \int_{-\pi}^{\pi} \frac{\left(\sin^2 \frac{n\lambda}{2}\right)}{\left(\sin^2 \frac{\lambda}{2}\right)} f(\lambda)\, d\lambda$$

$$\le n^2 \int_{-\pi}^{\pi} f(\lambda)\, d\lambda. \tag{2.1.6}$$

Hence $r = O(\log n)$. And further

$$\sigma_{kn}^2 = \mathrm{Var}S_{kn} = \sum_{j=1}^{k} E\xi_j^2 + 2\sum_{i\ne j} E\xi_i\xi_j + \sum_{i,j} E\xi_i\eta_j + \sum_{i,j} E\eta_i\eta_j. \tag{2.1.7}$$

By stationarity of the sequence, $E\xi_j^2 = \sigma_n^2 = \mathrm{Var}S_n$. From Lemma 1.2.8, we have

$$|E\xi_i\xi_j| \le 2\varphi(|i-j|r)^{1/2}\|\xi_i\|_2\|\xi_j\|_2 \le 2\varphi(r)^{1/2}\sigma_n^2 \tag{2.1.8}$$

for $i \neq j$. Using the Schwarz inequality and (1.2.6), we have

$$|E\xi_i\eta_j| \le \|\xi_i\|_2\|\eta_j\|_2 = \sigma_n\sigma_r = O(\sigma_n \log \sigma_n), \tag{2.1.9}$$

$$|E\eta_i\eta_j| \le \sigma_r^2 = O((\log \sigma_n)^2). \tag{2.1.10}$$

Inserting (2.1.8), (2.1.9) and (2.1.10) into (2.1.7) and noting $\varphi(r) = o(1)$ as $n \to \infty$, we obtain

$$\sigma_{kn}^2 = k\sigma_n^2 + o(\sigma_n^2),$$

which implies (2.1.5).

Next, we prove that the domain of $h(n)$ can be extended to R such that $h(x)$ is also slowly varying on R. Recalling (2.1.6), we define

$$\psi(x) = \int_{-\pi}^{\pi} \frac{\sin^2 \frac{x\lambda}{2}}{\sin^2 \frac{\lambda}{2}} f(\lambda)\, d\lambda,$$

$$h(x) = \psi(x)/x.$$

In order to show that $h(x)$ is slowly varying, it is enough to verify that for any $a > 0$

$$\lim_{x\to\infty} \psi(ax)/\psi(x) = a. \tag{2.1.11}$$

It is not difficult to know from the definition of ψ that

$$\psi(x) = \psi([x])(1 + o(1))$$

as $x \to \infty$. When a in (2.1.11) is an integer, we have

$$\frac{\psi(ax)}{\psi(x)} = \frac{[ax]h([ax])}{[x]h([x])}(1 + o(1)) = a(1 + o(1)).$$

Therefore, for $a = p/q$, where both p and q are integers, we obtain

$$\lim_{x\to\infty} \frac{\psi(ax)}{\psi(x)} = \lim_{x\to\infty} \frac{\psi(p\frac{x}{q})}{\psi(\frac{x}{q})} \frac{\psi(\frac{x}{q})}{\psi(q\frac{x}{q})} = \frac{p}{q} = a.$$

For any positive real number a, put

$$\psi_1(a) = \underline{\lim}_{x\to\infty} \frac{\psi(ax)}{\psi(x)}, \quad \psi_2(a) = \overline{\lim}_{x\to\infty} \frac{\psi(ax)}{\psi(x)}.$$

For any rational number a, $\psi_1(a) = \psi_2(a)$ by the above proof. Hence, it suffices to show that both $\psi_1(x)$ and $\psi_2(x)$ are continuous. Because

$$\begin{aligned}
&\left|\frac{\psi((a+\varepsilon)x) - \psi(ax)}{\psi(x)}\right| \\
&\le \frac{1}{\psi(x)}\left|\int_{-\pi}^{\pi} \frac{\sin^2 \frac{\varepsilon x\lambda}{2}}{\sin^2 \frac{\lambda}{2}} f(\lambda)\, d\lambda + \int_{-\pi}^{\pi} \frac{\sin \varepsilon x\lambda \sin ax\lambda}{2\sin^2 \frac{\lambda}{2}} f(\lambda)\, d\lambda\right| \\
&\le \frac{\psi(\varepsilon x)}{\psi(x)} + \left(\frac{\psi(\varepsilon x)}{\psi(x)}\right)^{\frac{1}{2}},
\end{aligned}$$

it is enough to show that $\psi_1(a)$ and $\psi_2(a)$ are continuous at $a = 0$. Using Property A4 about a slowly varying function(see Appendix A), for $\varepsilon > 0$ small enough, we have

$$\begin{aligned}\frac{\psi(\varepsilon x)}{\psi(x)} &= \frac{[\varepsilon x]h\left(\frac{[\varepsilon x]}{[x]}[x]\right)}{[x]h([x])}(1+o(1))\\ &\le \varepsilon^{1/2}(1+o(1))\end{aligned}$$

as $x \to \infty$. Hence both $\psi_1(a)$ and $\psi_2(a)$ are continuous at $a = 0$. Theorem 2.1.2 are proved.

Remark 2.1.1. In the proof of Theorem 2.1.2, φ-mixing property is only used to give the inequality

$$E\Big|S_n \sum_{j=n+p}^{m+n+p} X_j\Big| \le 2\varphi(p)^{1/2}\|S_n\|_2\|S_m\|_2.$$

Then, for α-mixing case, we have also

Theorem 2.1.3. *Let $\{X_n\}$ be a strictly stationary α-mixing sequence satisfying that $EX_1 = 0$, $EX_1^2 < \infty$, $\sigma_n^2 = ES_n^2 \to \infty$ and $\{S_n^2/\sigma_n^2,\ n \ge 1\}$ is integrable uniformly. Then the conclusions of Theorem 2.1.2 hold true.*

Proof. By the proof of Theorem 2.1.2 and Remark 2.1.1, it suffices to show the following facts.

1. $\sigma_n^2 \to \infty$;
2. for any $\varepsilon > 0$, there exist $p = p(\varepsilon)$, $N = N(\varepsilon)$ such that

$$\Big|ES_n \sum_{j=n+p}^{m+n+p} X_j\Big| \le \varepsilon\sigma_n\sigma_m \qquad \text{if } n, m \ge N(\varepsilon).$$

The first fact is an assumption of the theorem. Consider the latter. From uniform integrability of $\{S_n^2/\sigma_n^2\}$, for any $\varepsilon > 0$, there exists a $K > 0$ such that for p large enough,

$$\int_{S_n^2/\sigma_n^2 \ge K} S_n^2/\sigma_n^2\, dP < \frac{\varepsilon}{4}, \quad K\alpha(p) < \varepsilon/16.$$

Then, by Lemma 1.2.1, Schwarz's inequality and strict stationarity, we obtain

$$\left|ES_n\sum_{j=n+p}^{m+n+p}X_j\right|\Big/\sigma_n\sigma_m$$
$$\le\int_{|\frac{S_n}{\sigma_n}|<\sqrt{K},\,|\frac{S_{m+n+p}-S_{n+p-1}}{\sigma_m}|<\sqrt{K}}\left|\frac{S_n}{\sigma_n}\cdot\frac{S_{m+n+p}-S_{n+p-1}}{\sigma_m}\right|dP$$
$$+\int_{|\frac{S_n}{\sigma_n}|\ge\sqrt{K}}\sqrt{K}\left|\frac{S_n}{\sigma_n}\right|dP$$
$$+\int_{|\frac{S_{m+n-p}-S_{n+p-1}}{\sigma_m}|\ge\sqrt{K}}\sqrt{K}\left|\frac{S_{m+n+p}-S_{n+p-1}}{\sigma_m}\right|dP$$
$$+\int_{|\frac{S_n}{\sigma_n}|\ge\sqrt{K},\,|\frac{S_{m+n+p}-S_{n+p-1}}{\sigma_m}|\ge\sqrt{K}}\left|\frac{S_n}{\sigma_n}\cdot\frac{S_{m+n+p}-S_{n+p-1}}{\sigma_m}\right|dP$$
$$\le 4K\alpha(p)+\varepsilon/4+\varepsilon/4+\varepsilon/4\le\varepsilon.$$

For a ρ-mixing sequence, Peligrad (1982) showed the following general result. Denote $S_k(n)=S_{k+n}-S_k=\sum_{j=k+1}^{k+n}X_j$.

Theorem 2.1.4. *Let $\{X_n,\ n\ge 1\}$ be a ρ-mixing sequence of random variables with $EX_n=0$. Assume that*

(i) $\sup_n EX_n^2=\sigma_0^2<\infty$;

(ii) $ES_n^2\to\infty$ *as* $n\to\infty$;

(iii) $\lim_{n\to\infty}\dfrac{ES_k^2(n)}{ES_n^2}=1$ *uniformly in* k.

Then

$$ES_n^2=nh(n),$$

where $h(n)$ is a slowly varying function and its domain of definition can be extended to R such that $h(x)$ is also slowly varying on R. If, in addition, assume that

(iv) $\sum_{n=1}^{\infty}\rho(2^n)<\infty$,

then $ES_n^2/n\to\sigma^2>0$.

In order to prove Theorem 2.1.4, we need the following lemma.

Lemma 2.1.1. *Let $\{X_n,\ n\ge 1\}$ be a ρ-mixing sequence with $EX_n=0$. If condition (i) in Theorem 2.1.4 is satisfied, then, for natural numbers p,q,m with $p+q=m$,*

$$\begin{aligned}(1-\rho(n))(ES_{km}^2(p)+ES_{km+p}^2(q))-C_1&\\ \le ES_{km}^2(m)&\\ \le(1+\rho(n))(ES_{km}^2(p)+ES_{km+p}^2(q))+C_1,&\qquad(2.1.12)\end{aligned}$$

where k and n are positive integers and

$$C_1 = C_1(m,p,n) \le 20\sigma_0^2 n^2 + 12\sigma_0 n(\|S_{km}(p)\|_2 + \|S_{km+p}(q)\|_2),$$

and further

$$(1-\rho(n))^{1/2}\|S_{km}(p)\|_2 \le \|S_{km}(m)\|_2 + C_2 \tag{2.1.13}$$

where $C_2 \le 2\sigma_0 n$.

Proof. By the definition of ρ-mixing, we have

$$\begin{aligned}\Big|E(S_{km}(p) + S_{km+p+i}(q))^2 - (ES_{km}^2(p) + ES_{km+p+i}^2(q))\Big| \\ \le \rho(i)(ES_{km}^2(p) + ES_{km+p+i}^2(q)).\end{aligned} \tag{2.1.14}$$

Noting that $S_{km+p+n}(q) = S_{km+p}(q) - S_{km+p}(n) + S_{(k+1)m}(n)$, we obtain

$$\|S_{km+p+n}(q)\|_2 = \|S_{km+p}(q)\|_2 + \theta_1, \tag{2.1.15}$$

where $|\theta_1| \le 2\sigma_0 n$. Hence, from (2.1.14) and (2.1.15), it follows that

$$\begin{aligned}&(1-\rho(n))(ES_{km}^2(p) + ES_{km+p}^2(q) + \theta_2) \\ &\quad \le E(S_{km}(p) + S_{km+p+n}(q))^2 \\ &\quad \le (1+\rho(n))(ES_{km}^2(p) + ES_{km+p}^2(q) + \theta_2),\end{aligned} \tag{2.1.16}$$

where $|\theta_2| \le 4\sigma_0^2 n^2 + 4\sigma_0 n\|S_{km+p}(q)\|_2$. Write

$$S_{km}(m) = S_{km}(p) + S_{km+p}(n) + S_{km+p+n}(q) - S_{(k+1)m}(n).$$

Then

$$\|S_{km}(m)\|_2 = \|S_{km}(p) + S_{km+p+n}(q)\|_2 + \theta_3, \tag{2.1.17}$$

where $|\theta_3| \le 2\sigma_0 n$. Hence $ES_{km}^2(m) = E(S_{km}(p) + S_{km+p+n}(q))^2 + \theta_4$, where

$$|\theta_4| \le 12\sigma_0^2 n^2 + 4\sigma_0 n(\|S_{km}(p)\|_2 + \|S_{km+p}(q)\|_2).$$

Inserting it into (2.1.16) we obtain (2.1.12), where

$$\begin{aligned}C_1 &= \max(|(1-\rho(n))\theta_2 + \theta_4|, |(1+\rho(n))\theta_2 + \theta_4|) \\ &\le 20\sigma_0^2 n^2 + 12\sigma_0 n(\|S_{km}(p)\|_2 + \|S_{km+p}(q)\|_2).\end{aligned}$$

We turn to (2.1.13). (2.1.14) implies

$$(1-\rho(n))ES_{km}^2(p) \le E(S_{km}(p) + S_{km+p+n}(q))^2.$$

Then (2.1.13) is showed from (2.1.17).

Proof of Theorem 2.1.4. At first, we prove that for any $h \in \mathbb{N}$

$$\lim_{n\to\infty} ES_{hn}^2/ES_n^2 = h. \tag{2.1.18}$$

By (2.1.12) with $k = 0$, $m = hn$, $p = (h-1)n$, $q = n$, $n = [(ES_n^2)^{1/3}]$, we have

$$\begin{aligned}(1-\rho(n))&(ES_{(h-1)n}^2 + ES_{(h-1)n}^2(n)) - C_0 \\ &\leq ES_{hn}^2 \\ &\leq (1+\rho(n))(ES_{(h-1)n}^2 + ES_{(h-1)n}^2(n)) + C_0,\end{aligned}$$

where $C_0 = 20\sigma_0^2 n^2 + 12\sigma_0 n(\|S_{(h-1)n}\|_2 + \|S_{(h-1)n}(n)\|_2)$. Using conditions (ii) and (iii), (2.1.18) follows by induction on h. Therefore,

$$h(n) := ES_n^2/n$$

is a slowly varying function. Extend its domain by letting

$$h(t) = ES_{[t]}^2/t.$$

We show that

$$\lim_{n\to\infty} h((1-\varepsilon_n)n)/h(n) = 1, \tag{2.1.19}$$

where $\varepsilon_n \downarrow 0$ suah that $n\varepsilon_n$ are integers. By Property A4 (in Appendix),

$$\lim_{n\to\infty} \frac{h(n\varepsilon_n)}{h(n)}\varepsilon_n^\delta = 0. \tag{2.1.20}$$

Let $h_n = \max(h : hn\varepsilon_n < (1-\varepsilon_n)n)$. Note that

$$S_{n-n\varepsilon_n} = S_n - S_{h_n n\varepsilon_n}(n\varepsilon_n) + S_{h_n n\varepsilon_n}(p) - S_{h_n n\varepsilon_n + n\varepsilon_n}(p),$$

where $p = n - (h_n+1)n\varepsilon_n \leq n\varepsilon_n$. By (2.1.13) with $m = n\varepsilon_n$, $k = h_n$ and $h_n + 1$, we have

$$\begin{aligned}&|\|S_{n-n\varepsilon_n}\|_2 - \|S_n\|_2| \\ &\qquad \leq \|S_{h_n n\varepsilon_n}(n\varepsilon_n)\|_2 + \frac{1}{(1-\rho(i))^{1/2}}(\|S_{h_n n\varepsilon_n}(n\varepsilon_n)\|_2 \\ &\qquad\quad + \|S_{(h_n+1)n\varepsilon_n}(n\varepsilon_n)\|_2 + 4\sigma_0 i),\end{aligned}$$

where i is such that $\rho(i) < 1$. Dividing both sides of the above inquality by $\|S_n\|_2$ and using (iii) and (2.1.20), we obtain (2.1.19). For integer $k > 0$, there exists n_k such that for every $n \geq n_k$,

$$\left|\log\frac{h(nk)}{h(n)}\right| < \frac{1}{k} \tag{2.1.21}$$

since $h(n)$ is slowly varying. Without loss of generality, assume that n_k is strictly increasing on k. Let $t > 1$. For integer $n > 0$, define $k = q_n$ such that $n_k \le nt < n_{k+1}$. Then, (2.1.21) implies that

$$\lim_{n\to\infty} \log(h([nt]q_n)/h(nt)) = 0. \tag{2.1.22}$$

Put $p_n = [q_n t]$. Then $p_n = [kt] \ge k$. Hence, (2.1.21) also implies that

$$\lim_{n\to\infty} \log(h(np_n)/h(n)) = 0. \tag{2.1.23}$$

Moreover, from (2.1.19) we have

$$\lim_{n\to\infty} h([nt]q_n)/h(np_n) = 1.$$

Combining it with (2.1.22) and (2.1.23) yields

$$\lim_{n\to\infty} h(nt)/h(n) = 1.$$

Therefore, by Property A1 of Appendix

$$\lim_{x\to\infty} \frac{h(xt)}{h(x)} = \lim_{x\to\infty} \frac{h([x]t)}{h([x])} = 1$$

as required.

Now we consider the second part of the theorem. By (2.1.12) with $m = 2N$, $p = q = N$, $n = [N^{1/3}]$, we have

$$\begin{aligned}(1-&\rho([N^{1/3}]))(ES_{2kN}^2(N) + ES_{(2k+1)N}^2(N))(1 - \alpha_N)\\ &\le ES_{2kN}^2(2N)\\ &\le (1 + \rho([N^{1/3}]))(ES_{2kN}^2(N) + ES_{(2k+1)N}^2(N))(1 + \alpha_N)\end{aligned} \tag{2.1.24}$$

where

$$\alpha_N = \sup_k \frac{20\sigma_0^2 N^{2/3} + 12\sigma_0^2 N^{1/3}(\|S_{2kN}(N)\|_2 + \|S_{(2k+1)N}(N)\|_2)}{(1 - \rho([N^{1/3}]))(ES_{2kN}^2(N) + ES_{(2k+1)N}^2(N))}$$

and N_0 is so large that $\rho([N^{1/3}]) < 1$ for $N \ge N_0$. Conditions (i) and (iii) imply that

$$\alpha_N = O\Big(\frac{N^{2/3} + N^{1/3}\|S_N\|_2}{\|S_N\|_2^2}\Big).$$

By Property A2 of Appendix, for any $0 < \varepsilon < 1/6$

$$\lim_{N\to\infty} N^\varepsilon ES_N^2/N = \infty.$$

Hence $(ES_N^2)^{-1} = O(N^{-1+\varepsilon})$, and further,

$$\alpha_N = O(N^{-\frac{1}{6}+\varepsilon}). \tag{2.1.25}$$

Then (2.1.24) implies that for integers $r > p \geq N_0$ with $\rho([2^{N_0/3}]) < 1$,

$$\begin{aligned}\prod_{i=p}^{r-1}(1-\rho([2^{i/3}]))(1-\alpha_{2^i}) \sum_{i=0}^{2^{r-p}-1} ES^2_{i2^p}(2^p) \\ \leq ES^2_{2^r} \\ \leq \prod_{i=p}^{r-1}(1+\rho([2^{i/3}]))(1+\alpha_{2^i}) \sum_{i=0}^{2^{r-p}-1} ES^2_{i2^p}(2^p).\end{aligned} \tag{2.1.26}$$

By conditon (iv), $\sum_i \rho([2^{i/3}]) < \infty$. Moreover, (2.1.25) implies that $\sum_i \alpha(2^i) < \infty$. Therefore, from (2.1.26) we obtain

$$\lim_{r>p\to\infty} ES^2_{2^r} \Big/ \sum_{i=0}^{2^{r-p}-1} ES^2_{i2^p}(2^p) = 1.$$

Consequently, it follows by condition (iii) that

$$\lim_{r>p\to\infty} h(2^r)/h(2^p) = 1,$$

and further, $h(2^r)$ converges to a positive constant. Applying Property A3 of Appendix to $h(t)$ and $1/h(t)$, we obtain that $h(n)$ converges to the same limit as $h(2^r)$. Theorem 2.1.4 is proved.

For a strictly stationary ρ-mixing sequence, we also have the following result.

Theorem 2.1.5.(Ibragimov 1975) *Let $\{X_n, n \geq 1\}$ be a strictly stationary ρ-mixing sequence with $EX_1 = 0$, $EX_1^2 < \infty$ and $\sum_{n=1}^{\infty} \rho(2^n) < \infty$. Then $\{X_n\}$ possesses a continuous spectral density $f(\lambda)$ and*

$$\text{Var}S_n = 2\pi f(0)n + o(n) \quad \textit{as} \quad n \to \infty$$

if $f(0) \neq 0$.

For the proof of Theorem 2.1.5 we refer to Lemma 17 of Ibragimov and Rozanov (1978). They first show that $\{X_n\}$ possesses a bounded spectral density if $\sum_{n=1}^{\infty} \rho(2^n) < \infty$, and then show that

$$E_n(f) \leq 128 \max_{\lambda} f(\lambda) \sum_{k=0}^{\infty} \rho(2^{k-1}n),$$

where $E_n(f)$ denotes the error of best approximation of $f(\lambda)$ by a trigonometric polynomial of degree less than or equal to n on $[-\pi, \pi]$. With the help of this result, we may know that $f(\lambda)$ is a continuous function on $[-\pi, \pi]$. The rest of the proof can be completed by using Theorem 2.1.1.

2.2 Further inequalities

In order to show limit theorems for a mixing sequence, we often need some further inequalities besides the basic inequalities in Section 1.2.

The following extended *Ottaviani inequality* for an α-mixing sequence was given by Lin (1982).

Recall the notation $\mathcal{F}_a^b = \sigma(X_i, a \le i \le b)$ for a sequence $\{X_n\}$ of random variables.

Lemma 2.2.1. *Let $\{X_n, n \ge 1\}$ be an α-mixing sequence. For any given integers p, q and k, let ξ_j be $\mathcal{F}_{(j-1)(p+q)+1}^{jp+(j-1)q}$ measurable, $j = 1, 2, \cdots, k$. If*

$$P\{|\xi_{l+1} + \cdots + \xi_k| \le C\} \ge \frac{1}{2}, \quad l = 1, \cdots, k-1,$$

then

$$P\Big\{\max_{1\le l\le k} |\xi_1 + \cdots + \xi_l| > 2C\Big\} \le 2P\{|\xi_1 + \cdots + \xi_k| > C\} + 2k\alpha(q).$$

Proof. Let events

$$A = \Big\{\max_{1\le l\le k} |\xi_1 + \cdots + \xi_l| > 2C\Big\}, \quad B = \{|\xi_1 + \cdots + \xi_k| > C\},$$
$$A_1 = \{|\xi_1| > 2C\},$$
$$A_l = \Big\{\max_{1\le r\le l-1} |\xi_1 + \cdots + \xi_r| \le 2C,\ |\xi_1 + \cdots + \xi_l| > 2C\Big\},\ l = 2, \cdots, k,$$
$$B_l = \{|\xi_{l+1} + \cdots + \xi_k| \le C\},\ l = 1, \cdots, k-1,\ B_k = \Omega.$$

Then

$$A_iA_j = \emptyset\ (i \ne j), \quad A = \bigcup_{l=1}^k A_l, \quad \bigcup_{l=1}^k A_lB_l \subseteq B.$$

By the conditions of the lemma,

$$P(A_lB_l) \ge P(A_l)P(B_l) - \alpha(q) \ge \frac{1}{2}P(A_l) - \alpha(q),$$

and hence

$$P(B) \geq \sum_{l=1}^{k} P(A_l B_l) \geq \frac{1}{2} \sum_{l=1}^{k} P(A_l) - k\alpha(q)$$
$$= \frac{1}{2} P(A) - k\alpha(q)$$

as required.

The following lemmas all are about the bounds of the moments of partial sums. For a ρ-mixing sequence, earlier work was due to Peligrad (1982, 1987). Shao (1988b, 1989a,b) improved and generalized her results.

Lemma 2.2.2. *Let $\{X_n, n \geq 1\}$ be a ρ-mixing sequence with $EX_n = 0$, $EX_n^2 < \infty$ for each $n \geq 1$. Then for any $\varepsilon > 0$, there exists a $C = C(\varepsilon) > 0$ such that*

$$ES_k^2(n) \leq Cn \exp\Big\{(1+\varepsilon) \sum_{i=0}^{[\log n]} \rho(2^i)\Big\} \max_{k<i\leq k+n} EX_i^2$$

for each $k \geq 1$ and $n \geq 1$, where $S_k(n) = \sum_{i=k+1}^{k+n} X_i$.

Proof. Without loss of generality, assume $0 < \varepsilon < \frac{1}{4}$. Let C_n be a non-decreasing sequence of numbers such that

$$ES_k^2(n) \leq C_n n \max_{k<i\leq k+n} EX_i^2. \tag{2.2.1}$$

For $n \leq 2^{1/\varepsilon}$ we need only to take $C_n \geq 2^{1/\varepsilon}$ by the Minkowski inequality. Let $C_1 = 2^{1/\varepsilon}$. We suppose that C_m, $m = 1, \cdots, n-1$, are already defined as demanded in the lemma. Put $n_1 = [n/2]$, $n_2 = n - n_1$, $n_3 = [n^{1/(1+\varepsilon)}] + 1$. It is clear that

$$ES_k^2(n) = ES_k^2(n_1) + ES_{k+n_1}^2(n_2) + 2ES_k(n_1)S_{k+n_1}(n_2), \tag{2.2.2}$$

and further

$$\begin{aligned}
&|ES_k(n_1)S_{k+n_1}(n_2)| \\
&\quad \leq |ES_k(n_1)S_{k+n_1}(n_3)| + |ES_k(n_1)S_{k+n_1+n_3}(n_2-n_3)| \\
&\quad \leq \|S_k(n_1)\|_2 \|S_{k+n_1}(n_3)\|_2 \\
&\qquad + \rho(n_3)\|S_k(n_1)\|_2 \|S_{k+n_1+n_3}(n_2-n_3)\|_2 \\
&\quad \leq 2\|S_k(n_1)\|_2 \|S_{k+n_1}(n_3)\|_2 + \rho(n_3)\|S_k(n_1)\|_2 \|S_{k+n_1}(n_2)\|_2.
\end{aligned}$$

Inserting the above inequality into (2.2.2) and noting (2.2.1) we obtain

$$\begin{aligned} ES_k^2(n) &\le (ES_k^2(n_1) + ES_{k+n_1}^2(n_2))(1+\rho(n_3)) \\ &\quad + 4\|S_k(n_1)\|_2\|S_{k+n_1}(n_3)\|_2 \\ &\le C_{n_2}\cdot n(1+\rho(n_3)) \max_{k<i\le k+n} EX_i^2 + 4C_{n_1} n_1^{\frac{1}{2}} n_3^{\frac{1}{2}} \max_{k<i\le k+n} EX_i^2 \\ &\le C_{n_2}(1+\rho(n_2^{\frac{1}{1+\varepsilon}}) + 4n_2^{-\frac{\varepsilon}{2(1+\varepsilon)}}) n \max_{k<i\le k+n} EX_i^2, \end{aligned}$$

where

$$\rho(x) = (\rho(i+1)-\rho(i))(x-i)+\rho(i) \quad \text{if } i \le x < i+1.$$

Hence for $n \ge 2$, we define

$$C_n = C_{n_2}\Big(1+\rho(n_2^{\frac{1}{1+\varepsilon}}) + 4n_2^{-\frac{\varepsilon}{2(1+\varepsilon)}}\Big).$$

Obviously C_n is nondecreasing, and

$$\begin{aligned} C_{2^n} &= C_{2^{n-1}}(1+\rho(2^{\frac{n-1}{1+\varepsilon}}) + 4\cdot 2^{-\frac{\varepsilon(n-1)}{2(1+\varepsilon)}}) \\ &= C_1\prod_{i=0}^{n-1}\Big(1+\rho(2^{\frac{i}{1+\varepsilon}}) + 4\cdot 2^{-\frac{\varepsilon i}{2(1+\varepsilon)}}\Big) \\ &\le C_1 \exp\Big\{\sum_{i=o}^{n-1}\Big(\rho(2^{\frac{i}{1+\varepsilon}}) + 4\cdot 2^{-\frac{\varepsilon i}{2(1+\varepsilon)}}\Big)\Big\} \\ &\le C_1\exp\Big\{3 + \int_2^{n-1}\rho(2^{\frac{x}{1+\varepsilon}})\,dx + c_\varepsilon\Big\} \\ &\le C_1\exp\Big\{3+(1+\varepsilon)\int_{2/(1+\varepsilon)}^{n-1}\rho(2^x)\,dx + c_\varepsilon\Big\} \\ &\le C_1\exp\Big\{3+(1+\varepsilon)\sum_{i=1}^{n-1}\rho(2^i) + c_\varepsilon\Big\}, \end{aligned} \tag{2.2.3}$$

where $c_\varepsilon = 4/(1-2^{-\varepsilon/(2(1+\varepsilon))})$. Put $d_\varepsilon = 2^{1/\varepsilon}\exp(3+c_\varepsilon)$. We get

$$C_{2^n} \le d_\varepsilon \exp\Big\{(1+\varepsilon)\sum_{i=1}^{n-1}\rho(2^i)\Big\}.$$

For any n, there exists an m such that $2^m \le n < 2^{m+1}$. Using the monotonicity of C_n, it follows that

$$\begin{aligned} C_n \le C_{2^{m+1}} &\le d_\varepsilon\exp\Big\{(1+\varepsilon)\sum_{i=1}^{m}\rho(2^i)\Big\} \\ &\le d_\varepsilon\exp\Big\{(1+\varepsilon)\sum_{i=0}^{[\log n]}\rho(2^i)\Big\}. \end{aligned}$$

The lemma is proved.

Lemma 2.2.3. *Let $\{X_n, n \geq 1\}$ be a ρ-mixing sequence with $EX_n = 0, EX_n^2 < \infty$ for each $n \geq 1$. Suppose that*

$$ES_k^2(n)/\min_{k<i\leq k+n} EX_i^2 \to \infty \quad \text{as } n \to \infty \tag{2.2.4}$$

uniformly in k and

$$\max_{k<i\leq k+n} EX_i^2 \leq a \min_{k<i\leq k+n} EX_i^2 \quad \text{for some } a \geq 1. \tag{2.2.5}$$

Then, for any $\varepsilon > 0$, there exist $C' = C'(\varepsilon, \rho(\cdot), a) > 0$ and an integer N such that for each $k \geq 0$ and $n \geq N$

$$ES_k^2(n) \geq C' n \exp\Big\{-(1+\varepsilon) \sum_{i=0}^{[\log n]} \rho(2^i)\Big\} \min_{k<i\leq k+n} EX_i^2.$$

Proof. Without loss of generality assume that $0 < \varepsilon < 1/400$. Consequently,

$$1 - 5\varepsilon^2 > (3/2)^{-\varepsilon/6}.$$

Hence, noting $\rho(n) \to 0$ as $n \to \infty$, we have

$$1 - 5\varepsilon^2 - \rho(m_0) > (3/2)^{-\varepsilon/6}$$

for some large m_0. It is not hard to verify that $\exp\{2\sum_{i=0}^{[\log n]} \rho(2^i)\}$ is a slowly varying function. Then, by Lemma 2.2.2 and condition (2.2.4), there exists an n_0 such that for $n \geq n_0$

$$ES_k^2(n) \leq n^{1+\varepsilon^4} \max_{k<i\leq k+n} EX_i^2, \tag{2.2.6}$$

$$ES_k^2(n) \geq \frac{4am_0^2}{\varepsilon^4} \min_{k<i\leq k+n} EX_i^2. \tag{2.2.7}$$

When $n \geq 2n_0$, put $n_1 = [n/2]$, $n_2 = n - n_1$. Then

$$\begin{aligned}
ES_k^2(n) &= ES_k^2(n_1) + ES_{k+n_1}^2(n_2) \\
&\quad + 2ES_k(n_1)S_{k+n_1}(m_0) + 2ES_k(n_1)S_{k+n_1+m_0}(n_2 - m_0) \\
&\geq ES_k^2(n_1) + ES_{k+n_1}^2(n_2) - 2\|S_k(n_1)\|_2\|S_{k+n_1}(m_0)\|_2 \\
&\quad - 2\rho(m_0)\|S_k(n_1)\|_2\|S_{k+n_1+m_0}(n_2 - m_0)\|_2 \\
&\geq (1 - \rho(m_0))(ES_k^2(n_1) + ES_{k+n_1}^2(n_2)) \\
&\quad - 4\|S_k(n_1)\|_2\|S_{k+n_1}(m_0)\|_2 \\
&\geq (1 - \rho(m_0))(ES_k^2(n_1) + ES_{k+n_1}^2(n_2)) - 4\varepsilon^2 ES_k^2(n_1) \\
&\quad - \frac{4}{\varepsilon^2} m_0^2 \max_{k<i\leq k+n} EX_i^2 \\
&\geq (1 - 4\varepsilon^2 - \rho(m_0))(ES_k^2(n_1) + ES_{k+n_1}^2(n_2)) \\
&\quad - \frac{4}{\varepsilon^2} m_0^2 a \min_{k<i\leq k+n} EX_i^2 \\
&\geq (1 - 5\varepsilon^2 - \rho(m_0))(ES_k^2(n_1) + ES_{k+n_1}^2(n_2)) \\
&\geq (3/2)^{-\varepsilon/6}(ES_k^2(n_1) + ES_{k+n_1}^2(n_2)). \qquad (2.2.8)
\end{aligned}$$

We first show that for each $n \geq n_0$

$$ES_k^2(n) \geq C_2 n^{1-\varepsilon/6} \min_{k<i\leq k+n} EX_i^2, \qquad (2.2.9)$$

where $C_2 = 2am_0^2 n_0^{-1}\varepsilon^{-4}$. By (2.2.7), (2.2.9) holds for $n_0 \leq n < 2n_0$. When $n \geq 2n_0$, we assume that (2.2.9) is true for each positive integer less than n. Then it is also true for n. In fact, using (2.2.8) we obtain

$$\begin{aligned}
ES_k^2(n) &\geq (3/2)^{-\varepsilon/6} C_2(n_1^{1-\varepsilon/6} + n_2^{1-\varepsilon/6}) \min_{k<i\leq k+n} EX_i^2 \\
&\geq \left(\frac{3}{2}\right)^{-\varepsilon/6} C_2\left(\left(\frac{1}{3}\right)^{1-\varepsilon/6} + \left(\frac{2}{3}\right)^{1-\varepsilon/6}\right) \\
&\quad \cdot n^{1-\varepsilon/6} \min_{k<i\leq k+n} EX_i^2 \\
&\geq C_2 n^{1-\varepsilon/6} \min_{k<i\leq k+n} EX_i^2
\end{aligned}$$

as required.

Next, we turn to the assertion of the lemma. For $n \geq n_0^{1+\varepsilon}$, put $n_1 = [\frac{n}{2}]$, $n_2 = n - n_1$, $n_3 = [n^{1/(1+\varepsilon)}] + 1$. From (2.2.6) and (2.2.9), it follows that

$$\begin{aligned}
&ES_k^2(n)\\
&\quad= ES_k^2(n_1) + ES_{k+n_1}^2(n_2) + 2ES_k(n_1)ES_{k+n_1}(n_3)\\
&\qquad + 2ES_k(n_1)S_{k+n_1+n_3}(n_2-n_3)\\
&\quad\geq ES_k^2(n_1) + ES_{k+n_1}^2(n_2) - 4\|S_k(n_1)\|_2\|S_{k+n_1}(n_3)\|_2\\
&\qquad - 2\rho(n_3)\|S_k(n_1)\|_2\|S_{k+n_1}(n_2)\|_2\\
&\quad\geq (1-\rho(n_3))(ES_k^2(n_1) + ES_{k+n_1}^2(n_2))\\
&\qquad - 4(n_1n_3)^{(1+\varepsilon^4)/2} \max_{k<i\leq k+n} EX_i^2\\
&\quad\geq (1-\rho(n_3))(ES_k^2(n_1) + ES_{k+n_1}^2(n_2))\\
&\qquad - 4n^{1-(\varepsilon-3\varepsilon^4)/2(1+\varepsilon)} \max_{k<i\leq k+n} EX_i^2\\
&\quad\geq (1-\rho(n_3))(ES_k^2(n_1) + ES_{k+n_1}^2(n_2)) - 4an^{1-\varepsilon/5} \min_{k<i\leq k+n} EX_i^2\\
&\quad\geq (1-\rho(n_3))(ES_k^2(n_1) + ES_{k+n_1}^2(n_2)) - 8aC_2^{-1}n^{-\varepsilon/30}ES_k^2(n_1)\\
&\quad\geq (1-\rho(n_3) - n^{-\varepsilon/40})(ES_k^2(n_1) + ES_{k+n_1}^2(n_2)). \qquad (2.2.10)
\end{aligned}$$

The last inequality holds for $n \geq (8aC_2^{-1})^{120/\varepsilon}$. Put

$$n_o' = \max(n_0^{1+\varepsilon}, (8aC_2^{-1})^{120/\varepsilon}).$$

Let C_n be non-increasing so that

$$ES_k^2(n) \geq C_n n \min_{k<i\leq k+n} EX_i^2$$

for $n \geq n_0$. Then, by (2.2.10)

$$\begin{aligned}
ES_k^2(n) &\geq (1-\rho(n_3) - n_2^{-\varepsilon/40})C_{n_2} n \min_{k<i\leq k+n} EX_i^2\\
&\geq (1-\rho(n_2^{\frac{1}{1+\varepsilon}}) - n_2^{-\varepsilon/40})C_{n_2} n \min_{k<i\leq k+n} EX_i^2
\end{aligned}$$

for $n \geq n_0'$. Hence we can choose

$$C_n = C_{n_2}\Big(1-\rho(n_2^{\frac{1}{1+\varepsilon}}) - n_2^{-\varepsilon/40}\Big) \qquad (2.2.11)$$

for $n \geq n_0'$. It is easy to see that there exists an n_0'' such that for $n \geq n_0''$

$$1-\rho(n_2^{\frac{1}{1+\varepsilon}}) - n_2^{-\varepsilon/40} \geq \exp\Big\{-(1+\varepsilon)\Big(\rho(n_2^{\frac{1}{1+\varepsilon}}) + n_2^{-\varepsilon/40}\Big)\Big\}. \qquad (2.2.12)$$

Put $n_0^* = n_0' \vee n_0''$. In view of (2.2.7), we take $C_{n_0^*} = 4am_0^2/(\varepsilon^4 n_0^*)$. Obviously $\{C_n, n \geq 2n_0^*\}$ defined by (2.2.11) is non-increasing. From (2.2.11) and (2.2.12), we obtain that for $2^m > n_0^*$

$$\begin{aligned} C_{2^m} &= C_{2^{m-1}}(1 - \rho(2^{\frac{m-1}{1+\varepsilon}}) - 2^{-(m-1)\varepsilon/40}) \\ &\geq C_{2^{m-1}} \exp\{-(1+\varepsilon)(\rho(2^{\frac{m-1}{1+\varepsilon}}) + 2^{-(m-1)\varepsilon/40})\}, \end{aligned}$$

which implies that

$$\begin{aligned} C_{2^m} &\geq C_{n_0^*} \prod_{i=0}^{m-1} \exp\{-(1+\varepsilon)(\rho(2^{i/(1+\varepsilon)}) + 2^{-i\varepsilon/40})\} \\ &= C_{n_0^*} \exp\Big\{-(1+\varepsilon) \sum_{i=0}^{m-1} (\rho(2^{i/(1+\varepsilon)}) + 2^{-i\varepsilon/40})\Big\}. \end{aligned}$$

Similarly to (2.2.3), there exists a $d_\varepsilon > 0$ such that

$$\begin{aligned} &\exp\Big\{-(1+\varepsilon) \sum_{i=0}^{m-1} (\rho(2^{i/(1+\varepsilon)}) + 2^{-i\varepsilon/40})\Big\} \\ &\qquad\qquad \geq d_\varepsilon \exp\Big\{-(1+\varepsilon)^2 \sum_{i=0}^{m} \rho(2^i)\Big\}. \end{aligned}$$

Therefore

$$C_{2^m} \geq d_\varepsilon C_{n_0^*} \exp\Big\{-(1+\varepsilon)^2 \sum_{i=0}^{m} \rho(2^i)\Big\}.$$

For arbitrary $n > n_0^*$, there exists an m such that $2^m \leq n < 2^{m+1}$. By monotonicity of C_n, we find that

$$\begin{aligned} C_n \geq C_{2^{m+1}} &\geq d_\varepsilon C_{n_0^*} \exp\Big\{-(1+\varepsilon)^2 \sum_{i=0}^{m+1} \rho(2^i)\Big\} \\ &\geq \frac{1}{8} d_\varepsilon C_{n_0^*} \exp\Big\{-(1+\varepsilon)^2 \sum_{i=0}^{[\log n]} \rho(2^i)\Big\}, \end{aligned}$$

and hence we arrive at the assertion of the lemma.

Sometimes we need the bounds of moments of higher than two orders.

Lemma 2.2.4. *Let* $\{X_n, n \geq 1\}$ *be a* ρ*-mixing sequence with* $EX_n = 0$, $\sup_n E|X_n|^{2+\delta} < \infty$ *for some* $0 < \delta < 1$ *and*

$$\sum_{n=1}^{\infty} \rho(2^n) < \infty. \tag{2.2.13}$$

Then there exists a $C = C(\delta, \rho(\cdot)) > 0$ *such that for each* $n \geq 1$

$$\begin{aligned}\sup_{k\geq 1} E|S_k(n)|^{2+\delta} \leq & C\Big\{n^{1+\delta/2}(\sup_{k\geq 1} EX_k^2)^{1+\delta/2} \\ & + n\exp\Big\{(C\log n)^{\delta/(2+\delta)}\Big\}\sup_{k\geq 1} E|X_k|^{2+\delta}\Big\}.\end{aligned}$$

Proof. It is not difficult to verify that

$$\begin{aligned}(1+x)^{2+\delta} &\leq 1 + (2+\delta)^2(x + x^{1+\delta}) + x^{2+\delta} \\ &\leq 1 + 9(x + x^{1+\delta}) + x^{2+\delta}\end{aligned} \tag{2.2.14}$$

for $x \geq 0$. Put

$$a_m = \sup_{k\geq 1} \|S_k(m)\|_{2+\delta}, \quad \sigma_m = \sup_{k\geq 1} \|S_k(m)\|_2.$$

Obviously,

$$\|S_k(2m)\|_{2+\delta} \leq \|S_k(m) + S_{k+m+[m^{1/5}]}(m)\|_{2+\delta} + 2m^{1/5}a_1.$$

By (2.2.14), putting $m_1 = m + [m^{1/5}]$, we have

$$\begin{aligned}E|S_k(m) + S_{k+m_1}(m)|^{2+\delta} \leq & 2a_m^{2+\delta} + 9E|S_k(m)|^{1+\delta}|S_{k+m_1}(m)| \\ & + 9E|S_k(m)||S_{k+m_1}(m)|^{1+\delta}.\end{aligned}$$

Moreover, by the Schwarz inequality and Lemma 1.2.7, we have

$$\begin{aligned}& E|S_k(m)|^{1+\delta}|S_{k+m_1}(m)| \\ & \quad \leq \|S_k(m)\|_{2+\delta}^{\delta}\|S_k(m)S_{k+m_1}(m)\|_{(2+\delta)/2} \\ & \quad \leq a_m^{\delta}\{\sigma_m^{2+\delta} + 4\rho([m^{1/5}])a_m^{2+\delta}\}^{2/(2+\delta)} \\ & \quad \leq a_m^{\delta}\sigma_m^2 + 4\rho^{2/(2+\delta)}([m^{1/5}])a_m^{2+\delta}.\end{aligned}$$

Similarly

$$E|S_k(m)||S_{k+m_1}(m)|^{1+\delta} \leq a_m^{\delta}\sigma_m^2 + 4\rho^{2/(2+\delta)}([m^{1/5}])a_m^{2+\delta}.$$

Combining these inequalities yields that

$$\begin{aligned}& E|S_k(m) + S_{k+m_1}(m)|^{2+\delta} \\ & \quad \leq 2a_m^{2+\delta} + 18(a_m^{\delta}\sigma_m^2 + 4\rho^{2/(2+\delta)}([m^{1/5}])a_m^{2+\delta}) \\ & \quad \leq \Big\{[2(1 + 36\rho^{2/(2+\delta)}([m^{1/5}]))]^{1/(2+\delta)}a_m + 18\sigma_m\Big\}^{2+\delta},\end{aligned}$$

which implies that

$$a_{2m} \le \left\{2(1 + 36\rho^{2/(2+\delta)}([m^{1/5}]))\right\}^{1/(2+\delta)} a_m + 18\sigma_m + 2m^{1/5}a_1. \quad (2.2.15)$$

Noting monotonecity of $\rho(n)$ and condition (2.2.13), we have

$$\rho(n) \le c/\log n,$$

here, and in the sequel, c stands for a positive constant, which may take different values at different places. Hence, applying Lemma 2.2.2, we obtain

$$\begin{aligned} a_{2^r} &\le \{2(1 + 36\rho^{2/(2+\delta)}([2^{(r-1)/5}]))\}^{1/(2+\delta)} a_{2^{r-1}} \\ &\quad + 18\sigma_{2^{r-1}} + 2\cdot 2^{(r-1)/5}a_1 \\ &\le 2^{(r-1)/(2+\delta)} \prod_{i=0}^{r-1} (1 + 36\rho^{2/(2+\delta)}([2^{i/5}]))^{1/(2+\delta)} a_1 \\ &\quad + c\sigma_1 \sum_{i=0}^{r-1} 2^{i/2} \prod_{j=i+1}^{r-1} \left\{2(1 + 9\rho^{2/(2+\delta)}([2^{j/5}]))\right\}^{1/(2+\delta)} \\ &\quad + 2a_1 \sum_{i=0}^{r-1} 2^{i/5} \prod_{j=i+1}^{r-1} \left\{2(1 + 9\rho^{2/(2+\delta)}([2^{j/5}]))\right\}^{1/(2+\delta)} \\ &\le C2^{r/2}\sigma_1 + 2^{r/(2+\delta)} \exp(Cr)^{\delta/(2+\delta)} a_1. \end{aligned} \quad (2.2.16)$$

This implies the conclusion of the lemma.

Similarly, by finer estimation, Shao (1989a) showed the following results, whose proof will not be presented here.

Lemma 2.2.5. *Let $\{X_n, n \ge 1\}$ be a ρ-mixing sequence with*

$$EX_n = 0, \quad \sup_n E|X_n|^{2+\delta} < \infty \qquad \text{for some } \delta \ge 0.$$

Then, for any $\varepsilon > 0$, there exists a $C = C(\delta, \rho(\cdot), \varepsilon) > 0$, such that for each $n \ge 2$

$$\begin{aligned} E|S_k(n)|^{2+\delta} \le C\Big\{&\Big(n\exp\Big\{(1+\varepsilon)\sum_{i=0}^{[\log n]} \rho(2^i)\Big\} \max_{k<i\le k+n} EX_i^2\Big)^{1+\delta/2} \\ &+ n\exp\Big\{C\sum_{i=0}^{[\log n]} \rho^{2/(2+\delta)}(2^i)\Big\} \max_{k<i\le k+n} E|X_i|^{2+\delta}\Big\}. \end{aligned}$$

Lemma 2.2.6. *Let $\{X_n, n \geq 1\}$ be a ρ-mixing sequence with*

$$EX_n = 0, \quad E|X_n|^q < \infty,\ q \geq 2, \quad ES_k^2(n) \leq nh(n) \max_{k<i\leq k+n} EX_i^2.$$

Suppose that there exists a function $h(n)$ such that for every $k \geq 0$, $n \geq 1$ and there exist a positive integer n_0 and a constant $0 < \theta < 2^{1-2/(q\wedge 3)}$ such that

$$\max(h([n/2]),\, h(n-[n/2])) \leq \theta h(n)$$

for $n \geq n_0$. Furthermore, when $q > 3$ assume that there exists a $C > 0$ such that

$$h(n) \geq \frac{1}{C}\exp\Big\{-C\sum_{i=0}^{[\log n]} \rho^{2/q}(2^i)\Big\}.$$

Then there exists a constant $K = K(q, n_0, \theta, C, \rho(\cdot))$, such that for every $k \geq 0$, $n \geq 1$

$$E|S_k(n)|^q \leq K\Big\{(nh(n)\max_{k<i\leq k+n} EX_i^2)^{q/2} + n\exp\Big\{K\sum_{i=0}^{[\log n]} \rho^{2/q}(2^i)\Big\}\max_{k\leq i\leq k+n} E|X_i|^q\Big\}.$$

Next, we turn our attention to a φ-mixing sequence. Peligrad (1985) showed the following *inequality of tail probability* (see Shao 1988a).

Lemma 2.2.7. *Let $\{X_n, n \geq 1\}$ be a φ-mixing sequence, $0 < \eta < 1$. Suppose that there exists an integer p, $1 \leq p \leq n$, a number $A > 0$ such that*

$$\varphi(p) + \max_{p\leq i\leq n} P\{|S_n - S_i| \geq A\} \leq \eta. \tag{2.2.17}$$

Then, for any $a \geq 0$, $b \geq 0$, we have

$$P\Big\{\max_{1\leq i\leq n}|S_i| \geq a + A + b\Big\} \leq \frac{1}{1-\eta}P\{|S_n| \geq a\} + \frac{1}{1-\eta}P\Big\{\max_{1\leq i\leq n}|X_i| \geq \frac{b}{p-1}\Big\}. \tag{2.2.18}$$

$$P\{|S_n| \geq a + A + b\} \leq \eta P\Big\{\max_{1\leq i\leq n}|S_i| \geq a\Big\} + P\Big\{\max_{1\leq i\leq n}|X_i| \geq \frac{b}{p}\Big\}. \tag{2.2.19}$$

Proof. Put $E_i = \{\max_{1\le j<i}|S_j| < a+A+b \le |S_i|\}$. Then

$$P\Big\{\max_{1\le i\le n}|S_i| \ge a+A+b\Big\}$$

$$\le P(|S_n|\ge a) + \sum_{i=1}^{n-1} P\Big(E_i\cap\{|S_n-S_i|\ge A+b\}\Big)$$

$$\sum_{i=1}^{n-1} P\Big(E_i\cap\{|S_n-S_i|\ge A+b\}\Big)$$

$$\le \sum_{i=1}^{n-p-1} P\Big(E_i\cap\{|S_{i+p-1}-S_i|\ge b\}\Big)$$

$$+ \sum_{i=1}^{n-p-1} P\Big(E_i\cap\{|S_n-S_{i+p-1}|\ge A\}\Big)$$

$$+ \sum_{i=n-p}^{n-1} P\Big(E_i\cap\{|S_n-S_i|\ge A+b\}\Big)$$

$$\le \sum_{i=1}^{n-1} P\Big(E_i\cap\Big\{\max_{1\le j\le n}|X_j|\ge \frac{b}{p-1}\Big\}\Big)$$

$$+ \sum_{i=1}^{n-p-1} P(E_i)\Big(P\{|S_n-S_{i+p-1}|\ge A\}+\varphi(p)\Big)$$

$$\le P\Big\{\max_{1\le j\le n}|X_j|\ge \frac{b}{p-1}\Big\} + \eta P\Big\{\max_{1\le i\le n}|S_i|\ge a+A+b\Big\},$$

where condition (2.2.17) is used in the last inequality. Consequently, (2.2.18) is proved.

As for (2.2.19), putting $E_i' = \{\max_{1\le j<i}|S_j| < a \le |S_i|\}$ and noting

$$|S_n - S_{j+p-1}| \ge ||S_n| - |S_{j-1}| - p\max_{1\le i\le n}|X_i|| \text{for} 1\le j\le n-p,$$

we have

$$P\{|S_n|\ge a+A+b\}$$

$$\le P\Big\{|S_n|\ge a+A+b, \max_{1\le i\le n-p}|S_i|\ge a, \max_{1\le i\le n}|X_i|\le \frac{b}{p}\Big\}$$

$$+ P\Big\{\max_{1\le i\le n}|X_i| > \frac{b}{p}\Big\}$$

$$\le \sum_{i=1}^{n-p} P(E_i'\cap\{|S_n-S_{i+p-1}|>A\}) + P\Big\{\max_{1\le i\le n}|X_i|\ge \frac{b}{p}\Big\}$$

$$\le \eta P\Big\{\max_{1\le i\le n}|S_i|\ge a\Big\} + P\Big\{\max_{1\le i\le n}|X_i|\ge \frac{b}{p}\Big\}$$

as required.

Lemma 2.2.8 is due to Shao and Lu (1986).

Lemma 2.2.8. *Let* $\{X_n,\, n \ge 1\}$ *be a* φ*-mixing sequence with* $EX_n = 0$ *and* $\sup_n E|X_n|^{2+\delta} < \infty$ *for some* $\delta > 0$. *Suppose that*

$$\sup_k ES_k^2(n) \le Mn \sup_k EX_k^2 \quad \text{for some} M > 0. \tag{2.2.20}$$

Then there exists a $C = C(\delta,\, M,\, \varphi(\cdot)) > 0$ *such that for each* $n \ge 1$

$$\sup_k E|S_k(n)|^{2+\delta} \le Cn^{1+\delta/2} \sup_k E|X_k|^{2+\delta}.$$

Proof. It is easy to see that for $r \ge 1$ and $x \ge 0$

$$(1+x)^r \le \sum_{k=0}^{[r]} \binom{r}{k} x^k + \delta_r x^r, \tag{2.2.21}$$

where $\delta_r = 1$ if r is not an integer, otherwise $\delta_r = 0$. We now prove the lemma by induction on $r := 2+\delta$. Assume that the lemma holds for $l \le [r], r$ being non-integer. Denoting $a_m = \sup_k \|S_k(m)\|_r$, from (2.2.21) we obtain

$$\begin{aligned} &E|S_k(m) + S_{k+m+k_0}(m)|^r \\ &\quad \le E|S_k(m)|^r + E|S_{k+m+k_0}(m)|^r \\ &\qquad + \textstyle\sum_{j=1}^{[r]} \binom{r}{j} E|S_k(m)|^j |S_{k+m+k_0}(m)|^{r-j} \\ &\quad \le \Big(2 + 2\textstyle\sum_{j=1}^{[r]} \binom{r}{j} \varphi^{1/r}(k_0)\Big) a_m^r \\ &\qquad + \textstyle\sum_{j=1}^{[r]} \binom{r}{j} E|S_k(m)|^j E|S_{k+m+k_0}(m)|^{r-j} \\ &\quad =: I_1 + I_2. \end{aligned} \tag{2.2.22}$$

By the induction hypothesis, we have

$$\begin{aligned} I_2 &\le \textstyle\sum_{j=1}^{[r]} \binom{r}{j} (E|S_k(m)|^{[r]})^{j/[r]} (E|S_{k+m+k_0}(m)|^{[r]})^{(r-j)/[r]} \\ &\le (m^{[r]/2} \sup_k E|X_k|^{[r]})^{r/[r]} \le c m^{r/2} a_1^r. \end{aligned}$$

Substituting the above inequality into (2.2.22), we obtain

$$a_{2m} \le \Big(2 + 2\sum_{j=1}^{[r]} \binom{r}{j} \varphi^{1/r}(k_0)\Big)^{1/r} a_m + c m^{1/2} a_1.$$

Now choosing a sufficiently large k_0 and proceeding as in the proof of Lemma 2.2.4, we conclude that the lemma holds in this case and similarly we have the lemma for $[r]+1$. This proves the lemma.

Using Lemma 2.2.7, Shao (1988a) proved the following Lemma.

Lemma 2.2.9. *Let $\{X_n, n \geq 1\}$ be a φ-mixing sequence satisfying (2.2.17) and $q > 0$ satisfying $\eta 4^q < 1 - \eta$. Then*

$$E \max_{1\leq i\leq n} |S_i|^q \leq (1-\eta-\eta 4^q)^{-1}\Big\{(8A)^q + 2(4p)^q E \max_{1\leq i\leq n} |X_i|^q\Big\},$$

where η, p, A are defined in Lemma 2.2.7.

Proof. By Lemma 2.2.7, we have for $x \geq 8A$

$$\begin{aligned}
&P\Big\{\max_{1\leq i\leq n}|S_i| \geq x\Big\} \\
&\quad\leq \frac{1}{1-\eta}\Big(P\Big\{|S_n| \geq \frac{5}{8}x\Big\} + P\Big\{\max_{1\leq i\leq n}|X_i| \geq \frac{x}{4p}\Big\}\Big) \\
&\quad\leq \frac{1}{1-\eta}\Big(\eta P\Big\{\max_{1\leq i\leq n}|S_i| \geq \frac{x}{4}\Big\} + 2P\Big\{\max_{1\leq i\leq n}|X_i| \geq \frac{x}{4p}\Big\}\Big).
\end{aligned}$$

Hence for any $B > 8A$

$$\begin{aligned}
&\int_0^B qy^{q-1}P\Big\{\max_{1\leq i\leq n}|S_i| \geq y\Big\}\,dy \\
&\quad\leq \int_0^{8A} qy^{q-1}P\Big\{\max_{1\leq i\leq n}|S_i| \geq y\Big\}\,dy \\
&\qquad + \frac{\eta}{1-\eta}\int_{8A}^B qy^{q-1}P\Big\{\max_{1\leq i\leq n}|S_i| \geq \frac{y}{4}\Big\}\,dy \\
&\qquad + \frac{2}{1-\eta}\int_{8A}^B qy^{q-1}P\Big\{\max_{1\leq i\leq n}|X_i| \geq \frac{y}{4p}\Big\}\,dy \\
&\quad\leq (8A)^q + \frac{\eta}{1-\eta}4^q\int_0^B qy^{q-1}P\Big\{\max_{1\leq i\leq n}|S_i| \geq y\Big\}\,dy \\
&\qquad + \frac{2(4p)^q}{1-\eta}\int_0^\infty qy^{q-1}P\Big\{\max_{1\leq i\leq n}|X_i| \geq y\Big\}\,dy.
\end{aligned}$$

which implies that

$$\begin{aligned}
&\int_0^B qy^{q-1}P\Big\{\max_{1\leq i\leq n}|S_i| \geq y\Big\}\,dy \\
&\quad\leq \Big(1-\frac{\eta 4^q}{1-\eta}\Big)^{-1}\Big((8A)^q + \frac{2(4p)^q}{1-\eta}\int_0^\infty qy^{q-1}P\Big\{\max_{1\leq i\leq n}|X_i| \geq y\Big\}\,dy\Big) \\
&\quad\leq (1-\eta-\eta 4^q)^{-1}\Big((8A)^q + 2(4p)^q E\max_{1\leq i\leq n}|X_i|^q\Big).
\end{aligned}$$

Letting $B \to \infty$ yields the assertion of the lemma.

A similar result is

Lemma 2.2.10. *Let $\{X_n, n \geq 1\}$ be a φ-mixing sequence. Suppose that there exists an array $\{c_{kn}\}$ of positive numbers such that*

$$\max_{1 \leq i \leq n} ES_k^2(i) \leq c_{kn}. \tag{2.2.23}$$

Then, for any $q \geq 2$, there exists a $C = C(q, \varphi(\cdot))$ such that

$$E \max_{1 \leq i \leq n} |S_k(i)|^q \leq C\Big(c_{kn}^{q/2} + E \max_{k < i \leq k+n} |X_i|^q\Big). \tag{2.2.24}$$

Proof. Take $\eta = 4^{-2q}$, $A^2 = 2c_{kn}/\eta$.There exists a p_0 such that $\varphi(p_0) \leq \eta/2$ since $\varphi(p) \to 0$ as $p \to \infty$. Using (2.2.23) we can verify that (2.2.17) is satisfied. Hence, we get (2.2.24) from Lemma 2.2.9.

Part II Weak Convergence

In this part, we investigate weak convergence of probability measures (or distributions) of normalized sums of the form

$$\frac{1}{B_n}\sum_{j=1}^{n} X_j - A_n. \tag{II1}$$

We have found a series of this kind of results for an independent sequence (cf. e.g. Petrov 1971 and Billingsley 1968). A natural question is what about a weakly dependent sequence. Only assumption of weak dependence is not enough for weak convergence. For instance, let $\{\xi_n,\ n \geq 1\}$ be a sequence of i.i.d. random variables with a common characteristic function $f(t)$, and let

$$X_n = \xi_{n+1} - \xi_n.$$

Then $\{X_n,\ n \geq 1\}$ is a strictly stationary sequence and satisfies any mixing conditions mentioned in §1.1. The sum

$$\sum_{k=1}^{n} X_k = \xi_{n+1} - \xi_1 \tag{II2}$$

has the characteristic function $|f(t)|^2$ for all n. It is reasonable to introduce some restrictions to make the variance of the sum $\sum_{k=1}^{n} X_k$ increase when n is increasing. Hence we assume always that B_n in (II1) tends to infinity as $n \to \infty$.

First of all, we state the so-called *Bernstein's blocking technique* utilized frequently in showing limit theorems for mixing random variables. Let positive integers $p = p(n)$, $q = q(n)$ and $k = k(n)$ with $1 \leq p \leq n$, $q =$

$o(p)$, $k = [n/(p+q)]$,[1] and let

$$\xi_j = \sum_{i=(j-1)(p+q)+1}^{jp+(j-1)q} X_i, \qquad \eta_j = \sum_{i=jp+(j-1)q+1}^{j(p+q)} X_i, \; j = 1, \cdots, k,$$

$$\eta_{k+1} = \sum_{i=k(p+q)+1}^{n} X_i. \tag{II3}$$

Then

$$S_n = \sum_{j=1}^{k} \xi_j + \sum_{j=1}^{k+1} \eta_j. \tag{II4}$$

By weak dependence, $\xi_1, \xi_2, \cdots, \xi_k$ are asymptotically independent as $q = q(n)$ is large enough. On the other hand, the sum $\sum_{j=1}^{k+1} \eta_j$ is negligible, compared with S_n by noting $q = o(p)$. Consequently, the Bernstein method allows us to consider the sums of mixing random variables as independent sums.

Using this method, by the procedure similar to that for an independent sequence, for α-mixing sequence, we may prove the following theorem about the class of possible limit distributions of sums.

Theorem II1. *Let $\{X_n, n \geq 1\}$ be a strictly stationary α-mixing sequence, $\{A_n\}$ and $\{B_n\}$ two sequences of real numbers with $B_n \to \infty$ as $n \to \infty$. Suppose that the distribution function $F_n(x)$ of the sum*

$$\frac{1}{B_n} \sum_{k=1}^{n} X_k - A_n$$

converges weakly to a distribution function $F(x)$. Then $F(x)$ is stable with some exponent a. Moreover,

$$B_n = n^{1/a} h(n),$$

where $h(n)$ is a slowly varying function with positive integer argument.

[1] In this book, the sign $[\cdot]$ sometimes denotes the greatest integer part or at other times denotes brackets. It will be clear from the context.

Chapter 3 Weak Convergence for α-mixing Sequences

3.1 Necessary and sufficient conditions for the CLT

For an α-mixing sequence $\{X_n, n \geq 1\}$, Ibragimov (1959, 1962) first gave necessary and sufficient conditions for the *central limit theorem* (*CLT*). In this chapter we always assume that $\{X_n, n \geq 1\}$ is a strictly stationary α-mixing sequence unless special indication. Put $S_n = \sum_{j=1}^n X_j$, $\sigma_n^2 = \mathrm{Var} S_n$.

Theorem 3.1.1. *Suppose that $EX_1 = 0$ and $EX_1^2 < \infty$. Then in order that $\{X_n\}$ obeys the CLT and $\lim_{n\to\infty} \sigma_n^2 = \infty$, it is necessary that*

(1) $\sigma_n^2 = nh(n)$, where $h(x)$ is a slowly varying function of the continuous variable $x > 0$,

(2) for any pair of sequences $p = p(n)$, $q = q(n)$ satisfying that as $n \to \infty$.

(a) $p \to \infty$, $q \to \infty$, $q = o(p)$, $p = o(n)$,

(b) $n^{1-\beta} q^{1+\beta} p^{-2} \to 0$ for all $\beta > 0$,

(c) $np^{-1}\alpha(q) \to 0$, and

$$\lim_{n\to\infty} \frac{n}{p\sigma_n^2} \int_{|x|>\varepsilon\sigma_n} x^2 \, dF_p(x) = 0 \tag{3.1.1}$$

for any $\varepsilon > 0$, where $F_p(x) = P(S_p < x)$. Conversely, if (1) holds and if (3.1.1) is satisfied for some choice of the functions p and q satsifying the given conditions, then the CLT is satisfied.

We do not prove this theorem here. For its proof we refer to Ibragimov and Linnik's book (1971).

A simpler necessary and sufficient condition for the CLT was given by Denker (1985).

Theorem 3.1.2. *Suppose that $EX_1 = 0$, $EX_1^2 < \infty$ and $\sigma_n^2 \to \infty$ as $n \to \infty$. Then in order that $\{X_n\}$ obeys the CLT, it is necessary and sufficient that $\{S_n^2/\sigma_n^2,\ n \geq 1\}$ is integrable uniformly.*

Proof. Necessity. Suppose that

$$S_n/\sigma_n \xrightarrow{d} N(0,1),$$

where $N(0,1)$ stands for a standard normal variable. Therefore, for any $\varepsilon > 0$ there exists a $K > 0$ such that

$$\lim_{n\to\infty} \int_{|S_n/\sigma_n|>K} S_n^2/\sigma_n^2\, dP = \int_{|N|>K} N^2\, dP < \varepsilon,$$

which implies uniform integrability of $\{S_n^2/\sigma_n^2,\ n \geq 1\}$.

Sufficiency. Suppose that $\{S_n^2/\sigma_n^2,\ n \geq 1\}$ is integrable uniformly. By Theorem 2.1.3, we have

$$\sigma_n^2 = nh(n), \tag{3.1.2}$$

where $h(x)$ is a slowly varying function on $[1, \infty)$. Assume that p and q are the functions satisfying conditions (a) and (b) in Theorem 3.1.1. Furthermore, we choose p and q that satisfy condition (c) in Theorem 3.1.1 and

$$n^2p^{-2}\sigma_q^2\sigma_p^{-2} = n^2p^{-3}qh(q)/h(p) \to 0 \quad \text{as } n \to \infty. \tag{3.1.3}$$

For any given $\delta > 0$, there exists a $K > 0$ such that

$$\int_{|S_n|>K\sigma_n} S_n^2\, dP < \delta\sigma_n^2 \tag{3.1.4}$$

for each $n \geq 1$. By Property A4 of a slowly varying function,

$$\frac{\sigma_p^2}{\sigma_n^2} = \frac{ph(p)}{nh(n)} = o(1) \quad \text{as}\, n \to \infty,$$

which implies that $K\sigma_p < \varepsilon\sigma_n$ for all large n. Hence

$$\begin{aligned}\frac{n}{p}\int_{|S_p|>\varepsilon\sigma_n} S_p^2\, dP &\leq \frac{n}{p}\int_{|S_p|>K\sigma_p} S_p^2\, dP \\ &< \frac{n}{p}\delta\sigma_p^2 < 2\delta\sigma_n^2.\end{aligned}$$

The last inequality is due to (II4), condition (c) and (3.1.3). Consequently, condition (3.1.1) is satisfied. From Theorem 3.1.1, sufficiency is proved.

It is well known that boundedness of $\{E|S_n/\sigma_n|^{2+\delta}, n \geq 1\}$ implies uniform integrability of $\{S_n^2/\sigma_n^2, n \geq 1\}$. The former requires existence of higher order moments. This explains that the moment conditions imposed on $\{S_n\}$ are important for the CLT. A related result is given by Dehling, Denker and Philipp (1986).

Theorem 3.1.3. *Suppose that $EX_1 = 0$, $EX_1^2 = 1$, $\sigma_n^2 = nh(n)$, where $h(n)$ is a slowly varying function. Then in order that the distributions of $\{S_n/\sigma_n, n \geq 1\}$ tend to the standard normal distribution $\Phi(x)$, it is necessary and sufficient that*

$$\limsup_{n\to\infty} \sigma_n/E|S_n| \leq \sqrt{\pi/2}. \tag{3.1.5}$$

In order to prove this theorem, we need the following lemma.

At first, we introduce some notations. Let integer p and real number g satisfy

$$2 \leq g \leq \alpha^{-1/4}(\sigma_p^{1/4}) \wedge \sigma_p^{1/4},$$

where $\alpha(x) = \alpha([x])$, and let

$$v^2 = \sigma_p^{-2} \int_{g^{1/2}<|S_p|/\sigma_p\leq g} S_p^2\, dP, \tag{3.1.6}$$

$$u^2 = \int_{|S_p|\leq g\sigma_p} S_p^2\, dP, \tag{3.1.7}$$

$$r = [g^2 d] \tag{3.1.8}$$

where d satisfies $2g^{-1/2} \vee v^2 \leq d < 1$,

$$n = r(p + [\sigma_p^{1/4}]) \text{ and} \tau^2 = ru^2. \tag{3.1.9}$$

Lemma 3.1.1. *Suppose that $EX_1 = 0$ and $EX_1^2 = 1$. We have*

$$\begin{aligned} &|E\exp(itS_n/\tau) - \exp(-t^2/2)| \\ &\quad\leq 2d + |t|\sigma_p d^{1/2}/u + |t|^3\sigma_p/(ug^{1/4}) + |t|^3\sigma_p^3 v/u^3 \\ &\qquad + 4\alpha^{1/2}(\sigma_p^{1/4}) + t^4/g + t^2\sigma_p^2/(u^2 g). \end{aligned}$$

Proof. Note that $u \leq \sigma_p$. Hence, for $|t| > r^{1/2}$, we have

$$|t|^3\sigma_p/(ug^{1/4}) \geq r^{3/2}/g^{1/4} \geq ((g^2 d - 1)g^{-1/6})^{3/2} \geq g^2.$$

Then the conclusion of Lemma 3.1.1 holds obviously.

Now we assume that $|t| \le r^{1/2}$. Let $q = [\sigma_p^{1/4}]$ and

$$\xi_j = \sum_{i=(j-1)(p+q)+1}^{jp+(j-1)q} X_i,$$

$$\eta_j = \sum_{i=jp+(j-1)q+1}^{j(p+q)} X_i, \qquad j = 1, \cdots, r.$$

Recalling the definition of n, we can write

$$S_n = \sum_{j=1}^{r} \xi_j + \sum_{j=1}^{r} \eta_j =: S_n' + S_n''.$$

By the Minkowski inequality

$$ES_n''^2 \le r^2 \sigma_p^{1/2} \le g^4 \sigma_p^{1/2} \le \sigma_p^{3/2}.$$

Hence

$$\begin{aligned}
&|E\exp(itS_n/\tau) - E\exp(itS_n'/\tau)| \\
&\quad \le |E\exp(itS_n''/\tau) - 1| \\
&\quad \le t^2 ES_n''^2/\tau^2 \le t^2\sigma_p^{3/2}/(u^2 r) \\
&\quad \le t^2\sigma_p^2/(u^2 g).
\end{aligned} \tag{3.1.10}$$

Moreover, by Lemma 1.2.1, we have

$$\begin{aligned}
&|E\exp(itS_n'/\tau) - (E\exp(itS_p/\tau))^r| \\
&\quad \le 4r\alpha(\sigma_p^{1/4}) \le 4\alpha^{1/2}(\sigma_p^{1/4}).
\end{aligned} \tag{3.1.11}$$

We now estimate $|E\exp(itS_p/\tau) - (1 - t^2/(2r))|$. By the Chebyshev inequality we have

$$\Big|\int_{|S_p|>g\sigma_p} \exp(itS_p/\tau)\, dP\Big| \le g^{-2} \le d/r. \tag{3.1.12}$$

By Taylor's theorem

$$\begin{aligned}
&\Big|\int_{|S_p|\le g\sigma_p} \exp(itS_p/\tau)\, dP - (1 - t^2/(2r))\Big| \\
&\le \Big|P(|S_p| \le g\sigma_p) + \frac{it}{\tau}\int_{|S_p|\le g\sigma_p} S_p\, dP \\
&\quad - \frac{t^2}{2\tau^2}\int_{|S_p|\le g\sigma_p} S_p^2\, dP - \Big(1 - \frac{t^2}{2r}\Big)\Big| \\
&\quad + \frac{|t|^3}{\tau^3}\int_{|S_p|\le g\sigma_p} |S_p|^3\, dP.
\end{aligned} \tag{3.1.13}$$

Similarly to (3.1.12),

$$|1 - P(|S_p| \le g\sigma_p)| \le g^{-2} \le d/r. \tag{3.1.14}$$

Noting $ES_p = 0$, we obtain

$$\tau^{-1}\Big|\int_{|S_p|\le g\sigma_p} S_p\, dP\Big| = \tau^{-1}\Big|\int_{|S_p|>g\sigma_p} S_p\, dP\Big| \le \sigma_p/\tau g \le \sigma_p d^{1/2}/(ru). \tag{3.1.15}$$

It is clear that

$$\frac{t^2}{2\tau^2}\int_{|S_p|\le g\sigma_p} S_p^2\, dP = \frac{t^2}{2r}. \tag{3.1.16}$$

The cubic term in (3.1.13) is estimated as follows.

$$\begin{aligned}\tau^{-3}&\int_{g^{1/2}\sigma_p<|S_p|\le g\sigma_p} |S_p|^3\, dP\\ &\le \tau^{-3} g\sigma_p^3 v^2\\ &\le \sigma_p^3 v/(ru^3)\end{aligned} \tag{3.1.17}$$

and

$$\begin{aligned}\tau^{-3}&\int_{|S_p|\le g^{1/2}\sigma_p} |S_p|^3\, dP\\ &\le \tau^{-3} g^{1/2}\sigma_p u^2\\ &\le \sigma_p g^{1/2}/(ur^{3/2})\\ &\le \sigma_p/(rug^{1/4}).\end{aligned} \tag{3.1.18}$$

Hence substituting (3.1.14)-(3.1.18) into (3.1.13) we obtain by (3.1.12)

$$|E\exp(itS_p/\tau) - (1 - t^2/(2r))| \le \eta/r,$$

where

$$\eta = 2d + |t|\sigma_p d^{1/2}/u + |t|^3\sigma_p/(ug^{1/4}) + |t|^3\sigma_p^3 v/u^3.$$

Note that $|a^r - b^r| \le r|a-b|$ for $|a| \le 1$, $|b| \le 1$. We obtain for $|t| < r^{1/2}$

$$|E\exp(itS_n'/\tau) - (1 - t^2/(2r))^r| \le \eta + 4\alpha^{1/2}(\sigma_p^{1/4}) \tag{3.1.19}$$

by (3.1.11). Moreover

$$|\exp(-t^2/2) - (1 - t^2/(2r))^r| \le \frac{1}{4}t^4 r^{-1} \quad \text{for } |t| < r^{1/2},$$

since $|e^x - (1+x)| \le x^2$ for $|x| \le \frac{1}{2}$. Consequently, the lemma follows from (3.1.19), (3.1.10).

Proof of Theorem 3.1.3. Necessity. If the distribution of S_n/σ_n tends to $\Phi(x)$, then for any $a > 0$

$$\liminf_{n\to\infty} E|S_n|/\sigma_n \ge \int_{-a}^{a} \frac{|x|}{\sqrt{2\pi}} e^{-x^2/2}\, dx,$$

which implies (3.1.5).

Sufficiency. Let $\rho_n = \sqrt{\pi/2}E|S_n|$. If we can show that

$$S_n/\rho_n \xrightarrow{d} N(0,1) \quad \text{as } n \to \infty, \tag{3.1.20}$$

then for any $\alpha > 0$

$$\begin{aligned}\int_{|N|\le\alpha} N^2 dP &= \lim_{n\to\infty} \rho_n^{-2} \int_{|S_n|/\rho_n \le \alpha} S_n^2\, dP \\ &\le \limsup_{n\to\infty} \sigma_n^2/\rho_n^2 \le 1\end{aligned}$$

by (3.1.5). Letting $\alpha \to \infty$ yields that $\sigma_n/\rho_n \to 1$ as $n \to \infty$. Hence (3.1.20) implies $S_n/\sigma_n \xrightarrow{d} N(0,1)$.

Now we are going to prove (3.1.20). At first, for each $n \in \mathbb{N}$ we show that there is an infinite sequence $Q \subset \mathbb{N}$ and real numbers τ_n, $n \in Q$ such that

$$S_n/\tau_n \xrightarrow{d} N(0,1) \quad n \to \infty,\ n \in Q. \tag{3.1.21}$$

For this purpose we prove that there exist a sequence $\{g(p),\, p \ge 1\}$ and a monotone sequence $\{c(p),\, p \ge 1\}$ with the following properties:

$$g(p) \to \infty,\ c(p) \to 0 \quad \text{as } p \to \infty, \tag{3.1.22}$$

$$g(p) \le \alpha^{-1/4}(\sigma_p^{1/4}) \wedge \sigma_p^{1/4}, \tag{3.1.23}$$

$$v^2(p) := \sigma_p^{-2} \int_{g(p)^{1/2} < |S_p|/\sigma_p \le g(p)} S_p^2\, dP \to 0 \quad \text{as } p \to \infty \tag{3.1.24}$$

and

$$2g(p)^{-1/2} \vee v^2(p) \le c(p) < 1. \tag{3.1.25}$$

We first choose a sequence $\{z(p),\, p \ge 1\}$ with

$$\lim_{p\to\infty} z(p) = \infty, \quad z(p) \le \alpha^{-1/4}(\sigma_p^{1/4}) \wedge \sigma_p^{1/4}. \tag{3.1.26}$$

Next, we choose a sequence $\{i(p),\, p \ge 1\}$ such that

$$i(p) \to \infty,\ 2^{-i(p)} \log z(p) \to \infty \quad \text{as } p \to \infty. \tag{3.1.27}$$

Fix $p \in \mathbb{N}$. Since the intervals $I_i(p) := (z(p)^{2^{-i-1}}, z(p)^{2^{-i}}]$, $0 \le i < i(p)$, are disjoint, there exists an integer $k = k(p)$ with $0 \le k \le i(p)$ such that

$$\sigma_p^{-2} \int_{|S_p|/\sigma_p \in I_k(p)} S_p^2 \, dP \le 1/i(p). \tag{3.1.28}$$

Let

$$g(p) = z(p)^{2^{-k(p)}}. \tag{3.1.29}$$

Then $g(p) \to \infty$ as $p \to \infty$ by (3.1.27). Because of (3.1.26)–(3.1.28), (3.1.23) and (3.1.24) are satisfied. Since $2g(p)^{-1/2} \vee v^2(p) \to 0$ as $p \to \infty$ we can choose $\{c(p)\}$ with $c(p) \downarrow 0$ and satisfying (3.1.25).

With these choices of $\{g(p)\}$ and $\{c(p)\}$, we define $u(p)$, $r(p)$ and $n(p)$ by (3.1.7), (3.1.8) and (3.1.9) respectively. Put $Q = \{n(p),\, p \ge 1\}$ and define τ_n^2, $n \in Q$ by (3.1.9). Since

$$\begin{aligned} \sigma_p^{-1} E|S_p| &= \sigma_p^{-1} \int_{|S_p|/\sigma_p \le g(p)} |S_p| \, dP + \sigma_p^{-1} \int_{|S_p|/\sigma_p > g(p)} |S_p| \, dP \\ &\le \sigma_p^{-1} u(p) + g(p)^{-1}, \end{aligned} \tag{3.1.30}$$

we have, by (3.1.5), for p large enough

$$u(p)/\sigma_p \ge \gamma/2, \tag{3.1.31}$$

where $\gamma = \inf\{E|S_p|/\sigma_p,\, p \ge 1\} > 0$. Lemma 3.1.1 now implies (3.1.21).

Next we show that

$$\lim_{n \to \infty, n \in Q} \tau_n/\rho_n = 1. \tag{3.1.32}$$

To see this we choose a sequence $\{b(m),\, m \ge 1\}$ with

$$\begin{aligned} &\lim_{m \to \infty} b(m) = \infty, \\ &\lim_{m \to \infty} \sup\{|h(tm)/h(m) - 1|,\, 1 \le t \le b(m)\} = 0. \end{aligned} \tag{3.1.33}$$

This is possible. Indeed, by the Karamata theorem (see Appendix Theorem A1) there exists an increasing sequence $\{m_k,\, k \ge 2\}$ such that

$$\left| \sup_{1 \le t \le k} h(tm)/h(m) - 1 \right| \le \frac{1}{k}, \quad m \ge m_k.$$

Then $b(\cdot)$ defined by $b(m) = k$ for $m_k < m \le m_{k+1}$ has the desired properties. Of course, we can assume that $\{z(p),\, p \ge 1\}$ is chosen so that in addition to (3.1.26) we have $z(p) \le b(p)^{1/2}/2$. Then for all large p we have, by (3.1.9) and $\sigma_p^2 \le p^2$,

$$\frac{\sigma^2(n(p))}{r(p)\sigma_p^2} = \frac{r(p)(p+[\sigma_p^{1/4}])h(r(p)(p+[\sigma_p^{1/4}]))}{r(p)ph(p)}$$
$$= (1+O(p^{-1/2}))\frac{h(r(p)(p+[\sigma_p^{1/4}]))}{h(p)} = 1+o(1)$$

by (3.1.33). Thus, by (3.1.9) and (3.1.30) we have for p large enough

$$E(S_{n(p)}/\tau_{n(p)})^2 = \sigma_{n(p)}^2/\tau_{n(p)}^2 \le 2\sigma_p^2/u(p)^2 \le 8/\gamma^2. \tag{3.1.34}$$

Hence $\{S_n/\tau_n,\ n \in Q\}$ is uniformly integrable and thus, by (3.1.21),

$$\lim_{n\to\infty, n\in Q} E|S_n|/\tau_n = E|N| = \sqrt{2/\pi}.$$

This proves (3.1.32).

If $Q = \mathbb{N}$, (3.1.21) and (3.1.32) imply

$$S_n/\rho_n \xrightarrow{d} N(0,1) \quad \text{as}\, n \to \infty. \tag{3.1.35}$$

Consider the case of $Q \subset \mathbb{N}$. Assume that $\{z(p),\ p \ge 1\}$ constructed in the proof of (3.1.21) satisfies

$$z(p) \le \min(\alpha(\sigma_p^{1/4})^{-1/16},\ \sigma_p^{1/16},\ p^{1/4},\ b(p)^{1/8}/2) \tag{3.1.36}$$

and

$$z(p) \le z(q) \le z(p)^{3/2},\ p \le q \le p^2. \tag{3.1.37}$$

Such a sequence can be constructed as follows: First choose an increasing sequence $y(p)$ satisfying (3.1.36). By induction on k define $z(p) = y(p_k) \wedge z(p_{k-1})^{3/2}$ for $p = p_k = 2^{2^k}, p_k+1, \cdots, p_{k+1}-1$. Let $\{l(n),\ n \in Q\}$ and $\{j(n),\ n \in Q\}$ be two arbitrary sequences of real numbers tending to infinity and with $l(n) < j(n)$, $n \in Q$.

Recalling condition (3.1.5) and noting $E|S_n| \le \sigma_n$, we have

$$\sqrt{2/\pi} \le \limsup_{n\to\infty} \sigma_n/\rho_n \le 1. \tag{3.1.38}$$

Moreover, we have shown that $S_n/\rho_n \xrightarrow{d} N(0,1)$, $n \to \infty$, $n \in Q$. Consequently, we obtain, for any $\alpha > 0$

$$\int_{|N|\le\alpha} N^2\, dP = \lim_{n\in Q} \int_{|S_n|/\rho_n\le\alpha} S_n^2/\rho_n^2\, dP$$
$$\le \liminf_{n\in Q} \rho_n^{-2} \int_{|S_n|/\sigma_n \le l(n)} S_n^2\, dP,$$

which implies by letting $\alpha \to \infty$

$$1 \leq \liminf_{n\in Q} \rho_n^{-2} \int_{|S_n|/\sigma_n \leq l(n)} S_n^2 dP$$
$$\leq \limsup_{n\in Q} \rho_n^{-2} \int_{|S_n|/\sigma_n \leq j(n)} S_n^2 dP \leq 1. \tag{3.1.39}$$

(3.1.38) and (3.1.39) imply

$$\lim_{n\in Q} \sigma_n^{-2} \int_{l(n)<|S_n|/\sigma_n \leq j(n)} S_n^2\, dP = 0. \tag{3.1.40}$$

We shall apply Lemma 3.1.1 again. To prepare for it we set

$$l(n) = l(n(p)) = z(n(p)) \quad \text{if } n = n(p),\, p \geq 1 \tag{3.1.41}$$

and

$$j(n) = j(n(p)) = \min(\alpha^{-1/4}(\sigma_n^{1/4}),\, \sigma_n^{1/4},\, b(n)^{1/2}/2)$$
$$\text{if}\, n = n(p),\, p \geq 1. \tag{3.1.42}$$

For p large enough, we have by (3.1.22), (3.1.29) and (3.1.36)

$$n(p) \leq g^2(p)c(p)(p + p^{1/4}) \leq z^2(p)p \leq p^{1/2}p < p^2. \tag{3.1.43}$$

By (3.1.36), (3.1.42) and (3.1.41)

$$j(n) \geq z^4(n) = l^4(n) > l(n), \quad n \in Q. \tag{3.1.44}$$

Since, by (3.1.41)

$$w^2(n) := \sigma_n^{-2} \int_{l^{1/2}(n)<|S_n|/\sigma_n \leq j(n)} S_n^2\, dP \to 0, \quad n \in Q, \tag{3.1.45}$$

we can choose a nonincreasing sequence $\{d(n),\, n \in Q\}$ such that

$$\lim_{n\in Q} d(n) = 0 \quad \text{and} \quad d(n) \geq 2h(n)^{-1/2} \wedge w^2(n),\, n \in Q. \tag{3.1.46}$$

Let $Q = \{n_k,\, k \geq 1\}$ be arranged in increasing order and let J_k be the interval

$$J_k = [n_k l^2(n_k) d(n_k),\, n_k j^2(n_k) d(n_k)].$$

We show that there exists a k_0 such that

$$J_k \cap J_{k+1} \neq \emptyset,\, k \geq k_0. \tag{3.1.47}$$

Obviously $n(n_k) = r(n_k)(n_k + [\sigma(n_k)^{1/4}]) \in Q$. As n_{k+1} is the smallest member of Q bigger than n_k we must have for large k

$$n_{k+1} \le n(n_k) \le n_k z^2(n_k) < n_k^2$$

by (3.1.43). Hence, by (3.1.41) and (3.1.37), the left endpoint of J_{k+1} does not exceed

$$\begin{aligned} n_{k+1} l^2(n_{k+1}) d(n_{k+1}) &\le n_k z^2(n_k) z^2(n_{k+1}) \\ &\le n_k z^2(n_k) z^2(n_k^2) \le n_k z^5(n_k) \end{aligned}$$

for large k. On the other hand, by (3.1.46) and (3.1.44), the right endpoint of J_k is bigger than

$$n_k j^2(n_k) d(n_k) \ge n_k j^2(n_k) l^{-1/2}(n_k) \ge n_k z(n_k)^{15/2}$$

for large k. Since $z(n_k) \to \infty$ we obtain (3.1.47). Let $m \ge \min\{l, l \in J_{k_0}\}$. Then there is a $k \ge k_0$ such that $m \in J_k$. Thus we have for some $g \in [l(n_k), j(n_k)]$ and some $|\theta| \le 2$

$$\begin{aligned} m = g^2 d(n_k) n_k &= [g^2 d(n_k)](n_k + [\sigma^{1/4}(n_k)]) + \theta n_k \\ &=: M_k + \theta n_k. \end{aligned} \tag{3.1.48}$$

Now by (3.1.8), M_k is of the form (3.1.9) and hence we can apply Lemma 3.1.1. Put $p = n_k$ and $d = d(n_k)$. By (3.1.31), $u(n_k)/\sigma(n_k) \ge \frac{1}{2}\gamma > 0$. Now $g \ge l(n_k) \to \infty$ and $\alpha(\sigma^{1/4}(n_k)) \to 0$. Finally, by (3.1.45) and noting $l(n_k) \le g \le j(n_k)$, we have

$$v^2(n_k) := \sigma^{-2}(n_k) \int_{g^{1/2} < |S_{n_k}|/\sigma(n_k) \le g} S_{n_k}^2 \, dP \le w^2(n_k) \to 0.$$

Hence, by Lemma 3.1.1,

$$S_{M_k}/\tau(M_k) \xrightarrow{d} N(0,1). \tag{3.1.49}$$

Since $|\theta| \le 2$ we have by (3.1.33) for large k

$$\frac{ES^2_{|\theta| n_k}}{\sigma^2(n_k)} \le \frac{|\theta| n_k h(|\theta| n_k)}{n_k h(n_k)} \le 4.$$

Consequently

$$E(S_m - S_{M_k})^2 = ES^2_{|\theta| n_k} \le 4\sigma^2(n_k). \tag{3.1.50}$$

Denoting $r^*(n_k) = [g^2 d(n_k)]$ we obtain by (3.1.33) for large k

$$\frac{\sigma^2(M_k)}{r^*(n_k)\sigma^2(n_k)} \geq \frac{r^*(n_k)n_k h(r^*(n_k)(n_k + [\sigma^{1/4}(n_k)]))}{r^*(n_k)n_k h(n_k)} \geq \frac{1}{2}$$

since $r^*(n_k) \leq g^2 d(n_k) \leq j^2(n_k)d(n_k) \leq b(n_k)/2$ by (3.1.41). Hence, from (3.1.50) and as $r^*(n_k) \to \infty$, we have

$$E(S_m - S_{M_k})^2/\sigma^2(M_k) \to 0 \quad \text{as } k \to \infty. \tag{3.1.51}$$

In the same way as (3.1.34) one can prove

$$\sigma^2(M_k)/\tau^2(M_k) \leq 8/\gamma^2. \tag{3.1.52}$$

Put

$$\tau_m = \tau(M_k), \quad \text{if } m \text{ and } M_k \text{ are as in (3.1.47).}$$

Then by (3.1.49), (3.1.51) and (3.1.52)

$$S_m/\tau_m \xrightarrow{d} N(0,1) \quad \text{as } m \to \infty. \tag{3.1.53}$$

The sequence $\{S_m/\tau_m, m \geq 1\}$ is uniformly integerable in view of (3.1.51) and (3.1.52). We obtain (3.1.35) from (3.1.53) and (3.1.32).

With the help of it and similarly to (3.1.39) we have $\sigma_n/\rho_n \to 1$ as $n \to \infty$. This completes the proof of Theorem 3.1.3.

3.2 Sufficient conditions for the CLT and WIP

In the last section, we give some necessary and sufficient conditions for the CLT. But it is not easy to verify them. In this section, we shall give some sufficient conditions for the CLT and *weak invariance principle* (*WIP*). Rosenblatt (1956) first gave sufficient conditions for the CLT. After that many authors (e.g. Ibragimov 1962) have discussed this subject and have obtained some further results. One of the best is due to Herrndorf (1984, 1985). The following theorem is attributed to Gordin (1969) and has been restated and proved by Hall and Heyde (1980) (Corollary 5.3(ii)) via approximating S_n by a naturally related martingale with stationary ergodic differences. Its proof will not been presented here.

Theorem 3.2.1. *Suppose that* $EX_1 = 0$, $E|X_1|^{2+\delta} < \infty$ *for some* $\delta > 0$ *and*

$$\sum_{n=1}^{\infty} \alpha(n)^{\delta/(2+\delta)} < \infty. \tag{3.2.1}$$

Then

$$\sigma^2 := EX_1^2 + 2\sum_{j=2}^{\infty} EX_1X_j < \infty \tag{3.2.2}$$

and, if $\sigma \neq 0$,

$$S_n/\sigma\sqrt{n} \xrightarrow{d} N(0,1). \tag{3.2.3}$$

Remark 3.2.1. For the case of bounded variables, i.e. $\delta = \infty$, condition (3.2.1) is reduced to

$$\sum_{n=1}^{\infty} \alpha(n) < \infty. \tag{3.2.4}$$

Remark 3.2.2. Davydov (1973) gave two examples which pointed out that the rate of $\alpha(n)$ in Theorem 3.2.1 could not be improved in certain senses.

Example 3.2.1. For any $\delta > 0$ and $\varepsilon > 0$, there is a strictly stationary countable-state Markov chain $\{X_n,\ n \geq 1\}$ with $EX_1 = 0$, $E|X_1|^{2+\delta} < \infty$ such that

(i) $\alpha(n) = o(n^{-(1-\varepsilon)(1+2/\delta)})$ as $n \to \infty$,

(ii) $\mathrm{Var}S_n \approx n^d$ [1] for some $1 < d < 2$,

(iii) S_n is attracted to a symmetric stable law with exponent α, $1 < \alpha < 2$.

Example 3.2.2. For any $\varepsilon > 0$ there exists a strictly stationary countable-state Markov chain $\{X_n,\ n \geq 1\}$ with $EX_1 = 0$, $|X_1| < c_0$ a.s. for some $c_0 < \infty$ such that $\alpha(n) = o(n^{-(1-\varepsilon)})$ as $n \to \infty$ and properties (ii) and (iii) in Example 3.2.1 hold.

We now investigate the weak invariance principles. Define random elements on $D[0,1]$ as follows:

$$W_n(t) = S_{[nt]}/\sigma_n, \quad 0 \leq t \leq 1.$$

Convergence theory of probability measures tells us that the key to the proof for weak invariance principle lies in verification of *tightness*. One of the following conditions is sufficient for tightness (cf. Billingsley 1968, Section 16).

(1) For any $\varepsilon > 0$, $\eta > 0$, there exist a δ, $0 < \delta < 1$, and an integer n_0 such that, for $0 \leq t \leq 1$

$$\frac{1}{\delta}P\Big\{\sup_{t \leq s \leq t+\delta} |W_n(s) - W_n(t)| \geq \varepsilon\Big\} \leq \eta, \quad n \geq n_0. \tag{3.2.5}$$

[1] $a \approx b$ means $\lim a/b = 1$.

(2) For any $\varepsilon > 0$, there exists a $\lambda > 1$ and an integer n_0 such that

$$P\Big\{\max_{1\le i\le n} |S_i| \ge \lambda\sigma_n\Big\} \le \varepsilon/\lambda^2, \quad n \ge n_0. \tag{3.2.6}$$

Davydov (1968) first generalized the CLT to the invariance principle for bounded variables, but condition (3.2.4) was strengthened as $\sum_{n=1}^{\infty} \alpha^{1/2}(n) < \infty$. Oodaira and Yoshihara (1972) sharpened his result and obtained the invariance principle under the conditions for the CLT basically.

Theorem 3.2.2. *Suppose that $EX_1 = 0$, $|X_1| < c_0 < \infty$. If (3.2.4) is satisfied and $\alpha(n) \le c/n \log n$. Then, if $\sigma > 0$,*

$$W_n \Longrightarrow W,$$

i.e. W_n weakly converges to a Wiener process W with $\sigma_n = \sigma\sqrt{n}$.

Proof. By Remark 3.2.1, it follows that $W_n(t)$ converges to $W(t)$ in distribution for any t, $0 \le t \le 1$. By the Cramer-Wold method, it is easy to see that for any given $0 \le t_1 < t_2 < \cdots < t_k \le 1$, $(W_n(t_1), \cdots, W_n(t_k))$ converges to $(W(t_1), \cdots, W(t_k))$ in distribution.

We now show the tightness of $\{W_n\}$. It is enough by (3.2.6) to prove that

$$P\Big\{\max_{1\le i\le n} |S_i| \ge 3\lambda\sigma\sqrt{n}\Big\} \le \varepsilon/\lambda^2, \quad n \ge n_0. \tag{3.2.7}$$

Put $p = [\sqrt{n}/(\log n)^{3/8}]$, $k = [n/p]$. We have

$$P\{|X_1| + \cdots + |X_{2p}| \ge \lambda\sigma\sqrt{n}\} = 0 \tag{3.2.8}$$

for large n by boundedness of $\{X_n\}$. Moreover, using Lemma 1.2.1 and condition (3.2.4) we have uniform integrability of $\{S_n^2/n,\ n \ge 1\}$. Therefore, for any $\varepsilon > 0$, there exists a $\lambda > 1$ such that for each $i \ge 1$

$$P\{|S_i| \ge \lambda\sigma\sqrt{i}\} \le \varepsilon/3\lambda^2. \tag{3.2.9}$$

Put $E_j = \{\max_{1\le i<j}|S_i| < 3\lambda\sigma\sqrt{n} \le |S_j|\}$. We have

$$\begin{aligned}
&P\Big\{\max_{1\le i\le n}|S_i| \ge 3\lambda\sigma\sqrt{n}\Big\}\\
&\le P\{|S_n| \ge \lambda\sigma\sqrt{n}\} + P\Big(\bigcup_{j=1}^{n}\{E_j \bigcap (|S_n - S_j| \ge 2\lambda\sigma\sqrt{n})\}\Big)\\
&\le P\{|S_n| \ge \lambda\sigma\sqrt{n}\}\\
&\quad + \sum_{i=0}^{k-2} P\Big(\bigcup_{j=1}^{p}\Big\{E_{ip+j}\bigcap(|S_n - S_{ip+j}| \ge 2\lambda\sigma\sqrt{n})\Big\}\Big)\\
&\quad + \sum_{j=(k-1)p+1}^{n} P\Big\{|S_n - S_j| \ge 2\lambda\sigma\sqrt{n}\Big\}\\
&\le P\{|S_n| \ge \lambda\sigma\sqrt{n}\}\\
&\quad + \sum_{i=0}^{k-2} P\Big(\bigcup_{j=1}^{p}\Big\{E_{ip+j}\bigcap(|S_n - S_{(i+2)p}| \ge \lambda\sigma\sqrt{n})\Big\}\Big)\\
&\quad + \sum_{j=1}^{p} P\Big\{|S_{(i+2)p} - S_{ip+j}| \ge \lambda\sigma\sqrt{n}\Big\}\\
&\quad + \sum_{j=(k-1)p+1}^{n} P\Big\{|X_1| + \cdots + |X_{n-j}| \ge 2\lambda\sigma\sqrt{n}\Big\}\\
&\le P\{|S_n| \ge \lambda\sigma\sqrt{n}\}\\
&\quad + \sum_{i=0}^{k-2} P\Big\{\Big(\bigcup_{j=1}^{p} E_{ip+j}\Big)\bigcap\Big(|S_n - S_{(i+2)p}| \ge \lambda\sigma\sqrt{n}\Big)\Big\}\\
&\quad + 2nP\{|X_1| + \cdots + |X_{2p}| \ge \lambda\sigma\sqrt{n}\}\\
&=: I_1 + I_2 + I_3. \qquad (3.2.10)
\end{aligned}$$

By (3.2.8), $I_3 = 0$. (3.2.9) implies that $I_1 \le \varepsilon/3\lambda^2$. Since $\cup_{j=1}^{p} E_{ip+j} \in \mathcal{F}_0^{(i+1)p}$, $(|S_n - S_{(i+2)p}| \ge \lambda\sigma\sqrt{n}) \in \mathcal{F}_{(i+2)p+1}^{\infty}$, we obtain

$$\begin{aligned}
I_2 &\le \sum_{i=o}^{k-2} P\Big\{\bigcup_{j=1}^{p} E_{ip+j}\Big\}P\{|S_n - S_{(i+2)p}| \ge \lambda\sigma\sqrt{n}\} + k\alpha(p)\\
&\le \varepsilon/3\lambda^2 + k\alpha(p)
\end{aligned}$$

by (3.2.9) again. From $\alpha(n) \le c/n\log n$, it follows that

$$k\alpha(p) \le \frac{cn}{n(\log n)^{-3/4}\log(n^{1/2}(\log n)^{-3/8})} \to 0 \quad \text{as} \quad n \to \infty.$$

Inserting these estimations into (3.2.10) yields (3.2.7). The proof of Theorem 3.2.2 is completed.

Some authors have discussed and extended this theorem. The general result was given by Herrndorf (1985). He removed the assumption of stationarity and made the moment condition more flexible. Denote

$$\begin{aligned}\mathcal{G} = \{g(x) : [0, \infty) \to [0, \infty),\ g(x) \text{ is convex}; g(0) = 0,\\ g(x)/x^2 \text{is non-decreasing, } \lim_{x\to\infty} g(x)/x^2 = \infty\}.\end{aligned}$$

For every $g \in \mathcal{G}$ we define the inverse $\operatorname{inv} g : (0, \infty) \to (0, \infty)$ by $g(\operatorname{inv} g(x)) = x$ and $f_g : [0, \infty) \to [0, \infty)$ by $f_g(0) = 0$ and

$$f_g(x) = (\operatorname{inv} g(1/x))^2 x \quad \text{for } x > 0.$$

Theorem 3.2.3. *Let $\{X_n, n \geq 1\}$ be an α-mixing sequence with $EX_n = 0$, $EX_n^2 < \infty$ for all $n \geq 1$ and*

$$ES_n^2/n \to \sigma^2 \quad \text{as } n \to \infty \tag{3.2.11}$$

for some $\sigma > 0$. If there exists a $g \in G$ such that

$$\sup_{n\geq 1} Eg(|X_n|) < \infty, \qquad \sum_{n=1}^{\infty} f_g(\alpha(n)) < \infty, \tag{3.2.12}$$

then $W_n \Longrightarrow W$.

The proof of Theorem 3.2.3 needs the following lemmas.

Lemma 3.2.1. *Let $\xi_1, \cdots, \xi_n$ be random variables. Put*

$$\begin{aligned}\alpha = \max_{1\leq k\leq n-1} \sup\{|P(AB) - P(A)P(B)| :\ A \in \sigma(\xi_1, \cdots, \xi_k),\\ B \in \sigma(\xi_{k+1}, \cdots, \xi_n)\}.\end{aligned}$$

Then for any $\varepsilon > 0$,

$$\begin{aligned}P\Big\{\max_{1\leq k\leq n}\Big|\sum_{j=1}^{k} \xi_j\Big| > 2\varepsilon\Big\}\\ \leq \frac{P\{|\sum_{j=1}^n \xi_j| > \varepsilon\} + n\alpha}{\min_{1\leq k\leq n-1} P\{|\sum_{j=k+1}^n \xi_j| \leq \varepsilon\}}.\end{aligned} \tag{3.2.13}$$

Proof. Put

$$A_1 = \{|\xi_1| > 2\varepsilon\},$$
$$A_k = \Big\{\Big|\sum_{j=1}^{k}\xi_j\Big| > 2\varepsilon,\ \Big|\sum_{j=1}^{l}\xi_j\Big| \le 2\varepsilon,\ 1 \le l \le k-1\Big\}, \qquad 1 < k \le n.$$
$$B_k = \Big\{\Big|\sum_{j=k+1}^{n}\xi_j\Big| < \varepsilon\Big\},\ 1 \le k < n,\ B_n = \Omega,\ C = \Big\{\Big|\sum_{j=1}^{n}\xi_j\Big| > \varepsilon\Big\}.$$

It is easy to see that $\cup_{k=1}^{n} A_k B_k \subset C$ and

$$|P(A_k B_k) - P(A_k)P(B_k)| \le \alpha.$$

Hence

$$P(C) \ge \sum_{k=1}^{n} P(A_k B_k) \ge \min_{1\le k\le n} P(B_k) \sum_{k=1}^{n} P(A_k) - n\alpha. \tag{3.2.14}$$

Note that

$$P\Big\{\max_{1\le k\le n}\Big|\sum_{j=1}^{k}\xi_j\Big| > 2\varepsilon\Big\} = \sum_{k=1}^{n} P(A_k).$$

Combining it with (3.2.14) yields (3.2.13).

Lemma 3.2.2. *Let $\{X_n, n \ge 1\}$ be as in Theorem 3.2.3, and satisfy*

$$\sup_{n\ge 1,\, m\ge 0} E(S_{m+n} - S_m)^2/n < \infty. \tag{3.2.15}$$

Assume that there exist positive integers $p = p(n)$, $q = q(n)$ such that as $n \to \infty$

$$p = o(n),\ q = o(p),\ np^{-1}\alpha(q) = o(1), \tag{3.2.16}$$
$$n^{-1} \sum_{1\le i,\, j\le n,\, |i-j|>q} |EX_iX_j| \to 0, \tag{3.2.17}$$

$$p^{-1} \max_{0\le m\le n-p} E\{(S_{m+p}-S_m)^2 I(|S_{m+p}-S_m| > \varepsilon n^{1/2})\} \to 0 \quad \text{for any } \varepsilon > 0. \tag{3.2.18}$$

Then the CLT holds.

If, moreover, for any $\varepsilon > 0$

$$np^{-1} \max_{0\le m\le n-p} P\Big\{\max_{1\le r\le p} |S_{m+r} - S_m| > \varepsilon\sqrt{n}\Big\} \to 0 \tag{3.2.19}$$

then $W_n \Longrightarrow W$ with $\sigma_n = \sigma\sqrt{n}$.

Proof. Denote $k = [n/(p+q)]$. $k \to \infty$ as $n \to \infty$ by (3.2.16). Put

$$\xi_j = \sum_{i=j(p+q)+1}^{j(p+q)+p} X_i,$$

$$\eta_j = \sum_{i=j(p+q)+p+1}^{(j+1)(p+q)} X_i, \quad 0 \le j \le k-1,$$

$$\eta_k = \sum_{i=k(p+q)+1}^{n} X_i,$$

$$S_n' = \sum_{j=0}^{k-1} \xi_j, \quad S_n'' = \sum_{j=0}^{k} \eta_j.$$

In order to prove the CLT, it suffices to show that as $n \to \infty$

(a) $ES_n''^2/n \to 0$,

(b) $\sum_{0\le i<j\le k-1} |E\xi_i\xi_j|/n \to 0$,

(c) $\left|E\exp(itS_n' - \prod_{j=0}^{k-1} E\exp(it\xi_j)\right| \to 0$ uniformly in $t \in (-\infty, \infty)$,

(d) $\sum_{j=0}^{k-1} E(\xi_j^2 I(|\xi_j| > \varepsilon\sigma n^{1/2}))/n \to 0$ for any $\varepsilon > 0$.

From Lemma 1.2.3 and (3.2.15) we have

$$\begin{aligned} ES_n''^2 &\le \sum_{j=0}^{k} E\eta_j^2 + 2\sum_{i=1}^{n} \sum_{i+p<j\le n} |EX_iX_j| \\ &\le c(kq+p+q) + 20n \sum_{i=p+1}^{\infty} f_g(\alpha(i)) \sup_{j\le n} \|X_j\|_g^2. \end{aligned}$$

Now (a) follows from (3.2.12) and (3.2.16). By the same way, we obtain (b). As to (c), we have by Lemma 1.2.1 and (3.2.16)

$$\left|E\exp(itS_n') - \prod_{j=0}^{k-1} E\exp(it\xi_j)\right| \le 16k\alpha(q) \to 0 \quad \text{as } n \to \infty.$$

Finally, (d) follows from (3.2.18) immediately. Thus the CLT holds true for $\{X_n\}$.

Let $0 \le t_1 < \cdots < t_k \le 1$ be given. We wish to show

$$(W_n(t_1), \cdots, W_n(t_k)) \to (W(t_1), \cdots, W(t_k)) \quad \text{as} \quad n \to \infty \tag{3.2.20}$$

in distribution. From (3.2.11) and the CLT, $W_n(t)$ converges to $W(t)$ in distribution for any t, $0 \le t \le 1$. Therefore $\{W_n(t_1), \cdots, W_n(t_k)\}$ is tight by Prohorov's classical characterization of tightness. Let Q be the limit distribution of some subsequence of $\{W_n(t_1), \cdots, W_n(t_k)\}$, and π_{t_i} the mapping that carries the point $x = (x_1, \cdots, x_k)$ of R^k to the point x_{t_i} of R. According to the CLT, the marginal distribution $Q\pi_{t_i}^{-1}$ are normal with variance t_i. Take $r_n = q/n$, then $r_n \to 0$ and $\alpha_n(q) \to 0$, where

$$\begin{aligned}\alpha_n(q) = \sup\{|P(AB) - P(A)P(B)| :\\ &A \in \sigma(X_i,\, 1 \le i \le m),\\ &B \in \sigma(X_i,\, m+q \le i \le n,\, 1 \le m \le n-q\}.\end{aligned}$$

Using (3.2.15) one obtains $E(W_n(t_i + r_n) - W_n(t_i))^2 \to 0$ as $n \to \infty$. Hence $Q(\pi_{t_1}, \pi_{t_2} - \pi_{t_1}, \cdots, \pi_{t_k} - \pi_{t_{k-1}})^{-1}$ is the limit distribution of some subsequence of $(W_n(t_1),\, W_n(t_2) - W_n(t_1 + r_n), \cdots, W(t_k) - W_n(t_{k-1} + r_n))$. Now $\alpha_n([nr_n] - 1) \to 0$ implies that $\pi_{t_1}, \pi_{t_2} - \pi_{t_1}, \cdots, \pi_{t_k} - \pi_{t_{k-1}}$ are independent under Q. Therefore Q is the distribution of $(W(t_1), \cdots, W(t_k))$. This argument proves (3.2.20).

Finally, we have to prove the tightness of the sequence $\{W_n\}$. It suffices to show a version of (3.2.5), i.e., for any $\varepsilon > 0$, $\eta > 0$ there exist a δ, $0 < \delta < 1$, and an integer n_0 such that

$$\sum_{k=0}^{[1/\delta]} P\Big\{\max_{[nk\delta] < r \le [n(k+1)\delta]} |S_r - S_{[nk\delta]}| > \varepsilon\sigma\sqrt{n}\Big\} < \eta, \quad n \ge n_0. \tag{3.2.21}$$

Let $k \in \{0, \cdots, [1/\delta]\}$ be fixed, $m = m(n) = [([n(k+1)\delta] - [nk\delta])/(p+q)]$,

$$\xi_j = \sum_{i=[nk\delta]+j(p+q)+1}^{[nk\delta]+j(p+q)+p} X_i,\; \eta_j = \sum_{i=[nk\delta]+j(p+q)+p+1}^{[nk\delta]+(j+1)(p+q)} X_i, \quad j = 0, 1, \cdots, m-1.$$

Then we have

$$\begin{aligned}&P\Big\{\max_{[nk\delta] < r \le [n(k+1)\delta]} |S_r - S_{[nk\delta]}| > \varepsilon\sigma\sqrt{n}\Big\}\\ &\le (m+1) \max_{0 \le l \le n-(p+q)} P\Big\{\max_{1 \le r \le p+q} |S_{l+r} - S_l| > \varepsilon\sigma\sqrt{n}/3\Big\}\\ &\quad + P\Big\{\max_{0 \le r \le m-1}\Big|\sum_{j=0}^{r} \xi_j\Big| > \varepsilon\sigma\sqrt{n}/3\Big\}\\ &\quad + P\Big\{\max_{0 \le r \le m-1}\Big|\sum_{j=0}^{r} \eta_j\Big| > \varepsilon\sigma\sqrt{n}/3\Big\}\\ &=: I_1 + I_2 + I_3.\end{aligned} \tag{3.2.22}$$

From (3.2.19), it follows that as $n \to \infty$

$$I_1 \le cnp^{-1} \max_{0 \le l \le n-(p+q)} P\Big\{\max_{1 \le r \le p+q} |S_{l+r} - S_l| > \varepsilon\sigma\sqrt{n}/3\Big\} \to 0.$$

By Lemma 1.2.3, (3.2.15) and (3.2.12), we obtain

$$\begin{aligned}
&\max_{J\subset\{0,\cdots,m-1\}} E\Big(\sum_{j\in J}\eta_j\Big)^2/(\sigma^2 n)\\
&\le \sum_{0\le j\le m-1} E\eta_j^2/(\sigma^2 n) + 2\sum_{1\le i\le n}\sum_{i+p\le j\le n}|EX_iX_j|/(\sigma^2 n)\\
&\le cmq/(\sigma^2 n) + 20\sigma^{-2}\sum_{j>p} f_g(\alpha(j))\sup_{j\le n}\|X_j\|_g^2 \to 0.
\end{aligned}$$

Similarly

$$\begin{aligned}
&\max_{J\subset\{0,\cdots,m-1\}} E\Big(\sum_{j\in J}\xi_j\Big)^2/(\sigma^2 n)\\
&\le \sum_{0\le j\le m-1} E\xi_j^2/(\sigma^2 n) + 2\sum_{1\le i\le n}\sum_{i+q\le j\le n}|EX_iX_j|/(\sigma^2 n)\\
&\le cmp/(\sigma^2 n) + 20\sigma^{-2}\sum_{j>q} f_g(\alpha(j))\sup_{j\le n}\|X_j\|_g^2.
\end{aligned}$$

Using the definition of m and (3.2.12), we obtain that the last expression converges to $c\sigma^{-2}\delta$ as $n\to\infty$. Choose $\delta_0(\varepsilon)>0$ such that

$$(18/\varepsilon)^2 c\sigma^{-2}\delta_0(\varepsilon) < \frac{1}{2}.$$

From now on we shall assume $\delta<\delta_0(\varepsilon)$. By the Chebyshev inequality we have

$$\min_{0\le r\le m-2} P\Big\{\Big|\sum_{j=r+1}^{m-1}\xi_j\Big| \le \varepsilon\sigma\sqrt{n}/6\Big\} \ge \frac{1}{2}$$

for large n. Now we apply Lemma 3.2.1 and obtain

$$I_2 \le 2P\Big\{\Big|\sum_{j=0}^{m-1}\xi_j\Big| > \varepsilon\sigma\sqrt{n}/6\Big\} + 2m\alpha_n(q+1).$$

(3.2.16) implies $m\alpha_n(q+1)\to 0$ as $n\to\infty$. Since

$$E\Big(\sum_{j=0}^{m-1}\eta_j\Big)^2/(\sigma^2 n)\to 0,$$

$$E\Big(S_{[n(k+1)\delta]} - S_{[nk\delta]} - \sum_{j=0}^{m-1}(\xi_j+\eta_j)\Big)^2/(\sigma^2 n)\to 0 \ \text{ as } n\to\infty,$$

we obtain

$$\limsup_{n\to\infty} P\Big\{\max_{0\le r\le m-1}\Big|\sum_{j=0}^{r}\xi_j\Big| > \varepsilon\sigma\sqrt{n}/3\Big\}$$
$$\le 2\limsup_{n\to\infty} P\Big\{\Big|S_{[n(k+1)\delta]} - S_{[nk\delta]}\Big| > \varepsilon\sigma\sqrt{n}/7\Big\}.$$

Applying the Chebyshev inequality and Lamma 3.2.1 again yields

$$\limsup_{n\to\infty} P\Big\{\max_{0\le r\le m-1}\Big|\sum_{j=0}^{r}\eta_j\Big| > \varepsilon\sigma\sqrt{n}/3\Big\}$$
$$\le 2\limsup_{n\to\infty} P\Big\{\Big|\sum_{j=0}^{m-1}\eta_j\Big| > \varepsilon\sigma\sqrt{n}/6\Big\} + 2\limsup_{n\to\infty} m\alpha_n(p+1) = 0.$$

Summing up these results for $k = 0, \cdots, [1/\delta]$, we obtain

$$\limsup_{n\to\infty}\sum_{k=0}^{[1/\delta]} P\Big\{\max_{[nk\delta]<r\le[n(k+1)\delta]}\Big|S_r - S_{[nk\delta]}\Big| > \varepsilon\sigma\sqrt{n}\Big\}$$
$$\le 2\sum_{k=0}^{[1/\delta]}\limsup_{n\to\infty} P\Big\{\Big|S_{[n(k+1)\delta]} - S_{[nk\delta]}\Big| > \varepsilon\sigma\sqrt{n}/7\Big\}$$
$$\le 2(\frac{1}{\delta}+1)P\{|N(0,\delta)| > \varepsilon/7\} \to 0 \quad \text{as}\,\delta\to 0.$$

Thus Lemma 3.2.2 is proved.

Proof of Theorem 3.2.3.

Assume that some $g \in G$ satisfies condition (3.2.12). Then $K := \sup_{n\ge1}\|X_n\|_g < \infty$.

We note that the conditions (3.2.18) and (3.2.19) of Lemma 3.2.2 are both implied by

$$p^{-1}\max_{0\le m\le n-p} E\Big(\Big(\sum_{i=m+1}^{m+p}|X_i|\Big)^2 I\Big(\sum_{i=m+1}^{m+p}|X_i| > \varepsilon\sqrt{n}\Big)\Big) \to 0,$$
$$\text{as}\quad n\to\infty. \tag{3.2.23}$$

From the monotonicity of $g(x)/x^2$, the convexity of g and $Eg(|X|/\|X\|_g) \le 1$ if $0 < \|X\|_g < \infty$, we have

$$E\Big(\Big(\sum_{i=m+1}^{m+p}|X_i|\Big)^2 I\Big(\sum_{i=m+1}^{m+p}|X_i| > \varepsilon\sqrt{n}\Big)\Big)$$
$$\le Eg\Big(\sum_{i=m+1}^{m+p}|X_i|/Kp\Big)\varepsilon^2 n/g(\varepsilon\sqrt{n}/Kp)$$
$$\le \varepsilon^2 n/g(\varepsilon\sqrt{n}/Kp).$$

Thus (3.2.23) is implied by

$$p^{-1}n/g(\varepsilon\sqrt{n}/Kp) \to 0 \quad \text{as}\, n \to \infty. \tag{3.2.24}$$

From Lemma 1.2.3 we have for $m \geq 0$, $n \geq 1$

$$E(S_{m+n} - S_m)^2 \leq n \sup_{j\geq 1} EX_j^2 + 2nK^2 \sum_{k=1}^{\infty} f_g(\alpha(k)).$$

Since $\sup_{j\geq 1} EX_j^2 < \infty$, we obtain (3.2.15). The proof of Theorem 3.2.3 will be completed by constructing sequences $p(n)$, $q(n)$ such that (3.2.16), (3.2.17) and (3.2.24) are fulfilled. Since $k \to f_g(\alpha(k))$ is non-increasing, the assumption $\sum_{k=1}^{\infty} f_g(\alpha(k)) < \infty$ implies $f_g(\alpha(k))k \to 0$ as $k \to \infty$. Therefore we can choose $a: [0, \infty) \to [0, \infty)$ continuous and strictly decreasing such that

$$f_g(a(x))x \to 0 \quad \text{as}\, x \to \infty, \tag{3.2.25}$$

$$a(x) \geq \alpha(k) \quad \text{for all integers } k \geq x. \tag{3.2.26}$$

Since $x \to x \operatorname{inv} g(1/a(x))$ homeomorphically maps $(0, \infty)$ on $(0, \infty)$, we can define $x(n) \in (0, \infty)$ by

$$x(n) \operatorname{inv} g(1/a(x(n))) = n^{1/4}.$$

Then $x(n) \to \infty$ as $n \to \infty$. (3.2.25) implies that there exists an $L = L(n) \geq n^{-1/4}$ with $L(n) \to 0$ as $n \to \infty$ such that

$$L(n)^{-2} \sup_{t\geq x(n)} (\operatorname{inv} g(1/a(t)))^2 a(t)t \to 0 \quad \text{as}\, n \to \infty. \tag{3.2.27}$$

Define $y = y(n) \in (0, \infty)$ by

$$y(n) \operatorname{inv}g(1/a(y(n))) = L(n)n^{1/2}.$$

Clearly $y(n) \geq x(n)$ and $y(n) = o(n^{1/2})$. Put $q(n) = \min\{j,\, j \geq y(n)\}$ and choose a sequence $p(n)$ such that as $n \to \infty$

$$p(n)/y(n) \to \infty, \quad p(n)/n \to 0, \qquad L(n)p(n)/y(n) \to 0. \tag{3.2.28}$$

Now $p = p(n) = o(n)$, $q = q(n) = o(p)$ hold true. Since $q(n) \to \infty$, (3.2.17) can be obtained from the assumption of the theorem by an application of Lemma 1.2.3. Using (3.2.26), the definition of $y(n)$ and (3.2.27) we obtain

$$\begin{aligned} nq^{-1}\alpha(q) &\leq ny^{-1}\alpha(q) \leq ny^{-1}a(y) \\ &= L^{-2}(\operatorname{inv} g(1/a(y)))^2 a(y)y \to 0 \quad \text{as}\, n \to \infty, \end{aligned} \tag{3.2.29}$$

which and $q = o(p)$ imply $np^{-1}\alpha(q) = o(1)$. For all large n we have $p \geq y$ and $y\varepsilon/KpL \geq 1$ by (3.2.28). For such n we obtain by the convexity of g and the definition of $y(n)$:

$$\begin{aligned} p^{-1}n/g(\varepsilon\sqrt{n}/Kp) &\leq y^{-1}npLK/(\varepsilon yg(L\sqrt{n}/y)) \\ &= y^{-1}na(y)pLK/(\varepsilon y). \end{aligned}$$

This expression tends to 0 by (3.2.28) and (3.2.29). Hence (3.2.24) holds. The proof of Theorem 3.2.3 is completed.

The following corollaries are immediate. Let $g(x) = x^{2+\delta}$ for some $\delta > 0$ in Theorem 3.2.3.

Corollary 3.2.1. *Let $\{X_n, n \geq 1\}$ be an α-mixing sequence with*

$$EX_n = 0,$$

$$\sup_n E|X_n|^{2+\delta} < \infty, \ \sum_{n=1}^{\infty} \alpha(n)^{\delta/(2+\delta)} < \infty \quad \text{for some} \quad \delta > 0.$$

Suppose that condition (3.2.11) is satisfied. Then

$$W_n \Longrightarrow W.$$

Corollary 3.2.2. *Let $\{X_n, n \geq 1\}$ be an α-mixing sequence with $EX_n = 0$. If*

$$\sup_n EX_n^2|\log|X_n||^a < \infty, \quad \sum_{n=1}^{\infty}|\log \alpha(n)|^{-a} < \infty \tag{3.2.30}$$

for some $a > 0$, particularly

$$\sup_n EX_n^2|\log|X_n||^a < \infty, \ \alpha(n) = O(b^{-n}) \tag{3.2.31}$$

for some $a > 1$ and $b > 1$, then

$$W_n \Longrightarrow W.$$

Herrndorf (1985) has given two examples, which show that $a > 1$ cannot be replaced by $a = 1$ in (3.2.31). The constructions of the examples are omitted here.

Remark 3.2.3. Doukhan, Massart and Rio (1994) discussed the functional CLT for α-mixing sequence, via the α-mixing function $\alpha(t)$ and the tail distribution function of $|X_1|$, they gave another sufficient condition as follows:

Let $\{X_n, n \geq 1\}$ be a strictly stationary α-mixing sequence of centered random variables satisfying

$$\int_0^1 \mathrm{inv}\alpha(u)[\mathrm{inv}G(u)]^2 du < \infty \tag{3.2.32}$$

where $G(u) = P(|X_1| > u)$. Then the series $\sigma^2 = \sum_{n=1}^{\infty} \mathrm{Cov}(X_1, X_n)$ is absolutely convergent, and

$$Z_n/\sigma \Longrightarrow W,$$

where $Z_n(t) = \frac{1}{\sqrt{n}} \sum_{k=1}^{[nt]} X_k$.

Particularly, if $|X_1| \leq c < \infty$, condition (3.2.32) is equivalent to (3.2.4). So, in this case, Theorem 3.2.2 is recovered.

3.3 The CLT and WIP when the variance is infinite

The CLT for a mixing sequence with possibly infinite variance was first discussed by Lin (1981, 1982). He proved the following theorem under the conditions comparable with that in the case of independent sums. Assume also that $\{X_n,\, n \geq 1\}$ is a strictly stationary α-mixing sequence.

Theorem 3.3.1. *Suppose that $EX_1 = 0$ and the following conditions are satisfied:*

(i) There exist two sequences of positive integers $p = p(n)$ and $q = q(n)$ satisfying

$$p = o(n),\ q = o(p),\ k\alpha(q) \to 0 \quad \text{as } n \to \infty, \tag{3.3.1}$$

where $k = k(n) = [n/(p+q)]$.

(ii) There exists a sequence of constants $\{C_n\}$ with $C_n \uparrow \infty$ and

$$k \int_{|X_1|>C_k} dP = o\Big(\frac{1}{p}\Big), \tag{3.3.2}$$

$$C_k^2 \Big/ \Big(k \int_{|X_1|<C_k} X_1^2\, dP\Big) = o\Big(\frac{1}{p}\Big), \tag{3.3.3}$$

$$\frac{\int_{|X_1|<C_k, |X_i|<C_k} X_1 X_i \, dP}{\int_{|X_1|<C_k} X_1^2 \, dP} = o\Big(\frac{1}{p}\Big) \quad \text{as } n \to \infty \tag{3.3.4}$$

uniformly in i. Then, $\{X_n\}$ *obeys the CLT.*

If moreover there exists a constant $a > 0$ *such that* $p \leq k^a$ *and*

$$\lim_{n\to\infty} k^{2+a}\alpha(q) = 0, \tag{3.3.5}$$

then there are constants $B_n > 0$ *such that* $\{W_n(t) = S_{[nt]}/B_n, 0 \leq t \leq 1\}_{n\geq 1}$ *weakly converges to W.*

Proof. Define ξ_j, $j = 0, 1, \cdots, k-1$, and η_j, $j = 0, 1, \cdots, k$ as in Lemma 3.2.2. From (3.3.2) and (3.3.3) we obtain

$$C_k^2 \int_{|X_1|>C_k} dP \Big/ \int_{|X_1|<C_k} X_1^2 \, dP = o(p^{-2}). \tag{3.3.6}$$

Hence

$$\begin{aligned} 0 &\leq \int_{|\xi_0|<pC_k} \xi_0^2 \, dP - \int_{\cap_{j=1}^p (X_j|<C_k)} \xi_0^2 \, dP \\ &= \int_{(|\xi_0|<pC_k)\cap(\cup_{j=1}^p(|X_j|\geq C_k))} \xi_0^2 \, dP \\ &\leq p^3 C_k^2 \int_{|X_1|\geq C_k} dP = o\Big(p \int_{|X_1|<C_k} X_1^2 \, dP\Big), \end{aligned}$$

i.e.,

$$\begin{aligned} \int_{|\xi_0|<pC_k} \xi_0^2 \, dP &= \int_{\cap_{j=1}^p(|X_j|<C_k)} \Big(\sum_{j=1}^p X_j^2 + 2 \sum_{1\leq i<j\leq p} X_i X_j\Big) dP \\ &\quad + o\Big(p \int_{|X_1|<C_k} X_1^2 \, dP\Big). \end{aligned} \tag{3.3.7}$$

Furthermore, from (3.3.6) again

$$\begin{aligned} 0 &\leq \int_{|X_j|<C_k} X_j^2 \, dP - \int_{\cap_{i=1}^p(|X_i|<C_k)} X_j^2 \, dP \\ &= \int_{(|X_j|<C_k)\cap \cup_{j\neq i<p}(|X_i|\geq C_k)} X_j^2 \, dP \\ &\leq p C_k^2 \int_{|X_1|\geq C_k} dP = o\Big(p^{-1} \int_{|X_1|<C_k} X_1^2 \, dP\Big). \end{aligned}$$

Therefore

$$\begin{aligned} &\sum_{j=1}^p \int_{\cap_{i=1}^p(|X_i|<C_k)} X_j^2 \, dP \\ &\qquad = p \int_{|X_1|<C_k} X_1^2 \, dP + o\Big(\int_{|X_1|<C_k} X_1^2 \, dP\Big). \end{aligned} \tag{3.3.8}$$

Similarly

$$\int_{\cap_{j=1}^{p}(|X_j|<C_k)} X_i X_j \, dP$$
$$= \int_{|X_i|<C_k,\, |X_j|<C_k} X_i X_j \, dP + o\Big(p^{-1} \int_{|X_1|<C_k} X_1^2 \, dP\Big).$$

Using condition (3.3.4), we obtain

$$\sum_{1\le i<j\le p} \int_{\cap_{j=1}^{p}(|X_j|<C_k)} X_i X_j \, dP = o\Big(p \int_{|X_1|<C_k} X_1^2 \, dP\Big). \tag{3.3.9}$$

Now (3.3.7), (3.3.8) and (3.3.9) imply

$$\int_{|\xi_0|<pC_k} \xi_0^2 \, dP = (1+o(1))p \int_{|X_1|<C_k} X_1^2 \, dP. \tag{3.3.10}$$

Moreover,

$$\int_{|\xi_0|>pC_k} dP \le \int_{\cup_{j=1}^{p}(|X_j|>C_k)} dP \le p \int_{|X_1|>C_k} dP. \tag{3.3.11}$$

Then using conditions (3.3.3) and (3.3.2) we obtain

$$\frac{k}{(pC_k)^2} \int_{|\xi_0|<pC_k} \xi_0^2 \, dP \to \infty, \tag{3.3.12}$$

$$k \int_{|\xi_0|>pC_k} dP \to 0. \tag{3.3.13}$$

According to (3.3.10), $\int_{|\xi_0|<pC_k} \xi_0^2 \, dP \to \infty$ as $n \to \infty$. Imitating the proof of Theorem 35.1 in Gnedenko and Kolmogorov (1954), we have

$$\Big(\int_{|\xi_0|<pC_k} \xi_0 \, dP\Big)^2 = o\Big(\int_{|\xi_0|<pC_k} \xi_0^2 \, dP\Big),$$

i.e., (3.3.12) is equivalent to

$$\frac{k}{(pC_k)^2}\Big\{\int_{|\xi_0|<pC_k} \xi_0^2 \, dP - \Big(\int_{|\xi_0|<pC_k} \xi_0 \, dP\Big)^2\Big\} \to \infty. \tag{3.3.14}$$

Let ξ_j', $j = 0, 1, \cdots, k-1$, be independent random variables with the same distribution as ξ_0. By checking the proof of Theorem 26.4 in Gnedenko and Kolmogorov (1954), under conditions (3.3.13) and (3.3.14), there exists a sequence of positive constants $B_k^{(p)}$ such that

$$\frac{1}{B_k^{(p)}} \sum_{j=0}^{k-1} \xi_j' \xrightarrow{d} N(0,\, 1) \quad \text{as}\, n \to \infty. \tag{3.3.15}$$

In fact, we can take

$$\begin{aligned} B_k^{(p)^2} &= k\Big\{\int_{|\xi_0|<pC_k} \xi_0^2\,dP - \Big(\int_{|\xi_0|<pC_k} \xi_0\,dP\Big)^2\Big\} \\ &= (1+o(1))k\int_{|\xi_0|<pC_k} \xi_0^2\,dP \\ &= (1+o(1))kp\int_{|X_1|<C_k} X_1^2\,dP \end{aligned} \tag{3.3.16}$$

by (3.3.10). Using Lemma 1.2.1 and (3.3.1) we have

$$\Big|E\exp\Big(it\sum_{j=0}^{k-1}\xi_j/B_k^{(p)}\Big) - \prod_{j=1}^{k-1} E\exp(it\xi'_j/B_k^{(p)})\Big| \le 4k\alpha(q) \to 0. \tag{3.3.17}$$

Combining it with (3.3.15) yields

$$\frac{1}{B_k^{(p)}}\sum_{j=0}^{k-1}\xi_j \xrightarrow{d} N(0,\,1) \quad \text{as}\, n\to\infty. \tag{3.3.18}$$

Repeat the above discussion for η_j, $j = 0, 1, \cdots, k-1$. Then there exists a $B_k^{(q)} > 0$ such that

$$\frac{1}{B_k^{(q)}}\sum_{j=0}^{k-1}\eta_j \xrightarrow{d} N(0,\,1) \quad \text{as}\, n\to\infty. \tag{3.3.19}$$

Similarly to (3.3.16), $B_k^{(q)^2} = (1+o(1))kq\int_{|X_1|<C_k} X_1^2 dP$. Therefore

$$\Big(B_k^{(q)}/B_k^{(p)}\Big)^2 = (1+o(1))q/p = o(1). \tag{3.3.20}$$

Thus

$$\frac{1}{B_k^{(p)}}\sum_{j=0}^{k-1}\eta_j \xrightarrow{P} 0 \quad \text{as } n\to\infty. \tag{3.3.21}$$

Moreover, noting $n - k(p+q) \le p+q$ and using (3.3.16) and (3.3.3) we have for any $\varepsilon > 0$

$$\begin{aligned} P\{|\eta_k/B_k^{(p)}| > \varepsilon\} &\le (p+q)E|X_1|/\varepsilon B_k^{(p)} \\ &\le 2pE|X_1|\Big/\varepsilon\Big(kp\int_{|X_1|<C_k} X_1^2\,dP\Big)^{1/2} = o(1/C_k) \to 0. \end{aligned} \tag{3.3.22}$$

Let $B_n = B_k^{(p)}$. Then (3.3.18), (3.3.21) and (3.3.22) imply

$$S_n/B_n \xrightarrow{d} N(0,\,1) \quad \text{as}\, n\to\infty.$$

Now we turn to the weak invariance principle. Put $k_a = [k^{2+a}]$. For any positive integer N there exists an n such that $k_a(p+q) \leq N < (k_a + 1)(p+q)$. Define ξ_j and η_j as above, however their indices are extended to $j = 0, 1, \cdots, k_a - 1$. Recalling the proof of (3.3.18), under the condition $k_a \alpha(q) \to 0$, there are constants $B_m^{(p)} > 0$ $(m = k,\ k+1, \cdots, k_a)$ such that

$$\frac{1}{B_m^{(p)}} \sum_{j=0}^{m-1} \xi_j \xrightarrow{d} N(0,\ 1) \quad \text{as}\, n \to \infty, \tag{3.3.23}$$

where $B_m^{(p)}$ are the normed constants related to the independent random variables with the same distribution as ξ_0 (for fixed p). If we denote the sequence of constants corresponding to $\{\xi_j,\ j = 0, 1, \cdots, m\}$ in Theorem 26.4 of Gnedenko and Kolmogorov (1954) by $\{C_m'\}$, then, similarly to (3.3.16),

$$B_m^{(p)^2} = (1 + o(1)) mp \int_{|X_1| < C_m'} X_1^2 \, dP. \tag{3.3.24}$$

And similarly to (3.3.19),

$$B_{k_a}^{(q)-1} \sum_{j=0}^{k_a - 1} \eta_j \xrightarrow{d} N(0,\ 1) \quad \text{as}\, n \to \infty. \tag{3.3.25}$$

Define $B_N = B_{k_a}^{(p)}$, put

$$S_{[Nt]} = \sum_{j=0}^{[(k_a-1)t]} \xi_j + \sum_{j=0}^{[(k_a-1)t]} \eta_j + \eta_{N,t},$$

where $\eta_{N,t} = X_{([(k_a-1)t]+1)(p+q)+1} + \cdots + X_{[Nt]}$, with the number of terms

$$[Nt] - ([(k_a - 1)t] + 1)(p+q) \leq (k_a + 1)(p+q)t - (k_a - 1)(p+q)t \leq 2(p+q).$$

Define

$$W_N'(t) = B_N^{-1} \sum_{j=0}^{[(k_a-1)t]} \xi_j, \quad W_N''(t) = B_N^{-1} \Big(\sum_{j=0}^{[(k_a-1)t]} \eta_j + \eta_{Nt} \Big).$$

Recalling (3.3.17), we know that the normed sums of $\xi_j(\eta_j)$ and the normed sums of $\xi_j'(\eta_j')$, where $\xi_j', j = 1, \cdots, k_a - 1, (\eta_j', j = 1, \cdots, k_a - 1)$ are independent, with the common normed constants, have the same convergence.

Hence by Theorem 26.2 in Gnedenko and Kolmogorov (1954), (3.3.23) implies that for any $\varepsilon > 0$, $0 \leq t \leq 1$,

$$[mt] \int_{|x| \geq \varepsilon} dF_m^{(p)}(x) \to 0,$$

$$[mt]\Big\{\int_{|x|<\varepsilon} x^2\, dF_m^{(p)}(x) - \Big(\int_{|x|<\varepsilon} x\, dF_m^{(p)}(x)\Big)^2\Big\} \to t,$$

where $F_m^{(p)}(x)$ is the distribution of $\xi_0/B_m^{(p)}$, i.e.

$$[mt]\int_{|y|\geq \varepsilon/\sqrt{t}} dF_m^{(p)}(\sqrt{t}y) \to 0,$$

$$[mt]\Big\{\int_{|y|<\varepsilon/\sqrt{t}} y^2\, dF_m^{(p)}(\sqrt{t}y) - \Big(\int_{|y|<\varepsilon/\sqrt{t}} y\, dF_m^{(p)}(\sqrt{t}y)\Big)^2\Big\} \to 1.$$

Then by the same theorem cited just now we obtain

$$W_N'(t) \xrightarrow{d} W(t). \tag{3.3.26}$$

Similarly, we can also show

$$W_N'(t) - W_N'(s) \xrightarrow{d} W(t) - W(s) \quad \text{for } 0 \leq s < t \leq 1. \tag{3.3.27}$$

An analogue of (3.3.25) for $\{\eta_j\}$ is

$$\frac{1}{B_{k_a}^{(q)}} \sum_{j=0}^{[(k_a-1)t]} \eta_j \xrightarrow{d} W(t). \tag{3.3.28}$$

With the help of the rusult similar to (3.3.20), (3.3.27) implies

$$\frac{1}{B_N} \sum_{j=0}^{[(k_a-1)t]} \eta_j \xrightarrow{P} 0$$

uniformly in t. Imitating (3.3.22) we also have

$$\eta_{N,t}/B_N \xrightarrow{P} 0$$

uniformly in t. Consequently

$$W'' \xrightarrow{P} 0.$$

Thus we investigate W_N' instead of W_N.

At first we consider convergence of finit-dimensional distributions of W_N'. The convergence of one-dimensional distribution has been given by (3.3.26). For the two-dimensional case we need to prove

$$(W_N'(s),\, W_N'(t) - W_N'(s)) \xrightarrow{d} (W(s),\, W(t) - W(s)), \quad 0 \leq s < t \leq 1,$$

in other words, to show that for any Borel sets A_1 and A_2

$$P\{W'_N(s) \in A_1, W'_N(t) - W'_N(s) \in A_2\} \\ \to P\{W(s) \in A_1\}P\{W(t) - W(s) \in A_2\}. \quad (3.3.29)$$

According to the α-mixing property, we have

$$\Big|P\{W'_N(s) \in A_1, W'_N(t) - W'_N(s) \in A_2\} \\ - P\{W'_N(s) \in A_1\}P\{W'_N(t) - W'_N(s) \in A_2\}\Big| \\ \leq \alpha(q).$$

Combining it with (3.3.26) and (3.3.27) yields (3.3.32). Three or more dimension case can be treated in the same way, and hence the finite-dimensional distributions converge properly.

Finally, we prove tightness of W'_N. By (3.2.5), it suffices to show that for any ε, $\eta > 0$ there exist a δ, $0 < \delta < 1$ and an integer n_0 such that for $0 \leq t \leq 1$

$$P\{\sup_{t\leq s\leq t+\delta} |W'_N(s) - W'_N(t)| \geq \varepsilon\} \leq \delta\eta, \quad n \geq n_0.$$

Equivalently,

$$P\Big\{\max_{0\leq j\leq 2k_a\delta} \Big|\sum_{i=0}^{j} \xi_j\Big| \geq \varepsilon B_N\Big\} \leq \delta\eta, \quad n \geq n_0. \quad (3.3.30)$$

By the CLT, we have

$$P\Big\{\Big|\sum_{i=0}^{[2k_a\delta]} \xi_i\Big| \geq \varepsilon B_N/2\Big\} \to P\{|W(2\delta)| \geq \varepsilon/2\} \\ = P\{|W(1)| \geq \varepsilon/(2\sqrt{2\delta})\} \leq \frac{16\delta}{\varepsilon^3}\sqrt{2\delta}E|W(1)|^3 \leq \frac{\delta\eta}{4}$$

provided that δ is small enough. Hence, there exists an n_0 such that

$$P\Big\{\Big|\sum_{i=0}^{[2k_a\delta]} \xi_i\Big| \geq \varepsilon B_N/2\Big\} \leq \frac{\delta\eta}{2}, \quad n \geq n_0.$$

Therefore, if we prove that for large N and every j, $0 \leq j \leq 2k_a\delta - 1$,

$$P\Big\{\Big|\sum_{i=0}^{j} \xi_i\Big| \leq \varepsilon B_N/2\Big\} \geq \frac{1}{2}, \quad (3.3.31)$$

it follows from Lemma 2.2.1 that we have

$$P\Big\{\max_{0\le j\le 2k_a\delta}\Big|\sum_{i=0}^{j}\xi_i\Big|\ge \varepsilon B_N\Big\}\le 2P\Big\{\Big|\sum_{i=0}^{[2k_a\delta]}\xi_i\Big|\ge \varepsilon B_N/2\Big\}\le \delta\eta.$$

Now it remains to show (3.3.31). First we consider the case of $j\le k$. Using (3.3.24) and noting $p\le k^a$, we have

$$\begin{aligned}P\Big\{\Big|\sum_{i=0}^{j}\xi_i\Big|\ge \varepsilon B_N/2\Big\}&\le 2jpE|X_1|/\varepsilon B_N\\ &\le 3kpE|X_1|/\varepsilon\Big(k_a p\int_{|X_1|<C'_{k_a}}X_1^2\,dP\Big)^{1/2}\to 0.\end{aligned}$$

Assume $k<j\le 2k_a\delta$. Recall (3.3.23). By Theorem II1, we can write

$$B_j^{(p)^2}=jh^{(p)}(j).$$

With the help of the property of a slowly varying function,

$$\limsup_{N\to\infty}\frac{B_j^{(p)^2}}{B_N^2}=\limsup_{N\to\infty}\frac{jh^{(p)}(j)}{k_a h^{(p)}(k_a)}\le 2\delta.$$

Therefore, there exists an n_1 such that $B_N^2/B_j^{(p)^2}\ge 1/3\delta$ for $n\ge n_1$. Let δ be small enough so that

$$P\{|W(1)|\ge \varepsilon/2\sqrt{3\delta}\}<1/4.$$

Then

$$\begin{aligned}P\Big\{\Big|\sum_{i=0}^{j}\xi_i\Big|\ge \varepsilon B_N/2\Big\}&\le P\Big\{\Big|\sum_{i=0}^{j}\xi_i\Big|\ge \varepsilon B_{j+1}^{(p)}/2\sqrt{3\delta}\Big\}\\ &\to P\{|W(1)|\ge \varepsilon/2\sqrt{3\delta}\}<1/4,\end{aligned}$$

which implies (3.3.31). The proof of Theorem 3.3.1 is completed.

Chapter 4 Weak Convergence for ρ-mixing Sequences

Ibragimov (1975) first showed that the CLT and the WIP hold true under some conditions for a strictly stationary ρ-mixing sequence of random variables, i.e.,

Theorem 4.0.1. *Let $\{X_n, n \geq 1\}$ be a strictly stationary ρ-mixing sequence of random variables with $EX_1 = 0, EX_1^2 < \infty$ and*

(i) $\sigma_n^2 = ES_n^2 \to \infty$,

(ii) $\sum_{n=1}^{\infty} \rho(2^n) < \infty$.

Then the distridution of S_n/σ_n converges to $\Phi(x)$.

Theorem 4.0.2. *Let $\{X_n, n \geq 1\}$ be a strictly stationary ρ-mixing sequence of random variables with $EX_1 = 0$,*

$$E|X_1|^{2+\delta} < \infty \quad \textit{for some} \quad \delta > 0$$

and condition (i) is satisfied. Then with $W_n(t) = S_{[nt]}/\sigma_n$, $0 \leq t \leq 1$,

$$W_n \Longrightarrow W.$$

Peligrad (1982) gave a functional form under 2-th moment condition, however, the mixing rate is more restrictive, namely $\sum \rho^{1/2}(2^n) < \infty$. Peligrad (1986) suggested five open problems, one of which is to prove the weak invariance principle under the same sufficient conditions for the CLT. Shao (1988b) gave a positive answer, which will be introduced in Section 4.1.

When the moments of order $s > 2$ is finite, Bradley (1984) raised the problem of finding the slowest permissible ρ-mixing rates to assure the CLT under more flexible moment assumptions. Suppose that $\{X_n, n \geq 1\}$

is a strictly stationary ρ-mixing sequence of random variables with $EX_1 = 0, \sigma_n^2 = ES_n^2 \to \infty$ and $EX_1^2 g(X_1) < \infty$, where $g : [0,\infty) \to [0,\infty)$ satisfies:

both $g(x)$ and $x^\delta/g(x)$ are non-decreasing functions for some $\delta > 0$.

Bradley asked whether the CLT holds true when

$$\exp(d \sum_{i=1}^{[logn]} \rho(2^i)) = O(g(n^{1/2}))$$

for any $d > 0$. Peligrad (1987) proved a more meticulous result.

Theorem 4.0.3. *Let $\{X_n, n \geq 1\}$ and $g(x)$ be as above. If*

$$\exp\Big((2+\varepsilon^*) \sum_{i=1}^{[logn]} \rho(2^i)\Big) = O(g(n^{1/2}))$$

for some $o < \varepsilon^* < 1$, then $S_n/\sigma_n \to N(0,1)$ in distribution as $n \to \infty$.

Peligrad (1987) conjectured that $W_n \Longrightarrow W$ under the same conditions as in Theorem 4.0.3. Shao (1989b) gave a positive answer, which will be discussed in Section 4.2. In Shao (1989a), he gave a more general result for the non-stationary sequence, which will be discussed in Section 4.3. Some invariance principles for stationary ρ-mixing sequences with infinite variance will also be introduced in Section 4.4.

In the study of invariance principles for dependent random variables, the following basic result due to Billingsley (1968) is used frequently. Let

$$\mathcal{B} = \sigma\{(-\infty, x), -\infty < x < \infty\}$$

be the Borel σ-field of R.

Theorem 4.0.4. *Let $\{W_n, n \geq 1\}$ be a sequence of random elements in $D[0,1]$ satisfying the following conditions:*

(i) $\{W_n(t), 0 \leq t \leq 1\}$ has asymptotically independent increments, i.e., for any given $B_i \in \mathcal{B}, i = 1, \cdots, r$ and $0 \leq s_1 \leq t_1 < \cdots < s_r \leq t_r \leq 1$ we have

$$\lim_{n\to\infty} \{P(W_n(t_i) - W_n(s_i) \in B_i, i = 1, \cdots, r) \\ - \prod_{i-1}^{r} P(W_n(t_i) - W_n(s_i) \in B_i)\} = 0,$$

(ii) $\{W_n^2(t), n \geq 1\}$ *is uniformly integrable for each* t,

(iii) $EW_n(t) \to 0, EW_n^2(t) \to t$ *as* $n \to \infty$,

(iv) *for any* $\varepsilon, \eta > 0$ *there exist a* $\delta > 0$ *and a positive integer* n_0 *such that*

$$P\{w(W_n, \delta) \geq \varepsilon\} \leq \eta \quad n \geq n_0,$$

where $w(x, \delta) = \sup_{|s-t|<\delta} |x(s) - x(t)|$. *Then* $W_n \Longrightarrow W$.

4.1 The WIP when the moments of order 2 are finite

Shao (1988b) proved the following theorem.

Theorem 4.1.1. *Let* $\{X_n, n \geq 1\}$ *be a* ρ*-mixing sequence of random variables with* $EX_n = 0, EX_n^2 < \infty$ *and*

(i) $\lim_{n\to\infty} ES_n^2/n = \sigma^2 > 0$,

(ii) $\{X_n^2, n \geq 1\}$ *is uniformly integrable,*

(iii) $\sum_{n=1}^{\infty} \rho(2^n) < \infty$.

Then

$$W_n \Longrightarrow W,$$

where

$$W_n(t) = S_{[nt]}/\sigma\sqrt{n}, \qquad 0 \leq t \leq 1.$$

The proof of Theorem 4.1.1 will need the following lemmas.

Lemma 4.1.1.(Peligrad 1982) *Let* $\{X_n, n \geq 1\}$ *be an* α*-mixing sequence of random variables with* $EX_n = 0, EX_n^2 < \infty$, *satisfying*

(i) *of Theorem 4.1.1 and*

(ii)' *for each* $t \in [0,1], \{W_n^2(t), n \geq 1\}$ *is uniformly integrable,*

(iv) *for any* $\varepsilon > 0$ *there exist a real number* $\lambda > 2$ *and a positive integer* n_0 *such that for every* $n \geq n_0$ *and all* $k \geq 1$

$$P\{\max_{1\leq i\leq n} |S_k(i)| \geq \lambda\sigma\sqrt{n}\} \leq \varepsilon/\lambda^2. \tag{4.1.1}$$

Then $W_n \Longrightarrow W$.

Lemma 4.1.1 is a corollary of Theorem 4.0.4.

Lemma 4.1.2.(Moricz 1982) *Let* $\{X_n, n \geq 1\}$ *be a sequence of random variables. Suppose that the non-negative function* $f(k, m)$ *satisfies*

$$f(k, m) + f(k+m, l) \leq f(k, m+l) \tag{4.1.2}$$

for $k \geq 0, m \geq 1, l \geq 1$ *and the function* $g(t,s)$ *is non-decreasing for each arguments. If*

$$E|S_k(m)|^r \leq f(k,m)g^r(f(k,m),m) \qquad r \geq 1, \tag{4.1.3}$$

then

$$E \max_{1\leq m\leq n} |S_k(m)|^r \leq \frac{5}{2} f(k,n)\Big\{\sum_{i=0}^{[\log n]} g\Big(\frac{f(k,n)}{2^i}, \Big[\frac{n}{2^i}\Big]\Big)\Big\}^r. \tag{4.1.4}$$

The proof of Lemma 4.1.2 will not be presented here.

Proof of Theorem 4.1.1. We need only to check the conditions $(ii)'$ and (iv) in Lemma 4.1.1.

1) We prove that $\{S_k^2(n)/n, n \geq 1, k \geq 0\}$ is uniformly integrable. Let $N > 0$ be specified later on. Denote

$$\begin{aligned} X_i^N &= X_i I(|X_i| < N) - EX_i I(|X_i| < N), \\ \overline{X}_i^N &= X_i - X_i^N, \\ S_k^N(n) &= \sum_{i=k+1}^{k+n} X_i^N, \\ \overline{S}_k^N(n) &= \sum_{i=k+1}^{k+n} \overline{X}_i^N, \\ E_\alpha U &= \int_{U>\alpha} U dP. \end{aligned}$$

It is obvious that

$$S_k(n) = S_k^N(n) + \overline{S}_k^N(n),$$

$$E_\alpha S_k^2(n)/n \leq 4E_{\alpha/4}(S_k^N(n))^2/n + 4E(\overline{S}_k^N(n))^2/n. \tag{4.1.5}$$

From Lemma 2.2.2 it follows that for every n

$$\sup_{k\geq 1} E(\overline{S}_k^N(n))^2/n \leq K \sup_{k\geq 1} E(\overline{X}_k^N)^2$$

for some $K > 0$. Since $\{X_n^2, n \geq 1\}$ is uniformly integrable, for any given $\varepsilon > 0$ we have N such that $\sup_{k\geq 1} E(\overline{X}_k^N)^2 \leq \varepsilon/8K$. Thus for each $n \geq 1$ and $k \geq 1$

$$E(\overline{S}_k^N(n))^2/n \leq \varepsilon/8. \tag{4.1.6}$$

On the other hand, from Lemma 2.2.4, there exists a constant $K_1 = K_1(N, \delta, \rho) > 0$ such that for every n

$$\sup_{k\geq 1} E|S_k^N(n)|^{2+\delta}/n^{1+\delta/2} \leq K_1.$$

Then for large α we obtain

$$E_{\alpha/4}(S_k^N(n))^2/n \le \frac{c}{\alpha^{\delta/2}} E_{\alpha/4}|S_k^N(n)|^{2+\delta}/n^{1+\delta/2} < \frac{\varepsilon}{8}. \tag{4.1.7}$$

Inserting (4.1.6), (4.1.7) into (4.1.5) we prove that $\{S_k^2(n)/n, n \ge 1, k \ge 1\}$ is uniformly integrable.

2) We show (4.1.1). Let

$$p = [\exp(2C\log^{1/3} n)], \qquad r = [n/p],$$

where the constant C is specified in Lemma 2.2.4 corresponding to $\delta = 1$. Denote

$$\begin{aligned}
X_i^n &= X_i I(|X_i| < n^{1/2}/p) - EX_i I(|X_i| < n^{1/2}/p),\\
\overline{X}_i^n &= X_i - X_i^n,\\
S_k^n(l) &= \sum_{j=k+1}^{k+l} X_j^n, & \overline{S}_k^n(l) &= \sum_{j=k+1}^{k+l} \overline{X}_j^n,\\
Y_i &= \sum_{j=k+1+(2i-1)r}^{k+2ir} \overline{X}_j^n, & i &= 1, 2, \cdots, p_1 := [p/2],\\
Z_i &= \sum_{j=k+1+2ir}^{k+(2i+1)r} \overline{X}_j^n, & i &= 0, 1, \cdots, p_2 := [(p-1)/2],\\
T_l(i) &= \sum_{j=l+1}^{l+i} Y_j, & \overline{T}_l(i) &= \sum_{j=l}^{l+i} Z_j,\\
T(i) &= T_0(i), & \overline{T}(i) &= \overline{T}_0(i).
\end{aligned}$$

It is obvious that $S_k(i) = S_k^n(i) + \overline{S}_k^n(i)$. Without loss of generality assume that $\sigma = 1$.

We first prove

$$P\Big\{\max_{1\le i\le n} |S_k^n(i)| \ge \lambda n^{1/2}\Big\} \le \varepsilon/6\lambda^2 \tag{4.1.8}$$

for large n. Form Lemma 2.2.4 with $\delta = 1$ it follows that

$$\begin{aligned}
\max_{1\le k\le n} |ES_k^n(i)|^3 &\le C\{n^{3/2}\sigma_0^3 + n\exp(C\log^{1/3} n)a_0\}\\
&\le cn^{3/2},
\end{aligned}$$

where $\sigma_0^2 = \sup_n E(X_i^{(n)})^2$, $a_0 = \sup_n E|X_i^{(n)}|^3$. Using Lemma 4.1.2 we have

$$\sup_k E \max_{1\le i\le n} |S_k^n(i)|^3 = O(n^{3/2}).$$

That is to say, $\{\max_{1\le i\le n}|S_k^n(i)|^2/n, n\ge 1, k\ge 1\}$ is uniformly integrable. Therefore for any $\varepsilon>0$ there exist $\lambda\ge 2$ and large n_0 such that (4.1.8) is satisfied for $n\ge n_0$.

Next, we show

$$P\{\max_{1\le i\le n}|\overline{S}_k^n(i)|\ge 5\lambda n^{1/2}\}\le 5\varepsilon/6\lambda^2. \tag{4.1.9}$$

Note that the left hand side of (4.1.9) does not exceed

$$\begin{aligned}&P\Big\{\max_{1\le i\le p_1}|T(i)|\ge 2\lambda n^{1/2}\Big\}+P\Big\{\max_{1\le i\le p_2}|\overline{T}(i)|\ge 2\lambda n^{1/2}\Big\}\\&\quad+\sum_{j=0}^{p}P\Big\{\max_{1\le i\le r}|\overline{S}_{k+jr}^n(i)|\ge \lambda n^{1/2}\Big\}\\&=: I_1+I_2+I_3.\end{aligned} \tag{4.1.10}$$

Since $\{X_n^2, n\ge 1\}$ is uniformly integrable, without loss of generality we can assume that $\sup_{n\ge 1}\|X_n\|_2\le 1$. Thus we have

$$\begin{aligned}&P\Big\{\max_{1\le i\le r}|\overline{S}_{k+jr}^n(i)|\ge \lambda n^{1/2}\Big\}\\&\le P\Big\{\sum_{i=k+1+jr}^{k+(j+1)r}(|\overline{X}_i^n|-E|\overline{X}_i^n|)\ge \lambda n^{1/2}/2\Big\}\\&=O\Big(\frac{r}{n\lambda^2}\sup_m E(\overline{X}_m^n)^2\Big).\end{aligned}$$

Then by condition (ii) again

$$I_3=O\Big(\frac{pr}{\lambda^2 n}\sup_m E(\overline{X}_m^n)^2\Big)\le\frac{\varepsilon}{6\lambda^2}. \tag{4.1.11}$$

Now we estimate I_1. Denote

$$\begin{aligned}&\mathcal{G}_i=\sigma(Y_1,\cdots,Y_i), && u_i=E(Y_i|\mathcal{G}_{i-1}),\\&U_l(i)=\textstyle\sum_{j=l+1}^{l+i}u_j, && T_l^*(i)=T_l(i)-U_l(i),\\&U(i)=U_0(i), && T^*(i)=T(i)-U(i).\end{aligned}$$

It is easy to see that

$$\begin{aligned}I_1&\le P\Big\{\max_{1\le i\le p_1}|T^*(i)|\ge \lambda n^{1/2}\Big\}+P\Big\{\max_{1\le i\le p_1}|U(i)|\ge \lambda n^{1/2}\Big\}\\&=: I_{11}+I_{12}.\end{aligned} \tag{4.1.12}$$

Since $\{T^*(i), i = 1, \cdots, p_1\}$ is a martingale, for large n

$$I_{11} \le \varepsilon/6\lambda^2. \tag{4.1.13}$$

In order to estimate I_{12}, we prove that there exists a constant C_0, which does not depend on l, i, k and n, such that

$$EU_l^2(i) \le C_0 i r \rho^2(r)(\log 2i)^2. \tag{4.1.14}$$

From Lemma 2.2.2 there exists a constant $C_1 > 0$, which does not depend on l, i, k and n, such that

$$ET_l^2(i) \le C_1 i r. \tag{4.1.15}$$

Using induction for i, we can show that (4.1.14) holds for $C_0 = C_1/(\log \frac{3}{2})^2$. From the definition of ρ-mixing, we have

$$EU_l^2(1) = Eu_{l+1}^2 = EY_{l+1}u_{l+1} \le \rho(r)\|Y_{l+1}\|_2\|u_{l+1}\|_2.$$

Combining it with (4.1.15) implies (4.1.14) for $i = 1$.

For $i \ge 2$, assume that (4.1.14) holds for $j < i$. Put $i_1 = [i/2], i_2 = i - i_1$. We have

$$\begin{aligned} EU_l^2(i) &= EU_l^2(i_1) + EU_{l+i_1}^2(i_2) + 2EU_l(i_1)U_{l+i_1}(i_2) \\ &\le EU_l^2(i_1) + EU_{l+i_1}^2(i_2) + 2\rho(r)\|U_l(i_1)\|_2\|T_{l+i_1}(i_2)\|_2. \end{aligned}$$

By the assumption of induction and (4.1.15), we obtain

$$\begin{aligned} EU_l^2(i) &\le C_0 i_1 r \rho^2(r)(\log(2i_1))^2 + C_0 i_2 r \rho^2(r)(\log(2i_2))^2 \\ &\quad + 2\rho^2(r) r i_1^{1/2} i_2^{1/2} C_0^{1/2} C_1^{1/2} \log(2i_1) \\ &\le C_0 i r \rho^2(r)\Big((\log(2i_2))^2 + 2\Big(\log \frac{3}{2}\Big)\log(2i_2)\Big) \\ &\le C_0 i r \rho^2(r)(\log(2i))^2. \end{aligned}$$

This proves that (4.1.14) holds for every i. From (4.1.14) and Lemma 4.1.2 we obtain

$$E \max_{1 \le i \le p_1} U^2(i) \le c\, p_1 r(\log(2p_1))^4 \rho^2(r) \le c n (\log n)^{-1/2}.$$

Thus for large n there exists a $\lambda \ge 2$ such that

$$I_{12} \le \varepsilon/6\lambda^2. \tag{4.1.16}$$

Combining it with (4.1.13) yields

$$I_1 \le \varepsilon/3\lambda^2. \tag{4.1.17}$$

By the same way we have

$$I_2 \le \varepsilon/3\lambda^2. \tag{4.1.18}$$

From (4.1.11), (4.1.17) and (4.1.18) it follows that (4.1.9) holds true. This proves Theorem 4.1.1.

From Theorem 4.1.1 and Theorem 2.1.5, we have the following corollary immediately.

Corollary 4.1.1. *Let $\{X_n, n \ge 1\}$ be a strictly stationary ρ-mixing sequence of random variables with $EX_1 = 0, EX_1^2 < \infty$ and $\sum_{n=1}^{\infty} \rho(2^n) < \infty$. If $\sigma_n^2 = ES_n^2 \to \infty$, then*

$$\lim_{n\to\infty} \sigma_n^2/n = \sigma^2.$$

Moreover if $\sigma > 0$, then

$$W_n \Longrightarrow W,$$

where $W_n(t) = S_{[nt]}/\sigma\sqrt{n}, \quad 0 \le t \le 1$.

4.2 The WIP when moments of higher than two orders

Let $\{X_n, n \ge 1\}$ be a strictly stationary ρ-mixing sequence of random variables with $EX_1 = 0, EX_1^2 < \infty$ and $\sigma_n^2 = ES_n^2 \to \infty (n \to \infty)$. Let $g : [0, \infty) \to [0, \infty)$ be a non-decreasing function and for some $0 < \delta < 1, x^\delta/g(x)$ be also a non-decreasing function.

Theorem 4.2.1.(Peligrad 1987, Shao 1989b) *Let $\{X_n, n \ge 1\}$ be as above and satisfy*

(i) $EX_1^2 g(|X_1|) < \infty$,

(ii) $\exp((2+\varepsilon^*)\sum_{k=1}^{[\log n]} \rho(2^k)) = O(g(n^{1/2}))$

for some $0 < \varepsilon^* < 1$. Then

$$W_n \Longrightarrow W.$$

By some simple computations we have the following corollaries:

Corollary 4.2.1. *Let $\{X_n, n \geq 1\}$ be as above. Suppose that for some $\varepsilon > 0$ and $a > 0$*

$$EX_1^2(\log^+ |X_1|)^{2a/(1-\varepsilon)} < \infty$$

and

$$\rho(n) \leq a/\log n$$

for every n sufficiently large. Then

$$W_n \Longrightarrow W.$$

Corollary 4.2.2. *Let $\{X_n, n \geq 1\}$ be as above. Suppose that for some $0 < \beta < 1, \varepsilon > 0$ and $a > 0$*

$$EX_1^2 \exp\Big(\frac{2a(1+\varepsilon)}{1-\beta}(2\log^+ |X_1|)^{1-\beta}\Big) < \infty$$

and

$$\rho(n) \leq a/(\log n)^\beta$$

for every n sufficiently large. Then

$$W_n \Longrightarrow W.$$

Corollary 4.2.3. *Let $\{X_n, n \geq 1\}$ be as above. Soppose that for some $r > 0, \varepsilon > 0$ and $a > 0$*

$$EX_1^2 \exp\Big(\frac{4a\log^+ |X_1|}{(1-\varepsilon)(\log^+\log^+ |X_1|)^r}\Big) < \infty$$

and

$$\rho(n) \leq a/(\log\log n)^r$$

for every n sufficiently large. Then

$$W_n \Longrightarrow W.$$

The proof of Theorem 4.2.1 will need the following lemma, which is an immediate consequence of Theorem 3.1.2 and Theorem 8.4 of Billingsley (1968).

Lemma 4.2.1. *Let $\{X_n, n \geq 1\}$ be as above. In order that W_n weakly converges to W it is necessary and sufficient that $\{S_n^2/\sigma_n^2, n \geq 1\}$ is uniformly integrable and for any given $\varepsilon > 0$ there exists a $\lambda > 1$ such that*

$$P\Big\{\max_{1\leq i\leq n} |S_i| > \lambda\sigma_n\Big\} \leq \varepsilon/\lambda^2. \tag{4.2.1}$$

Lemma 4.2.2.(Peligrad 1987) *Let $\{X_n, n \geq 1\}$ be as above with $EX_1^4 < \infty$. Then for any given $\varepsilon > 0$ there exists a $C = C(\varepsilon, \rho(\cdot))$ such that for every $n \geq 1$*

$$ES_n^4 \leq C(n^{1+\varepsilon}EX_1^4 + \sigma_n^4).$$

Proof. Denote $a_m = \|S_m\|_4$. It is obvious that

$$a_{2m} \leq \|S_m + S_{k+m}(m)\|_4 + 2ka_1.$$

Using the Schwarz inequality and by the definition of ρ-mixing we have

$$\begin{aligned}
&E|S_m + S_{k+m}(m)|^4 \\
&\quad\leq 2a_m^4 + 6E|S_mS_{k+m}(m)|^2 + 8a_m^2(E|S_mS_{k+m}(m)|^2)^{1/2} \\
&\quad\leq 2(1 + 7\rho^{1/2}(k))a_m^4 + 8a_m^2\sigma_m^2 + 6\sigma_m^4 \\
&\quad\leq (2^{1/4}(1 + 7\rho^{1/2}(k))^{1/4}a_m + 2\sigma_m)^4.
\end{aligned}$$

It follows that

$$a_{2m} \leq 2^{1/4}(1 + 7\rho^{1/2}(k))^{1/4}a_m + 2\sigma_m + 2ka_1.$$

Let $0 < \varepsilon < 1/3$ and k be large enough such that $1 + 7\rho^{1/2}(k) \leq 2^\varepsilon$. By the recurrence method for every integer $r \geq 1$ we have

$$a(2^r) \leq 2^{r(1+\varepsilon)/4}a_1 + 2\sum_{i=1}^{r} 2^{(i-1)(1+\varepsilon)/4}(\sigma(2^{r-i}) + ka_1).$$

Whence

$$a(2^r) \leq c(2^{r(1+\varepsilon)/4}a_1 + \sigma(2^r)).$$

This implies the conclusion of the lemma.

Proof of Theorem 4.2.1.

If $\sum \rho(2^n) < \infty$, it follows from Theorem 2.1.4 that the conditions of Theorem 4.1.1 are satisfied. Therefore the conclusion of Theorem 4.2.1 holds true. We shall treat here the case when $\sum \rho(2^n) = \infty$. At this time we have that $g(x) \to \infty$ as $x \to \infty$ by condition (ii). Without loss of generality, we can assume that

$$\rho(n) \geq (\log n)^{-1}(\log\log n)^{-2} \tag{4.2.2}$$

for every large n.

1) We first prove that $\{S_n^2/\sigma_n^2, n \geq 1\}$ is uniformly integrable. It is easy to see that the condition (ii) implies

$$g(n^{1/2}) > \exp\Big(2\sum_{i=1}^{[\log n]} \rho(2^i)/(1-\varepsilon^*)\Big) \tag{4.2.3}$$

for large n. Put

$$T = \operatorname{inv} g\Big(\exp\Big(2\sum_{i=1}^{[(1-\varepsilon)\log n]} \rho(2^i)/(1-\varepsilon^*)\Big)\Big), \tag{4.2.4}$$

where $0 < \varepsilon < \varepsilon^* < 1$. Denote

$$\begin{aligned}
X_{i1} &= X_i I(|X_i| \leq T) - EX_i I(|X_i| \leq T),\\
X_{i2} &= X_i I(|X_i| > T) - EX_i I(|X_i| > T),\\
S_{n1} &= \sum_{i=1}^n X_{i1}, \qquad S_{n2} = \sum_{i=1}^n X_{i2},\\
\sigma_{n1}^2 &= \operatorname{Var} S_{n1}, \qquad \sigma_{n2}^2 = \operatorname{Var} S_{n2}.
\end{aligned}$$

By Lemma 2.2.2 and noting that the function $g(x)$ is non-decreasing, we have

$$\begin{aligned}
\sigma_{n2}^2 \leq & C_1 \frac{n}{g(T)} EX_1^2 g(|X_1|) I(|X_1| > T)\\
& \times \exp\Big(\sum_{i=1}^{[(1-\varepsilon)\log n]} \rho(2^i)/(1-\varepsilon)\Big),
\end{aligned}$$

where $C_1 = C_1(\varepsilon)$. From the definition of T and Lemma 2.2.3 we get

$$\sigma_{n2}^2 \leq \frac{C_1}{C'} \sigma_n^2 EX_1^2 g(|X_1|) I(|X_1| > T). \tag{4.2.5}$$

Obviously $\operatorname{inv} g(x) \to \infty$ $(x \to \infty)$. It follows from (4.2.4) and (4.2.5) that

$$\sigma_{n2} = o(\sigma_n), \tag{4.2.6}$$

which implies

$$\sigma_{n1}/\sigma_n \to 1 \qquad \text{as } n \to \infty. \tag{4.2.7}$$

By Lemma 4.2.2 for the sequence $\{X_{n1}, n \geq 1\}$, there exists a constant $K_1 = K_1(\rho(\cdot), \varepsilon)$ such that for every $n \geq 1$

$$E|S_{n1}|^4 \leq K_1(n^{1+\varepsilon/2}T^2EX_1^2 + \sigma_{n1}^4). \tag{4.2.8}$$

From Lemma 2.2.3 and noting that $\exp(d\sum_{i=1}^{[\log n]}\rho(2^i))$ is a slowly varying function ($n \to \infty$), it follows that there exists a $C = C(\rho(\cdot), \varepsilon)$ such that for every $n \geq 1$

$$\sigma_n^4 \geq Cn^{2-\varepsilon/2}.$$

Thus by (4.2.7) and (4.2.8) we have

$$E(|S_{n1}|/\sigma_n)^4 \leq K_2(T^2/n^{1-\varepsilon} + 1),$$

where $K_2 = K_2(\rho(\cdot), \varepsilon)$.

From (4.2.3) it follows that for large n

$$g([n^{1-\varepsilon}]^{1/2}) \geq \exp\Big(2\sum_{i=1}^{[(1-\varepsilon)\log n]}\rho(2^i)/(1-\varepsilon^*)\Big).$$

Combining it with (4.2.4) we obtain that $T/n^{(1-\varepsilon)/2}$ is bounded by 1. Therefore

$$\sup_n E(|S_{n1}|/\sigma_n)^4 < \infty. \tag{4.2.9}$$

Then, from (4.2.6) and (4.2.9), we prove that $\{S_n^2/\sigma_n^2, n \geq 1\}$ is uniformly integrable.

2) Next we show that for any $\varepsilon > 0$ there exists a $\lambda > 1$ such that

$$P\Big\{\max_{1\leq i\leq n}|S_i| \geq 6\lambda\sigma_n\Big\} \leq 6\varepsilon/\lambda^2. \tag{4.2.10}$$

Denote

$$T_1 = \exp\Big(\frac{40}{\delta}\sum_{i=1}^{[\log n]}\rho^{2/(2+\delta)}(2^i)\Big), \qquad J = n^{1/2}/T_1, \tag{4.2.11}$$

$$X_{i1} = X_iI(|X_i| \leq J) - EX_iI(|X_i| \leq J),$$
$$X_{i2} = X_iI(|X_i| > J) - EX_iI(|X_i| > J),$$
$$S_{n1}(k) = \sum_{i=1}^{k}X_{i1}, \quad S_{n2}(k) = \sum_{i=1}^{k}X_{i2},$$
$$\sigma_{n1}^2(k) = ES_{n1}^2(k), \quad \sigma_{n2}^2(k) = ES_{n2}^2(k).$$

Obviously $S_k = S_{n1}(k) + S_{n2}(k)$ and

$$P\Big\{\max_{1\le i\le n}|S_i| \ge 6\lambda\sigma_n\Big\}$$
$$\le P\Big\{\max_{1\le i\le n}|S_{n1}(i)| \ge \lambda\sigma_n\Big\} + P\Big\{\max_{1\le i\le n}|S_{n2}(i)| \ge 5\lambda\sigma_n\Big\}.$$

We first note that

$$\begin{aligned}\log T_1 &= \frac{40}{\delta}\sum_{i=1}^{[\log n]}\rho^{2/(2+\delta)}(2^i)\\ &\le \frac{40}{\delta}\rho^{-\delta/(2+\delta)}\Big(\frac{n}{T_1^2}\Big)\sum_{i=1}^{[\log n/T_1^2]}\rho(2^i)\\ &\quad+\frac{90}{\delta}\rho^{2/(2+\delta)}\Big(\frac{n}{T_1^2}\Big)\log T_1.\end{aligned}$$

Hence we have for every n sufficiently large

$$\log T_1 \le \frac{50}{\delta}\rho^{-\delta/(2+\delta)}\Big(\frac{n}{T_1^2}\Big)\sum_{i=1}^{[\log(n/T_1^2)]}\rho(2^i) \tag{4.2.12}$$

and

$$\sum_{i=1}^{[\log n]}\rho(2^i) \le \Big(1+\frac{\varepsilon^*}{12}\Big)\sum_{i=1}^{[\log(n/T_1^2)]}\rho(2^i). \tag{4.2.13}$$

From this and by condition (ii) and the fact that $g(x)$ is non-decreasing we have

$$g(J) \ge \exp\Big(\frac{2+\varepsilon^*}{1+\varepsilon^*/12}\sum_{i=1}^{[\log n]}\rho(2^i)\Big), \tag{4.2.14}$$

and by Lemma 2.2.2, Lemma 2.2.3 and (4.2.14) for every $k \le n$ and n sufficiently large

$$\begin{aligned}\sigma_{n2}^2(k) &\le CkEX_1^2I(|X_1|>J)\exp\Big(\sum_{i=1}^{[\log n]}\Big(1+\frac{\varepsilon^*}{4}\Big)\rho(2^i)\Big)\\ &\le \frac{CC'^{-1}\sigma_k^2EX_1^2g(|X_1|)}{g(J)}\exp\Big(\Big(2+\frac{3\varepsilon^*}{4}\Big)\sum_{i=1}^{[\log n]}\rho(2^i)\Big)\\ &\le CC'^{-1}\sigma_k^2EX_1^2g(|X_1|)\exp\Big(\frac{-\varepsilon^*}{52}\sum_{i=1}^{[\log n]}\rho(2^i)\Big).\end{aligned}$$

From this and noting that $\sum \rho(2^i) = \infty$, we deduce that

$$\max_{1\le k\le n} \frac{\sigma_{n2}(k)}{\sigma_k} = o(1) \quad \text{as } n \to \infty.$$

Whence it is easy to see that for $k = 1, 2, \cdots, n$ and n sufficiently large

$$\sigma_{n1}^2(k) \le 2\sigma_k^2. \tag{4.2.15}$$

By Lemma 2.2.4 and (4.2.15)

$$E|S_{n1}(k)|^{2+\delta} \le c\Big(\sigma_k^{2+\delta} + kE|X_1|^{2+\delta} I(|X_1| < J) \exp\Big(30 \sum_{i=1}^{[\log n]} \rho(2^i)\Big)\Big).$$

From this and by Lemma 2.2.2, conditions (i), (ii), (4.2.11), (4.2.14) and Lemma 4.1.2 we see

$$\begin{aligned} &E \max_{1\le k\le n} |S_{n1}(k)|^{2+\delta} \\ &\le c\Big(\sigma_n^{2+\delta} + n(\log n)^{2+\delta} E|X_1|^{2+\delta} I(|X_1| < J) \\ &\quad \times \exp\Big(30 \sum_{i=1}^{[\log n]} \rho^{2/(2+\delta)}(2^i)\Big)\Big) \\ &\le c\Big(\sigma_n^{2+\delta} + \frac{n^{2+\delta}(\log n)^{2+\delta} EX_1^2 g(|X_1|)}{g(J)T_1^\delta} \\ &\quad \times C\exp\Big(35 \sum_{i=1}^{[\log n]} \rho^{2/(2+\delta)}(2^i)\Big)\Big) \\ &\le c\sigma_n^{2+\delta}(1 + EX_1^2 g(|X_1|)). \end{aligned}$$

Thus there exists a constant $\lambda > 1$ such that for every n sufficiently large

$$P\Big\{\max_{1\le k\le n} |S_{n1}(k)| \ge \lambda\sigma_n\Big\} \le \varepsilon/\lambda^2. \tag{4.2.16}$$

We now estimate $P\{\max_{1\le k\le n} |S_{n2}(k)| \ge 5\lambda\sigma_n\}$. Let

$$p = \exp\Big(\frac{50}{\delta} \sum_{i=1}^{[\log n]} \rho^{2/(2+\delta)}(2^i)\Big),$$

$$r = \Big[\frac{n}{p}\Big], \quad p_1 = \Big[\frac{p}{2}\Big], \quad p_2 = \Big[\frac{p-1}{2}\Big].$$

Put

$$
\begin{aligned}
&Y_i = \sum_{j=1+(2i-1)r}^{2ir} X_{j2}, && i = 1, 2, \cdots, p_1;\\
&Z_i = \sum_{j=1+2ir}^{(2i+1)r} X_{j2}, && i = 0, 1, \cdots, p_2;\\
&T_1(i) = \sum_{j=1}^{i} Y_j, && T_2(i) = \sum_{j=0}^{i} Z_j.
\end{aligned}
$$

Noting that $\{X_{j2}, 1 \le j \le n\}$ is stationary we have

$$
\begin{aligned}
&P\Big\{\max_{1\le k\le n} |S_{n2}(k)| \ge 5\lambda\sigma_n\Big\}\\
&\le P\Big\{\max_{1\le k\le p_1} |T_1(k)| \ge 2\lambda\sigma_n\Big\}\\
&\quad + P\Big\{\max_{1\le k\le p_2} |T_2(k)| \ge 2\lambda\sigma_n\Big\}\\
&\quad + (p+1)P\Big\{\max_{1\le k\le r} |S_{n2}(k)| \ge \lambda\sigma_n\Big\}\\
&=: I_1 + I_2 + I_3.
\end{aligned}
$$

In terms of Lemmas 2.2.2 and 2.2.3 we have for every n sufficiently large

$$
\begin{aligned}
&P\Big\{\max_{1\le k\le r} |S_{n2}(k)| \ge \lambda\sigma_n\Big\}\\
&\le P\Big\{\sum_{i=1}^{r}(|X_{n2}(i)| - E|X_{n2}(i)|) \ge \lambda\sigma_n - 2\sum_{i=1}^{r} E|X_{n2}(i)|\Big\}\\
&\le P\Big\{\sum_{i=1}^{r}(|X_{n2}(i)| - E|X_{n2}(i)|) \ge \lambda\sigma_n - 2r\frac{EX_1^2 g(|X_1|)}{g(J)J}\Big\}\\
&\le P\Big\{\sum_{i=1}^{r}(|X_{n2}(i)| - E|X_{n2}(i)|) \ge \lambda\sigma_n/2\Big\}\\
&\le 4Cr\sigma_n^{-2}\exp\Big((1+\varepsilon^*/4)\sum_{i=1}^{[\log n]}\rho(2^i)\Big)EX_1^2 I(|X_1| > J)\cdot\lambda^{-2}.
\end{aligned}
$$

Whence by (4.2.14)

$$
\begin{aligned}
I_3 &\le 4Cn\sigma_n^{-2}\exp\Big((1+\varepsilon^*/4)\sum_{i=1}^{[\log n]}\rho(2^i)\Big)EX_1^2 I(|X_1| > J)\cdot\lambda^{-2}\\
&\le \frac{4C}{C'\lambda^2 g(J)}\exp\Big((2+3\varepsilon^*/4)\sum_{i=1}^{[\log n]}\rho(2^i)\Big)EX_1^2 g(|X_1|)\\
&\le \frac{4C}{C'\lambda^2}\exp\Big(-\frac{\varepsilon^*}{52}\sum_{i=1}^{[\log n]}\rho(2^i)\Big)EX_1^2 g(|X_1|)\\
&\le \varepsilon/\lambda^2. \qquad\qquad (4.2.17)
\end{aligned}
$$

In order to establish the estimation of I_1, let

$$\mathcal{F}_0 = (\Omega, \emptyset), \quad \mathcal{F}_k = \sigma(X_i, 1 \le i \le 2rk);$$
$$u_k = E(Y_k|\mathcal{F}_{k-1}), \quad k = 1, 2, \cdots, p_1;$$
$$U_i(k) = \sum_{j=i+1}^{i+k} u_j, \quad T^*(k) = T_1(k) - U_0(k).$$

Obviously

$$\begin{aligned} I_1 &\le P\Big\{\max_{1\le i\le p_1} |T^*(i)| \ge \lambda\sigma_n\Big\} + P\Big\{\max_{1\le i\le p_1} |U_0(i)| \ge \lambda\sigma_n\Big\} \\ &=: I_{11} + I_{12}. \end{aligned}$$

Because $\{T^*(i), i = 1, 2, \cdots, p_1\}$ is a martingale, we have

$$I_{11} \le \frac{16}{\lambda^2\sigma_n^2} \sum_{i=1}^{p_1} EY_i^2.$$

In a way somewhat similar to the estimation of I_3 we also have for any $\lambda > 1$ and for every n sufficiently large

$$I_{11} \le \epsilon/\lambda^2. \tag{4.2.18}$$

Finally, we shall prove by induction on k that for every i, k, n,

$$\begin{aligned} EU_i^2(k) \le& Ck\rho^2(r)\log^2(2k)EX_1^2 I(|X_1| > J) \\ &\cdot r \cdot \exp\Big((1+\varepsilon^*/4)\sum_{i=1}^{[\log n]} \rho(2^i)\Big). \end{aligned} \tag{4.2.19}$$

When $k = 1$, by the definition of ρ-mixing

$$\begin{aligned} EU_i^2(1) &= EE^2(Y_{i+1}|\mathcal{F}_i) = E(Y_{i+1}E(Y_{i+1}|\mathcal{F}_i)) \\ &\le \rho(r)\|Y_{i+1}\|_2 \cdot \|E(Y_{i+1}|\mathcal{F}_i)\|_2, \end{aligned}$$

thus (4.2.19) is true for $k = 1$ and for every $i + 1 \le p_1$ by Lemma 2.2.2. When $k \ge 2$, assume (4.2.19) holds for every integer less than k. Put $k_1 = [k/2], k_2 = k - k_1$, then

$$\begin{aligned} EU_i^2(k) &= EU_i^2(k_1) + EU_{i+k_1}^2(k_2) + 2EU_i(k_1)U_{i+k_1}(k_2) \\ &= EU_i^2(k_1) + EU_{i+k_1}^2(k_2) + 2EU_i(k_1) \sum_{j=i+k_1+1}^{i+k} Y_j \\ &\le EU_i^2(k_1) + EU_{i+k_1}^2(k_2) \\ &\quad + 2\|U_i(k_1)\|_2 \cdot \|\sum_{j=i+k_1+1}^{i+k} Y_j\|_2 \rho(r). \end{aligned}$$

By induction hypothesis and Lemma 2.2.2

$$
\begin{aligned}
EU_i^2(k) &\le C(k_1 \log^2 2k_1 + k_2 \log^2 2k_2 + 2(k_1k_2)^{1/2} \log 2k_1) \\
&\quad \cdot \rho^2(r) \cdot r \cdot \exp\Big(\Big(1+\frac{\varepsilon^*}{4}\Big) \sum_{j=1}^{[\log n]} \rho(2^j)\Big) EX_1^2 I(|X_1| > J) \\
&\le Ck(\log 2k)^2 r \cdot \exp\Big(\Big(1+\frac{\varepsilon^*}{4}\Big) \sum_{j=1}^{[\log n]} \rho(2^j)\Big) \\
&\quad \cdot EX_1^2 I(|X_1| > J) \cdot \rho^2(r),
\end{aligned}
$$

which proves that (4.2.19) holds.

From (4.2.19) we obtain by Lemma 4.1.2

$$
\begin{aligned}
&E \max_{1\le i\le p_1} U_0^2(i) \\
&\le 3Crp_1\rho^2(r)\log^4(2p_1) \\
&\quad \cdot \exp\Big((1+\varepsilon^*/4) \sum_{j=1}^{[\log n]} \rho(2^j)\Big) EX_1^2 I(|X_1| > J) \\
&\le \frac{3C\sigma_n^2\rho^2(n/p_1)\log^4(2p_1)}{C'g(J)} \\
&\quad \cdot \exp\Big((2+3\varepsilon^*/4) \sum_{j=1}^{[\log n]} \rho(2^j)\Big) EX_1^2 g(|X_1|) \\
&\le \frac{3C\sigma_n^2\rho^2(n/p_1)\log^4(2p_1)}{C'} \\
&\quad \cdot \exp\Big(-\frac{\varepsilon^*}{52} \sum_{j=1}^{[\log n]} \rho(2^j)\Big) EX_1^2 g(|X_1|).
\end{aligned}
$$

By (4.2.12)

$$
\rho^2\Big(\frac{n}{p_1}\Big)\log^4(2p_1) \le \Big(\frac{50}{\delta}\Big)^4 \rho^{2/3}\Big(\frac{n}{T_1^2}\Big) \sum_{j=1}^{[\log n]} \rho(2^j),
$$

hence we finally get that for any $\lambda > 1$ and for every n sufficiently large

$$
I_{12} \le \varepsilon/\lambda^2.
$$

Therefore

$$
I_1 \le 2\varepsilon/\lambda^2. \tag{4.2.20}
$$

Similarly, we have

$$I_2 \leq 2\varepsilon/\lambda^2. \tag{4.2.21}$$

(4.2.10) now follows from (4.2.16), (4.2.17) and (4.2.20), (4.2.21). Theorem 4.2.1 is proved.

4.3 A generalized result when moments of higher than two orders

Shao (1989a) gave a generalized result of Theorem 4.2.1, where the condition of strict stationarity was removed.

Theorm 4.3.1. *Let $\{X_n, n \geq 1\}$ be a ρ-mixing sequence of random variables with $EX_n = 0, EX_n^2 \geq c > 0$. Suppose that $g : [0,\infty) \to [0,\infty)$ is a non-decreasing function and $x^\delta/g(x), 0 < \delta < 1$, is also non-decreasing. If the following conditions are satisfied:*

(i) $\{X_n^2 g(|X_n|), n \geq 1\}$ *is uniformly integrable,*

(ii) $\sigma_n^2 := ES_n^2 = nh(n)$, *where $h(n)$ is a slowly varying function,*

(iii) $\sup_{m\geq 0, n\geq 1} ES_m^2(n)/\sigma_n^2 < \infty$,

(iv) $\lim_{n\to\infty} \inf_{m\geq 0} ES_m^2(n) = \infty$,

(v) $\exp\Big((2+\varepsilon^*)\sum_{k=1}^{[\log n]} \rho(2^k)\Big) = O(g(n^{1/2}))$, for some $0 < \varepsilon^* < 1$,

then W_n weakly converges to W.

The proof of Theorem 4.3.1 needs the following lemmas.

Lemma 4.3.1 . *Let $f(k,m)$ be a non-negative function satisfying (4.1.2). Suppose that there exist $\alpha > 0, r \geq 1$ such that*

$$E|S_k(l)|^r \leq f(k,l)\{f^\alpha(k,l)w_1(f(k,l)) + w_2(f(k,l))\}, \tag{4.3.1}$$

where w_2 is a non-negative non-decreasing function, w_1 is a non-negative function with

$$\max_{0\leq s\leq t} s^\beta w_1(s) \leq at^\beta w_1(t) \tag{4.3.2}$$

for some $a > 0, 0 < \beta < \alpha$ and any $t > 0$. Then we have

$$\begin{aligned} &E\max_{1\leq i\leq n}|S_k(i)|^r \\ &\leq 2^{r+1} f(k,n)\{a(1-(\frac{1}{2})^{(\alpha-\beta)/r})^{-r} f^\alpha(k,n)w_1(f(k,n)) \\ &\quad + w_2(f(k,n))\log^r(2n)\}. \end{aligned} \tag{4.3.3}$$

Proof. From (4.1.2) and (4.3.2), we have for $l \le n$

$$f^{\beta}(k,l)w_1(f(k,l)) \le af^{\beta}(k,n)w_1(f(k,n)).$$

Therefore, by (4.3.1), for every $k \ge 0, 1 \le l \le n$

$$E|S_k(l)|^r \le f(k,l)\{f^{\alpha-\beta}(k,l) \cdot af^{\beta}(k,n)w_1(f(k,n)) + w_2(f(k,n))\}.$$

It follows from Lemma 4.1.2 and the monotonicity of $w_2(\cdot)$ that

$$\begin{aligned}
&E \max_{1\le l\le n} |S_k(l)|^r \\
&\le \frac{5}{2}f(k,n)\Big\{\sum_{i=0}^{[\log n]} w_2^{1/r}(f(k,n)) \\
&\quad + (af^{\beta}(k,n)w_1(f(k,n)))^{1/r}\sum_{i=0}^{[\log n]}\Big(\frac{f(k,n)}{2^i}\Big)^{(\alpha-\beta)/r}\Big\}^r \\
&\le \frac{5}{2}f(k,n)\Big\{w_2^{1/r}(f(k,n))\log(2n) \\
&\quad + (af^{\alpha}(k,n)w_1(f(k,n)))^{1/r}\Big(1-\Big(\frac{1}{2}\Big)^{(\alpha-\beta)/r}\Big)^{-1}\Big\}^r \\
&\le 5\cdot 2^{r-2}f(k,n)\Big\{af^{\alpha}(k,n)w_1(f(k,n))\Big(1-\Big(\frac{1}{2}\Big)^{(\alpha-\beta)/r}\Big)^{-r} \\
&\quad + w_2(f(k,n))\log^r(2n)\Big\}.
\end{aligned}$$

This completes the proof of Lemma 4.3.1.

Lemma 4.3.2. *Let $\{X_n, n \ge 1\}$ be a ρ-mixing sequence with $EX_n = 0$, and $q_1, q_2 \ge 2$. Suppose that the non-negative function $h(n)$ satisfies:*

$$\max\Big(h\Big(\Big[\frac{n}{2}\Big]\Big), h\Big(n-\Big[\frac{n}{2}\Big]\Big)\Big) \le \theta h(n) \tag{4.3.4}$$

for every $n \ge 1$ and some $0 < \theta < 2^{1-\frac{2}{q_1\wedge 3}}$, and if $q_1 > 3$

$$h(n) \ge \frac{1}{a}\exp\Big\{-a\sum_{i=0}^{[\log n]}\rho^{2/q_1}(2^i)\Big\}, \tag{4.3.5}$$

$$\max_{1\le i\le n} i^{\alpha}h(i) \le an^{\alpha}h(n)$$

for some $a > 0$ and some α, $0 < \alpha < q_1 - 2$. And suppose that integers $1 \le k \le n$, $l \ge 0$ and numbers $x > 0, 0 < B \le A < \infty$ satisfy

$$4n \max_{l<i\le l+n} E|X_i|I(|X_i| \ge A) \le x, \tag{4.3.6}$$

$$48k \max_{l<i\le l+n} E|X_i|I(|X_i| \ge B) \le x, \tag{4.3.7}$$

$$\mathrm{Var}\Big(\sum_{i=j+1}^{j+n} X_i I(|X_i| \le B)\Big) \le nh(n) \max_{j<i\le j+n} EX_i^2 I(|X_i| \le B). \tag{4.3.8}$$

Then for any given $\varepsilon > 0$, there exists a $K = K(\varepsilon, q_1, q_2, a, \theta, \alpha, \rho(\cdot))$ such that

$$\begin{aligned}
P\Big\{\max_{1\le i\le n}|S_l(i)| \ge x\Big\}
&\le \sum_{i=l+1}^{l+n} P(|X_i| \ge A) \\
&\quad + K\Big\{x^{-q_1}\Big[(nh(n)\max_{l<i\le l+n} EX_i^2 I(|X_i| \le B))^{q_1/2} \\
&\quad + n\exp\Big\{K\sum_{i=0}^{[\log n]} \rho^{2/q_1}(2^i)\Big\}\log^{q_1}(2n) \\
&\quad \times \max_{l<i\le l+n} E|X_i|^{q_1} I(|X_i| \le B)\Big] \\
&\quad + x^{-q_2}\Big[\Big(n\exp\Big\{(1+\varepsilon)\sum_{i=0}^{[\log n]}\rho(2^i)\Big\} \\
&\quad \times \max_{l<i\le l+n} EX_i^2 I(B < |X_i| < A)\Big)^{q_2/2} \\
&\quad + n\exp\Big\{K\sum_{i=0}^{[\log n]}\rho^{2/q_2}(2^i)\Big\} \\
&\quad \times \max_{l<i\le l+n} E|X_i|^{q_2} I(B < |X_i| < A)\Big] \\
&\quad + x^{-2}n\rho^2(k)\log^4[\frac{n}{k}] \cdot \exp\Big\{(1+\varepsilon)\sum_{i=0}^{[\log n]}\rho(2^i)\Big\} \\
&\quad \times \max_{l<i\le l+n} EX_i^2 I(B < |X_i| < A)\Big\}.
\end{aligned}$$

Proof. For simplicity, we assume that $X_n, n \ge 1$ have a common distribution. Denote

$$\begin{aligned}
X_{i1} &= X_i I(|X_i| \le B) - EX_i I(|X_i| \le B),\\
X_{i2} &= X_i I(B < |X_i| < A) - EX_i I(B < |X_i| < A),\\
X_{i3} &= X_i I(|X_i| \ge A) - EX_i I(|X_i| \ge A),
\end{aligned}$$

$$S_{lm}(i) = \sum_{j=l+1}^{l+i} X_{jm}, \qquad m = 1, 2, 3.$$

It is easy to see that

$$\begin{aligned}
&P\{\max_{1\le i\le n} |S_l(i)| \ge x\}\\
&\le P\{\max_{1\le i\le n} |S_{l1}(i)| \ge \frac{x}{4}\} + P\{\max_{1\le i\le n} |S_{l2}(i)| \ge \frac{x}{4}\}\\
&\quad + P\{\max_{1\le i\le n} |S_{l3}(i)| \ge \frac{x}{2}\}\\
&=: I_1 + I_2 + I_3.
\end{aligned} \tag{4.3.9}$$

By (4.3.6), we have

$$\begin{aligned}
I_3 &\le P\Big\{\sum_{i=l+1}^{l+n} |X_i| I(|X_i| \ge A) \ge \frac{x}{2} - \sum_{i=l+1}^{l+n} E|X_i| I(|X_i| \ge A)\Big\}\\
&\le P\Big\{\sum_{i=l+1}^{l+n} |X_i| I(|X_i| \ge A) \ge \frac{x}{4}\Big\}\\
&\le \sum_{i=l+1}^{l+n} P(|X_i| \ge A).
\end{aligned} \tag{4.3.10}$$

By Lemma 2.2.6, there exists a $K_1 = K_1(q_1, a, \theta, \rho(\cdot))$ such that

$$\begin{aligned}
&E\Big|\sum_{i=j+1}^{j+n} X_{i1}\Big|^{q_1}\\
&\qquad \le K_1\Big\{(nh(n)EX_1^2 I(|X_1| \le B))^{q_1/2}\\
&\qquad\quad + n\exp\Big[K_1 \sum_{i=0}^{[\log n]} \rho^{2/q_1}(2^i)\Big] E|X_1|^{q_1} I(|X_1| \le B)\Big\}.
\end{aligned}$$

It follows from Lemma 4.3.1 that there exists a $K_2 = K_2(K_1, \alpha, q_1)$ such that

$$\begin{aligned} E\max_{1\le i\le n}|S_{l1}(i)|^{q_1} &\le K_2\Big\{(nh(n)EX_1^2 I(|X_1|\le B))^{q_1/2} \\ &\quad + n\exp\Big\{K_2\sum_{i=0}^{[\log n]}\rho^{2/q_1}(2^i)\Big\} \\ &\quad \cdot(\log(2n))^{q_1}E|X_1|^{q_1}I(|X_1|\le B)\Big\}. \end{aligned}$$

Then we have

$$\begin{aligned} I_1 &\le K_2 x^{-q_1}\Big\{(nh(n)EX_1^2 I(|X_1|\le B))^{q_1/2} \\ &\quad + n\exp\Big\{K_2\sum_{i=0}^{[\log n]}\rho^{2/q_1}(2^i)\Big\}(\log(2n))^{q_1} \\ &\quad \cdot E|X_1|^{q_1}I(|X_1|\le B)\Big\}. \end{aligned} \tag{4.3.11}$$

In order to estimate I_2, put

$$\begin{aligned} Y_i &= \sum_{j=2ik+l+1}^{(2i+1)k+l} X_{j2}, \quad i=0,1,\cdots,p_1, \\ Z_i &= \sum_{j=(2i+1)k+l+1}^{2(i+1)k+l} X_{j2}, \quad i=0,1,\cdots,p_2, \end{aligned}$$

where $p_1=[(\frac{n}{k}-1)/2], p_2=[(\frac{n}{k}-2)/2]$. Denote

$$W_i=\sum_{j=0}^{i}Y_j, \qquad W_i^*=\sum_{j=0}^{i}Z_j.$$

It is easy to see that

$$\begin{aligned} I_2 &\le P\Big\{\max_{0\le i\le p_1}|W_i|\ge\frac{x}{12}\Big\}+P\Big\{\max_{0\le i\le p_2}|W_i^*|\ge\frac{x}{12}\Big\} \\ &\quad + P\Big\{\max_{0\le i\le[n/k]}\ \max_{ik+1\le j<(i+1)k}\Big|\sum_{\nu=l+ik+1}^{l+j}X_{\nu 2}\Big|\ge\frac{x}{12}\Big\} \\ &=: I_{21}+I_{22}+I_{23}. \end{aligned} \tag{4.3.12}$$

From the condition (4.3.7) and Lemma 2.2.5 it follows that

$$\begin{aligned}
I_{23} \le{}& 2\left[\frac{n}{k}\right] \max_{0\le i\le [n/k]} P\Big\{ \sum_{j=l+ik+1}^{l+(i+1)k} (|X_j| I(B<|X_j|<A) \\
&- E|X_j| I(B<|X_j|<A)) \ge \frac{x}{12} \\
&- 2\sum_{j=l+ik+1}^{l+(i+1)k} E|X_j| I(B<|X_j|<A)\Big\} \\
\le{}& \frac{2n}{k} \max_{0\le i\le [n/k]} P\Big\{ \sum_{j=l+ik+1}^{l+(i+1)k} (|X_j| I(B<|X_j|<A) \\
&- E|X_j| I(B<|X_j|<A)) \ge x/24\Big\} \\
\le{}& C\frac{n}{k} x^{-q_2}\Big\{\Big(k\exp\Big((1+\varepsilon)\sum_{i=0}^{[\log k]} \rho(2^i)\Big) EX_1^2 I(B<|X_1|<A)\Big)^{q_2/2} \\
&+ k\exp\Big\{C\sum_{i=0}^{[\log k]} \rho^{2/q_2}(2^i)\Big\} E|X_1|^{q_2} I(B<|X|<A)\Big\} \\
\le{}& Cx^{-q_2}\Big\{\Big(n\exp\Big\{(1+\varepsilon)\sum_{i=0}^{[\log n]} \rho(2^i)\Big\} EX_1^2 I(B<|X_1|<A)\Big)^{q_2/2} \\
&+ n\exp\Big\{C\sum_{i=0}^{[\log n]} \rho^{2/q_2}(2^i)\Big\} E|X_1|^{q_2} I(B<|X_1|<A)\Big\}.
\end{aligned}$$

Next, we estimate I_{21}. The estimation of I_{22} can be obtained by the same way. Denote $\mathcal{F}_{-1} = \{\emptyset, \Omega\}$.

$$\mathcal{F}_i = \sigma(X_j : j \le l + (2i+1)k), \qquad i = 0, 1, \cdots, p_1.$$

Put

$$\begin{aligned}
U_i &= Y_i - E(Y_i|\mathcal{F}_{i-1}), & G_i &= \textstyle\sum_{j=0}^{i} U_j, \\
H_i &= \textstyle\sum_{j=1}^{i} E(Y_j|\mathcal{F}_{j-1}), & i &= 0, 1, \cdots, p_1.
\end{aligned}$$

It is easy to see that

$$\begin{aligned}
I_{21} &\le P\{\max_{1\le i\le p_1} |G_i| \ge x/24\} + P\{\max_{1\le i\le p_1} |H_i| \ge x/24\} \\
&=: I_{21}(1) + I_{21}(2).
\end{aligned} \tag{4.3.13}$$

Since $\{U_i, \mathcal{F}_i, i \ge 1\}$ is a martingale difference sequence, by the maximum value inequality of Brown (1971), we have

$$\begin{aligned}I_{21}(1) &\le \frac{24}{x}E|G_{p_1}|I(|G_{p_1}|\ge \frac{x}{48})\\ &\le \frac{48}{x}\Big\{E|W_{p_1}|I\Big(|W_{p_1}|\ge \frac{x}{96}\Big)\\ &\quad + E|H_{p_1}|I\Big(|H_{p_1}|\ge \frac{x}{96}\Big)\Big\}\\ &\le (96/x)^{q_2}E|W_{p_1}|^{q_2}+(96/x)^2E|H_{p_1}|^2. \end{aligned}\tag{4.3.14}$$

From Lemma 2.2.5 it follows that

$$\begin{aligned}&E|W_{p_1}|^{q_2}\\ &\le C\Big\{\Big(n\exp\Big[(1+\varepsilon)\sum_{i=0}^{[\log n]}\rho(2^i)\Big]EX_1^2I(B<|X_1|<A)\Big)^{q_2/2}\\ &\quad + n\exp\Big[C\sum_{i=0}^{[\log n]}\rho^{2/q_2}(2^i)\Big]E|X_1|^{q_2}I(B<|X_1|<A)\Big\}. \end{aligned}\tag{4.3.15}$$

We prove below by induction that there exists a constant K' such that

$$\begin{aligned}&E\Big(\sum_{j=i+1}^{i+m}E(Y_i|\mathcal{F}_{j-1})\Big)^2\\ &\quad\le K'mk\rho^2(k)(\log(2m))^2\\ &\qquad\times\exp\Big[(1+\varepsilon)\sum_{j=0}^{[\log m]}\rho(2^j)\Big]EX_1^2I(B<|X_1|<A). \end{aligned}\tag{4.3.16}$$

Indeed, it follows from Lemma 2.2.1 that there exists a constant C' such that for $mk\le n$

$$\begin{aligned}E\Big(\sum_{j=i+1}^{i+m}Y_j\Big)^2 &\le C'mk\exp\Big[(1+\varepsilon)\sum_{j=0}^{[\log n]}\rho(2^j)\Big]\\ &\quad\times EX_1^2I(B<|X_1|<A). \end{aligned}\tag{4.3.17}$$

When $m=1$, by the definition of ρ-mixing

$$E(E(Y_{i+1}|\mathcal{F}_i))^2=E(Y_{i+1}E(Y_{i+1}|\mathcal{F}_i))\le\rho(k)\|Y_{i+1}\|_2\|E(Y_{i+1}|\mathcal{F}_i)\|_2,$$

that is

$$E(E(Y_{i+1}|\mathcal{F}_i))^2\le\rho^2(k)EY_{i+1}^2.$$

Let $K'=C'/\log^2(3/2)$, it follows that (4.3.16) holds true for $m=1$.

Suppose that (4.3.16) holds true for the integers less than m. Now let us show that (4.3.16) holds true for m. Denote $m_1 = [m/2], m_2 = m - m_1$, we have

$$\begin{aligned}
&E\Big(\sum_{j=i+1}^{i+m} E(Y_j|\mathcal{F}_{j-1})\Big)^2 \\
&= E\Big(\sum_{j=i+1}^{i+m_1} E(Y_j|\mathcal{F}_{j-1})\Big)^2 + E\Big(\sum_{j=i+m_1+1}^{i+m} E(Y_j|\mathcal{F}_{j-1})\Big)^2 \\
&\quad + 2E\Big(\sum_{j=i+1}^{i+m_1} E(Y_j|\mathcal{F}_{j-1})\Big)\Big(\sum_{j=i+m_1+1}^{i+m} E(Y_j|\mathcal{F}_{j-1})\Big) \\
&\le E\Big(\sum_{j=i+1}^{i+m_1} E(Y_j|\mathcal{F}_{j-1})\Big)^2 + E\Big(\sum_{j=i+m_1+1}^{i+m} E(Y_j|\mathcal{F}_{j-1})\Big)^2 \\
&\quad + 2\rho(k)E\Big\|\sum_{j=i+1}^{i+m_1} E(Y_j|\mathcal{F}_{j-1})\Big\|_2\Big\|\sum_{j=i+m_1+1}^{i+m} Y_j\Big\|_2,
\end{aligned}$$

which, by the inductive assumption and (4.3.17), implies

$$\begin{aligned}
&E\Big(\sum_{j=i+1}^{i+m} E(Y_j|\mathcal{F}_{j-1})\Big)^2 \\
&\le K'\{m_1\log^2(2m_1) + m_2\log^2(2m_2) \\
&\quad + 2\Big(\log\frac{3}{2}\Big)(m_1m_2)^{1/2}\log(2m_1)\} \\
&\quad \times \rho^2(k)k\exp\Big\{(1+\varepsilon)\sum_{j=0}^{[\log n]}\rho(2^j)\Big\}EX_1^2I(B<|X_1|<A) \\
&\le K'mk\rho^2(k)\log^2(2m) \\
&\quad \times \exp\Big\{(1+\varepsilon)\sum_{j=0}^{[\log n]}\rho(2^j)\Big\}EX_1^2I(B<|X_1|<A).
\end{aligned}$$

This proves (4.3.16). Moreover, it follows from (4.3.16) and Lemma 4.3.1 that there exists a constant K'' such that

$$\begin{aligned}
E\max_{0\le i\le p_1} H_i^2 \le K''p_1k\rho^2(k)(\log(2p_1))^4\exp\Big\{(1+\varepsilon)\sum_{j=0}^{[\log n]}\rho(2^j)\Big\} \\
\cdot EX_1^2I(B<|X_1|<A). \qquad (4.3.18)
\end{aligned}$$

Thus we have

$$I_{21}(2) \le 288K''x^{-2}n\rho^2(k)\log^4[n/k]$$
$$\cdot\exp\Big\{(1+\varepsilon)\sum_{j=0}^{[\log n]}\rho(2^j)\Big\}EX_1^2I(B<|X_1|<A). \quad (4.3.19)$$

From (4.3.13), (4.3.14), (4.3.15), (4.3.18) and (4.3.19) it follows that

$$I_{21} \le K_3x^{-q_2}\Big\{\Big(n\exp\Big\{(1+\varepsilon)\sum_{j=0}^{[\log n]}\rho(2^j)\Big\}EX_1^2I(B<|X_1|<A)\Big)^{q_2/2}$$
$$+n\exp\Big\{K_3\sum_{j=0}^{[\log n]}\rho^{2/q_2}(2^j)\Big\}E|X_1|^{q_2}I(B<|X_1|<A)\Big\}$$
$$+K_3x^{-2}n\rho^2(k)\log^4\Big[\frac{n}{k}\Big]$$
$$\times\exp\Big\{(1+\varepsilon)\sum_{j=0}^{[\log n]}\rho(2^j)\Big\}EX_1^2I(B<|X_1|<A)$$

for some $K_3 > 0$. The proof of Lemma 4.3.2 is completed.

Lemma 4.3.3. *Let $0<\delta\le 1$. Suppose that the non-negative function $h(n)$ satisfies the following conditions: there exist integer $n_0>0$, $0<\theta<2^{\delta/(2+\delta)}$, $0<\delta'<\delta$ and $a>0$ such that for any $n\ge n_0$*

$$h\Big(\Big[\frac{n}{2}\Big]\Big)\vee h\Big(n-\Big[\frac{n}{2}\Big]\Big)\le\theta h(n), \quad (4.3.20)$$

$$\max_{1\le i\le n} i^{\delta'}h(i)\le an^{\delta'}h(n). \quad (4.3.21)$$

Let $\{X_n, n\ge1\}$ be a ρ-mixing sequence with $EX_n=0, EX_n^2<\infty$ and for every $k\ge0, n\ge1$

$$ES_k^2(n)\le nh(n)\max_{k<i\le k+n}EX_i^2. \quad (4.3.22)$$

Then for any given $\varepsilon>0$ there exists a $K=K(\varepsilon,\delta,\delta',n_0,a,\rho(\cdot))$ such that *for any* $n \geq k \geq 0, l \geq 0$ *and* $B > 0$

$$
\begin{aligned}
&E \max_{1\leq i\leq n} |S_l(i)|^2 I(\max_{1\leq i\leq n} |S_l(i)| \geq x) \\
&\quad \leq K\Big\{x^{-\delta}\Big\{(nh(n) \max_{l<i\leq l+n} EX_i^2 + n\exp\Big\{(1+\varepsilon)\sum_{i=0}^{[\log n]} \rho(2^i)\Big\} \\
&\quad \times \max_{l<i\leq l+n} EX_i^2 I(|X_i| \geq B))^{\frac{2+\delta}{2}} \\
&\quad + n\exp\Big\{K \sum_{i=0}^{[\log n]} \rho^{2/(2+\delta)}(2^i)\Big\} \log^{2+\delta}(2n) \\
&\quad \times \max_{l<i\leq l+n} E|X_i|^{2+\delta} I(|X_i| < B)\Big\} \\
&\quad + n\exp\Big\{(1+\varepsilon)\sum_{i=0}^{[\log n]} \rho(2^i)\Big\}\Big(1 + \rho^2(k)\log^4\Big[\frac{n}{k}\Big]\Big) \\
&\quad \max_{l<i\leq l+n} EX_i^2 I(|X_i| \geq B) \\
&\quad + nk \max_{l<i\leq l+n} (E|X_i| I(|X_i| \geq B))^2\Big\}. \qquad (4.3.23)
\end{aligned}
$$

Proof. Denote

$$
X_{i1} = X_i I(|X_i| < B) - EX_i I(|X_i| < B), \qquad S_{l1}(i) = \sum_{j=l+1}^{l+i} X_{j1},
$$

$$
X_{i2} = X_i I(|X_i| \geq B) - EX_i I(|X_i| \geq B), \qquad S_{l2}(i) = \sum_{j=l+1}^{l+i} X_{j2}.
$$

It is easy to see that

$$
\begin{aligned}
&E \max_{1\leq i\leq n} S_l^2(i) I\Big(\max_{1\leq i\leq n} |S_l(i)| \geq x\Big) \\
&\leq 4E \max_{1\leq i\leq n} S_{l1}^2(i) I(\max_{1\leq i\leq n} |S_{l1}(i)| \geq x/2) + 4E \max_{1\leq i\leq n} S_{l2}^2(i) \\
&\leq 8x^{-\delta} E \max_{1\leq i\leq n} |S_{l1}(i)|^{2+\delta} + 4E \max_{1\leq i\leq n} S_{l2}^2(i) \\
&=: 8I_1 + 4I_2. \qquad (4.3.24)
\end{aligned}
$$

From Lemma 2.2.2 , for every $n \geq 1, l \geq 0$ we have

$$
ES_{l2}^2(n) \leq Cn\exp\Big\{(1+\varepsilon)\sum_{i=0}^{[\log n]} \rho(2^i)\Big\} \max_{l\leq i\leq l+n} EX_i^2 I(|X_i| \geq B),
$$

where $C = C(\varepsilon)$. Therefore, by (4.3.22) one obtains

$$\begin{aligned} ES_{l1}^2(n) &\le 2ES_l^2(n) + 2ES_{l2}^2(n) \\ &\le 2nh(n) \max_{l \le i \le l+n} EX_i^2 \\ &\quad + 2Cn \exp\Big\{(1+\varepsilon) \sum_{i=0}^{[\log n]} \rho(2^i)\Big\} \\ &\quad \times \max_{l \le i \le l+n} EX_i^2 I(|X_i| \ge B). \end{aligned} \tag{4.3.25}$$

It follows from Lemma 2.2.6 that there exists a K_1 such that for every $l \ge 0, n \ge 1$

$$\begin{aligned} &E|S_{l1}(n)|^{2+\delta} \\ &\le K_1 \Big\{ \Big(nh(n) \max_{l < i \le l+n} EX_i^2 \\ &\quad + n \exp\Big\{(1+\varepsilon) \sum_{i=0}^{[\log n]} \rho(2^i)\Big\} \max_{l < i \le l+n} EX_i^2 I(|X_i| \ge B)\Big)^{(2+\delta)/2} \\ &\quad + n \exp\Big\{K_1 \sum_{i=0}^{[\log n]} \rho^{2/(2+\delta)}(2^i)\Big\} \max_{l < i \le l+n} E|X_i|^{2+\delta} I(|X_i| \le B)\Big\}. \end{aligned}$$

By Lemma 4.3.1 we get

$$\begin{aligned} I_1 &\le x^{-\delta} c \Big\{ \Big(nh(n) \max_{l < i \le l+n} EX_i^2 \\ &\quad + n \exp\Big\{(1+\varepsilon) \sum_{i=0}^{[\log n]} \rho(2^i)\Big\} \max_{l < i \le l+n} EX_i^2 I(|X_i| > B)\Big)^{\frac{2+\delta}{2}} \\ &\quad + n \exp\Big\{(1+\varepsilon) \sum_{i=0}^{[\log n]} \rho^{2/(2+\delta)}(2^i)\Big\} \\ &\quad \cdot \max_{l < i \le l+n} E|X_i|^{2+\delta} I(|X_i| \le B) \log^{2+\delta}(2n)\Big\}. \end{aligned} \tag{4.3.26}$$

We estimate I_2 below. Denote

$$\begin{aligned} &Y_i = \textstyle\sum_{j=l+2ik+1}^{l+(2i+1)k} X_{j2}, \qquad && W_i = \textstyle\sum_{j=0}^{i} Y_j, \quad i = 0, 1, \cdots, p_1, \\ &Z_i = \textstyle\sum_{j=l+(2i+1)k+1}^{l+2(i+1)k} X_{j2}, && W_i^* = \textstyle\sum_{j=0}^{i} Z_j, \quad i = 0, 1, \cdots, p_2, \end{aligned}$$

where $p_1 = [(\frac{n}{k} - 1)/2]$, $p_2 = [\frac{n}{2k}] - 1$. We have

$$I_2 \le 8\Big\{E \max_{0\le i\le p_1} W_i^2 + E \max_{0\le i\le p_2} W_i^{*2} + E \max_{0\le i\le [n/k]} \max_{ik+1\le j\le (i+1)k} \Big|\sum_{v=l+ik+1}^{l+j} X_{v2}\Big|^2\Big\}$$
$$=: 8(I_{21} + I_{22} + I_{23}). \tag{4.3.27}$$

It is easy to see from the proof of Lemma 4.3.2 that

$$I_{21} + I_{22} \le c\, n \exp\Big\{(1+\varepsilon)\sum_{i=0}^{[\log n]} \rho(2^i)\Big\} \cdot \Big(1 + \rho^2(k)\log^4\Big[\frac{n}{k}\Big]\Big) \max_{l<i\le l+n} EX_i^2 I(|X_i| > B). \tag{4.3.28}$$

And from Lemma 2.2.2

$$\begin{aligned} I_{23} &\le \sum_{o\le i\le [n/k]} E \max_{ik+1\le j\le (i+1)k} \Big|\sum_{v=l+ik+1}^{l+j} X_{v2}\Big|^2 \\ &\le 32 \sum_{0\le i\le [n/k]} \Big(E\Big|\sum_{v=l+ik+1}^{l+(i+1)k} \{|X_v| I(|X_v| \ge B) - E|X_v| I(|X_v| \ge B)\}\Big|^2 \\ &\quad + k^2 \max_{l+ik+1\le j\le l+(i+1)k} (E|X_j| I(|X_j| \ge B)^2)\Big) \\ &\le c\Big\{n \exp\Big\{(1+\varepsilon)\sum_{i=0}^{[\log n]} \rho(2^i)\Big\} \max_{l<i\le l+n} EX_i^2 I(|X_i| \ge B) \\ &\quad + nk \max_{l<i\le l+n} (E|X_i| I(|X_i| \ge B))^2\Big\}. \end{aligned} \tag{4.3.29}$$

Inserting (4.3.26)-(4.3.29) into (4.3.24) we prove (4.3.23), as desired.

Proof of Theorem 4.3.1.

By the condition (ii), we need only to show that

(a) $\{S^2_{[nt]}/\sigma_n^2, n \ge 1\}$ is uniformly integrable for any $0 \le t \le 1$.

(b) For any given $\varepsilon > 0$, there exist a $\lambda > 1$ and an integer n_0 such that for $n \ge n_0, 0 \le k \le n\lambda^2/\varepsilon$ we have

$$P\Big\{\max_{1\le i\le n}\Big|\sum_{j=k+1}^{k+i} X_j\Big| \ge \lambda\sigma_n\Big\} \le \varepsilon/\lambda^2. \tag{4.3.30}$$

To this end, we need only to show that for any given $\varepsilon > 0$ there exist a $\lambda > 1$ and an integer n_0 such that for every $n \geq n_0, l \geq 0$

$$E \max_{1\leq i\leq n} S_l^2(i) I(\max_{1\leq i\leq n} |S_l(i)| \geq \lambda(nh(n))^{1/2})/nh(n) \leq \varepsilon. \qquad (4.3.31)$$

Without loss of generality we assume that for $n \geq 16$

$$\rho(n) \geq 1/(\log n(\log\log n)^2). \qquad (4.3.32)$$

In fact, if put $\rho^*(n) = \rho(n) \vee (\log n)^{-1}(\log\log n)^{-2}$, it is easy to check that $\rho^*(n)$ satisfies condition (v) also.

By conditions (i), (ii), (iii) and Lemma 4.3.3, there exists a constant K such that for every $n, 1 \leq k \leq n, B > 0$ and $\lambda > 0$, we have

$$\begin{aligned}
&E \max_{1\leq i\leq n} S_l^2(i) I(\max_{1\leq i\leq n} |S_l(i)| \geq \lambda(nh(n))^{1/2})/nh(n) \\
&\leq K\Big\{\Big(\{nh(n) + n\exp\Big(\Big(1+\frac{\varepsilon}{4}\Big)\sum_{i=0}^{[\log n]} \rho(2^i)\Big) \\
&\quad \times \max_{l<i\leq l+n} EX_i^2 I(|X_i| \geq B)\}^{\frac{2+\delta}{2}} \\
&\quad + n\exp\Big(K\sum_{i=0}^{[\log n]} \rho^{2/(2+\delta)}(2^i)\Big)(\log(2n))^{2+\delta} \\
&\quad \times \max_{l<i\leq l+n} E|X_i|^{2+\delta} I(|X_i| < B)\Big)\Big/(\lambda^\delta (nh(n))^{\frac{2+\delta}{2}}) \\
&\quad + n\exp\Big((1+\frac{\varepsilon}{4})\sum_{i=0}^{[\log n]} \rho(2^i)\Big) \max_{l<i\leq l+n} EX_i^2 I(|X_i| \geq B) \\
&\quad \times \Big(1 + \rho^2(k)\log^4\Big[\frac{n}{k}\Big]\Big)/nh(n) \\
&\quad + nk \max_{l<i\leq l+n} (E|X_i| I(|X_i| \geq B))^2/nh(n)\Big\} \\
&=: K(I_1 + I_2 + I_3). \qquad (4.3.33)
\end{aligned}$$

Put

$$\begin{aligned}
&T = \exp\Big\{\frac{3K}{\delta}\sum_{i=0}^{[\log n]} \rho^{2/(2+\delta)}(2^i)\Big\}, \\
&B = n^{1/2}/T, \qquad k = [n/T^2] + 1.
\end{aligned}$$

We first estimate I_1. Note that

$$\begin{aligned}\log T &= \frac{3K}{\delta}\sum_{i=0}^{[\log n]}\rho^{2/(2+\delta)}(2^i)\\ &\le \frac{3K}{\delta}\rho^{-\delta/(2+\delta)}(\frac{n}{T^2})\sum_{i=0}^{[\log nT^{-2}]}\rho(2^i)\\ &\quad+\frac{3K}{\delta}\rho^{2/(2+\delta)}(\frac{n}{T^2})\sum_{i=[\log nT^{-2}]+1}^{[\log n]}1\\ &\le \frac{3K}{\delta}\rho^{-\delta/(2+\delta)}(\frac{n}{T^2})\sum_{i=0}^{[\log nT^{-2}]}\rho(2^i)\\ &\quad+\frac{7K}{\delta}\rho^{2/(2+\delta)}(\frac{n}{T^2})\log T.\end{aligned}$$

Therefore for large n we have

$$\log T \le \frac{3K}{\delta}\Big(1+\frac{\varepsilon}{24}\Big)\rho^{-\delta/(2+\delta)}\Big(\frac{n}{T^2}\Big)\sum_{i=0}^{[\log nT^{-2}]}\rho(2^i) \tag{4.3.34}$$

and

$$\sum_{i=0}^{[\log n]}\rho(2^i)\le\Big(1+\frac{\varepsilon}{12}\Big)\sum_{i=0}^{[\log nT^{-2}]}\rho(2^i). \tag{4.3.35}$$

By condition (v), there exists a $C_1>0$ such that for every $n\ge 1$

$$g(n^{1/2})\ge C_1\exp\Big\{(2+\varepsilon)\sum_{i=0}^{[\log n]}\rho(2^i)\Big\}. \tag{4.3.36}$$

Combining it with (4.3.35) and condition (v), we obtain

$$g(B)\ge C_1\exp\Big\{\frac{2+\varepsilon}{1+\varepsilon/12}\sum_{i=0}^{[\log n]}\rho(2^i)\Big\} \tag{4.3.37}$$

for large n. By Lemma 2.2.3 and condition (iii), there exists a constant $C_2>0$ such that for large n

$$ES_n^2\ge C_2 n\exp\Big\{-\Big(1+\frac{\varepsilon}{4}\Big)\sum_{i=0}^{[\log n]}\rho(2^i)\Big\},$$

and hence, by condition (ii), we have

$$h(n) \geq C_2 \exp\Big\{-\Big(1+\frac{\varepsilon}{4}\Big)\sum_{i=0}^{[\log n]} \rho(2^i)\Big\}. \tag{4.3.38}$$

By the monotonicity of $g(x)$ and $x^\delta/g(x)$, we obtain

$$\max_{i\geq 1} EX_i^2 I(|X_i| \geq B) \leq \frac{1}{g(B)} \max_{i\geq 1} EX_i^2 g(|X_i|) I(|X_i| \geq B), \tag{4.3.39}$$

$$\max_{i\geq 1} E|X_i|^{2+\delta} I(|X_i| \leq B) \leq \frac{B^\delta}{g(B)} \max_{i\geq 1} EX_i^2 g(|X_i|). \tag{4.3.40}$$

Combining (4.3.37), (4.3.39),(4.3.40) and (4.3.32) together implies

$$\begin{aligned} &n\exp\Big\{\Big(1+\frac{\varepsilon}{4}\Big)\sum_{i=0}^{[\log n]} \rho(2^i)\Big\} \max_{i\geq 1} EX_i^2 I(|X_i| \geq B) \\ &\leq \frac{nh(n)}{C_1 C_2} \max_{i\geq 1} EX_i^2 g(|X_i|) I(|X_i| \geq B), \end{aligned} \tag{4.3.41}$$

$$\begin{aligned} &n\exp\Big\{K\sum_{i=0}^{[\log n]} \rho^{2/(2+\delta)}(2^i)\Big\} \log^{2+\delta}(2n) \max_{i\geq 1} E|X_i|^{2+\delta} I(|X_i| < B) \\ &\leq n\exp\Big\{2K\sum_{i=0}^{[\log n]} \rho^{2/(2+\delta)}(2^i)\Big\} B^\delta \max_{i\geq 1} EX_i^2 g(|X_i|)/g(B) \\ &\leq n^{(2+\delta)/2'} \max_{i\geq 1} EX_i^2 g(|X_i|) \exp\Big\{-K\sum_{i=0}^{[\log n]} \rho^{2/(2+\delta)}(2^i)\Big\} \\ &\leq (nh(n))^{(2+\delta)/2} \max_{i\geq 1} EX_i^2 g(|X_i|) \end{aligned} \tag{4.3.42}$$

for large n. We have $\max_{i\geq 1} EX_i^2 g(|X_i|) < \infty$ by condition (i). Therefore we obtain

$$I_1 \leq \varepsilon/(3K) \tag{4.3.43}$$

for large n provided that λ is large enough.

We now estimate I_2. From (4.3.34) and the definition of k, we have

$$\begin{aligned} &(\log[\frac{n}{k}])^4 \rho^2(k) \leq 2(\log T)^4 \rho^2(k) \\ &\leq \Big(\frac{3K}{\delta}\Big)^4 \Big(1+\frac{\varepsilon}{24}\Big)^4 \Big(\rho(\frac{n}{T^2})\Big)^{2-\frac{4\delta}{2+\delta}} \Big(\sum_{i=0}^{[\log n]} \rho(2^i)\Big)^4 \\ &\leq \Big(\frac{3K}{\delta}\Big)^4 \Big(\sum_{i=0}^{[\log n]} \rho(2^i)\Big)^4. \end{aligned} \tag{4.3.44}$$

By (4.3.44),(4.3.39),(4.3.37) and (4.3.38), we get

$$
\begin{aligned}
& n \exp\Big\{\Big(1+\frac{\varepsilon}{4}\Big)\sum_{i=0}^{[\log n]} \rho(2^i)\Big\} \\
& \max_{i\ge 1} E X_i^2 I(|X_i| \ge B)\Big(1+\rho^2(k)\log^4\Big[\frac{n}{k}\Big]\Big) \\
& \quad \le \frac{nh(n)}{g(B)h(n)} \exp\Big\{\Big(1+\frac{\varepsilon}{4}\Big)\sum_{i=0}^{[\log n]} \rho(2^i)\Big\} \\
& \quad \cdot \Big(1+(\frac{3K}{\delta})^4\Big(\sum_{i=0}^{[\log n]} \rho(2^i)\Big)^4\Big) \max_{i\ge 1} E X_i^2 g(|X_i|) I(|X_i| \ge B) \\
& \quad \le \frac{nh(n)}{C_1 C_2} \exp\Big\{-\frac{\varepsilon}{8}\sum_{i=0}^{[\log n]} \rho(2^i)\Big\} \\
& \quad \cdot \Big(1+\Big(\frac{3K}{\delta}\Big)^4\Big(\sum_{i=0}^{[\log n]} \rho(2^i)\Big)^4\Big) \max_{i\ge 1} E X_i^2 g(|X_i|) I(|X_i| \ge B) \\
& \quad \le cnh(n) \max_{i\ge 1} E X_i^2 g(|X_i|) I(|X_i| \ge B), \qquad (4.3.45)
\end{aligned}
$$

where the following result is used:

$$\Big(\sum_{i=0}^{[\log n]} \rho(2^i)\Big)^4 \exp\Big\{-\frac{\varepsilon}{8}\sum_{i=0}^{[\log n]} \rho(2^i)\Big\} = O(1).$$

Combining (4.3.45) with condition (i), we have

$$I_2 \le \varepsilon/3K. \qquad (4.3.46)$$

Finally, we consider I_3. Note that

$$\max_{i\ge 1} E|X_i| I(|X_i| \ge B) \le \frac{1}{Bg(B)} \max_{i\ge 1} E X_i^2 g(|X_i|) I(|X_i| \ge B).$$

Hence it follows from (4.3.37) and (4.3.38) that

$$
\begin{aligned}
& kn \max_{i\ge 1} (E|X_i| I(|X_i| \ge B))^2 \\
& \quad \le \frac{kn}{B^2 g^2(B)} \max_{i\ge 1} (E X_i^2 g(|X_i|) I(|X_i| \ge B))^2 \\
& \quad \le \frac{2nh(n)}{C_1^2 C_2} \max_{i\ge 1} (E X_i^2 g(|X_i|) I(|X_i| \ge B))^2.
\end{aligned}
$$

By condition (i), we have

$$I_3 \leq \varepsilon/3K \tag{4.3.47}$$

for large n. Inserting (4.3.43),(4.3.46) and (4.3.47) into (4.3.33) we prove (4.3.31). The proof of Theorem 4.3.1 is completed.

From Theorem 4.3.1, we have the following corollary immediately.

Corollary 4.3.1. *Let $\{X_n, n \geq 1\}$ be a strictly stationary ρ-mixing sequence of random variables with $EX_1 = 0, EX_1^2 < \infty, \sigma_n^2 = ES_n^2 \to \infty$. If one of the following conditions is statisfied:*

(i) $EX_1^2 g(|X_1|) < \infty$, *and for some* $0 < \varepsilon < 1$,

$$\exp\Big((2+\varepsilon)\sum_{k=1}^{[\log n]} \rho(2^j)\Big) = O(g(n^{1/2})),$$

(ii) for some $\varepsilon > 0, a > 0, EX_1^2(\log|X_1|)^{2a/(1-\varepsilon)} < \infty$ *and* x $\rho(n) \leq a/\log n$,

(iii) for some $0 < \beta < 1, \varepsilon > 0, a > 0$,

$$EX_1^2 \exp\Big\{\frac{2a(1+\varepsilon)}{1-\beta}(2\log|X_1|)^{1-\beta}\Big\} < \infty$$

and

$$\rho(n) \leq a/(\log n)^\beta,$$

(iv) for some $r > 0, \varepsilon > 0, a > 0$,

$$EX_1^2 \exp\Big(\frac{4a(1+\varepsilon)\log|X_1|}{(\log\log|X_1|)^r}\Big) < \infty$$

$$\rho(n) \leq a(\log\log n)^{-r},$$

then

$$W_n \Longrightarrow W.$$

4.4 The WIP when the variance is infinite

Bradley (1988) established a CLT for a strictly stationary ρ-mixing sequence with infinite variance. Shao (1993a) showed a WIP under the same hypothesis.

Theorem 4.4.1. *Let $\{X_n, n \geq 1\}$ be a strictly stationary ρ-mixing sequence of non-degenerate random variables with $EX_1 = 0$. Suppose that*

(i) $H(x) := EX_1^2 I(|X_1| \leq x)$ *is slowly varying as* $x \to \infty$,

(ii) $\rho(1) < 1$,

(iii) $\sum_{n=1}^{\infty} \rho(2^n) < \infty$.

Then there exists a sequence $\{A_n, n \geq 1\}$ of positive numbers with $A_n \to \infty$ as $n \to \infty$, such that

$$W_n \Longrightarrow W$$

where $W_n(t) = S_{[nt]}/A_n, \quad 0 \leq t \leq 1.$

Remark 4.4.1. In fact, Shao (1993a) showed a more general result as follows: Let $g : (-\infty, \infty) \to [0, \infty)$ be a non-decreasing continuous even function and $x^\delta / g(x)$ is non-decreasing for any $\delta > 0$ and x large enough. Let

$$e(x, \varepsilon) = \exp\Big\{\varepsilon \sum_{i=1}^{[\log x]} \rho(2^i)\Big\}, \quad x(\delta) = \exp\Big\{\sum_{i=0}^{[\log x]} \rho^{1-\delta}(2^i)\Big\}.$$

Suppose that conditions (i) and (ii) in Theorem 4.4.1 are satisfied and

(i)′ $G(x) := EX_1^2 g(X_1) I(|X_1| \leq x)$ is slowly varying as $x \to \infty$,

(iv) $G(x)e(x^2, 2+\varepsilon) = O(H(x)g(x))$ for some $0 < \varepsilon < 1$,

(v) $g(x) = O(g(x/x(\delta)))$ or $G(x) = O(G(x/x(\delta)))$ for some $0 < \delta < 1$ as $x \to \infty$.

Then we also have the conclusion of Theorem 4.4.1.

Obviously, conditions (i) and (iii) imply condition (i)′, (iv) and (v) by taking $g(x) \equiv 1$. It is well-known that the mixing rate (iii) is essentially sharp, even in the case of a finite second moment. However the following example is interesting: Let X_1 have the density function $p(x) = a(1 + |x|^3)^{-1}$ for $x \in R^1$, where $a^{-1} = \int_{-\infty}^{\infty} (1+|x|^3)^{-1} dx$. Let $g(x) = \exp(\log(1+|x|^3)^\alpha)$ for some $0 < \alpha < 1$. It is easy to see that as $x \to \infty$

$$H(x) \sim 2a \log(1 + |x|^3)/3,$$
$$G(x) \sim 2a(\log(1 + |x|^3))^{1-\alpha} g(x)/3\alpha.$$

If $\rho(n) \leq \alpha/(5 \log n)$, we can easily verify that the conditions in Remark 4.4.1 are satisfied but the condition (iii) in Theorem 4.4.1 fails. Hence condition (iii) may be not essentially sharp in some particular case of infinite variance, even of finite variance.

In order to prove Theorem 4.4.1, we introduce some notations. Let M^* be a positive integer such that

$$\sup_{x>0} H(x)/x^2 > 1/M^*. \tag{4.4.1}$$

For each $n \geq M^*$ define

$$t_n = \sup\{x > 0 : H(x)/x^2 \geq 1/n\}. \tag{4.4.2}$$

It is clear that $t_n \to \infty$ monotonically as $n \to \infty$. Note by a trivial argument that

$$t_n^2 = nH(t_n) \qquad \text{for} \quad n \geq M^*.$$

By condition (i), for any $0 < \varepsilon < 1/2$ and large n

$$n^{1-\varepsilon} \leq t_n^2 \leq n^{1+\varepsilon}. \tag{4.4.3}$$

We need a few properties of these t_n's.

Lemma 4.4.1. *If condition (iii) is satisfied, for any* $0 < a < 1$,

$$t_{[na]}^2 / t_n^2 \longrightarrow a \qquad \text{as } n \to \infty. \tag{4.4.4}$$

Proof. (4.4.2) implies

$$t_n^2 H(t_{[na]})/(t_{[na]}^2 H(t_n)) \longrightarrow 1/a \qquad \text{as } n \to \infty. \tag{4.4.5}$$

We show that there is a $M > 0$ such that

$$\limsup_{n\to\infty} t_n^2 / t_{[na]}^2 \leq M. \tag{4.4.6}$$

In fact, if not, there is a subsequence n_k such that $\lim_{k\to\infty} t_{n_k}^2 / t_{[n_k a]}^2 = \infty$. Then using Property A5 of a slowly varying function (see Appendix) we obtain

$$\lim_{k\to\infty} t_{n_k}^2 H(t_{[n_k a]})/(t_{[n_k a]}^2 H(t_{[n_k a]} t_{n_k} / t_{[n_k a]})) = \infty,$$

which is contrary to (4.4.5). By (4.4.2) and (4.4.6), it follows that

$$n/[na] \leq t_n^2 / t_{[na]}^2 \leq nH(Mt_{[na]})/([na]H(t_{[na]})),$$

which implies (4.4.4).

Lemma 4.4.2.

$$\lim_{n\to\infty} nP(|X_1| > t_n) = 0$$

and

$$\lim_{n\to\infty} (n/H(t_n))^{1/2} E|X_1| I(|X_1| > t_n) = 0.$$

The proof of this lemma can be found in Bradley (1988) and will be not presented here.

Proof of Theorem 4.4.1.

For some $\varepsilon > 0$, put $C_1 = C\exp\Big\{(1+\varepsilon)\sum_{i=0}^{[\log n]}\rho(2^i)\Big\}$, where C is defined in Lemma 2.2.2 and $C_2 = C'\exp\Big\{-(1+\varepsilon)\sum_{i=0}^{[\log n]}\rho(2^i)\Big\}$, where C' is defined in Lemma 2.2.3. For $n \geq M^*$ define

$$X_k^{(n)} = X_k I(|X_k| \leq t_n) - EX_k I(|X_k| \leq t_n), \qquad k \geq 1 \tag{4.4.7}$$

and

$$S_m^{(n)} = X_1^{(n)} + \cdots + X_m^{(n)}, \qquad m \geq 1.$$

By the definitions of C_1 and C_2 we have

$$C_2 m E{X_1^{(n)}}^2 \leq E(S_m^{(n)})^2 \leq C_1 m E(X_1^{(n)})^2. \tag{4.4.8}$$

Put $A_n(m) = \|S_m^{(n)}\|_2$ and $A_n = A_n(n)$. Note that $EX_1 = 0$, it is easy to verify that

$$E(X_1^{(n)})^2 \sim H(t_n) \qquad \text{as} \quad n \to \infty. \tag{4.4.9}$$

Therefore there exist $0 < C_2' < C_1' < \infty$ such that

$$C_2' n H(t_n) \leq A_n^2 \leq C_1' n H(t_n). \tag{4.4.10}$$

Next we formulate the proof in two steps.

Step 1. We prove that $S_n^{(n)}/A_n \xrightarrow{d} N(0,1)$ in distribution as $n \to \infty$. It suffices to show that

$$\lim_{n\to\infty} E\{\exp(itS_n^{(n)}/A_n)\} = \exp(-t^2/2) \qquad \text{for any} \quad t \in R. \tag{4.4.11}$$

Since the case of $t = 0$ is trivial, it needs only to prove the above equality for $t \neq 0$. Fix $t \neq 0$. Let J be a positive integer specified later on. Define p^* and L^* to be positive integers such that

$$\frac{4C_1}{C_2 p^*}\sum_{j=1}^{\infty} 2^{J-j/2} \leq \frac{\varepsilon^2}{4}, \tag{4.4.12}$$

$$|(1-t^2/(2j))^j - \exp(-t^2/2)| \leq \varepsilon/3 \qquad \text{for each} \quad j \geq 2^{L^*}. \tag{4.4.13}$$

Let $N^* \geq M^*$ be a positive integer such that

$$N^* \geq 2p^* \cdot 2^{L^*}, \tag{4.4.14}$$

$$E(X_1^{(n)})^2 > 0 \qquad \text{for each} \quad n \geq N^*, \tag{4.4.15}$$

and

$$\frac{128p^{*2}t^2}{C_2 nE(X_1^{(n)})^2}(EX_1 I(|X_1| \le t_n))^2$$
$$+ \frac{400p^{*4}(t^2 \vee |t|^3)}{C_2 \wedge C_2^{3/2}} E\Big\{ \frac{X_1^2 I(|X_1| \le t_n)}{nE(X_1^{(n)})^2} \wedge \frac{|X_1|^3 I(|X_1| \le t_n)}{(nE(X_1^{(n)})^2)^{3/2}} \Big\}$$
$$\le \varepsilon/(3n) \qquad \text{for each} \quad n \ge N^*. \tag{4.4.16}$$

Here, (4.4.16) can be justified by (4.4.9).

Let $N \ge N^*$ be an arbitrary but fixed integer. Then to prove (4.4.11) it suffices to show that

$$|E\exp(itS_N^{(N)}/A_N) - \exp(-t^2/2)| \le \varepsilon. \tag{4.4.17}$$

Referring to (4.4.14), let L be the positive integer such that

$$p^* \le N/2^L \le 2p^*.$$

Note that $L \ge L^*$. Let p be the positive integer such that

$$p2^L \le N < (p+1)2^L. \tag{4.4.18}$$

It is easy to see that

$$p^* \le p < 2p^*. \tag{4.4.19}$$

Let us partition $\mathbb{N}$ into disjoint blocks of consecutive integers, leaving no gaps between the blocks. The order of the blocks is $G(1), G(2), \cdots$, with

$$\mathrm{Card}G(j) = \begin{cases} p, & \text{if} \quad j \quad \text{is odd;} \\ [2^{J+l/2}], & \text{where} \quad l \quad \text{is such integer that} \quad j/2^l \\ & \quad \text{is an odd integer if } j \text{ is even.} \end{cases} \tag{4.4.20}$$

Henceforth we shall deal only with the blocks $G(1), G(2), \cdots, G(2^{L+1}-1)$. For each $l = 1, \cdots, L$, there are exactly 2^{L-l} integers $j \in \{1, 2, \cdots, 2^{L+1}-1\}$ such that $j/2^l$ is an odd integer. Therefore

$$\mathrm{Card}\{G(2) \cup G(4) \cup \cdots \cup G(2^{L+1}-2)\} = \sum_{l=1}^{L} 2^{L-l}[2^{J+l/2}]. \tag{4.4.21}$$

Hence by (4.4.18)

$$\begin{aligned} N &\le 2^L p + \sum_{l=1}^{L} 2^{L-l} \le 2^L p + \sum_{l=1}^{L} 2^{L-l}[2^{J+l/2}] \\ &= \mathrm{Card}\{G(1) \cup G(2) \cup \cdots \cup G(2^{L+1}-1)\} \\ &\le N + \sum_{l=1}^{L} 2^{L-l}[2^{J+l/2}]. \end{aligned} \tag{4.4.22}$$

For each $j = 1, 2, \cdots, 2^{L+1} - 1$ define

$$U(j) = \sum_{k \in G(j)} X_k^{(N)}.$$

And further, for even integers j_1 and j_2 such that $0 \le j_1 < j_2 \le 2^{L+1}$, define

$$V(j_1, j_2) = U(j_1 + 1) + U(j_1 + 3) + \cdots + U(j_2 - 1).$$

If $1 \le j \le 2^l - 1$, then for the integer m such that $j/2^m$ is an odd integer, we have that $m < l$ and hence $(2^l + j)/2^m$ is an odd integer, and hence $\mathrm{Card}G(2^l + j) = \mathrm{Card}G(j)$ by (4.4.20). Consequently, if we denote $u = \mathrm{Card}\{G(1) \cup G(2) \cup \cdots \cup G(2^l)\}$ and use the notation $u + G = \{u + g : g \in G\}$ for sets G of positive integers, we have (by induction on j) that $G(2^l + j) = u + G(j)$ for $j = 1, 2, \cdots, 2^l - 1$. In particular, if we denote

$$G = G(1) \cup G(3) \cup \cdots \cup G(2^l - 1)$$

and

$$G^* = G(2^l + 1) \cup G(2^l + 3) \cup \cdots \cup G(2^{l+1} - 1),$$

then $G^* = u + G, V(2^l, 2^{l+1}) = \sum_{k \in G^*} X_k^{(N)} = \sum_{k \in G} X_{k+u}^{(N)}$ and $V(0, 2^l) = \sum_{k \in G} X_k^{(N)}$. The stationarity of the sequence $\{X_k^{(N)}, k \ge 1\}$ implies the following useful property:

For each $l = 1, \cdots, L$, $V(0, 2^l)$ and $L(2^l, 2^{l+1})$ have the same distribution. (4.4.23)

Hence by a simple calculation for each $l = 1, \cdots, L$,

$$\begin{aligned} &2(1 - \rho([2^{J+l/2}]))EV(0, 2^l)^2 \\ &\quad \le EV(0, 2^{l+1})^2 \\ &\quad \le 2(1 + \rho([2^{J+l/2}]))EV(0, 2^l)^2. \end{aligned} \tag{4.4.24}$$

Moreover for each $l = 1, \cdots, L$

$$\begin{aligned} &|E \exp\{itV(0, 2^{l+1})\} - (E \exp\{itV(0, 2^l)\})^2| \\ &\quad \le \rho([2^{J+l/2}])E|\exp\{itV(0, 2^l)\} - 1|^2 \\ &\quad \le \rho([2^{J+l/2}])t^2 EV(0, 2^l)^2 \\ &\quad \le \rho([2^{J+l/2}])t^2 C_1 2^{l-1} EU(1)^2. \end{aligned} \tag{4.4.25}$$

In what follows, it should be kept in mind that $E(S_m^{(N)})^2 > 0$ for all $m \geq 1$ by (4.4.15), the fact $N \geq N^*$ and (4.4.8). By (4.4.24) and induction

$$2^L\Big\{\prod_{l=1}^{L}(1-\rho([2^{J+l/2}]))\Big\}EU_1^2 \\ \leq EV(0,2^{L+1})^2 \leq 2^L\Big\{\prod_{l=1}^{L}(1+\rho([2^{J+l/2}]))\Big\}EU_1^2$$

and hence

$$1-\varepsilon/2 \leq \|V(0,2^{L+1})\|_2/(2^{L/2}\|U_1\|_2) \leq 1+\varepsilon/2 \tag{4.4.26}$$

provided J is large enough. By (4.4.21) and (4.4.8)

$$E(U(2)+U(4)+\cdots+U(2^{L+1}-2))^2 \\ \leq C_1\Big(\sum_{l=1}^{L}2^{L+J-l/2}\Big)E(X_1^{(n)})^2.$$

Also by (4.4.22), $U(1)+U(2)+\cdots+U(2^{L+1}-1)-S_N^{(N)}$ is the sum of at most $[\sum_{l=1}^{L}2^{L+J-l/2}]$ distinct $X_k^{(N)}$'s, and hence

$$E(U(1)+U(2)+\cdots+U(2^{L+1}-1)-S_N^{(N)})^2 \\ \leq C_1\Big(\sum_{l=1}^{L}2^{L+J-l/2}\Big)E(X_1^{(N)})^2.$$

Consequently, using (4.4.8), (4.4.12) and (4.4.19)

$$\begin{aligned}
&\|V(0,2^{L+1})-S_N^{(N)}\|_2 \\
&\quad\leq \|U(1)+U(2)+\cdots+U(2^{L+1}-1)-S_N^{(N)}\|_2 \\
&\qquad+\|U(2)+U(4)+\cdots+U(2^{L+1}-2)\|_2 \\
&\quad\leq 2C_1^{1/2}\Big(\sum_{l=1}^{L}2^{L+J-l/2}\Big)^{1/2}\|X_1^{(N)}\|_2 \\
&\quad\leq 2C_1^{1/2}2^{L/2}\Big(\sum_{l=1}^{L}2^{J-l/2}\Big)^{1/2}\|U(1)\|_2/(C_2p)^{1/2} \\
&\quad\leq 2^{L/2}\varepsilon\|U(1)\|_2/2.
\end{aligned} \tag{4.4.27}$$

We now come back to (4.4.17). By (4.4.26) and (4.4.27) we have

$$\begin{aligned}
&\left|1-\frac{A_N}{2^{L/2}\|U(1)\|_2}\right| \\
&\le \left|1-\frac{\|V(0,2^{L+1})\|_2}{2^{L/2}\|U(1)\|_2}\right| + \left|\frac{\|V(0,2^{L+1})\|_2 - \|S_N^{(N)}\|_2}{2^{L/2}\|U(1)\|_2}\right| \\
&\le \frac{\varepsilon}{2} + \frac{\|V(0,2^{L+1}) - S_N^{(N)}\|_2}{2^{L/2}\|U(1)\|_2} \\
&\le \varepsilon. \qquad (4.4.28)
\end{aligned}$$

(4.4.28) and (4.4.27) together imply that (4.4.17) is equivalent to

$$D := |E\exp\{itV(0,2^{L+1})/(2^{L/2}\|U(1)\|_2\} - \exp(-t^2/2)| \le \varepsilon. \qquad (4.4.29)$$

Obviously

$$\begin{aligned}
D \le\ & |E\exp\{itV(0,2^{L+1})/(2^{L/2}\|U(1)\|_2)\} \\
& - (E\exp\{i(t/2^{L/2})U(1)/\|U(1)\|_2\})^{2^L}| \\
& + |(E\exp\{i(t/2^{L/2})U(1)/\|U(1)\|_2\})^{2^L} - (1-(1/2)t^2/2^L)^{2^L}| \\
& + |(1-(1/2)t^2/2^L)^{2^L} - \exp(-t^2/2)| \\
=:\ & e_1 + e_2 + e_3.
\end{aligned}$$

Using (4.4.25) and the elementary inequality

$$\left|\prod_{k=1}^m y_k - \prod_{k=1}^m z_k\right| \le \sum_{k=1}^m |y_k - z_k|, \qquad (4.4.30)$$

where $y_1, \cdots, y_m, z_1 \cdots, z_m$ are complex numbers in the closed unit disc, we have

$$\begin{aligned}
&\left|\left(E\exp\{iTV(0,2^{l+1})\}\right)^{2^{L-l}} - \left(E\exp\{iTV(0,2^l)\}\right)^{2^{L-l+1}}\right| \\
&\le 2^{L-l}\rho([2^{J+l/2}])T^2C_1 2^{l-1}EU(1)^2
\end{aligned}$$

for any T. Hence by induction

$$\begin{aligned}
&|E\exp\{iTV(0,2^{L+1})\} - (E\exp\{iTV(0,2)\})^{2^L}| \\
&\le 2^L T^2 C_1 EU(1)^2 \sum_{l=1}^{L} \rho([2^{J+l/2}]).
\end{aligned}$$

Letting $T = t/(2^{L/2}\|U(1)\|_2)$ and keeping in mind that $U(1) = V(0,2)$, we have

$$e_1 \le t^2 C_1 \sum_{l=1}^{L} \rho([2^{J+l/2}]) \le \varepsilon/3 \qquad (4.4.31)$$

provided the constant J is large enough. In order to estimate e_2, define the event

$$F_k = \{|X_k^{(N)}| = \max_{1\le j\le p} |X_j^{(N)}|\}, k = 1, \cdots, p.$$

Put $s = t/(2^{L/2}\|U(1)\|_2)$ for simplicity. By (4.4.8) and (4.4.18),

$$s^2 \le t^2/(2^L C_2 p E(X_1^{(N)})^2) \le 2t^2/(C_2 N E(X_1^{(N)})^2).$$

Now we need a fact that for any real numbers x and r,

$$|x-r|^2 \wedge |x-r|^3 \le 4r^2 + 8(x^2 \wedge |x|^3).$$

Using this fact and (4.4.19), (4.4.16) and (4.4.18) we have

$$\begin{aligned}
&E(|sU(1)|^2 \wedge |sU(1)|^3) \\
&\le \sum_{k=1}^{p} EI(F_k)(|spX_k^{(N)}|^2 \wedge |spX_k^{(N)}|^3) \\
&\le p^4 E(|sX_1^{(N)}|^2 \wedge |sX_1^{(N)}|^3) \\
&\le p^4\{4s^2(EX_1 I(|X_1| \le t_N))^2 \\
&\quad + 8E\{(s^2 X_1^2 I(|X_1| \le t_N)) \wedge (|s|^3 |X_1|^3 I(|X_1| \le t_N))\}\} \\
&\le (2p^*)^4\Big\{\frac{8t^2}{C_2 N E(X_1^{(N)})^2}(EX_1 I(|X_1| \le t_N))^2 \\
&\quad + 8E\Big\{\frac{2t^2 X_1^2 I(|X_1| \le t_N)}{C_2 N E(X_1^{(N)})^2} \wedge \frac{2^{3/2}|t|^3 |X_1|^3 I(|X_1| \le t_N)}{C_2^{3/2} N^{3/2}(E(X_1^{(N)})^2)^{3/2}}\Big\}\Big\} \\
&\le \frac{128 p^{*4} t^2}{C_2 N E(X_1^{(N)})^2}(EX_1 I(|X_1| \le t_n))^2 \\
&\quad + \frac{400 p^{*4}(t^2 \vee |t|^3)}{C_2 \wedge C_2^{3/2}} E\Big\{\frac{X_1^2 I(|X_1| \le t_N)}{N E(X_1^{(N)})^2} \wedge \frac{|x_1|^3 I(|X_1| \le t_N)}{N^{3/2}(E(X_1^{(N)})^2)^{3/2}}\Big\} \\
&\le \varepsilon/(3N) \le \varepsilon/(3\cdot 2^L).
\end{aligned}$$

Hence noting (4.4.30) we obtain

$$\begin{aligned}
e_2 &\le 2^L |E\exp(isU(1)) - (1 - \frac{1}{2}s^2 EU(1)^2)| \\
&\le 2^L E(|sU(1)|^2 \wedge |sU(1)|^3) \le \varepsilon/3. \qquad (4.4.32)
\end{aligned}$$

Here we use the inequality about a characteristic function (cf. p.331 in Bradley 1988).

As for e_3, by (4.4.13) and noting $L \geq L^*$, it is clear that

$$e_3 \leq \varepsilon/3. \tag{4.4.33}$$

(4.4.31), (4.4.32) and (4.4.33) together imply (4.4.29). This completes the proof of the CLT.

Step 2. Now we show that

$$W_n \Longrightarrow W \qquad \text{as} \quad n \to \infty. \tag{4.4.34}$$

By Lemma 4.4.2 and (4.4.10) we have

$$\lim_{n\to\infty} P\{\sup_{0\leq t\leq 1} |S_{[nt]} - S_{[nt]}^{(n)}| > \varepsilon A_n\} = 0 \qquad \text{for any} \quad \varepsilon > 0.$$

Hence, in order to prove the theorem it suffices (cf. Theorem 4.1 in Billingsley 1968) to show that

$$W_n^* \Longrightarrow W \qquad \text{as} \quad n \to \infty \tag{4.4.35}$$

where $W_n^*(t) = S_{[nt]}^{(n)}/A_n$. By Theorem 4.0.4, it is enough to prove that for any $0 < t \leq 1$

$$A_n^2([nt])/A_n^2 \longrightarrow t, \qquad \text{as} \quad n \to \infty \tag{4.4.36}$$

$$\{W_n^*(t)^2, n \geq 1\} \quad \text{is uniformly integrable,} \tag{4.4.37}$$

and there exists a constant $\lambda > 1$ for any $\varepsilon > 0$ such that for all large n

$$P\Big\{\max_{0\leq i\leq n} |S_i^{(n)}| \geq \lambda A_n\Big\} < \varepsilon/\lambda^2. \tag{4.4.38}$$

We prove that

$$A_n^2([mt])/A_n^2(m) \longrightarrow t \tag{4.4.39}$$

as $m \to \infty$ uniformly in $n \geq M^*$, which implies (4.4.36). At first consider the case of $t = 1/p$ where $p \geq 2$ is an integer. Let $q = [m/p]$,

$$Y_i = \sum_{j=iq+1}^{(i+1)q} X_j^{(n)}, \qquad i = 0, 1, \cdots, p-1,$$

$$Y_p = \sum_{j=pq+1}^{m} X_j^{(n)}.$$

Note that

$$A_n^2(m) = pA_n^2(q) + \sum_{i \neq j} EY_iY_j + EY_p^2.$$

For $i \neq j$ and integer $k \geq 1$, by the Minkowski inequality

$$|EY_iY_j| \leq 2\|Y_i\|_2\|S_k^{(n)}\|_2 + \rho(k)\|Y_i\|_2\|Y_j\|_2.$$

Hence by (4.4.8)

$$\begin{aligned} |A_n^2(m)/A_n^2(q) - p| &\leq (p+1)^2(\|S_k^{(n)}\|_2/A_n(q) + \rho(k) + \|Y_p\|_2^2/A_n^2(q)) \\ &\leq (p+1)^2((k^2+p^2)m^{-1/3} + \rho(k)) \qquad (4.4.40) \end{aligned}$$

for every m large enough. Choose k such that $(p+1)^2\rho(k) < \varepsilon/2$. Then for m large enough

$$|A_n^2(m)/A_n^2([m/p]) - p| < \varepsilon$$

uniformly in $n \geq M^*$. Therefore as $m \to \infty$ uniformly in $n \geq M^*$

$$A_n^2(m)/A_n^2([m/p]) \longrightarrow p. \qquad (4.4.41)$$

If t is a rational number, that is, $t = q/p$ for some integers p and q with $q < p$, then

$$\begin{aligned} A_n^2([mq/p])/A_n^2(m) &= \frac{A_n^2([mq/p])}{A_n(mq)} \cdot \frac{A_n^2(mq)}{A_n^2(m)} \\ &\longrightarrow q/p = t \qquad (4.4.42) \end{aligned}$$

as $m \to \infty$ uniformly in $n \geq M^*$ by (4.4.41).

If t is a irrational number, then for any given $0 < \varepsilon < 1/2$, take rational number $t_1 > 0$ such that

$$\varepsilon/4 < t - t_1 < \varepsilon/2.$$

By the Minkowski inequality

$$|A_n([mt]) - A_n([mt_1])| \leq A_n([mt] - [mt_1]). \qquad (4.4.43)$$

Let $p = [m/([mt] - [mt_1])]$. Then

$$\begin{aligned} \frac{1}{2}\varepsilon^{-1} &< m/(mt - mt_1 + 1) - 1 \\ &\leq p \leq m/(mt - mt_1 - 1) < 5\varepsilon^{-1} \end{aligned}$$

for $m > 20/\varepsilon$. Similarly to the proof of (4.4.40),

$$\begin{aligned}(p-(p+1)^2\rho(k))A_n^2([mt]-[mt_1]) \\ \le A_n^2(m)+(p+1)^2A_n^2([mt]-[mt_1])A_n^2(k) \\ +(p+1)^2E(X_1^{(n)})^2.\end{aligned}$$

Take k such that $\rho(k) < \varepsilon/24$. Then

$$\begin{aligned}&A_n^2([mt]-[mt_1])/A_n^2(m)\\ &\le 6p^{-1}(1+3(p+1)^4(A_n^2(k)+E(X_1^{(n)})^2)/A_n^2(m))\\ &\le 13\varepsilon \qquad (4.4.44)\end{aligned}$$

provided m is large enough. Combining (4.4.44) with (4.4.43) and (4.4.42) yields (4.4.39). Hence (4.4.36) is proved.

Now turn to (4.4.37). We have proved in Step 1 that

$$S_{[nt]}^{(n)}/A_{[nt]} \xrightarrow{d} N(0,1) \qquad \text{as } n \to \infty$$

for any $0 < t \le 1$. From (4.4.8) and (4.4.4) we have

$$\begin{aligned}&E(S_{[nt]}^{(n)} - S_{[nt]}^{([nt])})^2 A_{[nt]}^{-2}\\ &= A_{[nt]}^{-2}\operatorname{Var}\Big(\sum_{i=1}^{[nt]} X_i I(t_{[nt]} < |X_i| \le t_n)\Big)\\ &\le C_2C_1^{-1}E(X_1^2 I(t_{[nt]} < |X_1| \le t_n))/E(X_1^2 I(|X_1| \le t_{[nt]}))\\ &= C_2C_1^{-1}(H(t_n)-H(t_{[nt]}))/H(t_{[nt]})\\ &\longrightarrow 0, \qquad \text{as} \quad n\to\infty.\end{aligned}$$

It follows that

$$A_n([nt])/A_{[nt]} \longrightarrow 1, \qquad \text{as } n \to \infty.$$

Therefore $S_{[nt]}^{(n)}/A_n([nt]) \xrightarrow{d} N(0,1)$ as $n \to \infty$. By a well-known result on uniform integrability (e.g. cf. Theorem 5.4 of Billingsley 1968), $(S_{[nt]}^{(n)}/A_n([nt]))^2$ is uniformly integrable and so is $(S_{[nt]}^{(n)}/A_n)^2$ by (4.4.36).

Finally we prove (4.4.38).

Let $l_n = \exp\Big\{\sum_{i=0}^{[\log n]}\rho(2^i)^{2/3}\Big\}$. Define

$$\begin{aligned}X_{i1}^{(n)} &= X_i I(|X_i| \le t_n/l_n) - EX_i I(|X_i| \le t_n/l_n),\\ X_{i2}^{(n)} &= X_i I(t_n/l_n < |X_i| \le t_n) - EX_i I(t_n/l_n < |X_i| \le t_n),\\ S_{k1}^{(n)} &= \sum_{i=1}^{k} X_{i1}^{(n)} \quad \text{and} \quad S_{k2}^{(n)} = \sum_{i=1}^{k} X_{i2}^{(n)}.\end{aligned}$$

Obviously.

$$\begin{aligned}
&P\Big\{\max_{1\le i\le n}|S_i^{(n)}|\ge 6\lambda A_n\Big\}\\
&\le P\Big\{\max_{1\le i\le n}|S_{i1}^{(n)}|\ge \lambda A_n\Big\}+P\Big\{\max_{1\le i\le n}|S_{i2}^{(n)}|\ge 5\lambda A_n\Big\}\\
&=: p_1+p_2. \qquad (4.4.45)
\end{aligned}$$

Note that for any integer $K>0$

$$\sum_{i=1}^{[\log n]}\rho^{4/5}(2^i)\le K+\rho(2^K)^{2/15}\sum_{i=1}^{[\log n]}\rho(2^i)^{2/3}.$$

Hence, when $l_n\to\infty$ as $n\to\infty$,

$$\sum_{i=1}^{[\log n]}\rho^{4/5}(2^i)=o\Big(\sum_{i=1}^{[\log n]}\rho^{2/3}(2^i)\Big).$$

Therefore, from Lemma 2.2.5, (4.4.2) and Property $A4$, it follows that

$$\begin{aligned}
E|S_{k1}^{(n)}|^{5/2}&\le C\Big\{k^{5/4}(E(X_{11}^{(n)})^2)^{5/4}\\
&\qquad+k\exp\Big\{C\sum_{i=1}^{[\log k]}\rho^{4/5}(2^i)\Big\}E|X_{11}^{(n)}|^{5/2}\Big\}\\
&\le C\Big\{(kH(t_n/l_n))^{5/4}\\
&\qquad+k\exp\Big\{C\sum_{i=1}^{[\log k]}\rho^{4/5}(2^i)\Big\}(t_n/l_n)^{1/2}H(t_n/l_n)\Big\}\\
&\le c\{(kH(t_n)/l_n)^{5/4}+kn^{1/4}H(t_n)^{5/4}/l_n\}.
\end{aligned}$$

Using Lemma 4.1.2 we have

$$E\max_{1\le k\le n}|S_{k1}^{(n)}|^{5/2}\le c\{(nH(t_n))^{5/4}/l_n\}(\log n)^{5/2}.$$

Without loss of generality we assume that $\rho(2^i)\ge 1/(i\log^2 i)$. Then, recalling (4.4.10) we obtain

$$E\max_{1\le k\le n}|S_{k1}^{(n)}|^{5/2}\le c(nH(t_n))^{5/4}\le cA_n^{5/2}. \qquad (4.4.46)$$

Whence there exists a $\lambda>1$ such that for each n large enough

$$p_1\le \varepsilon/\lambda^2. \qquad (4.4.47)$$

We now estimate p_2. Let

$$r_1 = [n/l_n], \quad r_2 = [n/l_n^2], \quad r = r_1 + r_2,$$
$$d_1 = [(n - r_1)/r], \qquad d_2 = [n/r],$$
$$Y_i = \sum_{j=ir+1}^{ir+r_1} X_{j2}^{(n)}, \quad i = 0, 1, \cdots, d_1,$$
$$Z_i = \sum_{j=ir+r_1+1}^{(i+1)r} X_{j2}^{(n)}, \quad i = 0, 1, \cdots, d_2,$$
$$T_i(1) = \sum_{j=0}^{i} Y_j \quad \text{and} \quad T_i(2) = \sum_{j=0}^{i} Z_j,$$
$$Y_i^* = \sum_{j=ir+1}^{ir+r_1} (|X_j| I(t_n/l_n < |X_j| \le t_n) - E|X_j| I(t_n/l_n < |X_j| \le t_n)), \qquad i = 0, 1, \cdots, d_1.$$

It is easy to see that

$$\begin{aligned} P\Big\{\max_{1 \le i \le n} |S_{i2}^{(n)}| \ge 5\lambda A_n\Big\} &\le P\Big\{\max_{1 \le i \le d_1} |T_i(1)| \ge 2\lambda A_n\Big\} + P\Big\{\max_{1 \le i \le d_2} |T_i(2)| \ge 2\lambda A_n\Big\} \\ &\quad + P\Big\{\max_{1 \le i \le d_1} \sum_{j=ir+1}^{ir+r_1} |X_{j2}^{(n)}| \ge \frac{1}{2}\lambda A_n\Big\} \\ &\quad + 2l_n P\Big\{\max_{1 \le i \le r_2} |S_{i2}^{(n)}| \ge \frac{1}{2}\lambda A_n\Big\} \\ &=: I_1 + I_2 + I_3 + I_4. \end{aligned} \tag{4.4.48}$$

By (4.4.2) and (4.4.10), there is a $\lambda_0 > 0$ such that

$$r_1 E|X_1| I(t_n/l_n < |X_1| \le t_n) \le r_1 l_n t_n^{-1} H(t_n) \le \lambda_0 A_n.$$

Hence for $\lambda > 8\lambda_0$

$$I_3 \le P\Big\{\max_{1 \le i \le d_1} |Y_i^*| \ge \frac{1}{4}\lambda A_n\Big\} \tag{4.4.49}$$

and by Lemma 2.2.2 and (4.4.10)

$$\begin{aligned} I_4 &\le 2l_n P\Big\{\sum_{i=1}^{r_2}\Big(|X_i|I(t_n/l_n < |X_i| \le t_n) \\ &\qquad - E|X_i|I(t_n/l_n < |X_i| \le t_n)\Big) \ge \frac{1}{4}\lambda A_n\Big\} \\ &\le c\, l_n(\lambda A_n)^{-2} r_2 E X_i^2 I(t_n/l_n < |X_i| \le t_n) \\ &\le c\,(\lambda A_n)^{-2} n l_n^{-1} H(t_n) \\ &\le c\,\lambda^{-2} l_n^{-1} \le \varepsilon/\lambda^2 \end{aligned} \tag{4.4.50}$$

provided n is large enough.

In order to estimate I_1, let

$$\begin{aligned} &G_{-1} = (\Omega, \emptyset), \qquad G_k = \sigma(X_i, 1 \le i \le r_1 + kr), \\ &U_0(0) = 0, \quad U_i(k) = \sum_{j=1}^{k} E(Y_{j+i}|G_{j+i-1}), \quad k = 0, 1, \cdots, d_1, \\ &T^*(k) = T_k(1) - U_0(k). \end{aligned}$$

Obviously

$$\begin{aligned} I_1 &\le P\Big\{\max_{1\le i\le d_1} |T^*(i)| \ge \lambda A_n\Big\} + P\Big\{\max_{1\le i\le d_1} |U_0(i)| \ge \lambda A_n\Big\} \\ &=: I_1^{(1)} + I_1^{(2)}. \end{aligned} \tag{4.4.51}$$

Noting that $\{T^*(i), G_i, i = 0, 1, \cdots, d_1\}$ is a martingale and using the maximal inequality of martingale, we have

$$I_1^{(1)} \le 4(\lambda A_n)^{-2} E T^*(d_1)^2 I(|T^*(d_1)| \ge \lambda A_n). \tag{4.4.52}$$

We prove below that for every i, k and n, by induction on k

$$E U_i^2(k) \le C_1 k r_1 \rho(r_2)^2 \log^2(2k) E X_1^2 I(t_n/l_n < |X_1| \le t_n). \tag{4.4.53}$$

If $k = 1$, from the definition of ρ-mixing

$$E U_i^2(1) = E(Y_{i+1} E(Y_{i+1}|G_i)) \le \rho(r_2) \|Y_{i+1}\|_2 \|E(Y_{i+1}|G_i)\|_2.$$

Thus (4.4.53) is true for $k = 1$ by a version of (4.4.8). If $k \ge 2$, assume that (4.4.53) holds for every integer less than k. Put $k_1 = [k/2], k_2 = k - k_1$.

Then

$$\begin{aligned}
EU_i^2(k) &= EU_i^2(k_1) + EU_{i+k_1}^2(k_2) + 2EU_i(k_1)U_{i+k_1}(k_2) \\
&\le EU_i^2(k_1) + EU_{i+k_1}^2(k_2) \\
&\quad + 2\rho(r_2)\|U_i(k_1)\|_2 \Big\| \sum_{j=k_1+1}^{k} Y_{j+i} \Big\|_2 \\
&\le C_1\{k_1 \log^2(2k_1) + k_2 \log^2(2k_2) + 2k_1^{1/2}k_2^{1/2}\log(2k_2)\} \\
&\quad \cdot r_1\rho(r_2)^2 EX_1^2 I(t_n/l_n < |X_1| \le t_n) \\
&\le C_1 k r_1 \rho(r_2)^2 \log^2(2k) EX_1^2 I(t_n/l_n < |X_1| \le t_n)
\end{aligned}$$

by induction assumption. This proves (4.4.53). From it and Lemma 4.1.2, we have

$$\begin{aligned}
E \max_{1\le i\le d_1} U_0^2(i) &\le 3C_1 d_1 r_1 \rho(r_2)^2 \log^4(2d_1) H(t_n) \\
&\le cA_n^2 \rho(r_2)^2 \log^4(2l_n) \\
&\le cA_n^2 \rho(r_2)^2 \Big(\sum_{i=1}^{[\log n]} \rho(2^i)^{2/3} \Big)^4.
\end{aligned}$$

Also,

$$\begin{aligned}
\sum_{i=1}^{[\log n]} \rho(2^i)^{2/3} &\le \sum_{i=1}^{[\log r_2]} \rho(2^i)^{2/3} + \rho(r_2)^{2/3}\log(n/r_2) \\
&\le \rho(r_2)^{-1/3} \sum_{i=1}^{[\log r_2]} \rho(2^i) + 2\rho(r_2)^{2/3} \sum_{i=1}^{[\log n]} \rho(2^i)^{2/3}
\end{aligned}$$

from which it follows

$$\sum_{i=1}^{[\log n]} \rho(2^i)^{2/3} = O(\rho(r_2)^{-1/3}) \qquad \text{as} \quad n \to \infty.$$

Therefore we obtain

$$E \max_{1\le i\le d_1} U_0^2(i) \le cA_n^2 \rho(r_2)^{2/3} \tag{4.4.54}$$

and further

$$I_1^{(2)} \le \varepsilon/\lambda^2 \tag{4.4.55}$$

provided n is large enough.

For I_2, having analogue to (4.4.52) and (4.4.55), we can get that for large n

$$\begin{aligned} I_2 &\le \varepsilon/\lambda^2 + 4(\lambda A_n)^{-2} ET_{d_2}^2(2) \\ &\le \varepsilon/\lambda^2 + 4C_1(\lambda A_n)^{-2} d_2 r_2 E X_1^2 I(t_n/l_n < |X_1| \le t_n) \\ &\le \varepsilon/\lambda^2 + 4C_1(\lambda A_n)^{-2} n l_n^{-1} H(t_n) \\ &\le \varepsilon/\lambda^2 + c\lambda^{-2} l_n^{-1} \le 2\varepsilon/\lambda^2. \end{aligned} \tag{4.4.56}$$

Now we can come back to $I_1^{(1)}$.

$$\begin{aligned} &ET^*(d_1)^2 I(|T^*(d_1)| \ge \lambda A_n) \\ &\qquad \le 4ET_{d_1}^2(1) I(|T_{d_1}(1)| \ge \frac{1}{2}\lambda A_n) + 4EU_0^2(d_1) \\ &\qquad \le 36\Big(E(S_{n2}^{(n)})^2 I(|S_{n2}^{(n)}| \ge \lambda \frac{A_n}{6}) \\ &\qquad\quad + E\Big(\sum_{i=d_2 r}^{n} X_{i2}^{(n)}\Big)^2 + EU_0^2(d_1)\Big) \\ &\qquad \le 144\Big(E(S_n^{(n)})^2 I(|S_n^{(n)}| \ge \lambda \frac{A_n}{12}) \\ &\qquad\quad + E(S_{n1}^{(n)})^2 I(|S_{n1}^{(n)}| \ge \lambda \frac{A_n}{12}) \\ &\qquad\quad + ET_{d_2}^2(2) + EU_0^2(d_1) + E\Big(\sum_{i=d_2 r}^{n} X_{i2}^{(n)}\Big)^2\Big). \end{aligned}$$

By (4.4.10), (4.4.54) and a version of (4.4.8)

$$\begin{aligned} &A_n^{-1}\Big(ET_{d_2}^2(2) + EU_0^2(d_1) + E\Big(\sum_{i=d_2 r}^{n} X_{i2}^{(n)}\Big)^2\Big) \\ &\qquad \le c(n^{-1} d_2 r_2 + \rho(r_2)^{2/3}) \le c(l_n^{-1} + \rho(n^{1/2})^{2/3}) \\ &\qquad \le \varepsilon/2000 \end{aligned}$$

for large n. Using the uniform integrability of $\{(S_n^{(n)})^2/A_n^2, n \ge 1\}$, we find that

$$A_n^{-2} E(S_n^{(n)})^2 I(|S_n^{(n)}| \ge \lambda A_n/12) \le \varepsilon/2000$$

for each $n \ge 1$ provided λ is large enough. Moreover, by (4.4.46)

$$A_n^{-2} E(S_{n1}^{(n)})^2 I(|S_{n1}^{(n)}| \ge \lambda A_n/12) \le 4\lambda^{-1/2} A_n^{-5/2} E|S_{n1}^{(n)}|^{5/2} \le c\lambda^{-1/2}.$$

Whence we obtain that there is a constant λ_1 such that for any $\lambda > \lambda_1$ and large n

$$I_1^{(1)} \le \varepsilon/\lambda^2. \tag{4.4.57}$$

(4.4.55) and (4.4.57) together yield

$$I_1 \leq 2\varepsilon/\lambda^2. \tag{4.4.58}$$

Proceeding exactly as the proof of (4.4.58), we also have

$$I_3 \leq 2\varepsilon/\lambda^2 \tag{4.4.59}$$

for any large λ and n.

It follows from (4.4.45), (4.4.47), (4.4.48), (4.4.50), (4.4.56), (4.4.58), (4.4.59) that (4.4.38) holds, as desired. This completes the proof of Theorem 4.4.1.

Chapter 5 Weak Convergence for φ-mixing Sequences

The CLT of a φ-mixing sequence is one of the earlier results for the dependent random variables. Ibragimov (1959) gave the following two Propositions.

Proposition 5.0.1. Let $\{X_n, n \geq 1\}$ be a strictly stationary φ-mixing sequence of random variables with $EX_1 = 0$, $EX_1^2 < \infty$. If

$$\sum_{n=1}^{\infty} \varphi^{1/2}(n) < \infty,$$

then

$$\sigma^2 = EX_1^2 + 2\sum_{j=2}^{\infty} EX_1X_j$$

converges absolutely, and if the condition $\sigma > 0$ is added, then $S_n/\sigma\sqrt{n}$ converges in distribution to $N(0,1)$.

Proposition 5.0.2. Let $\{X_n, n \geq 1\}$ be a strictly stationary φ-mixing sequence of random variables with $EX_1 = 0, E|X_1|^{2+\delta} < \infty$ for some $\delta > 0$ and $\sigma_n^2 = ES_n^2 \to \infty$. Then S_n/σ_n converges in distribution to $N(0,1)$.

Since then, the CLT and the WIP for a φ-mixing sequence have ever been discussed by many authors. Ibragimov-Linnik and Iosifescu have raised the following conjectures:

Conjecture 1 (Ibragimov and Linnik 1971). Let $\{X_n, n \geq 1\}$ be a strictly stationary φ-mixing sequence with $EX_1 = 0$, $EX_1^2 < \infty$. If $\sigma_n^2 \to \infty$ as $n \to \infty$, then the CLT holds.

Conjecture 2 (Iosifescu 1977). Let $\{X_n, n \geq 1\}$ be as above. Then W_n weakly converges to W, where $W_n(t) = S_{[nt]}/\sigma_n$.

Since 1970's, some mathematicians have obtained many beneficial results around these conjectures. Herrndorf (1983b) showed that there exists a strictly stationary φ-mixing sequence with $\sigma_n^2 \to \infty, \liminf_{n\to\infty} \sigma_n^2/n = 0$, conjecture 2 does not hold. Peligrad (1985) proved that these two conjectures hold true under the additional assumption $\liminf\limits_{n\to\infty} \sigma_n^2/n > 0$. Thus, we can reduce the study of the above conjectures to that of the variances of the partial sums. In those two papers they have also given some sufficient conditions for the CLT and the WIP of a φ-mixing sequence when the moment of order 2 is finite. We shall introduce these in Section 5.1.

In Section 5.2, we will discuss the above two conjectures and a general conjecture which was raised by Peligrad (1990).

5.1 The WIP when the moments of order 2 are finite

Since $\rho(n) \le 2\varphi^{1/2}(n)$, we have the same conclusion as in Theorem 4.1.1 with $\sum_{n=1}^{\infty} \varphi^{1/2}(2^n) < \infty$ instead of $\sum_{n=1}^{\infty} \rho(2^n) < \infty$.

Herrndorf (1983) dropped the condition on the mixing rate.

Theorem 5.1.1. *Let $\{X_n,\ n \ge 1\}$ be a φ-mixing sequence with $EX_n = 0$, $EX_n^2 < \infty$ satisfying*

(i) $\sigma_n^2 = nh(n)$, *where $h(n)$ is slowly varying,*

(ii) $\lim_{n\to\infty} P\{\max_{1\le i\le n} |X_i| \ge \varepsilon\sigma_n\} = 0$ *for any $\varepsilon > 0$,*

(iii) $\{S_m^2(n)/\sigma_n^2,\ m \ge 0,\ n \ge 1\}$ *is uniformly integrable.*

Then

$$W_n \Longrightarrow W.$$

Proof. We are going to verify the conditions in Theorem 4.0.4. By the definition of φ-mixing,

$$\Big|P\Big\{\bigcap_{i=1}^{r} E_i\Big\} - \prod_{i=1}^{r} P\{E_i\}\Big| \le r\varphi([n\delta]) \to 0 \quad \text{as } n \to \infty,$$

where $E_i = \{W_n(t_i) - W_n(s_i) \in B_i\}, i = 1, \cdots, r$ are defined in Theorem 4.0.4, $\delta = \min_{2\le i\le r}(s_i - t_{i-1}) > 0$. And hence, $\{W_n, n \ge 1\}$ has asymptotically independent increments. Moreover, conditions (i) and (iii) imply the uniform integrability of $\{W_n^2(t), n \ge 1\}$ for each $t \ge 0$. Obviously $EW_n(t) = 0$ and $EW_n^2(t) \to t$ as $n \to \infty$ by condition (i). In order to show the tightness, we need the following lemma.

Lemma 5.1.1. *Let $\{X_n,\, n \geq 1\}$ be a φ-mixing sequence. For any given positive integer q and $a > 0$, $m \geq 0$, $r \geq q+1$, we have*

$$\Big(1 - \varphi(q) - \max_{q\leq j\leq r} P\{|S_{m+r} - S_{m+j}| > a\}\Big) \times P\Big\{\max_{1\leq j\leq r}|S_{m+j} - S_m| > 3a\Big\} \leq P\{|S_{m+r} - S_m| > a\} + P\Big\{(q-1)\max_{1\leq j\leq r}|X_{m+j}| > a\Big\}. \tag{5.1.1}$$

Proof. Denote $A_1 = \{|X_{m+1}| > 3a\}$,

$$A_j = \{|S_{m+j} - S_m| > 3a,\ |S_{m+i} - S_m| \leq 3a,\ 1 \leq i \leq j-1\}, \quad 2 \leq j \leq r,$$
$$B_j = \{|S_{m+r} - S_{m+j+q-1}| \leq a\}, \quad 1 \leq j \leq r-q,$$
$$B_j = \Omega, \quad r-q+1 \leq j \leq r, \qquad C = \{|S_{m+r} - S_m| > a\}.$$

It is clear that

$$\bigcup_{j=1}^{r} A_jB_j \subset C \bigcup \Big\{(q-1)\max_{1\leq j\leq r}|X_{m+j}| > a\Big\}.$$

Hence

$$P(C) + P\Big\{(q-1)\max_{1\leq j\leq r}|X_{m+j}| > a\Big\} \geq P\Big\{\bigcup_{j=1}^{r} A_jB_j\Big\} = \sum_{j=1}^{r} P(A_jB_j) \geq \Big\{\min_{1\leq j\leq r-q} P(B_j) - \varphi(q)\Big\}\sum_{j=1}^{r} P(A_j). \tag{5.1.2}$$

Note that

$$\sum_{j=1}^{r} P(A_j) = P\Big\{\max_{1\leq j\leq r}|S_{m+j} - S_m| > 3a\Big\},$$

$$\min_{1\leq j\leq r-q} P(B_j) + \max_{q\leq j\leq r} P\{|S_{m+r} - S_{m+j}| > a\} \geq 1.$$

Inserting these into (5.1.2) implies (5.1.1), as desired.

In order to prove that $\{W_n\}$ is tight, it needs only to show that

$$\lim_{\delta\downarrow 0,\, 1/\delta\in\mathrm{N}} \frac{1}{\delta} \max_{0\leq k\leq 1/\delta} \limsup_{n\to\infty} P\Big\{\sup_{k\delta\leq s\leq (k+1)\delta}|W_n(s) - W_n(k\delta)| > \varepsilon\Big\} = 0. \tag{5.1.3}$$

Choose a positive integer q so large that $\varphi(q) < 1$. For any given $\varepsilon > 0$ and $\delta > 0$, by condition (iii) we have

$$\begin{aligned}
&\sup_{m\geq 0}\sup_{1\leq j\leq n\delta} P\{|S_m(j)| > \varepsilon\sigma_n/3\} \\
&\quad \leq 9\varepsilon^{-2} \sup_{m\geq 0,\, j\geq 1} E(S_m^2(j)/\sigma_j^2) \sup_{1\leq j\leq n\delta} \sigma_j^2/\sigma_n^2 \\
&\quad = C(\varepsilon) \sup_{1\leq j\leq n\delta} \sigma_j^2/\sigma_n^2.
\end{aligned}$$

It follows from (i) and Property A4, that

$$\lim_{\delta\downarrow 0}\limsup_{n\to\infty}\sup_{1\leq j\leq n\delta} \sigma_j^2/\sigma_n^2 = 0.$$

Therefore, there exists a $\delta_0 = \delta_0(\varepsilon) > 0$ such that for any $0 < \delta \leq \delta_0(\varepsilon)$

$$\begin{aligned}
&\limsup_{n\to\infty}\sup_{m\geq 0}\sup_{1\leq j\leq n\delta} P\Big\{|S_m(j)| > \frac{1}{3}\varepsilon\sigma_n\Big\} \\
&\qquad\qquad \leq C(\varepsilon)\delta \leq \frac{1}{2}(1-\varphi(q)). \qquad (5.1.4)
\end{aligned}$$

Applying Lemma 5.1.1 for $m = [nk\delta]$, $r = [n(k+1)\delta] - [nk\delta]$, $3a = \varepsilon\sigma_n$, from (5.1.4) we obtain that for any $0 < \delta \leq \delta_0$ with $1/\delta \in \mathbb{N}$, $0 \leq k \leq 1/\delta$,

$$\begin{aligned}
&\frac{1}{2}(1-\varphi(q))\limsup_{n\to\infty} P\Big\{\sup_{k\delta\leq s\leq (k+1)\delta} |W_n(s) - W_n(k\delta)| > \varepsilon\Big\} \\
&\quad \leq \limsup_{n\to\infty} P\{|W_n((k+1)\delta) - W_n(k\delta)| > \varepsilon/3\} \\
&\qquad + \limsup_{n\to\infty} P\Big\{(q-1)\max_{1\leq j\leq n} |X_j| > \frac{1}{3}\varepsilon\sigma_n\Big\} \\
&\quad =: I_1 + I_2 \qquad (5.1.5)
\end{aligned}$$

For fixed q, it follows from (ii) that $I_2 = 0$. And from (i)

$$\lim_{n\to\infty} \sigma_{[n(k+1)\delta]-[nk\delta]}/\sigma_n = \delta^{1/2}.$$

Thus we have

$$\begin{aligned}
I_1 &\leq \limsup_{n\to\infty} P\{|S_{[n(k+1)\delta]} - S_{[nk\delta]}| > \varepsilon\sigma_{[n(k+1)\delta]-[nk\delta]}/4\sqrt{\delta}\} \\
&\leq 16\varepsilon^{-2}\delta \sup_{m\geq 0,\, n\geq 1} ES_m^2(n) I(|S_m(n)| > \varepsilon\sigma_n/(4\sqrt{\delta}))/\sigma_n^2.
\end{aligned}$$

Combining it with (5.1.5) and condition (iii) implies (5.1.3). The proof of Theorem 5.1.1 is completed.

Corollary 5.1.1. *Let $\{X_n n \geq 1\}$ be a φ-mixing sequence with $EX_n = 0$, $EX_n^2 < \infty$. If the conditions (i) and (ii) of Theorem 5.1.1 and*

(iv) $\sup\{ES_m^2(n)/\sigma_n^2,\ m \geq 0,\ n \geq 1\} < \infty$

are satisfied, and if $\{X_n n \geq 1\}$ obey the CLT, then $W_n \Longrightarrow W$.

Proof. From the CLT and (i), it follows that $S_{[nt]}/\sigma_n$ converges in distribution to $W(t)$ for every $t > 0$. Now let $0 < s < t$ be given. We shall show

$$(S_{[nt]} - S_{[ns]})/\sigma_n \xrightarrow{d} W(t) - W(s) \quad \text{as } n \to \infty. \tag{5.1.6}$$

It is obvious that $\{(S_{[ns]}/\sigma_n,\ S_{[nt]}/\sigma_n),\ n \geq 1\}$ is tight (see Billingsley 1968 p.41). Hence, it follows from Helly's theorem that there exists a probability measure Q on R^2 and a subsequence $\{n_k\}$ such that

$$(S_{[n_k s]}/\sigma_{n_k},\ S_{[n_k t]}/\sigma_{n_k}) \xrightarrow{d} Q \quad \text{as } k \to \infty.$$

Let $\pi_i : R^2 \to R$, $i = 1, 2$, denote the projections. Then

$$(S_{[n_k s]}/\sigma_{n_k},\ (S_{[n_k t]} - S_{[n_k s]})/\sigma_{n_k}) \xrightarrow{d} Q(\pi_1,\ \pi_2 - \pi_1)^{-1} \quad \text{as } k \to \infty.$$

Taking $p = p(n) \in \{0, 1, \ldots, [ns]\}$ such that $p(n) \to \infty$ and $p(n)/\sigma_n \to 0$ as $n \to \infty$, we have

$$\begin{aligned}(ES^2_{[ns]-p(n)}(p(n)))^{1/2}/\sigma_n \\ \leq p(n)\sigma_n^{-1} \sup_j (EX_j^2)^{1/2} \to 0 \quad \text{as } n \to \infty.\end{aligned}$$

Hence it follows that

$$\begin{aligned}(S_{[n_k s]-p(n_k)}/\sigma_{n_k}, (S_{[n_k t]} - S_{[n_k s]})/\sigma_{n_k}) \\ \xrightarrow{d} Q(\pi_1,\ \pi_2 - \pi_1)^{-1} \quad \text{as } k \to \infty.\end{aligned}$$

For any Borel sets $A, B \subset R$ with $Q(\pi_1,\ \pi_2 - \pi_1)^{-1}(\partial(A \times B)) = 0$, we obtain by an application of the mixing condition

$$\begin{aligned}&|Q(\pi_1,\ \pi_2 - \pi_1)^{-1}(A \times B) - Q\pi_1^{-1}(A)Q(\pi_2 - \pi_1)^{-1}(B)| \\ &= \lim_{k\to\infty} |P\{S_{[n_k s]-p(n_k)}/\sigma_{n_k} \in A,\ (S_{[n_k t]} - S_{[n_k s]})/\sigma_{n_k} \in B\} \\ &\quad - P\{S_{[n_k s]-p(n_k)}/\sigma_{n_k} \in A\}P\{(S_{[n_k t]} - S_{[n_k s]})/\sigma_{n_k} \in B\}| \\ &= 0.\end{aligned}$$

Hence π_1 and $\pi_2 - \pi_1$ are Q-independent. Since $Q\pi_1^{-1} = N(0, s)$ and $Q\pi_2^{-1} = N(0, t)$, it follows that $Q(\pi_2 - \pi_1)^{-1} = N(0, t - s)$. This proves (5.1.6).

The remainder of the proof needs only to check (5.1.3). From (5.1.6) it follows that

$$\limsup_{n\to\infty} P\{|W_n((k+1)\delta) - W_n(k\delta)| \geq \varepsilon/3\} \leq N(0,\delta)\{x : |x| \geq \varepsilon/3\}. \tag{5.1.7}$$

Furthermore

$$\begin{aligned}
&\frac{1}{\delta}N(0,\delta)\{x : |x| \geq \varepsilon/3\} \\
&= \frac{1}{\delta}\frac{1}{\sqrt{2\pi\delta}}\int_{|x|\geq\varepsilon/3} \exp\Big(-\frac{x^2}{2\delta}\Big)\,dx \\
&= \frac{2}{\delta}\Big(1 - \Phi\big(\frac{\varepsilon}{3\sqrt{\delta}}\big)\Big) \\
&\leq \frac{2}{\delta}\frac{3\sqrt{\delta}}{\varepsilon}\frac{1}{\sqrt{2\pi}}e^{-\varepsilon^2/18\delta} \to 0 \quad \text{as } \delta \to 0.
\end{aligned}$$

Now (5.1.5) and (5.1.7) imply (5.1.3). This completes the proof of Corollary 5.1.1.

Remark 5.1.1. The conditions (i) and (ii) of Theorem 5.1.1 are also necessary for the WIP.

Suppose that W_n weakly converges to W. For $\delta > 0$ and a function f let $w(f,\delta) = \sup\{|f(x) - f(y)| : 0 \leq x,\, y \leq 1,\, |x-y| < \delta\}$. For any $\varepsilon > 0$, the set $\{f : f \in D[0,1],\, w(f,\delta) \geq \varepsilon\}$ is closed with respect to the uniform topology. Hence

$$\begin{aligned}
\limsup_{n\to\infty} P\{\max_{1\leq i\leq n} |X_i| \geq \varepsilon\sigma_n\} &\leq \limsup_{n\to\infty} P\{w(W_n,\,\delta) \geq \varepsilon\} \\
&\leq P\{w(W,\,\delta) \geq \varepsilon\} \to 0 \quad \text{as}\, \delta \to 0.
\end{aligned}$$

This proves that (ii) holds true.

In order to prove (i), we show that $h(t) := \max(1, \sigma_{[t]}^2)/t, t > 0$, is slowly varying. Since $\sigma_n^2 \to \infty$ as $n \to \infty, h(t) = \sigma_{[t]}^2/t$ for large t. For $t \in [0,1]$ we have

$$S_{[nt]}/\sigma_{[nt]} \xrightarrow{d} N(0,1), \quad S_{[nt]}/\sigma_n \xrightarrow{d} N(0,t) \quad \text{as } n \to \infty.$$

Therefore we obtain for every $t \in [0,1]$

$$\sigma_{[nt]}^2/\sigma_n^2 \longrightarrow t \quad \text{as} \quad n \to \infty. \tag{5.1.8}$$

Moreover (ii) implies $X_n/\sigma_n \xrightarrow{P} 0$. Hence both S_n/σ_n and S_n/σ_{n+1} weakly converge to $N(0,1)$ and further $\sigma_n/\sigma_{n+1} \to 1$. Consequently for $t \in [0,1]$

$$\sigma_{[ts]}/\sigma_{[t[s]]} \longrightarrow 1 \quad \text{as} \quad s \to \infty. \tag{5.1.9}$$

From (5.1.8) and (5.1.9), we obtain $\lim_{s\to\infty} h(ts)/h(s) = 1$ for every $t \in [0,1]$. Hence h is slowly varying.

Now we can write the following Corollary for the strictly stationary case immediately.

Corollary 5.1.2. *Let $\{X_n, n \geq 1\}$ be a strictly stationary φ-mixing sequence with $EX_1 = 0, EX_1^2 < \infty$ and $\sigma_n^2 \to \infty$. Denote*

$$Y_n(t) = \frac{1}{\sigma_n}(S_{[nt]} + (nt - [nt])X_{[nt]+1}).$$

Then the following assertions are equivalent

(a) $W_n \Longrightarrow W$,

(b) $Y_n \Longrightarrow W$,

(c) *$\{X_n\}$ obeys the CLT and (ii) is fulfilled,*

(d) *$\{S_n^2/\sigma_n^2, n \geq 1\}$ is uniformly integrable and (ii) is fulfilled.*

Proof. Obviously, (a) and (b) are equivalent and (b) implies (c). From Theorem 5.4 of Billingsley (1968), it follows that (c) implies (d). Finally "(d) implies (a)" follows from Theorem 5.1.1 .

Peligrad (1985) weakened the conditions of Herrndorf (1983) and gave the following theorem.

Theorem 5.1.2. *Let $\{X_n, n \geq 1\}$ be a φ-mixing sequence with $EX_n = 0, EX_n^2 < \infty$. Suppose that the conditions (i), (iv) and the Lindeberg condition*

(v) $\lim_{n\to\infty} \frac{1}{\sigma_n^2} \sum_{i=1}^n EX_i^2 I(X_i^2 > \varepsilon\sigma_n^2) = 0$ for any $\varepsilon > 0$ *are satisfied. Then $W_n \Longrightarrow W$.*

Remark 5.1.2. If $\varphi^* = \lim_{n\to\infty} \varphi(n) < 1$, then (ii) is equivalent to

$$\sum_{i=1}^n P\{|X_i| > \varepsilon\sigma_n\} \longrightarrow 0 \qquad \text{for any}\, \varepsilon > 0 \tag{5.1.10}$$

and the Lindeberg condition (v) is equivalent to

$$E \max_{1\leq i\leq n} X_i^2/\sigma_n^2 \longrightarrow 0. \tag{5.1.11}$$

In fact, (5.1.10) implies (ii) obviously. On the other hand, by (ii) we can choose n_0 and p_0 such that

$$P\Big\{\max_{1\leq i\leq n} X_i^2 < \varepsilon\sigma_n^2\Big\} - \varphi(p_0) \geq a > 0 \qquad \text{for } n \geq n_0. \tag{5.1.12}$$

Therefore for every $x \ge \varepsilon > 0, n \ge n_0$ and $j = 0, 1, \cdots, p_0 - 1$ we have

$$\begin{aligned}
&P\Big\{\max_{1\le i\le n} X_i^2 \ge x\sigma_n^2\Big\} \\
&\quad \ge P\Big\{X_j^2 \ge x\sigma_n^2, \max_{p_0+j\le i\le n} X_i^2 < x\sigma_n^2\Big\} \\
&\quad\quad + P\Big\{X_{p_0+j}^2 \ge x\sigma_n^2, \max_{2p_0+j\le i\le n} X_i^2 < x\sigma_n^2\Big\} \\
&\quad\quad + \cdots + P\{X_{[(n-j)/p_0]p_0+j}^2 \ge x\sigma_n^2\} \\
&\quad \ge \sum_{0\le i\le[(n-j)/p_0]} P\{X_{ip_0+j}^2 \ge x\sigma_n^2\} \\
&\quad\quad \Big(P\Big\{\max_{1\le i\le n} X_i^2 < x\sigma_n^2\Big\} - \varphi(p_0)\Big) \\
&\quad \ge a \sum_{0\le i\le[(n-j)/p_0]} P\{X_{ip_0+j}^2 \ge x\sigma_n^2\},
\end{aligned}$$

which implies

$$\sum_{j=1}^{n} P\{X_j^2 > x\sigma_n^2\} \le \frac{p_0}{a} P\Big\{\max_{1\le j\le n} X_j^2 > x\sigma_n^2\Big\}. \tag{5.1.13}$$

Hence we can obtain (5.1.10) from (ii). The fact that (v) implies (5.1.11) follows from

$$\begin{aligned}
&E \max_{1\le i\le n} X_i^2/\sigma_n^2 \\
&\quad \le \varepsilon + E \max_{1\le i\le n} X_i^2 I(X_i^2 > \varepsilon\sigma_n^2)/\sigma_n^2 \qquad \text{for any } \varepsilon > 0.
\end{aligned}$$

Note that, (5.1.12) and further (5.1.13) hold under (5.1.11). Hence, (5.1.11) implies that (v) follows from the following well-known relation: for every positive integrable random variable X,

$$EXI(X > b) = bP(X > b) + \int_b^\infty P(X > x)dx. \tag{5.1.14}$$

Remark 5.1.3. Utev (1990) showed that for a φ-mixing sequence $\{X_n\}$, the Lindeberg condition (v) implies the CLT. Furthermore, Grin (1991) showed that for a stationary φ-mixing sequence $\{X_n\}$ with $\varphi(1) < 1$, there exists a sequence $\{A_n\}$ of numbers such that $B_n^{-1}S_n - A_n \to N(0,1)$, where $B_n = \sup\{z \ge 0 : Var(\sum_{i=1}^n X_i I(|X_i| \le z)) \ge z^2\}$, if and only if $\{B_n\}$ is regularly varying with the exponent $1/2$.

The proof of theorem 5.1.2 needs the following lemmas. First we state an analogue of Lemma 2.2.7.

Lemma 5.1.2. *Let $\{Y_n, n \geq 1\}$ be a sequence of random variables. Denote $T_n = \sum_{i=1}^n Y_i$. If for some $b > 0, p \in \mathbb{N}$ and $a_0 > 0$*

$$\varphi(p) + \max_{1 \leq i \leq n} P\{T_n - T_i| > ba_0/2\} \leq \eta < 1, \tag{5.1.15}$$

then for every $a \geq a_0$ and $n > p$ the following relations hold:

$$\begin{aligned} P\Big\{\max_{1 \leq i \leq n} |T_i| > (1+b)a\Big\} \leq & \frac{1}{1-\eta} P(|T_n| > a) \\ & + \frac{1}{1-\eta} P\Big\{\max_{1 \leq i \leq n} |Y_i| > \frac{ba}{2(p-1)}\Big\} \end{aligned} \tag{5.1.16}$$

and

$$\begin{aligned} P\{|T_n| > (1+2b)a\} \leq & \frac{\eta}{1-\eta} P\{|T_n| > a\} \\ & + \frac{1}{1-\eta} P\Big\{\max_{1 \leq i \leq n} |Y_i| > \frac{ba}{2p}\Big\}. \end{aligned} \tag{5.1.17}$$

For simplicity, denote

$$E_a X = EXI(X > a).$$

Lemma 5.1.3. *Let $\{Y_n, n \geq 1\}$ be a sequence of random variables satisfying (5.1.15). Then for every $A \geq a_0^2$ we have*

$$\begin{aligned} E_{(1+2b)^2 A} T_n^2 \leq & (1+2b)^2 \frac{\eta}{1-\eta} E_A T_n^2 \\ & + \Big(\frac{2p(1+2b)}{b}\Big)^2 \frac{1}{1-\eta} E_{A(b/2p)^2} \max_{1 \leq i \leq n} Y_i^2. \end{aligned}$$

Proof. By (5.1.14) and a change of variables one obtains

$$\begin{aligned} E_{(1+2b)^2 A} T_n^2 = & (1+2b)^2 A P\{T_n^2 > (1+2b)^2 A\} \\ & + (1+2b)^2 \int_A^\infty P\{T_n^2 > (1+2b)^2 y\} dy. \end{aligned}$$

The lemma follows by (5.1.17) and then (5.1.14) again.

Lemma 5.1.4. *Let $\{X_n, n \geq 1\}$ be a centered sequence such that $\varphi^* < 1/4$ and $\{\max_{1\leq i\leq n} EX_i^2/\sigma_n^2, n \geq 1\}$ is bounded. Then*

$$\left\{\max_{1\leq i\leq n} E(S_n - S_i)^2/\sigma_n^2, n \geq 1\right\}$$

is bounded as well.

Proof. Let p be an integer such that $\varphi(p) < 1/4$. We have

$$\max_{1\leq i\leq n} E(S_n - S_i)^2 \leq \max_{1\leq i<n-p} E(S_n - S_i)^2 + p^2 \max_{1\leq i\leq n} EX_i^2. \tag{5.1.18}$$

For every $i < n - p$ we also have

$$\big|\|S_n\|_2 - \|S_i + (S_n - S_{i+p})\|_2\big| \leq p \max_{1\leq i\leq n} \|X_i\|_2.$$

By Lemma 1.2.8, we have

$$\|S_i + (S_n - S_{i+p})\|_2 \geq (1 - 2\varphi^{1/2}(p))^{1/2}(\sigma_i^2 + E(S_n - S_{i+p})^2)^{1/2}.$$

Therefore

$$\sigma_i < (1 - 2\varphi^{1/2}(p))^{-1/2}\left(\sigma_n + p \max_{1\leq i\leq n} \|X_i\|_2\right)$$

for every $i < n - p$. Whence, from (5.1.18)

$$\max_{1\leq i\leq n} E(S_n - S_i)^2/\sigma_n^2 \leq 2 + 4(1 - 2\varphi^{1/2}(p))^{-1} + p^2(1 + 4(1 - 2\varphi^{1/2}(p))^{-1}) \max_{1\leq i\leq n} EX_i^2/\sigma_n^2$$

as desired.

Lemma 5.1.5. *Let $\{X_n, n \geq 1\}$ be a centered sequence with $\phi^* < 1/4$. Then $\{\max_{1\leq i\leq n} S_i^2/\sigma_n^2, n \geq 1\}$ is uniformly integrable if and only if $\{\max_{1\leq i\leq n} X_i^2/\sigma_n^2, n \geq 1\}$ is uniformly integrable.*

proof. First, because

$$P\left\{\max_{1\leq i\leq n} |X_i| > 2x\sigma_n\right\} \leq P\left\{\max_{1\leq i\leq n} |S_i| > x\sigma_n\right\} \tag{5.1.19}$$

for any $x > 0$, one of the implications follows by the relation (5.1.14).

We prove the part of "if", and hence, assume that $\{\max_{1\le i\le n} X_i^2/\sigma_n^2, n \ge 1\}$ is uniformly integrable. By the Chebyshev inequality and by Lemma 5.1.4 for any $b > 0$

$$\lim_{t\to\infty} \sup_n \max_{1\le i\le n} P\{(S_n - S_i)^2 > (b/2)^2 t\sigma_n^2\} = 0. \tag{5.1.20}$$

By $\phi^* < 1/4$ and (5.1.20), we can find some constants $b > 0, \eta < 1/2, p \in \mathbb{N}$ and $a_0 \in R$ such that

$$(1+2b)^2\eta/(1-\eta) < 1 \tag{5.1.21}$$

and

$$\varphi(p) + \max_{1\le i\le n} P\{(S_n - S_i)^2 > (b/2)^2 a_0^2 \sigma_n^2\} \le \eta \quad \text{for every } n \ge 1. \tag{5.1.22}$$

From Lemma 5.1.3 and (5.1.21), (5.1.22) it follows that

$$E_{(1+2b)^2 A}\Big(\frac{S_n^2}{\sigma_n^2}\Big) \le (1+2b)^2 \frac{\eta}{1-\eta} E_A\Big(\frac{S_n^2}{\sigma_n^2}\Big) + \Big(\frac{2p(1+2b)}{b}\Big)^2 \frac{1}{1-\eta} E_{A(b/2p)^2}\Big(\max_{1\le i\le n} \frac{X_i^2}{\sigma_n^2}\Big)$$

for any $A > a_0^2$ and every $n \ge 1$. Taking the supremum on n in this relation and noting that $\sup_n E_A(S_n^2/\sigma_n^2)$ is decreasing in A and that $\{\max_{1\le i\le n}(X_i^2/\sigma_n^2), n \ge 1\}$ is uniformly integrable, we obtain

$$\lim_{A\to\infty} \sup_n E_A\Big(\frac{S_n^2}{\sigma_n^2}\Big) \le (1+2b)^2 \frac{\eta}{1-\eta} \lim_{A\to\infty} \sup_n E_A\Big(\frac{S_n^2}{\sigma_n^2}\Big).$$

Whence it follows by (5.1.21), (5.1.22) that $\{S_n^2/\sigma_n^2, n \ge 1\}$ is uniformly integrable. This implies by (5.1.16) and (5.1.14) that $\{\max_{1\le i\le n} S_i^2/\sigma_n^2, n \ge 1\}$ is uniformly integrable.

Proof of Theorem 5.1.2. By the proof of Theorem 5.1.1, $\{W_n, n \ge 1\}$ also has asymptotically independent increments. By Remark 5.1.2 it follows that $\{\max_{1\le i\le n}(X_i^2/\sigma_n^2), n \ge 1\}$ is uniformly integrable under the Lindeberg condition. Whence $\{W_n^2(t), n \ge 1\}$ is uniformly integrable for each t by (i) and Lemma 5.1.5. Moreover, $EW_n(t) = 0$ and $EW_n^2(t) \to t$ as $n \to \infty$ by (i) again.

For the tightness condition of $\{W_n, n \ge 1\}$, from the proof of Theorem 8.3 of Billingsley (1968), it suffices to show

$$\lim_{\delta\to 0} \limsup_{n\to\infty} \sum_{i=0}^{[1/\delta]-1} P\Big\{\max_{i\delta\le s\le (i+1)\delta} |W_n(s) - W_n(i\delta)| > \varepsilon\Big\} = 0. \tag{5.1.23}$$

For every $0 \le i \le 1/\delta - 1$ denote

$$f_i = f_i(n, \delta, a) := \max_{in\delta \le j \le (i+1)n\delta} P\Big\{\Big| \sum_{k=j}^{(i+1)n\delta} X_k \Big| > \frac{b}{2} a^{1/2} \sigma_n \Big\}.$$

By the Chebyshev inequality we have

$$f_i \le \Big(\frac{2}{b}\Big)^2 \frac{1}{a} \max_{in\delta \le j \le (i+1)n\delta} E\Big(\sum_{k=j}^{(i+1)n\delta} X_k \Big)^2 / \sigma_n^2.$$

By (i), (iv) and the properties of a slowly varying function that follows from the Karamata representation (see Appendix) we obtain

$$\lim_{\delta \to 0} \limsup_{n \to \infty} \max_{0 \le i \le 1/\delta - 1} f_i = 0. \tag{5.1.24}$$

Choose p and b such that

$$\varphi(p)(1 + 2b)^2 / (1 - \varphi(p)) < 1$$

and choose δ_0 and n_0 such that for any $\delta < \delta_0$ and $n > n_0$

$$\varphi(p) + \max_{1 \le i \le 1/\delta} f_i = \eta'(n, \delta, a) := \eta' < 1. \tag{5.1.25}$$

From (5.1.25) and Lemma 5.1.3 we obtain for every $0 \le i \le 1/\delta - 1$

$$\begin{aligned} &E_{(1+2b)^2 a}\Big(\Big(\sum_{ni\delta \le j \le n(i+1)\delta} X_j \Big)^2 / \sigma_n^2 \Big) \\ &\le (1 + 2b)^2 \frac{\eta'}{1 - \eta'} E_a \Big(\frac{\sum_{ni\delta \le j \le n(i+1)\delta} X_j}{\sigma_n} \Big)^2 \\ &\quad + \Big(\frac{2p(1 + 2b)}{b} \Big)^2 \frac{1}{1 - \eta'} \sum_{ni\delta \le j \le n(i+1)\delta} E_{a(b/2p)^2} \Big(\frac{X_j^2}{\sigma_n^2} \Big). \end{aligned} \tag{5.1.26}$$

Noting conditions (i) and (iv), for fixed $\delta > 0$ we have

$$\begin{aligned} &\limsup_{n \to \infty} \sum_{i=0}^{[1/\delta]-1} E\Big(\sum_{ni\delta \le j \le n(i+1)\delta} X_j \Big)^2 / \sigma_n^2 \\ &\qquad = O\Big(\limsup_{n \to \infty} \sum_{i \le 1/\delta} \sigma_{[n\delta]}^2 / \sigma_n^2 \Big) = O(1). \end{aligned}$$

Denote

$$l(a) = \limsup_{\delta \to 0} \limsup_{n \to \infty} \sum_{i=0}^{[1/\delta]-1} E_a \Big(\sum_{ni\delta \le j \le n(i+1)\delta} X_j \Big)^2 / \sigma_n^2.$$

From (5.1.26), (5.1.24) and condition (v) we obtain

$$l((1+2b)^2 a) \le \frac{(1+2b)^2\varphi(p)}{1-\varphi(p)} l(a) \quad \text{for every} a > 0.$$

Since $l(a)$ is a decreasing function in a, and $[(1+2b)^2\varphi(p)]/(1-\varphi(p)) < 1$, we obtain $\lim_{a\to 0} l(a) = 0$. Hence $l(a) = 0$ for every $a > 0$, which implies

$$\limsup_{\delta\to 0}\limsup_{n\to\infty} \sum_{i=0}^{[1/\delta]-1} P\Big(\Big|\sum_{ni\delta\le j\le n(i+1)\delta} X_j\Big| > \varepsilon\sigma_n\Big) = 0$$

for any $\varepsilon > 0$. The relation (5.1.23) follows now by (5.1.16), (v) and (5.1.24). The proof of Theorem 5.1.2 is completed.

Particularly, by Theorem 5.1.2 we have the following corollaries.

Corollary 5.1.3. *Let $\{X_n, n \ge 1\}$ be a stationary φ-mixing sequence with $EX_1 = 0, EX_1^2 < \infty, \sigma_n^2 \to \infty$ and the Lindeberg condition (v) is satisfied. Then $W_n \Longrightarrow W$.*

Corollary 5.1.4. *Let $\{X_n, n \ge 1\}$ be a strictly stationary φ-mixing sequence with $EX_1 = 0, EX_1^2 < \infty$, $\sigma_n^2 \to \infty$ and for any $\varepsilon > 0$*

$$\lim_{n\to\infty} \frac{n}{\sigma_n^2} EX_1^2 I(X_1^2 > \varepsilon\sigma_n) = 0. \tag{5.1.27}$$

Then $W_n \Longrightarrow W$.

A special case of Corollary 5.1.4 is

Corollary 5.1.5. *Let $\{X_n, n \ge 1\}$ be a strictly stationary φ-mixing sequence with $EX_1 = 0, EX_1^2 < \infty$ and $\liminf_n \sigma_n^2/n > 0$. Then $W_n \Longrightarrow W$.*

Remark 5.1.4. Peligrad (1985) pointed out that in some cases the Lindeberg condition (v) also is necessary: If $\{X_n, n \ge 1\}$ is a φ-mixing sequence such that $W_n \Longrightarrow W$, $\sigma_n^2 \to \infty$ and $\varphi(1) < 1$, then the Lindeberg condition (v) is satisfied.

In fact, by Remark 5.1.1 we have $\sigma_i^2 = ih(i)$, where h is a slowly varying function on R^+, whence,

$$\Big\{\max_{1\le i\le n} E(S_n - S_i)^2/\sigma_n^2, n \ge 1\Big\}$$

is bounded. So there exists a $t_0 > 0$ such that for every n

$$\varphi(1) + \max_{1\le i\le n} P\{|S_n - S_i| > t_0\sigma_n\} \le c < 1.$$

Then by the proof of Lemma 2.2.7 for any $x > t_0^2$ and for each $n \in \mathbb{N}$, we have

$$P\Big\{\max_{1\le i\le n} S_i^2 > 4x\sigma_n^2\Big\} \le \frac{1}{1-c} P\{S_n^2 > x\sigma_n^2\}. \tag{5.1.28}$$

On the other hand, the weak convergence to W implies the uniform integrability of $\{S_n^2/\sigma_n^2, n \ge 1\}$. This fact together with (5.1.28) and (5.1.14) implies $\{\max_{1\le i\le n} S_i^2/\sigma_n^2, n \ge 1\}$ is uniformly integrable. By Lemma 5.1.5 $\{\max_{1\le i\le n} X_i^2/\sigma_n^2, n \ge 1\}$ is also uniformly integrable. From Remark 5.1.1 we have

$$\lim_{n\to\infty} P\Big\{\max_{1\le i\le n} |X_i| > \varepsilon\sigma_n\Big\} = 0 \qquad \text{for any } \varepsilon > 0.$$

Hence we obtain (5.1.11) since $\{\max_{1\le i\le n} X_i^2/\sigma_n^2, n \ge 1\}$ is uniformly integrable. By Remark 5.1.2, (v) is a necessary condition for the weak convergence to W.

5.2 The Ibragimov-Linnik-Iosifescu conjecture

We have showed in Corollary 5.1.5 that the Iosifescu conjecture is true under the assumption $\liminf \sigma_n^2/n > 0$. But Herrndorf (1983b) showed by an example that the Iosifescu conjecture does not hold true if $\liminf \sigma_n^2/n = 0$.

Example 5.2.1. Assume that $\{\eta_n, n \ge 1\}$ be a strictly stationary φ-mixing sequence with $E\eta_n = 0$, $E\eta_n^2 < \infty$, $\tau_n^2 = E\Big(\sum_{i=1}^n \eta_i\Big)^2 \to \infty$, $\liminf \tau_n^2/n = 0$. If $\{\eta_n, n \ge 1\}$ does not satisfies the WIP, then we can take $X_n = \eta_n$ for $n \in \mathbb{N}$. Suppose that $\{\eta_n, n \ge 1\}$ satisfies the WIP. Let $\{\alpha_n, n \ge 0\}$ be a sequence of independent identically distributed random variables, which are independent of $\{\eta_n\}$ and satisfy

$$\begin{aligned} &P\{\alpha_0 = a_k\tau_{n_k}\} = b_k n_k^{-1} \quad \text{for} k \in \mathbb{N}, \\ &P\{\alpha_0 = 0\} = 1 - \sum_k b_k n_k^{-1}, \end{aligned} \tag{5.2.1}$$

where the positive integers $n_1 < n_2 < n_3 < \cdots$, and the sequences of real numbers $\{a_k\}, \{b_k\}$ with $a_k \to \infty, b_k \to \infty$ such that

$$\sum_k b_k n_k^{-1} \le 1/2, \qquad \sum_k a_k^2 \tau_{n_k}^2 b_k n_k^{-1} < \infty. \tag{5.2.2}$$

(For instance, from $\tau_n^2/n \to 0$, we can choose n_k such that $\tau_{n_k}^2 n_k^{-1} < 2^{-2k-1}$ and n_k to be increasing, $a_k = k$, $b_k = 2^k$.) (5.2.1) and (5.2.2) imply $E\alpha_0^2 < \infty, P(\alpha_0 = 0) \geq 1/2$. Now, define

$$X_n = \eta_n + \alpha_n - \alpha_{n-1}. \tag{5.2.3}$$

Denote $S_n = \sum_{j=1}^n X_j$, $T_n = \sum_{j=1}^n \eta_j$. We have

$$S_n = T_n + \alpha_n - \alpha_0. \tag{5.2.4}$$

It is clear that $\{X_n, n \geq 1\}$ is a strictly stationary φ-mixing sequence with $EX_1 = 0, EX_1^2 < \infty$, and

$$\sigma_n^2/\tau_n^2 \to 1. \tag{5.2.5}$$

For n, $m \geq 1$ we can write

$$\begin{aligned}&\Big(\frac{S_{n+m} - S_m}{\sigma_n}\Big)^2\\ &\leq 2\Big(\frac{T_{n+m} - T_m}{\tau_n}\Big)^2\Big(\frac{\tau_n}{\sigma_n}\Big)^2 + 2\Big(\frac{\alpha_{n+m} - \alpha_m}{\sigma_n}\Big)^2.\end{aligned} \tag{5.2.6}$$

It is well-known that if $\{\eta_n\}$ satisfies the WIP, then $\{(T_{n+m}-T_m)^2/\tau_n^2, m \geq 0, n \geq 1\}$ is uniformly integrable. Thus, from (5.2.6), (5.2.5) and

$$\|(\alpha_{n+m} - \alpha_m)/\sigma_n\|_2 \leq 2\|\alpha_0\|_2/\sigma_n \to 0, \tag{5.2.7}$$

it follows that $\{(S_{n+m} - S_m)^2/\sigma_n^2,\ m \geq 0, n \geq 1\}$ is also uniformly integrable. Since $\{\eta_n\}$ satisfies the WIP and the random vector

$$\sigma_n^{-1}(S_{[nt_1]}, \cdots, S_{[nt_k]}) - \tau_n^{-1}(T_{[nt_1]}, \cdots, T_{[nt_k]}) \xrightarrow{P} 0$$

for any $0 \leq t_1 \leq \cdots \leq t_k \leq 1$, $(W_{[nt_1]}, \cdots, W_{[nt_k]})$ converges weakly to $W\pi_{t_1\cdots t_k}^{-1}$, where $W_n(t) = S_{[nt]}/\sigma_n$. We shall show that W_n does not weakly converge to W. For $t > 0$ we have

$$\begin{aligned}&P\Big\{\max_{1\leq i\leq n} |S_i| \geq t\sigma_n\Big\}\\ &\geq P\Big\{\max_{1\leq i\leq n} |\alpha_i - \alpha_0| \geq 2t\sigma_n\Big\} - P\Big\{\max_{1\leq i\leq n} |T_i| \geq t\sigma_n\Big\},\end{aligned} \tag{5.2.8}$$

$$\begin{aligned}&P\Big\{\max_{1\leq i\leq n} |\alpha_i - \alpha_0| \geq 2t\sigma_n\Big\}\\ &\geq P\Big\{\alpha_0 = 0, \max_{1\leq i\leq n} |\alpha_i| \geq 2t\sigma_n\Big\}\\ &\geq \frac{1}{2}P\Big\{\max_{1\leq i\leq n} |\alpha_i| \geq 2t\sigma_n\Big\}\\ &= \frac{1}{2}(1 - (P\{|\alpha_0| < 2t\sigma_n\})^n)\\ &\geq \frac{1}{2} - \frac{1}{2}\exp(-nP\{|\alpha_0| \geq 2t\sigma_n\}).\end{aligned} \tag{5.2.9}$$

Using (5.2.9), (5.2.5), $a_k \to \infty$, (5.2.1) and $b_k \to \infty$, we obtain

$$\limsup_{n\to\infty} P\Big\{\max_{1\le i\le n} |\alpha_i - \alpha_0| \ge 2t\sigma_n\Big\}$$
$$\ge \frac{1}{2} - \frac{1}{2}\lim_{k\to\infty} \exp(-n_k b_k/n_k) = \frac{1}{2}.$$

Since $\{\eta_n\}$ fulfills the WIP and (5.2.5), we can choose $t_0 > 0$ with

$$\limsup_{n\to\infty} P\{\max_{1\le i\le n} |T_i| > t_0\sigma_n\} \le 1/4.$$

Then (5.2.8) and (5.2.9) imply:

$$\limsup_{n\to\infty} P\Big\{\max_{1\le i\le n} |S_i| \ge t\sigma_n\Big\} \ge \frac{1}{4} \quad \text{for every} \quad t \ge t_0,$$

whence $\{X_n, n \ge 1\}$ does not satisfy the WIP.

Furthermore, Peligrad (1990) pointed out that conjectures 1 and 2 are not the most general results which one can expect for a φ-mixing sequence. After comprehensive survey of the relevant works, she showed that the following conjecture might be true.

Conjecture 3 (Peligrad 1990). Let $\{X_n, n \ge 1\}$ be a strictly stationary centered φ-mixing sequence satisfying:

$$H(x) := EX_1^2 I(|X_1| \le x) \quad \text{is slowly varying as} \quad x \to \infty \tag{5.2.10}$$

and $\varphi(1) < 1$. Then $\tilde{W}_n$ weakly converges to W, where

$$\tilde{W}_n(t) = S_{[nt]}/((\pi/2)^{1/2} b_n) \qquad 0 \le t \le 1,$$

$b_n = E|S_n|$.

Remark 5.2.1. There are at least two situations of interest when it is easy to verify (5.2.10). One is $EX_1^2 < \infty$, and the other is that $P(|X_1| > x)$ is regularly varying with the exponent -2, i.e.,

$$P(|X_1| > x) = 1/(x^2 h(x)), \tag{5.2.11}$$

where $h(x)$ is slowly varying as $x \to \infty$.

In fact, we can rewrite $P(|X_1| > x) = h(x)/x^2$. By the partial integration we have

$$H(x) = -\int_0^x y^2 dP(|X_1| > y)$$
$$= -h(x) + \int_0^x h(y) y^{-1} dy.$$

By the Karamata representation (see Theorem A1), we can choose $z_1 = z_1(x) < x$ such that

$$\limsup_{x \to \infty} \sup_{z_1 \le y \le x} h(y)/h(x) = 1,$$
$$\lim_{x \to \infty} x/z_1(x) = \infty.$$

So that

$$\int_0^x h(y)y^{-1}dy \ge \int_{z_1}^x h(y)y^{-1}dy > \frac{1}{2}h(x)\log(xz_1^{-1}).$$

Then

$$H(x) = (1 + o(1)) \int_0^x h(y)y^{-1}dy.$$

For any given $k > 0$ we have

$$\Big|\int_x^{kx} h(y)y^{-1}dy\Big| \le 2(\log k)h(x) = o\Big(\int_0^x h(y)y^{-1}dy\Big).$$

It follows that

$$\lim_{x \to \infty} H(kx)/H(x) = 1.$$

Peligrad (1990) proved that Conjecture 3 is true under condition (5.2.11).

Theorem 5.2.1. *Let $\{X_n, n \ge 1\}$ be a centered, strictly stationary φ-mixing sequence satisfying (5.2.11) and $\varphi(1) < 1$. Then*

$$\tilde{W_n} \Longrightarrow W.$$

The proof of Theorem 5.2.1 will not be presented here.

Chapter 6 Weak Convergence for Mixing Random Fields

There are two kinds of definitions of mixing dependence for a random field. One is a natural generalization from the classical case, the sequence of dependent random variables, to the dependent random field, which has been discussed by Bulinskii-Zurbenko (1981), Gorodezkii (1982, 1984), Bolthausen (1982), Nahapetian (1987), Bradley (1992), Donkhan and Guyon (1991), Guyon(1992) and Donkhan (1994), etc. Another is appeared in the study of set-index partial sum process for weakly dependent random fields, which has been discussed by Goldie and Greenwood (1986 a,b), Chen (1991) and Lu (1995), etc. We shall introduce the first case in Section 6.1 and the second case in Sections 6.2 and 6.3 respectively.

6.1 The CLT for mixing random fields

A natural generalization from an α-mixing sequence to an α-mixing random field has been discussed by some mathematicians.

A random field $\{\xi_{\mathbf{t}}, \mathbf{t} \in \mathbb{Z}^d\}, d \geq 1$, is said to be α_*-mixing, if

$$\begin{aligned}\alpha_{m,n}(r) = \sup\{&|P(AB) - P(A)P(B)| : A \in \sigma_U, B \in \sigma_V,\\ &U, V \subset \mathbb{Z}^d, |U| \leq m, |V| \leq n, d(U,V) \geq r\}\\ \to 0 \qquad &as \quad r \to \infty. \end{aligned} \tag{6.1.1}$$

where $d(U,V) = \inf\{d(\mathbf{t},\mathbf{s}) : \mathbf{t} \in U, \ \mathbf{s} \in V\}$, $d(\mathbf{t},\mathbf{s}) = \max_{1\leq i\leq d}|t_i - s_i|$, $m, n \in \mathbb{N} \cup \{\infty\}$, $\sigma_A = \sigma\{\xi_{\mathbf{t}}, \mathbf{t} \in A\}, |A|$ is the number of elements of A.

For some subsets $\Delta_j^{(n)}, j = 1, \cdots, k$, of d-dimensional cube $\mathbf{J}_n = [-n, n]^d$, denote

$$S(n,j) = \sigma_n^{-1} S_{\Delta_j^{(n)}}, \qquad \sigma_n^2 = \mathrm{Var} S_{\mathbf{J}_n},$$
$$S_I = \sum_{\mathbf{t} \in I} \xi_{\mathbf{t}}, \qquad I \subset \mathbb{Z}^d, \quad |I| < \infty,$$

$$M_k = \left| E \prod_{j=1}^{k} e^{itS(n,j)} - \prod_{j=1}^{k} Ee^{itS(n,j)} \right|,$$

$$L_k(\varepsilon, \delta) = \sum_{j=1}^{k} \int_{|S(n,j)| \geq \varepsilon} |S(n,j)|^{2+\delta} P(d\omega), \qquad \varepsilon, \ \delta \geq 0. \quad (6.1.2)$$

Nahapetian (1987) proved the following conclusion, which is a generalization of Theorem 3.2.1.

Theorem 6.1.1. *Let $\{X_{\mathbf{t}}, \mathbf{t} \in \mathbb{Z}^d\}$ be a strictly stationary α_*-mixing random field with $EX_{\mathbf{t}} = 0, E|X_{\mathbf{t}}|^{2+\delta} < \infty$ for some $\delta > 0$. If for some $\tau > 0$,*

(i) $\alpha_{m,n}(r) \leq f(m) n^{\tau} \alpha(r)$, where $f(m)$ is a non-negative function, $m \in \mathbb{N} \cup \{\infty\}$;

(ii) $\sum_{r=1}^{\infty} r^{d-1} \alpha^{\delta/(2+\delta)}(r) < \infty, \quad \alpha(r) = o(r^{-(2\tau+1)d}), \qquad r \to \infty$,

then

$$\sigma^2 = \sum_{\mathbf{t} \in \mathbb{Z}^d} EX_{\mathbf{0}} X_{\mathbf{t}} < \infty,$$

and if $\sigma \neq 0$, we have

$$S_{\mathbf{J}_n} / \sigma_n \xrightarrow{d} N(0,1). \quad (6.1.3)$$

The proof of Theorem 6.1.1 will need some lemmas. It is clear from the proof of Lemma 1.2.3 that we have

Lemma 6.1.1. *Let random variables X and Y be measurable with respect to σ-fields σ_U and σ_V respectively, $|U| \leq m$, $|V| \leq n$, $d(U,V) \geq r$, $E|X|^p < \infty$, $E|Y|^q < \infty$, $p, q > 1$, $p^{-1} + q^{-1} \leq 1$. Then*

$$|EXY - EXEY| \leq c(E|X|^p)^{1/p} (E|Y|^q)^{1/q} \alpha_{m,n}^{1-p^{-1}-q^{-1}}(r).$$

Particularly, if $|X| \leq C_1$ a.s., $|Y| \leq C_2$ a.s., we have

$$|EXY - EXEY| \leq cC_1 C_2 \alpha_{m,n}(r).$$

Lemma 6.1.2. *Let $\xi_1, \cdots, \xi_n$ be a sequence of random vectors, $|E\prod_{j=i}^n \times \xi_j| < \infty$, $i = 1, \cdots, n-1$, $|E\xi_i| \le 1$, $i = 1, \cdots, n$. Then*

$$\begin{aligned}
&\left|E\prod_{s=1}^n \xi_s - \prod_{s=1}^n E\xi_s\right| \\
&\le \sum_{i=1}^{n-1}\sum_{j=i+1}^{n}\Big|E(\xi_i-1)(\xi_j-1) \\
&\quad\times \prod_{s=j+1}^{n}\xi_s - E(\xi_i-1)E(\xi_j-1)\prod_{s=j+1}^{n}\xi_s\Big|.
\end{aligned} \tag{6.1.4}$$

Proof. Obviously, we have

$$\left|E\prod_{s=1}^n \xi_s - \prod_{s=1}^n E\xi_s\right| \le \sum_{i=1}^{n-1}\left|E\xi_i\prod_{j=i+1}^{n}\xi_j - E\xi_i E\prod_{j=i+1}^{n}\xi_j\right|. \tag{6.1.5}$$

Moreover

$$\begin{aligned}
&\left|E\xi_i\prod_{j=i+1}^{n}\xi_j - E\xi_i E\prod_{j=i+1}^{n}\xi_j\right| \\
&\le \Big|E(\xi_i-1)(\xi_{i+1}-1)\prod_{j=i+2}^{n}\xi_j \\
&\quad - E(\xi_i-1)E(\xi_{i+1}-1)\prod_{j=i+2}^{n}\xi_j\Big| \\
&\quad + \left|E(\xi_i-1)\xi_{i+2}\prod_{j=i+3}^{n}\xi_j - E(\xi_i-1)E\xi_{i+2}\prod_{j=i+3}^{n}\xi_j\right|.
\end{aligned}$$

By this recursive process, we obtain

$$\begin{aligned}
&\left|E\xi_i\prod_{j=i+1}^{n}\xi_j - E\xi_i E\prod_{j=i+1}^{n}\xi_j\right| \\
&\le \sum_{j=i+1}^{n}\Big|E(\xi_i-1)(\xi_j-1) \\
&\quad\times \prod_{s=j+1}^{n}\xi_s - E(\xi_i-1)E(\xi_j-1)\prod_{s=j+1}^{n}\xi_s\Big|,
\end{aligned} \tag{6.1.6}$$

inserting (6.1.6) into (6.1.5) yields (6.1.4).

Proof of Theorem 6.1.1.

Let $p = p(n)$, $q = q(n)$ be the positive integers such that

$$p, q \to \infty, \quad p = o(n),\ q = o(p) \text{as } n \to \infty.$$

Denote

$$\hat{k} = \hat{k}(n) = [2n/(p+q)],$$
$$I_n(i) = [-n + ip + iq, -n + (i+1)p + iq] \quad i = 0, 1, \cdots, \hat{k} - 1,$$
$$I_n = \bigcup_{j=0}^{\hat{k}-1} I_n(j), \quad I_n^d = \overbrace{I_n \times I_n \times \cdots \times I_n}^{d \text{times}}.$$

Then I_n^d consists of $\hat{k}^d$ d-dimensional cubes with side p. For given n, we denote these $k = \hat{k}^d$ d-dimensional cubes by $\Delta_j^{(n)}, j = 1, \cdots, k$, i.e.

$$I_n^d = \bigcup_{j=1}^{k} \Delta_j^{(n)}.$$

Put $A_n = \mathbf{J}_n \backslash I_n^d$. By Bernstein's blocking technique, we need only to prove

1) $\sigma_n^{-2} \mathrm{Var} S_{A_n} \to 0$,

2) $M_k \to 0$, $\sum_{j=1}^{k} \mathrm{Var} S(n,j) \to 1$ and for any given $\varepsilon > 0$

$$L_k(\varepsilon, \delta) \to 0, \quad \text{as } n \to \infty.$$

From $E|X_{\mathbf{t}}|^{2+\delta} < \infty$, $\sum_{r=1}^{\infty} r^{d-1} \alpha^{\delta/(2+\delta)}(r) < \infty$ and Lemma 6.1.1, it follows that σ^2 is finite and $\mathrm{Var} S_I \sim \sigma^2 |I|$ for any d-dimensional cube I with $|I| < \infty$. Therefore we have $\sigma_n^{-2} \mathrm{Var} S_{A_n} \to 0$ and

$$\sum_{j=1}^{k} \mathrm{Var} S(n,j) = k \sigma_n^{-2} \mathrm{Var} S_{\Delta_1^{(n)}}$$
$$= \frac{k}{(2n)^d \sigma^2 (1 + o(1))} p^d \sigma^2 (1 + o(1)) \to 1, \quad n \to \infty.$$

It also permit us to prove 2) only for the bounded random field (cf. Ibragimov and Linnik 1971, the proofs of Theorems 18.5.3 and 18.5.4).

Now we prove $M_k \to 0$. From (6.1.5) it follows that

$$
\begin{aligned}
M_k \le & \sum_{j=1}^{k-1} \Big| E e^{itS(n,j)} \prod_{m=j+1}^{k} e^{itS(n,m)} - E e^{itS(n,j)} E \prod_{m=j+1}^{k} e^{itS(n,m)} \Big| \\
\le & \sum_{j=1}^{k-1} \Big| E(e^{itS(n,j)} - 1 - itS(n,j) \\
& + \frac{t^2}{2} S(n,j)^2) \prod_{m=j+1}^{k} e^{itS(n,m)} \\
& - E(e^{itS(n,j)} - 1 - itS(n,j) + \frac{t^2}{2} S(n,j)^2) E \prod_{m=j+1}^{k} e^{itS(n,m)} \Big| \\
& + |t| \sum_{j=1}^{k-1} \Big| ES(n,j) \prod_{m=j+1}^{k} e^{itS(n,m)} \\
& - ES(n,j) E \prod_{m=j+1}^{k} e^{itS(n,m)} \Big| \\
& + \frac{t^2}{2} \sum_{j=1}^{k-1} \Big| ES(n,j)^2 \prod_{m=j+1}^{k} e^{itS(n,m)} \\
& - ES(n,j)^2 E \prod_{m=j+1}^{k} e^{itS(n,m)} \Big| \\
=: & T_1 + |t| T_2 + \frac{t^2}{2} T_3. \qquad (6.1.7)
\end{aligned}
$$

Take $p = o(n^{1/2})$, we have

$$
\begin{aligned}
T_1 & \le 2 \frac{|t|^3}{3!} \sum_{j=1}^{k-1} E|S(n,j)|^3 \\
& \le c|t|^3 \Big(\frac{n}{p}\Big)^d \sigma_n^{-3} E|S_{\Delta_1^{(n)}}|^3 \\
& \le c|t|^3 \Big(\frac{n}{p}\Big)^d (n^d)^{-3/2} p^d E S^2_{\Delta_1^{(n)}} \\
& \le c|t|^3 \Big(\frac{p}{n^{1/2}}\Big)^d \to 0, \quad n \to \infty. \qquad (6.1.8)
\end{aligned}
$$

Using the proof of inequality (6.1.4) we have

$$T_2 \le \sigma_n^{-1} \sum_{j=1}^{k-1} \sum_{\mathbf{s}\in\Delta_j^{(n)}} \Big| EX_{\mathbf{s}} \prod_{m=j+1}^{k} e^{itS(n,m)} - EX_{\mathbf{s}} E \prod_{m=j+1}^{k} e^{itS(n,m)} \Big|$$

$$\le \sigma_n^{-1} \sum_{j=1}^{k-1} \sum_{\mathbf{s}\in\Delta_j^{(n)}} \sum_{m=j+1}^{k} \Big| EX_{\mathbf{s}} (e^{itS(n,m)} - 1) \prod_{r=m+1}^{k} e^{itS(n,r)}$$

$$- EX_{\mathbf{s}} E(e^{itS(n,m)} - 1) \prod_{r=m+1}^{k} e^{itS(n,r)} \Big|. \tag{6.1.9}$$

Noting $|X_{\mathbf{s}}| \le C_0$, $|e^{itS(n,m)} - 1| \le cp^d/\sigma_n$ a.s. and applying Lemma 6.1.1 in (6.1.9) we obtain

$$T_2 \le cp^d \sigma_n^{-2} \sum_{j=1}^{k-1} \sum_{\mathbf{s}\in\Delta_j^{(n)}} \sum_{m=j+1}^{k} \alpha_{1,n^d}(d(\Delta_j^{(n)}, \Delta_m^{(n)}))$$

$$\le cp^{2d} f(1) n^{\tau d} \sigma_n^{-2} \sum_{j=1}^{k-1} \sum_{m=j+1}^{k} \alpha(d(\Delta_j^{(n)}, \Delta_m^{(n)}))$$

$$\le cp^{2d} n^{\tau d} p^{-d} \sum_{j=1}^{k} \alpha(d(\Delta_1^{(n)}, \Delta_j^{(n)}))$$

$$\le cp^d n^{\tau d} \sum_{l=1}^{\infty} \sum_{j:(l-1)q \le d(\Delta_1^{(n)}, \Delta_j^{(n)}) < lq} \alpha(d(\Delta_1^{(n)}, \Delta_j^{(n)}))$$

$$\le cp^d n^{\tau d} \sum_{l=1}^{\infty} l^{d-1} \alpha(lq).$$

From condition (ii) of Theorem 6.1.1 it follows that

$$T_2 \le cp^d n^{\tau d} q^{-(2\tau+1)d} \sum_{l=1}^{\infty} \frac{\beta(lq)}{l^{1+2\tau d}}, \tag{6.1.10}$$

where $\beta(n) = \alpha(n) n^{(2\tau+1)d}$. Since $\beta(n) \to 0$ as $n \to 0$ we can take $p = p(n) = o(n^{1/2})$, $p(n) \to \infty$ as $n \to \infty$ and $q = q(n) \to \infty$, $q = o(p)$ such that the right hand side of (6.1.10) approaches zero.

Now we estimate T_3.

$$\begin{aligned}
T_3 &\le \sigma_n^{-2}\sum_{j=1}^{k-1}\sum_{\mathbf{s},\mathbf{u}\in\Delta_j^{(n)}}\Big|EX_{\mathbf{s}}X_{\mathbf{u}}\prod_{m=j+1}^{k}e^{itS(n,m)} \\
&\qquad\qquad - EX_{\mathbf{s}}X_{\mathbf{u}}E\prod_{m=j+1}^{k}e^{itS(n,m)}\Big| \\
&\le \sigma_n^{-2}\sum_{j=1}^{k-1}\sum_{m=j+1}^{k}\sum_{\mathbf{s},\mathbf{u}\in\Delta_j^{(n)}}\Big|EX_{\mathbf{s}}X_{\mathbf{u}}(e^{itS(n,j+1)}-1)\prod_{m=j+2}^{k}e^{itS(n,m)} \\
&\qquad\qquad - EX_{\mathbf{s}}X_{\mathbf{u}}E(e^{itS(n,j+1)}-1)\prod_{m=j+2}^{k}e^{itS(n,m)}\Big| \\
&\le c\frac{p^{2d}}{n^d}\frac{p^d}{n^{d/2}}\sum_{j=1}^{k-1}\sum_{m=j+1}^{k}\alpha_{2,n^d}(d(\Delta_j^{(n)},\Delta_m^{(n)})) \\
&\le c\frac{p^{3d}}{n^{3d/2}}\Big(\frac{n}{p}\Big)^d f(2)n^{\tau d}\sum_{s=1}^{\infty}s^{d-1}\alpha(sq) \\
&\le c\frac{p^{2d}n^{\tau d}\beta(q)}{n^{d/2}q^{(2\tau+1)d}}\to 0,\quad \text{as}\quad n\to\infty. \qquad (6.1.11)
\end{aligned}$$

Combining (6.1.7)-(6.1.11) yields that $M_k\to 0$ as $n\to\infty$.

Finally, we prove $L_k(\varepsilon,\delta)\to 0$. Noting that the stationarity of $\{\xi_{\mathbf{t}}\}$, we have

$$\begin{aligned}
L_k(\varepsilon,\delta) &\le c\Big(\frac{n}{p}\Big)^d n^{-d(1+\delta/2)}\int_{|S_{\Delta_1^{(n)}}|\ge\varepsilon\sigma_n}|S_{\Delta_1^{(n)}}|^{2+\delta}P(d\omega) \\
&\le c\Big(n^{\delta/2}p\Big)^{-d}\int_{|S_{\Delta_1^{(n)}}|\ge\varepsilon\sigma_n}|S_{\Delta_1^{(n)}}|^{2+\delta}P(d\omega).
\end{aligned}$$

There exist $p=p(n)\to\infty$, $p(n)=o(n)$, and $q=q(n)\to\infty$, $q=o(p)$ as $n\to\infty$ such that the right hand of the above inequality approaches zero. This completes the proof of Theorem 6.1.1.

Remark 6.1.1. In the above definition of α_*-mixing, the positions of set U and V are symmetric, so the assumption:

$$\alpha_{m,n}(r)\le f(m)n^\tau\alpha(r)$$

in the condition (i) is not very reasonable. The random field $\{X_{\mathbf{t}},\mathbf{t}\in\mathbb{Z}^d\}$

is said to be α-mixing, if

$$\begin{aligned}\alpha(r) = \sup\{|P(AB) - P(A)P(B)| : A \in \sigma_U, B \in \sigma_V,\\ U, V \subset \mathbb{Z}^d, d(U,V) \geq r\}\\ \to 0 \qquad as \quad r \to \infty.\end{aligned} \tag{6.1.12}$$

A more general result was given by Lu (1995), which is a generalization of Theorem 3.2.3 to α-mixing random fields.

Theorem 6.1.2. *Let $\{X_{\mathbf{t}}, \mathbf{t} \in \mathbb{Z}^d\}$ be an α-mixing random field with $EX_{\mathbf{t}} = 0$, $EX_{\mathbf{t}}^2 < \infty$. If there exists a $g \in \mathcal{G}$ such that*

$$\sup_{\mathbf{t}} Eg(|X_{\mathbf{t}}|) < \infty, \qquad \sum_{r=1}^{\infty} r^{d-1} f_g(\alpha(r)) < \infty, \tag{6.1.13}$$

$$\lim_{n\to\infty} \mathrm{Var} S_{J_n}/n^d = \sigma^2 > 0. \tag{6.1.14}$$

Then

$$W_n \Longrightarrow W$$

where $W_n(\mathbf{t}) = S_{J[n\mathbf{t}]}/\sigma_n$, $\mathbf{0} \leq \mathbf{t} \leq \mathbf{1}$.

The proof of Theorem 6.1.2 is similar to that of Theorem 3.2.3.

Remark 6.1.2. A similar theorem for non-stationary random fields has been discussed by Guyon (1992). The central limit theorem for α-mixing random fields with continuous parameters has been discussed by Gorodezkii (1984), Zhurbenko (1984), etc, and the weak invariance principle for this case has also been given there.

Bradley (1992) proved the CLT of strictly stationary random fields under the "unrestricted ρ-mixing" condition and just finite or "barely infinite" second moments. No mixing rate is assumed. Let $\{\xi_{\mathbf{t}}, \ \mathbf{t} \in \mathbb{Z}^d\}$ be a strictly stationary random field. For any nonempty disjoint sets $S, D \subset \mathbb{Z}^d$, put

$$\rho(S, D) = \rho(\sigma(\xi_{\mathbf{k}}, \mathbf{k} \in S), \ \sigma(\xi_{\mathbf{k}}, \mathbf{k} \in D)).$$

For each $r \geq 1$, define

$$\rho^*(r) = \sup \ \rho(S, D),$$

where the sup is taken over all pairs of nonempty disjoint subsets $S, D \subset \mathbb{Z}^d$ such that $d(S, D) \geq r$. Denote

$$L := L^{(n)} = (l_1^{(n)}, \cdots, l_d^{(n)}) \in \mathbb{N}^d, \ S(\xi, L) = \sum_{\mathbf{1} \leq \mathbf{t} \leq \mathbf{L}} \xi_{\mathbf{t}}.$$

Bradley (1992) proved the following theorem.

Theorem 6.1.3. *Let $\{X_{\mathbf{t}}, \mathbf{t} \in \mathbb{Z}^d\}$ be a centered strictly stationary random field with $0 < EX_{\mathbf{0}}^2 < \infty$, $\rho^*(r) \to 0$ as $r \to \infty$ and the continuous positive spectral density $f(\cdot)$ on T^d, satisfying $f(1, \cdots, 1) > 0$, where T denotes the unit circle in the complex plane. Then as $\|L^{(n)}\| = l_1^{(n)} \cdots l_d^{(n)} \to \infty$ one has that $\|S(\xi, L)\|_2 \to \infty$ and $S(\xi, L)/\|S(\xi, L)\|_2 \xrightarrow{d} N(0,1)$.*

The proof of Theorem 6.1.3 will not be presented here.

6.2 Convergence of finite dimensional distributions

Denote

$$\begin{aligned}
\mathcal{J}_n &= \{\mathbf{j} = (j_1/n, j_2/n, \cdots, j_d/n) : j_1, j_2, \cdots, j_d \in \{1, 2, \cdots, n\}\}, \\
C_{n\mathbf{j}} &= (\mathbf{j} - n^{-1}\mathbf{1}, \mathbf{j}), \\
(\mathbf{a}, \mathbf{b}] &= \{(x_1, x_2, \cdots, x_d) : a_i < x_i \le b_i, i = 1, 2, \cdots, d\}, \\
\mathcal{I}^d &= \{(\mathbf{a}, \mathbf{b}] : \mathbf{a}, \mathbf{b} \in [0, 1]^d\}.
\end{aligned}$$

Let $\{\xi_{n\mathbf{j}},\ \mathbf{j} \in \mathcal{J}_n,\ n \ge 1\}$ be a triangular array field. It is easy to see that a random field $\{\xi_{\mathbf{t}},\ \mathbf{t} \in \mathbb{Z}^d\}$ on a d-dimensional integer lattice is a special case of a triangular array random field $\{\xi_{n\mathbf{j}},\ \mathbf{j} \in \mathcal{J}_n, n \ge 1\}$, if we write $\xi_{n\mathbf{j}} = \xi_{n\mathbf{j}}$, $n\mathbf{j} = (j_1, j_2, \cdots, j_d)$. Now from a random field $\{\xi_{n\mathbf{j}},\ \mathbf{j} \in \mathcal{J}_n, n \ge 1\}$ we form the set-indexed partial-sum process of the n-th level as

$$Z_n(A) = n^{-d/2} \sum_{\mathbf{j} \in \mathcal{J}_n} \frac{|A \cap C_{n\mathbf{j}}|}{|C_{n\mathbf{j}}|} (\xi_{n\mathbf{j}} - E\xi_{\mathbf{n},\mathbf{j}}) \quad A \in \mathcal{B}^d,\ n \in \mathbb{N}, \qquad (6.2.1)$$

where $|\cdot|$ is Lebesgue measure and $\mathcal{B}^d$ is the class of Borel sets of $[0, 1]^d$. For the random field $\{\xi_{\mathbf{t}}, \mathbf{t} \in \mathbb{Z}\}$, correspondingly, define

$$Z_n(A) = n^{-d/2} \sum_{\mathbf{t} \in \mathbb{Z}^d} |A \cap C_{\mathbf{t}}| (\xi_{\mathbf{t}} - E\xi_{\mathbf{t}}) \qquad (6.2.1')$$

where $A/n^d \in \mathcal{B}^d$, $\quad C_{\mathbf{t}} = (\mathbf{t} - \mathbf{1}, \mathbf{t}), \mathbf{t} = (t_1, t_2, \cdots, t_d), t_i \in \mathbb{Z}$.

In Sections 6.2 and 6.3, we prove the weak convergence of Z_n to a Wiener process W, restricting its domain of definition to a subset of $\mathcal{B}^d$

satisfying a metric-entropy bound. We also impose moment and mixing conditions on the $\{\xi_{n\mathbf{j}}\}$. Let $x > 0$.

Definition 6.2.1. The random field $\{\xi_{n\mathbf{j}},\ \mathbf{j} \in \mathcal{J}_n\}$ is said to be *α-mixing* if $\alpha(nx) \to 0$ as $n \to \infty$, where

$$\alpha(nx) = \sup_{\substack{I,J\subset\mathcal{J}_n \\ d(I,J)\geq x}} \ \sup_{\substack{A\in\sigma(\xi_{n\mathbf{j}},\mathbf{j}\in I) \\ B\in\sigma(\xi_{n\mathbf{j}},\mathbf{j}\in J)}} |P(AB) - P(A)P(B)|.$$

Definition 6.2.2. The random field $\{\xi_{n\mathbf{j}},\ \mathbf{j} \in \mathcal{J}_n,\}$ is said to be *ρ-mixing* if $\rho(nx) \to 0$ as $n \to \infty$, where

$$\rho(nx) = \sup_{\substack{I,J\subset\mathcal{J}_n \\ d(I,J)\geq x}} \ \sup_{\substack{X\in L_2(\sigma(\xi_{n\mathbf{j}},\mathbf{j}\in I)) \\ Y\in L_2(\sigma(\xi_{n\mathbf{j}},\mathbf{j}\in J))}} \frac{|\mathrm{Cov}(X,Y)|}{\sqrt{\mathrm{Var}X\,\mathrm{Var}Y}}$$

and $L_2(\mathcal{F})$ is the set of L_2 random variables measurable with respect to $\mathcal{F}$.

Definition 6.2.3. The random field $\{\xi_{n\mathbf{j}},\ \mathbf{j} \in \mathcal{J}_n\}$ is said to be *symmetric φ-mixing* if $\varphi(nx) \to 0$ as $n \to \infty$, where

$$\varphi(nx) = \sup_{\substack{I,J\subset\mathcal{J}_n \\ d(I,J)\geq x}} \ \sup_{\substack{A\in\sigma(\xi_{n\mathbf{j}},\mathbf{j}\in I) \\ B\in\sigma(\xi_{n\mathbf{j}},\mathbf{j}\in J) \\ P(A)P(B)>0}} \max(|P(A|B) - P(A)|, |P(B|A) - P(B)|).$$

Definition 6.2.4. The random field $\{\xi_{n\mathbf{j}},\ \mathbf{j} \in \mathcal{J}_n\}$ is said to be *absolutely regular* if $\beta(nx) \to 0$ as $n \to \infty$, where

$$\beta(nx) = \sup_{I,J\subset\mathcal{J}_n, d(I,J)\geq x} \|\mathcal{L}(\xi_{n\mathbf{j}}, \mathbf{j} \in I \cup J) - \mathcal{L}(\xi_{n\mathbf{j}}, \mathbf{j} \in I)\mathcal{L}(\xi_{n\mathbf{j}}, \mathbf{j} \in J)\|_{\mathrm{Var}}$$

$\mathcal{L}(\xi(\cdot))$ is the distribution law of $\{\xi(\cdot)\}$ and $\|\cdot\|_{\mathrm{Var}}$ is variation norm.

It is clear that

$$\alpha(nx) \leq \rho(nx) \leq 2\varphi(nx), \qquad \alpha(nx) \leq \beta(nx) \leq \varphi(nx). \tag{6.2.2}$$

Next, we introduce the metric entropy condition. We say that Borel sets A, B in $\mathcal{B}^d$ are equivalent if $|A\triangle B| = 0$, and denote the set of equivalence classes by $\mathcal{E}$. Define $d_L(A,B) = |A\triangle B|$, it can be proved that $d_L(\cdot,\cdot)$ is a metric on $\mathcal{E}$. The set $\mathcal{E}$ forms a complete metric space under d_L.

Definition 6.2.5. A subset $\mathcal{A}$ of $\mathcal{E}$ is called *totally bounded* with inclusion, if for every $\delta > 0$ there is a finite set $\mathcal{A}_\delta \subset \mathcal{E}$, such that for every $A \in \mathcal{A}$ there exist $A^+, A^- \in \mathcal{A}_\delta$ with $A^- \subset A \subset A^+$ and $|A^+ \setminus A^-| \leq \delta$.

Note that $\mathcal{A}_\delta$ is a δ-net with respect to d_L for $\mathcal{A}$.

Let $\mathcal{A}$ be a totally bounded subset of $\mathcal{E}$. Its closure $\overline{\mathcal{A}}$ is complete and totally bounded, hence compact. Let $C(\overline{\mathcal{A}})$ be the space of continous functions on $\overline{\mathcal{A}}$ with the sup norm $\|\cdot\|$. Because $\overline{\mathcal{A}}$ is compact, $C(\overline{\mathcal{A}})$ is separable. Thus $C(\overline{\mathcal{A}})$ is a complete, separable metric space. Let $CA(\overline{\mathcal{A}})$ be the set of everywhere additive elements of $C(\overline{\mathcal{A}})$, namely, elements f such that $f(A\cup B) = f(A) + f(B) - f(A\cap B)$ whenever $A, B, A\cup B, A\cap B \in \overline{\mathcal{A}}$. It can be shown that for fixed $\omega, Z_n(\cdot) \in CA(\overline{\mathcal{A}})$, i.e. Z_n are random elements of $CA(\overline{\mathcal{A}})$. A standard Wiener process on $\overline{\mathcal{A}}$ is a random element W of $CA(\overline{\mathcal{A}})$ whose finite dimensional laws are Gaussian with $EW(A) = 0, EW(A)W(B) = |A\cap B|$. In order that W should exist it is necessary (see Dudley 1973) that $\mathcal{A}$ satisfies a metric entropy condition.

Definition 6.2.6. Let $\mathcal{A}$ be a totally bounded subset of $\mathcal{E}$, $\mathcal{A}_\delta$ be the smallest δ-net of $\mathcal{A}$. Denote

$$N(\delta, \mathcal{A}) = \text{Card } \mathcal{A}_\delta, \qquad H(\delta) = \log N(\delta, \mathcal{A}).$$

$\mathcal{A}$ is said to satisfy a *metric entropy condition*, or a *convergent entropy integral*, if

$$\int_0^1 \left(\frac{H(\delta)}{\delta}\right)^{1/2} d\delta < \infty. \tag{6.2.3}$$

Define the *exponent of metric entropy* of $\mathcal{A}$, denoted by $r := \inf\{s,\ s > 0, H(\delta) = O(\delta^{-s}) \text{ as } \delta \to 0\}$. If $r < 1$, then (6.2.3) holds.

Remark 6.2.1. Some examples of classes of sets which satisfy the metric entropy condition are as follows:

If $\mathcal{C}^d$ denotes the convex subsets of $[0,1]^d$, then $r = (d-1)/2$ (see Dudley 1974).

If $\mathcal{I}^d = \{(\mathbf{a}, \mathbf{b}] : \mathbf{a}, \mathbf{b} \in [0,1]^d\}$ as above, then $r = 0$.

If $\mathcal{P}^{d,m}$ denotes the family of all polygonal regions of $[0,1]^d$ with no more than m vertices, then $r = 0$ (see Erickson 1981).

If $\mathcal{E}^d$ denotes the sets of all ellipsoidal regions in $[0,1]^d$, then $r = 0$ (see Gaeussler 1983).

For the Vapnik-Červonenkis class $\mathcal{V}$ that includes the above three examples, it is known that $N(\delta, \mathcal{V}, d_\lambda) = \text{Card } \mathcal{V}_\delta \leq c\delta^{-v}$ for some c and $v > 0$ (Dudley 1978).

When $\{\xi_{\mathbf{t}},\ \mathbf{t} \in \mathbb{Z}^d\}$ is independent, the weak convergence of Z_n to W has been studied by Bass and Pyke (1984, 1985), Alexander and Pyke

(1986), Lu (1992), etc. For the mixing random field $\{\xi_{n,\mathbf{j}}, \mathbf{j} \in \mathcal{J}_n, n \geq 1\}$, the weak convergence of Z_n to W was first discussed by Goldie and Greenwood (1986a, b). They proved the following theorem.

Theorem 6.2.1. *Assume that* $E\xi_{n,\mathbf{j}} = 0$, *and*

(i) for some $s > 2$, $\{|n^{d/2}\xi_{n,\mathbf{j}}|^s, \mathbf{j} \in \mathcal{J}_n, n \geq 1\}$ *is uniformly integrable;*

(ii) the exponent r *of metric entropy (with inclusion) of* $\mathcal{A}$ *satisfies* $r < 1$;

(iii) $\beta(nx) = O((nx)^b)$ *as* $nx \to \infty$, *the exponent* b *of absolute regularity satisfies* $b \geq ds/(s-2)$ *and* $b > d(1+r)/(1-r)$;

(iv) the symmetric φ*-mixing coefficients satisfy*

$$\sup_{n\geq 1}\sum_{j=1}^{\infty}\varphi^{1/2}(2^j n^{-1}) < \infty;$$

(v) for any null family $\{D_h, 0 < h < h_0\}$ *in* $\mathcal{I}^d$ *(a null family is a collection such that* $D_h \subseteq D_{h'}$ *for* $h \leq h'$ *and* $|D_h| = h$ *for each* h*),*

$$\lim_{h\downarrow 0}\limsup_{n\to\infty}\left|\frac{EZ_n^2(D_h)}{|D_h|} - 1\right| = 0.$$

Then Z_n *converges weakly in* $CA(\overline{\mathcal{A}})$ *to* W.

For a random field $\{\xi_{\mathbf{t}}, \mathbf{t} \in \mathbb{Z}^d\}$, the versions of Definitions 6.2.1, 6.2.2, 6.2.3 and 6.2.4 are as follows:

$$\alpha(x) = \sup_{\substack{I,J\subset\mathbb{Z}^d\\ d(I,J)\geq x}}\ \sup_{\substack{A\in\sigma(\xi_{\mathbf{i}},\mathbf{i}\in I)\\ B\in\sigma(\xi_{\mathbf{j}},\mathbf{j}\in J)}} |P(AB) - P(A)P(B)|,$$

$$\rho(x) = \sup_{\substack{I,J\subset\mathbb{Z}^d\\ d(I,J)\geq x}}\ \sup_{\substack{X\in L_2(\sigma(\xi_{\mathbf{i}},\mathbf{i}\in I))\\ Y\in L_2(\sigma(\xi_{\mathbf{j}},\mathbf{j}\in J))}} \frac{|\mathrm{Cov}(X,Y)|}{\sqrt{\mathrm{Var}X\,\mathrm{Var}Y}},$$

$$\beta(x) = \sup_{I,J\subset\mathbb{Z}^d,\ d(I,J)\geq x} \|\mathcal{L}(\xi_{\mathbf{j}}, \mathbf{j} \in I\cup J) - \mathcal{L}(\xi_{\mathbf{j}}, \mathbf{j} \in I)\mathcal{L}(\xi_{\mathbf{j}}, \mathbf{j}\in J)\|_{\mathrm{Var}},$$

$$\varphi(x) = \sup_{\substack{I,J\subset\mathbb{Z}^d\\ d(I,J)\geq x}}\ \sup_{\substack{A\in\sigma(\xi_{\mathbf{i}},\mathbf{i}\in I)\\ B\in\sigma(\xi_{\mathbf{j}},\mathbf{j}\in J) P(A)P(B)>0}} \times \max(|P(A|B) - P(A)|, |P(B|A) - P(B)|).$$

Corollary 6.2.1. *Let* $\{\xi_{\mathbf{t}}, \mathbf{t} \in \mathbb{Z}^d\}$ *be a strictly stationary real random field with* $E\xi(\mathbf{0}) = 0$. *Assume that*

(i) $E|\xi_{\mathbf{0}}|^s < \infty$ *for some* $s > 2$;
(ii) $\mathcal{A}$ *has exponent of metric entropy (with inclusion)* $r < 1$;
(iii) $\beta(x) = O(x^{-b})$ $(x \to \infty)$ *for some* $b > \max(ds/(s-2), d(1+r)/(1-r))$;
(iv) $\sum_{j=1}^{\infty} \rho^{1/2}(2^j) < \infty$;
(v) $\sum_{\mathbf{i}\in\mathbb{Z}^d} E\xi_{\mathbf{0}}\xi_{\mathbf{i}} = 1$.
Then Z_n *converges weakly in* $CA(\overline{\mathcal{A}})$ *to* W.

Dobrushin (1968) showed that the φ-mixing condition is not satisfied even for some simple examples of Gibbs random fields. Dobrushin and Nahapetian (1974) introduced the nonuniform φ-mixing condition.

Definition 6.2.7. The random field $\{\xi_{\mathbf{t}}, \mathbf{t} \in \mathbb{Z}^d\}$ is said to be *nonuniform φ-mixing*, if for $\Lambda_i \subset \mathbb{Z}^d$, $|\Lambda_i| < \infty, i = 1, 2$, there exists a nonnegative function $\varphi_{|\Lambda_1|}(\cdot)$ depending only on $|\Lambda_1|$, such that

$$\sup_{E\in\sigma(\Lambda_1), F\in\sigma(\Lambda_2), P(F)>0} |P(E|F) - P(E)| \le \varphi_{|\Lambda_1|}(d(\Lambda_1, \Lambda_2))$$

and $\varphi_{|\Lambda_1|}(x) \to 0$ as $x \to \infty$, where $|\Lambda|$ is the cardinality of Λ.

Chen (1991) gave a sufficient condition under which a sequence of partial-sum set-indexed processes with the nonuniform φ-mixing condition converges to a Brownian motion when the indexed set $\mathcal{A} = \mathcal{I}^d = \{(\mathbf{a}, \mathbf{b}], \mathbf{a}, \mathbf{b} \in [0,1]^d\}$.

Theorem 6.2.2. *Let* $\{\xi_{\mathbf{t}}, \mathbf{t} \in \mathbb{Z}^d\}$ *be a strictly stationary nonuniform φ-mixing random field and satisfy*

(i) there exists a non-negative function $\varphi(\cdot)$ *on* R^1, *such that for any* $\Lambda \subset \mathbb{Z}^d$, $|\Lambda| < \infty$, $\varphi_{|\Lambda|}(\cdot) \le |\Lambda|\varphi(\cdot)$, *and for some* $\delta > 0$

$$\limsup_{r\to\infty} (\varphi(r))^{1/2} r^{3d+4d/\delta} < \infty, \tag{6.2.4}$$

(ii) $E\xi_{\mathbf{0}} = 0$, $E|\xi_{\mathbf{0}}|^{2+\delta} < \infty$,
(iii)

$$0 < \sigma^2 := \sum_{\mathbf{t}\in\mathbb{Z}^d} \mathrm{Cov}(\xi_{\mathbf{0}}, \xi_{\mathbf{t}}) < \infty. \tag{6.2.5}$$

Then Z_n/σ *converges weakly in* $CA(\overline{\mathcal{A}})$ *to a Brownian motion as* $n \to \infty$.

By a direct calculation, Lu (1995) proved that the conclusion of Theorem 6.2.2 holds for a more general indexed set $\mathcal{A}$ (with the metric entropy exponent $r, 0 < r < 1$) and the condition for the rate of nonuniform φ-mixing was weakened.

Theorem 6.2.3. *Let* $\{\xi_{\mathbf{t}}, \mathbf{t} \in \mathbb{Z}^d\}$ *be a strictly stationary nonuniform* φ*-mixing random field, and satisfy*

(i) $\varphi_{|\Lambda|}(\cdot) \le |\Lambda|\varphi(\cdot)$ *and* $\varphi(x) = O(x^{-2d-1-2d/\delta})$ *for some* $\delta > 0$,

(ii) $E\xi_{\mathbf{0}} = 0, E|\xi_{\mathbf{0}}|^{2+\delta} < \infty$,

(iii) $\mathcal{A}$ *has a exponent of metric entropy (with inclusion)* $r < 1$. *Then* Z_n/σ *converges weakly in* $CA(\overline{\mathcal{A}})$ *to a Brownian motion, as* $n \to \infty$, *where* σ^2 *is defined as in (6.2.5).*

Remark 6.2.1. From Theorem 6.2.2 a uniform central limit theorem for certain Gibbs fields has been given in Chen (1991) which is also true for indexed set $\mathcal{A}$, if $\mathcal{A}$ has the exponent of metric entropy $r < 1$.

The proofs of Theorems 6.2.1 and 6.2.2 are omitted here.

The proof of Theorem 6.2.3 will need the following lemmas. Some lemmas for moment estimations are of independent interest.

A *slice* in R^d is a set

$$S(\mathbf{c}, a, \eta) = \{\mathbf{x} \in R^d : a \le \mathbf{c}'\mathbf{x} \le a + \eta\},$$

where $\mathbf{x} \in R^d$, $|\mathbf{c}| = 1$, $a \in R$, and $\eta > 0$. The *thickness* is η, the *direction* (of the normal to the two bounding hyperplanes) $\mathbf{c}$, and the *displacement* a. The slice splits a set $A \subseteq R^d$ into three parts, namely $A \cap S(\mathbf{c}, a, \eta)$ and two sets

$$A_+ = A \cap \{\mathbf{x} \in R^d : \mathbf{c}'\mathbf{x} > a + \eta\}, \quad A_- = A \cap \{\mathbf{x} \in R^d : \mathbf{c}'\mathbf{x} < a\}.$$

If A is measurable and $|A_+| = |A_-|$ we say the slice bisects A.

Lemma 6.2.1. *There exist* C_0, q, *depending only on* d, *such that for all* p *satisfying* $0 < p < 1/d$, *and for every measurable* $A \subset R^d$ *of finite measure, we can find a slice* S *that bisects* A, *has thickness* $(|A|/2)^p$, *and is such that*

$$|A \cap S| \le C_0(|A|/2)^{(q+pd)/(q+1)}. \tag{6.2.6}$$

The proof of Lemma 6.2.1 was given in Goldie and Greenwood (1986b).

Let $\{X_{\mathbf{i}}, \mathbf{i} \in \mathbb{Z}^d\}$ be a ρ-mixing random field with $EX_{\mathbf{i}} = 0$. Denote

$$\rho = \sum_{i=0}^{\infty} \rho(2^i), \qquad \sigma = \sup_{\mathbf{i}} \|X_{\mathbf{i}}\|_2,$$

$$Z(A) = \sum_{\mathbf{i} \in \mathbb{Z}^d} |A \cap C_{\mathbf{i}}| X_{\mathbf{i}}, \qquad A \in \mathcal{B}^d.$$

Lemma 6.2.2. *There exist constants a, b, depending only on d, such that*

$$\|Z(A)\|_2 \le a\, e^{b\rho}\sigma |A|^{1/2}, \qquad A \in \mathcal{B}^d. \tag{6.2.7}$$

Proof. Without loss of generality we assume $\sigma = 1$. Set

$$\sigma(h) = \sup_{A\in\mathcal{B}^d, |A|=h} \|Z(A)\|_2, \qquad \overline{\sigma}(h) = \sup_{h'\le h} \sigma(h').$$

Observe

$$\|Z(A)\|_2 \le \sum_{\mathbf{j}\in\mathbf{Z}^d} |A\cap C_{\mathbf{j}}|\|X_{\mathbf{j}}\|_2 \le \sum_{\mathbf{j}\in\mathbf{Z}^d} |A\cap C_{\mathbf{j}}| = |A|,$$

so

$$\sigma(h) \le h. \tag{6.2.8}$$

Take p in Lemma 6.2.1 to be such that the exponent $r = (q+pd)/(q+1)$ does not have any positive integer power equal to $1/2$. This is to avoid a minor technically later. Pick A of positive finite measure $2m$, say. The slice S in Lemma 6.2.1 is $S(\mathbf{c}, a, m^p)$ for some $\mathbf{c}, a$. We refer to $\{\mathbf{x} \in R^d : \mathbf{c}'\mathbf{x} > a + m^p\}$ and $\{\mathbf{x} \in R^d : \mathbf{c}'\mathbf{x} < a\}$ as the "sides" of S containing A_+ and A_-, respectively. Since $|A_+| = |A_-|$ we know

$$m \ge |A_+| = |A_-| \ge m - |A\cap S| \ge m - C_0 m^r.$$

We may find A'_+, such that A'_+ is in the side of S containing A_+, is disjoint from A_-, and has measure $|A'_+| = m - |A_+|$. Thus $|A'_+| \le C_0 m^r$. Let $A''_+ = A_+ \cup A'_+$; then $|A''_+| = m$. Similarly we construct A'_- and A''_- on the side of S containing A_-. Now if $\mathbf{x} \in A''_+$, $\mathbf{y} \in A''_-$ then

$$\|\mathbf{x}-\mathbf{y}\| \ge d^{-1/2}|\mathbf{x}-\mathbf{y}| \ge d^{-1/2}m^p.$$

Hence if $C_{\mathbf{i}}$ and $C_{\mathbf{j}}$ intersect A''_+ and A''_-, respectively, then

$$\|\mathbf{i}-\mathbf{j}\| \ge d^{-1/2}m^p - 2,$$

whence

$$\begin{aligned} &E(Z(A''_+) + Z(A''_-))^2 \\ &\quad\le (1+\rho(d^{-1/2}m^p - 2))(EZ^2(A''_+) + EZ^2(A''_-)) \\ &\quad\le (1+\rho(d^{-1/2}m^p - 2))2\sigma(m)^2. \end{aligned}$$

Since

$$Z(A) = Z(A''_+) + Z(A''_-) - Z(A'_+) - Z(A'_-) + Z(A\cap S) \tag{6.2.9}$$

the triangle inequality gives

$$\sigma(2m) \leq 2^{1/2}(1+\rho(d^{-1/2}m^p - 2))^{1/2}\sigma(m) + 3\overline{\sigma}(C_0 m^r).$$

Choose $h > 1$, then $h = 2^k m$ where $k \in \mathbb{N}$ and $1/2 < m \leq 1$, and we have

$$\sigma(m) \leq 1, \qquad \sigma(2^{j+1}m) \leq \alpha_j \sigma(2^j m) + \beta_j,$$

where

$$\alpha_j := 2^{1/2}(1+\rho(d^{-1/2}2^{jp}m^p - 2))^{1/2}, \qquad \beta_j := 3\overline{\sigma}(C_0 2^{jr}).$$

Iterating,

$$\sigma(h) \leq \prod_{j=0}^{k-1} \alpha_j + \sum_{j=0}^{k-1} \beta_j \prod_{i=j+1}^{k-1} \alpha_i. \tag{6.2.10}$$

Let $l \in \mathbb{N}$ satisfy $p > 1/l$. Now $d^{1/2}2^{jp}2^{-p} - 2 > 2^{j/l}$ for $j \geq j_0 \geq 1$, so

$$\begin{aligned}
\sum_{j=j_0}^{\infty} \rho(d^{-1/2}2^{jp}m^p - 2) &\leq \sum_{j=1}^{\infty} \rho(2^{j/l}) \\
&= \sum_{j=0}^{\infty}\sum_{i=1}^{l} \rho(2^j 2^{i/l}) \\
&\leq \sum_{j=0}^{\infty}\sum_{i=1}^{l} \rho(2^j) = l\rho.
\end{aligned}$$

Thus

$$\prod_{j=0}^{\infty} (1+\rho(d^{-1/2}2^{jp}m^p - 2))^{1/2} \leq 2^{j_0/2}e^{l\rho/2} =: a_1 e^{b_1 \rho}, \tag{6.2.11}$$

and (6.2.10) yields

$$\sigma(h) \leq a_1 e^{b_1\rho}\Big(2^{k/2} + 3\sum_{j=0}^{k-1} 2^{(k-1-j)/2}\overline{\sigma}(C_0 2^{jr})\Big). \tag{6.2.12}$$

Let ν be the integer such that $r^\nu < 1/2 < r^{\nu-1}$ (recall that $r^\nu = 1/2$ was excluded by choice of r). We shall use (6.2.12) iteratively, ν times, to obtain the result of the lemma,

$$\sigma(h) \leq a e^{b\rho} h^{1/2}. \tag{6.2.13}$$

Applying (6.2.8) to the right-hand side of (6.2.12), if $\nu = 1$ we find

$$\sigma(h) \le a_1 e^{b_1\rho} 2^{k/2}\Big\{1 + 3C_0(1 - 2^{-(1/2-r)})^{-1}\Big\},$$

and since $2^{k/2} \le 2^{1/2}h^{1/2}$ the proof is concluded. In the other case, i.e. when $r > 1/2$, we obtain

$$\sigma(h) \le a_1 e^{b_1\rho} 2^{k/2}\Big\{1 + \frac{3}{2}C_0(2^{r-1/2} - 1)^{-1}2^{kr}\Big\},$$

whence $\overline{\sigma}(h) \le a_1' e^{b_1\rho} h^r$ for some a_1'. Substituting this in (6.2.12), if $r^2 < 1/2$ we obtain (6.2.13) and otherwise $\sigma(h) \le a_2 e^{b_2\rho} h^{r^2}$. After a total of ν uses of (6.2.12) we obtain (6.2.13).

By the same way, we can prove the following lemma.

Lemma 6.2.3. *Suppose that* $\tau := \sup_{\mathbf{i}} \|X_{\mathbf{i}}\|_s < \infty$ *for* $s = 2 + \delta, 0 \le \delta \le 1$ *and*

$$\rho" = \sum_{i=1}^{\infty} \rho^{2/s}(2^i) < \infty.$$

Then for any $A \in \mathcal{B}^d$ *we have*

$$\|Z(A)\|_s \le ce^{c\rho''}\tau |A|^{1/2}. \tag{6.2.14}$$

Proof. First it follows from Lemma 6.2.2 that Lemma 6.2.3 holds true for $\delta = 0$. For the case of $0 < \delta \le 1$, let

$$\tau(m) = \sup_{A\in\mathcal{B}^d, |A|=m} \|Z(A)\|_s, \quad \overline{\tau}(m) = \sup_{m'\le m} \tau(m').$$

From (6.2.9), we obtain

$$\tau(2m) \le \|Z(A_+'') + Z(A_-'')\|_{2+\delta} + 3\overline{\tau}(C_0 m^r). \tag{6.2.15}$$

By using

$$|1 + x|^{2+\delta} \le 1 + 9|x| + 9|x|^{1+\delta} + |x|^{2+\delta}, \tag{6.2.16}$$

we have

$$\begin{aligned} &E|Z(A_+'') + Z(A_-'')|^{2+\delta} \\ &\quad \le 2\tau^{2+\delta}(m) + 9(E|Z(A_+'')||Z(A_-'')|^{1+\delta} \\ &\qquad + E|Z(A_+'')|^{1+\delta}|Z(A_-'')|). \end{aligned} \tag{6.2.17}$$

By Lemma 1.2.7, we have

$$\begin{aligned}&E|Z(A''_+)||Z(A''_-)|^{1+\delta}\\&\quad\le E|Z(A''_+)|E|Z(A''_-)|^{1+\delta}+4\rho^{\frac{2}{2+\delta}}(d^{-\frac{1}{2}}m^p-2)\tau^{2+\delta}(m)\\&\quad\le\left(1+4\rho^{\frac{2}{2+\delta}}(d^{-\frac{1}{2}}m^p-2)\right)\tau^{2+\delta}(m).\end{aligned}\tag{6.2.18}$$

Similarly we have

$$\begin{aligned}&E|Z(A''_+)|^{1+\delta}|Z(A''_-)|\\&\quad\le\left(1+4\rho^{\frac{2}{2+\delta}}(d^{-\frac{1}{2}}m^p-2)\right)\tau^{2+\delta}(m).\end{aligned}\tag{6.2.19}$$

Inserting (6.2.18), (6.2.19) into (6.2.17) yields

$$\|Z(A''_+)+Z(A''_-)\|_{2+\delta}\le 2 0^{\frac{1}{2+\delta}}\left(1+2\rho^{\frac{2}{2+\delta}}(d^{-\frac{1}{2}}m^p-2)\right)^{\frac{1}{2+\delta}}\tau(m).$$

Write $\rho_0=\rho^{2/(2+\delta)}(d^{-1/2}m^p-2)$, whence

$$\tau(2m)\le 20^{\frac{1}{2+\delta}}\left(1+2\rho_0\right)^{\frac{1}{2+\delta}}\tau(m)+3\overline{\tau}(C_0m^r).$$

Iterating, for $h=2^km, k\in\mathbb{N}, 1/2<m\le 1$ we have

$$\tau(h)\le\prod_{j=0}^{k-1}\alpha_j+\sum_{j=0}^{k-1}\beta_j\prod_{i=j+1}^{k-1}\alpha_i,$$

where

$$\alpha_j\le 20^{\frac{1}{2+\delta}}\left\{1+2\rho^{\frac{2}{2+\delta}}(d^{-\frac{1}{2}}2^{jp}m^p-2)\right\}^{\frac{1}{2+\delta}},\quad \beta_j=3\overline{\tau}(C_02^{jr}m^r).$$

The conditon $\rho''<\infty$ implies that

$$\sum_{j=j_0}^{\infty}\rho^{\frac{2}{2+\delta}}\left(d^{-\frac{1}{2}}2^{jp}m^p-2\right)<\infty,$$

where j_0 is defined as in Lemma 6.2.2. The remainder of the proof is the same as in the case of $\delta=0$. The proof of Lemma 6.2.3 is completed.

Denote

$$\rho'=\sum_{i=0}^{\infty}\rho^{1/2}(2^i),$$
$$g(y)=\sup_{\mathbf{i}}EX_{\mathbf{i}}^2I(|X_{\mathbf{i}}|>y).$$

Lemma 6.2.4. *Let $\{X_{\mathbf{i}},\ \mathbf{i} \in \mathbb{Z}^d\}$ be a ρ-mixing random field. If $\{X_{\mathbf{i}}^2,\ \mathbf{i} \in \mathbb{Z}^d\}$ is uniformly integrable and $\rho' < \infty$, then the set of random variables $\{Z^2(A)/|A|, A \in \mathcal{B}^d, |A| < \infty\}$ is uniformly integrable, i.e. there exists a constant C depending only on d, such that*

$$E\Big(\frac{Z^2(A)}{|A|} I\Big(\frac{Z^2(A)}{|A|} > y\Big)\Big) < \Big\{C(1 \wedge y^{-1}) + Cg(y^{1/4})\Big\} e^{C\rho'}. \quad (6.2.20)$$

Lemma 6.2.5. *Let $\{W(A), A \in \mathcal{B}^d\}$ be an additive processs that satisfying*

(i) $EW(C) = 0$ for any $C \in \mathcal{I}^d$,

(ii) $EW^2(C) = |C|$ for any $C \in \mathcal{I}^d$,

(iii) $W(C_1), \cdots, W(C_k)$ are independent whenever $C_1, \cdots, C_k \in \mathcal{I}^d$ and $d(C_i, C_j) > 0$ for $i \neq j$,

(iv) $\lim_{m\to\infty} \sum_{\mathbf{j}\in J_m} P\{|W(C_{m\mathbf{j}})| \geq \varepsilon\} = 0$ for any $\varepsilon > 0$.

Then W is a standard Wiener process on $\mathcal{R} = \mathcal{R}(\mathcal{I}^d)$ (the ring of all finite unions of elements of $\mathcal{I}^d$).

Lemma 6.2.6. *Let $\{Z_n(A),\ A \in \mathcal{R} = \mathcal{R}(\mathcal{I}^d),\ n \geq 1\}$ be a sequence of additive processes such that*

(i) $EZ_n(C) \to 0$ $(n \to \infty)$ for any $C \in \mathcal{I}^d$,

(ii) $EZ_n^2(C) \to |C|$ $(n \to \infty)$ for any $C \in \mathcal{I}^d$,

(iii) whenever $C_1, \cdots, C_k \in \mathcal{I}^d$ are such that $\rho(C_i, C_j) > 0$ for $i \neq j$ we have for all real $z_1, \cdots, z_k$ that

$$P\{\cap_{i=1}^k (Z_n(C_i) \leq z_i)\} - \prod_{i=1}^k P\{Z_n(C_i) \leq z_i\} \to 0, \quad \text{as } n \to \infty, \quad (6.2.21)$$

(iv) $\lim_{m\to\infty} \limsup_{n\to\infty} \sum_{\mathbf{j}\in J_m} P\{|Z_n(C_{m\mathbf{j}})| \geq \varepsilon\} = 0$ for any $\varepsilon > 0$,

(v) for each $C \in \mathcal{I}^d$, the set $\{Z_n^2(C), n \geq 1\}$ is uniformly integrable.

Then the finite dimensional distributions of Z_n converge weakly to the corresponding finite dimensional distributions of a Wiener process W on $\mathcal{R}$.

Lemma 6.2.7. *Let $\{Z_n\}$ be a sequence of additive processes on $\mathcal{B}^d$ such that the set $\{Z_n^2(A)/|A|, A \in \mathcal{B}^d, n \geq 1\}$ is uniformly integrable, and satisfying Lemma 6.2.6 (i), (ii), (iii). Then the finite dimensional distributions of Z_n converge weakly to the corresponding finite dimensional distributions of a Wiener process W on $\mathcal{B}^d$.*

The proofs of Lemmas 6.2.4–6.2.7 are given in Goldie and Greenwood (1986 a, b).

Theorem 6.2.4. *Let $\{\xi_{n,\mathbf{j}}\ \mathbf{j} \in \mathcal{J}_n\}$, be a ρ-mixing triangular array. The smoothed partial-sum processes Z_n are defined by (6.2.1). Suppose that*

(i) $E\xi_{n,\mathbf{j}} = 0$ *for any* $n \geq 1, \mathbf{j} \in \mathcal{J}_n$*;*

(ii) the set $\{n^d \xi^2_{n,\mathbf{j}}, n \geq 1, \mathbf{j} \in \mathcal{J}_n\}$ *is uniformly integrable;*

(iii) $\sup_n \sum_{j=1}^{\infty} \rho_n^{1/2}(n^{-1}2^j) < \infty$, *and Lemma 6.2.6 (ii).*

Then the finite dimensional distributions of Z_n on $\mathcal{B}^d$ converges weakly to the corresponding finite dimensional distributions of a Wiener process.

Proof. By Lemma 6.2.4 we have uniform integrability of the set $\{Z_n^2(A)/|A|, n \geq 1, A \in \mathcal{B}^d\}$. Because $\rho_n(x)$ is non-increasing in x, condition (iii) implies $\rho_n(x) \to 0 (n \to \infty)$ for each fixed x. Thus $\alpha_n(x) \to 0 (n \to \infty)$. Clearly the left hand side of (6.2.21) does not exceed $(k-1)\alpha_n(\theta)$, where $\theta > 0$ is the least separation distance between the sets $C_1, \cdots, C_k$. Hence Lemma 6.2.6 (iii) is satisfied. By Lemma 6.2.7, the proof of Theorem 6.2.4 is completed.

6.3 Tightness

Fisrt we present a lemma for nonuniform φ-mixing sequence, which is similar to Lemma 6.2.3.

Lemma 6.3.1. *Let $\{\xi_{\mathbf{t}},\ \mathbf{t} \in \mathbb{Z}^d\}, \{Z_n(A), A \in \mathcal{B}^d\}$ be as in Theorem 6.2.3 with $\varphi(x) = O(x^{-(d+1+\theta)})$, for some $\theta > 0$, instead of $\varphi(x) = O(x^{-(2d+1+2d/\delta)})$. Then for any $A \in \mathcal{B}^d$*

$$\|Z_n(A)\|_{2+\delta} \leq c\sigma_0 |nA|^{1/2} \tag{6.3.1}$$

where $\sigma_0^2 = E\xi_{\mathbf{0}}^2$.

Proof. First, we prove that for $\delta = 0$

$$\sigma(2m) \leq 2^{1/2}(1 + 2m^{1/2}\varphi^{1/2}(d^{-1/2}m^p - 2))^{1/2}\sigma(m) + 3\overline{\sigma}(cm^r),$$

where $\sigma(m) = \sup_{A \in \mathcal{B}^d, |A|=m} \|Z_n(A)\|_2$, $\overline{\sigma}(m) = \sup_{m' \leq m} \sigma(m')$. Take $0 < p < 1/d$ in the bisection lemma to be such that the exponent $r = (q+pd)/(q+1)$ does not have any positive power equal to $1/2$. For any given $h > 1$, we write $h = 2^k m, k \in \mathbb{N}, 1/2 < m \leq 1$. Note that $\sigma(m) \leq 1$, and for nonuniform φ-mixing, by the same discussion as in the proof of Lemma 6.2.3 we have

$$\sigma(2^{j+1}m) \le \alpha_j \sigma(2^j m) + \beta_j,$$

where

$$\alpha_j = 2^{1/2}(1 + 2 \cdot 2^{j/2} m^{j/2} \varphi(d^{-1/2} 2^{jp} m^p - 2))^{1/2},$$
$$\beta_j = 3\,\overline{\sigma}(c2^{js}).$$

Iterating,

$$\sigma(h) \le \prod_{j=0}^{k-1} \alpha_j + \sum_{j=0}^{k-1} \beta_j \prod_{i=j+1}^{k-1} \alpha_i. \tag{6.3.2}$$

Furthermore, we take $p, \frac{1}{d} > p > \frac{1}{d+1}$, such that $r = \frac{q+pd}{q+1}$ does not also have any positive power equal to $1/2$. There exists a $j_0 \ge 1$ such that $d^{1/2} 2^{jp} 2^{-p} > 2^{j/(d+1)}$ for $j \ge j_0$, so

$$\begin{aligned} c_\varphi &:= \sum_{j=j_0}^{\infty} 2^{j/2} \varphi^{1/2}(d^{-1/2} 2^{jp} m^p - 2) \\ &\le \sum_{j=1}^{\infty} 2^{j/2} \varphi^{1/2}(2^{j/(d+1)}) \\ &\le c \sum_{j=1}^{\infty} 2^{\frac{j}{2}} 2^{-\frac{1}{2} \cdot \frac{j}{d+1}(d+1+\theta)} \\ &= c \sum_{j=1}^{\infty} \left(2^{\theta/(2(d+1))}\right)^{-j} < \infty. \end{aligned}$$

Thus we obtain

$$\sigma(h) \le c e^{c\varphi} \left(2^{\frac{k}{2}} + 3 \sum_{j=0}^{k-1} 2^{\frac{k-j-1}{2}} \overline{\sigma}(c2^{jr})\right). \tag{6.3.3}$$

The remainder of the proof is the same as in the proof of Lemma 6.2.2. This proves that (6.3.1) holds for $\delta = 0$.

Consider the case of $0 < \delta \le 1$. Recall the notation of Lemma 6.2.2 and write

$$Z_n(A) = Z_n(A''_+) + Z_n(A''_-) - Z_n(A'_+) - Z_n(A'_-) + Z_n(A \cap S), \tag{6.3.4}$$

where $|A| = 2m, S$ is a slice of A, $|A''_+| = |A''_-| = m, A''_+$ and A''_- is situated on the different side of S, the separation distance $d(A''_+, A''_-) \ge$

$d^{-1/2}m^p, |A'_+|, |A'_-|$, and $A \cap S$ all do not exceed cm^r. Denote $\tau(h) = \sup_{|A|=h, A\in\mathcal{B}^d} \|Z_n(A)\|_{2+\delta}$, $\overline{\tau}(h) = \sup_{h'\le h} \tau(h')$. From (6.3.4),

$$\tau(2m) \le \|Z_n(A''_+) + Z_n(A''_-)\|_{2+\delta} + 3\overline{\tau}(cm^r). \tag{6.3.5}$$

By (6.2.16) we have

$$\begin{aligned} &E|Z_n(A''_+) + Z_n(A''_-)|^{2+\delta} \\ &\quad \le 2\tau^{2+\delta}(m) + 9(E|Z_n(A''_+)||Z_n(A''_-)|^{1+\delta} \\ &\qquad + E|Z_n(A''_+)|^{1+\delta}|Z_n(A''_-)|). \end{aligned} \tag{6.3.6}$$

By the property of nonuniform φ-mixing, we have

$$\begin{aligned} &E|Z_n(A''_+)||Z_n(A''_-)|^{1+\delta} \\ &\quad \le E|Z_n(A''_+)|E|Z_n(A''_-)|^{1+\delta} + 2\varphi^{\frac{1}{2+\delta}}_{|A''_+|}(d^{-\frac{1}{2}}m^p - 2)\tau^{2+\delta}(m) \\ &\quad \le \Big(1 + 2m^{\frac{1}{2+\delta}}\varphi^{\frac{1}{2+\delta}}(d^{-\frac{1}{2}}m^p - 2)\Big)\tau^{2+\delta}(m). \end{aligned} \tag{6.3.7}$$

Similarly we have

$$\begin{aligned} &E|Z_n(A''_+)|^{1+\delta}|Z_n(A''_-)| \\ &\quad \le \Big(1 + 2m^{\frac{1}{2+\delta}}\varphi^{\frac{1}{2+\delta}}(d^{-\frac{1}{2}}m^p - 2)\Big)\tau^{2+\delta}(m). \end{aligned} \tag{6.3.8}$$

Inserting (6.3.7), (6.3.8) into (6.3.6) yields

$$\|Z_n(A''_+) + Z_n(A''_-)\|_{2+\delta} \le c\Big(1 + m^{\frac{1}{2+\delta}}\varphi^{\frac{1}{2+\delta}}(d^{-\frac{1}{2}}m^p - 2)\Big)^{\frac{1}{2+\delta}}\tau(m).$$

Whence

$$\tau(2m) \le c\Big(1 + m^{\frac{1}{2+\delta}}\varphi^{\frac{1}{2+\delta}}(d^{-\frac{1}{2}}m^p - 2)\Big)^{\frac{1}{2+\delta}}\tau(m) + 3c\overline{\tau}(cm^r).$$

Iterating, for $h = 2^k m, k \in \mathbb{N}, 1/2 < m \le 1$ we have

$$\tau(h) \le \prod_{j=0}^{k-1}\alpha_j + \sum_{j=0}^{k-1}\beta_j \prod_{i=j+1}^{k-1}\alpha_i,$$

where

$$\alpha_j \le c\Big\{1 + 2^{\frac{j}{2+\delta}}m^{\frac{j}{2+\delta}}\varphi^{\frac{1}{2+\delta}}(d^{-\frac{1}{2}}2^{jp}m^p - 2)\Big\}^{\frac{1}{2+\delta}}.$$

Note

$$\begin{aligned} &\sum_{j=j_0}^{\infty} 2^{\frac{j}{2+\delta}}\varphi^{\frac{1}{2+\delta}}\Big(d^{-\frac{1}{2}}2^{jp}m^p - 2\Big) \\ &\quad \le c\sum_{j=1}^{\infty}\Big(2^{\varepsilon/(2(2+\delta)(d+1))}\Big)^{-j} < \infty. \end{aligned}$$

The remainder of the proof is the same as in the case of $\delta = 0$. The proof of Lemma 6.3.1 is completed.

Proof of tightness in Theorem 6.2.3. For $f \in C(\overline{\mathcal{A}})$ write the modulus of continuity

$$\omega_\delta(f) = \sup_{A,B\in\mathcal{A},\,|A\triangle B|\le\delta} |f(A) - f(B)|. \tag{6.3.9}$$

Then, since $\overline{\mathcal{A}}$ is compact, we can use a version of the Arzela-Ascoli theorem: a subset U of $C(\overline{\mathcal{A}})$ has a compact closure iff it is equibounded ($\sup_{f\in U}\sup_{A\in\overline{\mathcal{A}}}|f(A)| < \infty$) and equicontinuous ($\lim_{\delta\downarrow 0}\sup_{f\in U}\omega_\delta(f) = 0$). Using this, from Theorem 8.2 of Billingsley (1968) it follows that a sequence $\{Z_n\}$ of random elements of $C(\overline{\mathcal{A}})$ is relatively compact, i.e. every subsequence of $\{Z_n\}$ contains a weakly convergent subsequence iff

(a) for each element A of some countable dence set in $\mathcal{A}$, the family $\{Z_n(A), n \ge 1\}$ is tight, and

(b) for every $\lambda > 0$, $\lim_{\delta\downarrow 0}\limsup_{n\to\infty} P\{\omega(Z_n) > \lambda\} = 0$.

(a) follows from Theorem 6.2.4. We need only prove (b). Without loss of generality, we assume that $\sigma^2 = 1, E\xi_{\mathbf{t}} = 0$. For $0 \le u \le v < \infty$, define

$$\eta_{n\mathbf{j}}(u,v) = n^{-(d/2)}\xi_{n\mathbf{j}} I(u \le n^{d\delta/(2(1+\delta))} n^{-d/2}|\xi_{n\mathbf{j}}| < v) \qquad \mathbf{j} \in \mathcal{J}_n,$$
$$Z_n(A,u,v) = \sum_{\mathbf{j}\in\mathcal{J}_n} \frac{|A\cap C_{n\mathbf{j}}|}{|C_{n\mathbf{j}}|}(\eta_{n\mathbf{j}}(u,v) - E\eta_{n\mathbf{j}}(u,v)).$$

Lemma 6.3.2. *Suppose $\{\xi_{\mathbf{t}},\ \mathbf{t} \in \mathbb{Z}^d\}$ satisfies the conditions of Theorem 6.2.3. Then as $n \to \infty$, $U_n(I^d, a, \infty) \to 0$, a.s., $EU_n(I^d, a, \infty) \to 0$, where $I^d = [0,1]^d, a > 0$ and*

$$U_n(A,a,\infty) = \sum_{\mathbf{j}\in\mathcal{J}_n} \frac{|A\cap C_{n\mathbf{j}}|}{|C_{n\mathbf{j}}|}|\eta_{n\mathbf{j}}(a,\infty)|.$$

Proof. We use Bass's technique (1985). For $\mathbf{k} = (k_1,\cdots,k_d) \in \mathbb{Z}_+^d$, let $\mu(\mathbf{k}) = \max\{k_1,\cdots,k_d\}$ and $\Psi(i,a) = \sup\{k \in \mathbb{Z}_+ : ak^{d/(2(1+\delta))} < i+1\}$. It is well known that $\mathrm{Card}\{\mathbf{k} \in \mathbb{Z}_+^d : \mu(\mathbf{k}) = r\} \le Cr^{d-1}$, where C

is a positive constant depending on d.

$$\begin{aligned}
&\sum_{\mathbf{k}\in\mathbb{Z}_+^d} I\Big(i+1 > a\mu(\mathbf{k})^{d/(2(1+\delta))}\Big)\mu(\mathbf{k})^{-d/2} \\
&\quad= \sum_{r=1}^{\infty}\sum_{\mathbf{j}:\mu(\mathbf{j})=r} I\Big(i+1 > ar^{d/(2(1+\delta))}\Big)r^{-d/2} \\
&\quad= c\sum_{r=1}^{\infty} I\Big(i+1 > ar^{d/(2(1+\delta))}\Big)r^{(d-2)/2} \\
&\quad\le c(\Psi(i,a))^{d/2} \le ca^{-1}(i+1)^{1+\delta}.
\end{aligned}$$

Thus

$$\begin{aligned}
&\sum_{\mathbf{k}\in\mathbb{Z}_+^d} E|\xi_{\mathbf{k}}|I\Big(a(\mu(\mathbf{k}))^{d/(2(1+\delta))} \le |\xi_{\mathbf{k}}| < \infty\Big)(\mu(\mathbf{k}))^{-d/2} \\
&\quad\le \sum_{\mathbf{k}}\ \sum_{i+1>a(\mu(\mathbf{k}))^{d/(2(1+\delta))}} (i+1)P\{i < |\xi_{\mathbf{k}}| \le i+1\}(\mu(\mathbf{k}))^{-d/2} \\
&\quad\le \sum_{i=0}^{\infty}\Big[\sum_{\mathbf{k}} I\Big(i+1 > a(\mu(\mathbf{k}))^{d/(2(1+\delta))}\Big)(\mu(\mathbf{k}))^{-d/2}\Big] \\
&\qquad\cdot (i+1)P\{i < |\xi_{\mathbf{0}}| \le i+1\} \\
&\quad\le ca^{-1}\sum_{i=0}^{\infty}(i+1)^{2+\delta}P\{i < |\xi_{\mathbf{0}}| \le i+1\} \\
&\quad\le ca^{-1}E(|\xi_{\mathbf{0}}|+1)^{2+\delta} < \infty.
\end{aligned}$$

Then for any $\varepsilon > 0$ and almost sure ω, there exists an $n_1(\omega)$, such that

$$\sum_{\mathbf{k}:\mu(\mathbf{k})>n_1} \Big|\xi_{\mathbf{k}} I\Big(a\mu(\mathbf{k})^{d/(2(1+\delta))} \le |\xi_{\mathbf{k}}| < \infty\Big)\Big|/(\mu(\mathbf{k}))^{d/2} < \varepsilon.$$

From this,

$$U_n(I^d, a, \infty) \le \sum_{\mu(\mathbf{k})\le n_1} |\xi_{\mathbf{k}}|I\Big(an^{d/(2(1+\delta))} \le |\xi_{\mathbf{k}}| < \infty\Big)n^{-d/2}.$$

Thus, as $n \to \infty$, $U_n(I^d, a, \infty) \to 0$, a.s. Analogously, we can obtain

$$EU_n(I^d, a, \infty) \to 0.$$

Lemma 6.3.2 is proved.

In order to prove (b) we need only prove

$$\lim_{\nu\downarrow 0}\limsup_{n\to\infty} P\{\|Z_n\|_{\mathcal{A}_\nu} > \lambda\} = 0, \tag{6.3.10}$$

where $\mathcal{A}_\nu = \{A\backslash B : A, B \in \mathcal{A},\ |A\backslash B| \le \nu\}$. Since $Z_n(A) = Z_n(A,0,a) + Z_n(A,a,\infty)$, and

$$|Z_n(A,a,\infty)| \le U_n(I^d,a,\infty) + EU_n(I^d,a,\infty),$$

by Lemma 6.3.2, in order to prove (6.3.10), we need only prove

$$\lim_{\nu\downarrow 0}\limsup_{n\to\infty} P\{\|Z_n(A,0,a)\|_{\mathcal{A}_\nu} > \lambda\} = 0. \qquad (6.3.11)$$

Let $p_n = [n^{(2+\delta)/(2(1+\delta))}]$ and $m_n = n/(2p_n)$. We divide I^d in the following two ways: $C_{p_n,\mathbf{l}}, \mathbf{l} \in J_{p_n}$ and $C_{2p_n,\mathbf{l}}, \mathbf{l} \in J_{2p_n}$. There are 2^d $C_{2p_n,\mathbf{l}}$ in each $C_{p_n,\mathbf{l}}$. Denote by $I_{n,\mathbf{l},i}$ the ith $C_{2p_n\mathbf{j}}$ in $C_{p_n,\mathbf{l}}$, $\mathbf{l} \in J_{p_n}, \mathbf{j} \in J_{2p_n}$. Let

$$I_{n,i} = \bigcup_{\mathbf{l}\in J_{p_n}} I_{n,\mathbf{l},i}, \qquad i = 1,\cdots,2^d.$$

Then

$$Z_n(\cdot,0,a) = \sum_{i=1}^{2^d} Z_n(\cdot\cap I_{n,i},0,a).$$

Now in order to prove (6.3.11), we need only prove

$$\lim_{\nu\downarrow 0}\limsup_{n\to\infty} P\Big\{\|Z_n(A\cap I_{n,i},0,a)\|_{\mathcal{A}_\nu} > \lambda\Big\} = 0. \qquad (6.3.12)$$

Write

$$\begin{aligned} Z_n(A\cap I_{n,i},0,a) &= \sum_{\mathbf{l}\in J_{p_n}}\sum_{\mathbf{j}\in S(n,\mathbf{l},i)} \frac{|A\cap I_{n,\mathbf{l},i}\cap C_{n,\mathbf{j}}|}{|C_{n,\mathbf{j}}|}(\eta_{n\mathbf{j}}(0,a) - E\eta_{n,\mathbf{j}}(0,a)) \\ &=: \sum_{\mathbf{l}\in J_{p_n}} V_{n\mathbf{l}}(A\cap I_{n,i},0,a), \\ \eta_{n,\mathbf{j}}(0,a) &= n^{-(d/2)}\xi_{n\mathbf{j}} I(n^{d\delta/(2(1+\delta))}n^{-d/2}|\xi_{n\mathbf{j}}| < a), \quad \mathbf{j}\in\mathcal{J}_n, \end{aligned}$$

where $S(n,\mathbf{l},i) = \{\mathbf{j}\in J_n : I_{n,\mathbf{l},i}\cap C_{n,\mathbf{j}} \ne \emptyset\}$. Denote $V_{n\mathbf{l}} = V_{n\mathbf{l}}(A\cap I_{n,i},0,a)$. By the property of nonuniform φ-mixing, we have

$$\begin{aligned} Ee^{\alpha\sum_{\mathbf{k}\in J_{p_n}} V_{n\mathbf{k}}} \le& Ee^{\alpha V_{n\mathbf{l}}} Ee^{\alpha\sum_{\mathbf{k}\ne\mathbf{l},\mathbf{k}\in J_{p_n}} V_{n\mathbf{k}}} \\ &+ 2\varphi_{|S(n,\mathbf{l},i)|}\Big(\frac{n}{2p_n}\Big) Ee^{\alpha\sum_{\mathbf{k}\ne\mathbf{l}} V_{n\mathbf{k}}}\|e^{V_{n\mathbf{l}}}\|_\infty. \end{aligned} \qquad (6.3.13)$$

Note that $|V_{n\mathbf{l}}| \le 2a$ and $e^{\alpha V_{n\mathbf{l}}} \le 1 + \alpha V_{n\mathbf{l}} + \alpha^2 V_{n\mathbf{l}}^2$ when $\alpha a \le 1/4$. From Lemma 6.3.1 it follows that

$$EV_{n\mathbf{l}}^2 \le C|A\cap I_{n,\mathbf{l},i}| \qquad \text{for} \quad i = 1,2,\cdots,2^d.$$

Therefore for $\alpha a \le 1/4$

$$Ee^{\alpha V_{n\mathbf{l}}} \le e^{E\alpha^2 V_{n\mathbf{l}}^2} \le e^{C\alpha^2 |A\cap I_{n,\mathbf{1},i}|}, \tag{6.3.14}$$

Inserting (6.3.14) into (6.3.13) yields

$$Ee^{\alpha \sum_{\mathbf{l}\in J_{p_n}} V_{n\mathbf{l}}} \le wEe^{\alpha \sum_{\mathbf{k}\ne\mathbf{l},\, \mathbf{k}\in J_{p_n}} V_{n\mathbf{k}}},$$

where

$$\begin{aligned} w &\le e^{C\alpha^2|A\cap I_{n,\mathbf{1},i}|}\Big(1+2e^{1/2}|S(n,\mathbf{1},i)|\varphi\Big(\frac{n}{2p_n}\Big)\Big) \\ &\le e^{C\alpha^2|A\cap I_{n,\mathbf{1},i}|}\Big(1+2e^{1/2}\frac{n^d}{(2p_n)^d}\varphi\Big(\frac{n}{2p_n}\Big)\Big). \end{aligned}$$

Iterating this procedure, we have

$$\begin{aligned} Ee^{\alpha \sum_{\mathbf{l}\in J_{p_n}} V_{n\mathbf{l}}} &\le e^{C\alpha^2|A\cap I_{n,i}|}\Big(1+2e^{1/2}\frac{n^d}{(2p_n)^d}\varphi\Big(\frac{n}{2p_n}\Big)\Big)^{p_n^d} \\ &\le \exp\Big(c\alpha^2|A| + cn^d\Big(\frac{n}{p_n}\Big)^{-2d-2d/\delta-1}\Big) \\ &= \exp\Big(c\alpha^2|A| + cn^{-\frac{\delta}{2(1+\delta)}}\Big) \\ &\le c\,\exp(\alpha^2|A|) \end{aligned} \tag{6.3.15}$$

for large n.

Now we return to estimate the left hand side of (6.3.12). Since $0 \le r < 1$, we can take $s > 0$ such that $r < 1/(1+s)$. Set

$$\begin{aligned} \delta_j &= \nu/2^j, \qquad j = 0, 1, \cdots, \\ \lambda_j &= \lambda_0 e^{-j(1+r-r(2+s))/(2+s)} \qquad j = 1, 2, \cdots, \\ \lambda_0 &= \lambda(1 - 2^{-(1+r-r(2+s))/(2+s)}), \\ a_j &= e^{-j(1+r)/(2+s)}a, \qquad j = 1, 2, \cdots, \\ a &= c\nu^{1/(1+s)}, \qquad c_0 = (6E|\xi_0|^{2+s}/\lambda_0)^{1/(1+s)}. \end{aligned}$$

For any $A \in \mathcal{A}_0$ there exist $A_j, A_j^+ \in \mathcal{A}_0(\delta_j)$ such that $A_j \subseteq A \subseteq A_j^+$ and $|A_j^+ \backslash A_j| \le \delta_j$. Then

$$\begin{aligned} &Z_n(A\cap I_{n,i}, 0, a) \\ &\quad = Z_n(A_0\cap I_{n,i}, 0, a) + \sum_{j=0}^{\infty}\{Z_n(A_{j+1}\cap I_{n,i}, 0, a_j) - Z_n(A_j\cap I_{n,i}, 0, a_j)\} \\ &\qquad + \sum_{j=0}^{\infty}\{Z_n(A\cap I_{n,i}, a_j, a_{j-1}) - Z_n(A_j\cap I_{n,i}, a_j, a_{j-1})\}. \end{aligned}$$

So if $\|Z_n(\cdot \cap I_{n,i}, 0, a)\|_{\mathcal{A}_0}$ is to exceed λ, at least one of the following must hold:

(a) for some $A_0 \in \mathcal{A}_0(\delta_0)$, $\quad |Z_n(A_0 \cap I_{n,i}, 0, a)| > \lambda_0$;

(b) for some j, for some

$$A_j \in \mathcal{A}_0(\delta_j), A_{j+1} \in \mathcal{A}_0(\delta_{j+1}), |A_j \triangle A_{j+1}| \leq 2\delta_j,$$

$$|Z_n(A_{j+1} \cap I_{n,i}, 0, a_j) - Z_n(A_j \cap I_{n,i}, 0, a_j)| > 2\lambda_j;$$

(c) for some j, for some

$$A_j, A_j^+ \in \mathcal{A}_0(\delta_j),\ A_j \subseteq A \subseteq A_j^+,\ |A_j^+ \backslash A_j| \leq \delta_j,$$

$$|Z_n(A \cap I_{n,i}, a_j, a_{j-1}) - Z_n(A_j \cap I_{n,i}, a_j, a_{j-1})| > \lambda_j.$$

The number of pairs A_j, A_j^+ in $\mathcal{A}_0(\delta_j)$ is $\leq \exp(4H(\delta_j/2))$, while the number of pairs $A_j \in \mathcal{A}_0(\delta_j)$, $A_{j+1} \in \mathcal{A}_0(\delta_{j+1})$ is $\leq \exp(4H(\delta_{j+1}/2))$.

We have

$$P\{\|Z_n(\cdot \cap I_{n,i}, 0, a)\|_{\mathcal{A}_0} > \lambda\} \leq p_0 + \sum_{j=0}^{\infty} r_j + \sum_{j=1}^{\infty} s_j,$$

where

$$\begin{aligned} p_0 &\leq 2\exp\{2H(\delta_0/2)\} \max_{|A_0| \leq 2\delta_0} P\{|Z_n(A_0 \cap I_{n,i}, 0, a)| > \lambda_0\}, \\ r_j &\leq 4\exp\{4H(\delta_{j+1}/2)\} \max_{|A_{j+1} \triangle A_j| \leq 2\delta_j} (P\{|Z_n((A_{j+1} \backslash A_j) \cap I_{n,i}, 0, a)| > \lambda_j\} \\ &\qquad + P\{|Z_n((A_j \backslash A_{j+1}) \cap I_{n,i}, 0, a)| > \lambda_j\}), \\ s_j &\leq \exp\{4H(\delta_j/2)\} \max_{A_j \subseteq A_j^+, |A_j^+ \backslash A_j| \leq 2\delta_j} P\{ \sup_{A_j \subseteq A \subseteq A_j^+} |Z_n(A \cap I_{n,i}, a_j, a_{j-1}) \\ &\qquad - Z_n(A_j \cap I_{n,i}, a_j, a_{j-1})| > \lambda_j\}. \end{aligned}$$

By (6.3.15), taking $\alpha = 1/4a_0$, we have

$$\begin{aligned} p_0 &\leq 2\exp\{2H(\delta_0/2)\} \exp\Big(-\frac{\lambda_0}{4a_0} + c_0 \frac{\delta_0}{a_0^2}\Big) \\ &\leq 2\exp\Big\{c2^{r+1}\delta_0^{-r} - \frac{\lambda_0}{4a_0} + c_0 \frac{\delta_0}{a_0}\Big\}. \end{aligned}$$

Similarly

$$\begin{aligned} r_j &\leq 4\exp\Big\{c2^{r+1}\delta_j^{-r} - \frac{\lambda_j}{4a_j} + c_0 \frac{\delta_j}{a_j}\Big\}, \\ s_j &\leq \exp\Big\{c2^{r+1}\delta_j^{-r} - \frac{\lambda_j}{4a_j} + c_0 \frac{\delta_j}{a_j}\Big\}. \end{aligned}$$

Thus

$$
\begin{aligned}
p_0 + \sum_{j=0}^{\infty} r_j + \sum_{j=1}^{\infty} s_j
&\leq 6 \sum_{j=0}^{\infty} \exp\Big\{ c\delta_j^{-r} - \frac{\lambda_j}{4a_j} + c_0 \frac{\delta_j}{a_j} \Big\} \\
&\leq 6 \sum_{j=0}^{\infty} \exp\Big\{ \Big(c\nu^{-r} - c'\nu^{-\frac{1}{1+s}} + c\nu^{-\frac{1-s}{1+s}} 2^{-\frac{jsr}{2+s}} \Big) 2^{jr} \Big\}.
\end{aligned}
$$

Because of $r < 1/(1+s)$, the coefficient of 2^{ir} may be negative by choosing ν small enough, and (6.3.12) follows. Theorem 6.2.3 is proved.

Chapter 7 The Berry-Esseen Inequality and the Rate of Weak Convergence

It is well-known that the uniform estimation of the difference between the distribution function $F_n(x)$ of the normalized sum of the first n terms of the sequence of independent random variables $\{X_n, n \geq 1\}$ and the normal distribution function $\Phi(x)$ is given by the Esseen and the Berry-Esseen inequalities. Furthermore, there exists a succinct non-uniformly estimation

$$|F_n(x) - \Phi(x)| \leq A\rho_0/(\sqrt{n}(1 + |x|^3))$$

for the case of i.i.d. random variables $\{X_n\}$ with $E|X_1|^3 < \infty$, $EX_1 = 0$, where $\rho_0 = E|X_1|^3/\sigma^3, \sigma^2 = EX_1^2, F_n(x) = P\{S_n/\sigma\sqrt{n} < x\}$.

In this chapter, we shall give the uniform estimations for α-mixing and ρ-mixing sequences in Section 7.1

In Section 7.2 we shall discuss the Prohorov distance $L(P \circ W_n^{-1}, W)$ between the measure W_n generated by the partial sums processes $\{W_n(t),\ 0 \leq t \leq 1,\ n \geq 1\}$ and the Wiener measure W. We shall give the estimation of $L(P \circ W_n^{-1}, W)$ for a φ-mixing sequence.

7.1 Rate of convergence in distribution for α-mixing and ρ-mixing sequences

The proofs of the Esseen and the Berry-Esseen inequalities for the independent random variables are based on the following proposition (Petrov 1975):

Proposition 7.1.1. Let $F_n(x)$ be a distribution, $f_n(t)$ be the characteristic function of $F_n(x)$. Then for any given $T > 0$ and $b > 1/(2\pi)$ we have

$$\sup_x |F_n(x) - \Phi(x)| \le b\int_{-T}^{T}\left|\frac{f_n(t) - e^{-t^2/2}}{t}\right|dt + \gamma(b)\frac{1}{\sqrt{2\pi}T} \tag{7.1.1}$$

for some $\gamma(b) > 0$.

To obtain the estimation of $|f_n(t) - e^{-t^2/2}|$ is the key to the proof. It is simpler in the case of independent random variables. How to get the estimation of $|f_n(t) - e^{-t^2/2}|$ for the case of mixing dependent random variables ? A method was given by Tikhomirov (1980), a modification of which was given by Sunklodas (1984). The sketch of this method is as follows:

Let $\{X_n, n \ge 1\}$ be an α-mixing sequence with $EX_n = 0$. Denote

$$\sigma_n^2 = ES_n^2, \qquad Z_n = S_n/\sigma_n, \qquad F_n(x) = P(Z_n < x).$$

Assume that $1 \le h \le n-1$, $2 \le k \le n-1$ such that $2kh + 1 < n$. Put

$$\begin{aligned}
Y_j &= X_j/\sigma_n,\\
Z_j^{(0)} &= X_j/\sigma_n = Y_j,\\
Z_j^{(l)} &= \sum_{|p-j|\le lh} Y_p,\\
z_j^{(0)} &= Z_n,\\
z_j^{(l)} &= Z_n - Z_j^{(l)},\\
\xi_j^{(l)} &= \exp\{it(z_j^{(l-1)} - z_j^{(l)})\} - 1,\\
a_j^{(r-1)} &= E\Big(Y_j \prod_{l=1}^{r-1} \xi_j^{(l)}\Big),\\
\eta_j^{(r)} &= e^{-itZ_j^{(r)}} - 1,\\
f_n(t) &= Ee^{itZ_n},
\end{aligned} \tag{7.1.2}$$

for $j = 1, 2, \cdots, n$; $l = 1, 2, \cdots, r-1$; $r = 2, \cdots, k$. Since

$$f_n'(t) = i\sum_{j=1}^{n} EY_je^{itZ_n} = i\sum_{j=1}^{n} EY_je^{itz_j^{(0)}}$$

and

$$Ee^{itz_j^{(r)}} = E(\eta_j^{(r)} + 1)f_n(t) + E[(\eta_j^{(r)} - E\eta_j^{(r)})e^{itZ_n}]$$

for $j = 1, 2, \cdots, n$; $r = 2, \cdots, k$, the derivative of the characteristic function $f_n(t)$ has the representation :

$$\begin{aligned} f_n'(t) = & i\Big\{\sum_{j=1}^{n} a_j^{(1)} E(\eta_j^{(2)} + 1) + \sum_{r=3}^{k}\sum_{j=1}^{n} a_j^{(r-1)} E(\eta_j^{(r)} + 1)\Big\} f_n(t) \\ & + i\sum_{r=2}^{k}\sum_{j=1}^{n}\Big\{E(Y_j \prod_{l=1}^{r-1} \xi_j^{(l)} e^{itz_j^{(r)}}) - E(Y_j \prod_{l=1}^{r-1} \xi_j^{(l)}) E e^{itz_j^{(r)}}\Big\} \\ & + i\sum_{r=2}^{k}\sum_{j=1}^{n} a_j^{(r-1)} E\{(\eta_j^{(r)} - E\eta_j^{(r)}) e^{itZ_n}\} \\ & + i\sum_{j=1}^{n} EY_j e^{itz_j^{(1)}} + i\sum_{j=1}^{n} EY_j \prod_{l=1}^{k} \xi_j^{(l)} e^{itz_j^{(k)}}. \end{aligned} \tag{7.1.3}$$

If we can give the estimation for each term of the right hand side of (7.1.3) under some suitable conditions, then we get the differential equation

$$f_n'(t) = (-t + a(t)) f_n(t) + b(t). \tag{7.1.4}$$

By solving this differential equation, one obtains the estimation of $|f_n(t) - e^{-t^2/2}|$. Finally, from (7.1.1) the estamation of $\Delta_n = \sup_x |F_n(x) - \Phi(x)|$ follows.

Denote $s = 2 + \delta$. Assume that

$$d := \max_{1\le j\le n} E|X_j|^{2+\delta} < \infty, \qquad 0 < \delta \le 1, \tag{7.1.5}$$

$$\sigma_n^2 := ES_n^2 \ge c_0 n \qquad for \quad 0 < c_0 < \infty. \tag{7.1.6}$$

Sunklodas (1984) gave the following theorems.

Theorem 7.1.1. *Let $\{X_n, n \ge 1\}$ be an α-mixing sequence with $EX_n = 0$, and satisfy conditions (7.1.5), (7.1.6) and*

$$\alpha(n) \le K e^{-\lambda n} \qquad \lambda > 0,\ K > 0.$$

Then there exist $c_1 = c_1(K, \delta)$, $c_2 = c_2(K, \delta)$ such that for λ, $\lambda_1 \le \lambda \le \lambda_2$,

$$\Delta_n \le c_1 \frac{d}{c_0 \sigma_n^{\delta}} \Big(\frac{\log(\sigma_n / c_0^{1/2})}{\lambda}\Big)^{1+\delta}, \qquad n \ge 1, \tag{7.1.7}$$

where

$$\lambda_1 = c_2 (\log(\sigma_n / c_0^{1/2}))^b / n, \qquad b > 2(1+\delta)/\delta;$$
$$\lambda_2 = 4(2+\delta)\delta^{-1} \log(\sigma_n / c_0^{1/2}).$$

Theorem 7.1.2. *Let $\{X_n, n \geq 1\}$ be as in Theorem 7.1.1, but*

$$\alpha(n) \leq Kn^{-\mu}, \qquad \mu = 2\beta(1+\delta)(2+\delta)/\delta^2, \quad \beta > 1, \ K > 0.$$

Then there exists a $C = C(K, \beta, \delta)$ such that for every $n \geq 1$

$$\Delta_n \leq Cdc_0^{-(1+\beta+\delta)/(1+\beta)}\sigma_n^{-(\beta-1)\delta/(1+\beta)}. \tag{7.1.8}$$

Remark 7.1.1. Theorem 7.1.1 implies the Theorem 2 of Tikhomirov (1980). Theorem 4 of Tikhomirov (1980) also points out that if $E|X_1|^{4+\gamma} < \infty$, $\gamma > 0$, then

$$\Delta_n \leq cn^{-1/2}\log n. \tag{7.1.9}$$

The proofs of Theorems will use the following lemmas.

Lemma 7.1.1. *We have*

$$i\sum_{j=1}^{n} a_j^{(1)} = -t + \Theta 20(1+12\alpha)^{1/2}d^{2/s}\sigma_n(\alpha(h+1))^{(s-2)/2s}|t|/c_0^{3/2}$$
$$+\Theta(2^{3-s}/(s-1))d(2h+1)^{s-1}|t|^{s-1}/(c_0\sigma_n^{s-2}), \tag{7.1.10}$$

where $\alpha = \sum_{j=1}^{n}(\alpha(j))^{(s-2)/s}$, $s = 2+\delta$, $0 < \delta \leq 1$ and Θ is a constant with $|\Theta| \leq 1$.

Proof. Let τ be a random variable uniformly distributed on the set $\{1, 2, \cdots, n\}$, and independent of $\{X_1, X_2, \cdots, X_n\}$. It is easy to see that for $i = 0, 1, \cdots, k$ and any $b \geq 1$

$$E|Z_\tau^{(l)}|^b \leq (2hl+1)^b \sum_{j=1}^{n} E|Y_j|^b/n. \tag{7.1.11}$$

By the Hölder inequality and (7.1.11), we obtain

$$\sum_{j=1}^{n} E(|Y_j||Z_j^{(1)}|^{s-1}) \leq n\|Y_\tau\|_s\|Z_\tau^{(1)}\|_s^{s-1}$$
$$\leq (2h+1)^{s-1}\sum_{j=1}^{n} E|Y_j|^s. \tag{7.1.12}$$

Denote by $\widehat{z}_j^{(l)}$ the sum of those Y_p from $z_j^{(l)}$, for which $p < j - lh$, and by $\widetilde{z}_j^{(l)}$ the sum of those Y_p from $z_j^{(l)}$, for which $p > j + lh$. Thus $z_j^{(l)} = \widehat{z}_j^{(l)} +$

$\widetilde{z}_j{}^{(l)}$. From Lemma 1.2.4 it follows that

$$\begin{aligned}|EY_j\widehat{z}_j{}^{(l)}| &\le 10(\alpha(h+1))^{1-\frac{1}{2}-\frac{1}{s}}\|Y_j\|_s\|\widehat{z}_j{}^{(l)}\|_2\\ &\le 10(1+12\alpha)^{1/2}d^{2/s}(\alpha(h+1))^{(s-2)/2s}/(c_0^{1/2}\sigma_n).\end{aligned}$$

The same quantity also bounds $|EY_j\widetilde{z}_j{}^{(l)}|$ for $j=1,2,\cdots,n$. Since $\sum_{j=1}^n EY_jZ_n=1$, we have

$$\begin{aligned}\sum_{j=1}^n EY_jZ_j^{(1)} &= 1-\sum_{j=1}^n(EY_j\widehat{z}_j{}^{(1)}+EY_j\widetilde{z}_j{}^{(1)})\\ &= 1+\Theta 20(1+12\alpha)^{1/2}d^{2/s}\\ &\quad\times(\alpha(h+1))^{(s-2)/2s}\sigma_n/c_0^{3/2},\end{aligned} \tag{7.1.13}$$

where we have used (7.1.6). According to Taylor's formula

$$i\sum_{j=1}^n a_j^{(1)}=-\sum_{j=1}^n EY_jZ_j^{(1)}t+\Theta(2^{3-s}/(s-1))\sum_{j=1}^n E(|Y_j||Z_j^{(1)}|^{s-1})|t|^{s-1}.$$

Inserting (7.1.12) and (7.1.13) into this equation and using (7.1.6), we obtain (7.1.10).

Lemma 7.1.2. *For $|t|\le(\sigma_n/32d^{1/s}(h+1))=:T_1$, $j=1,2,\cdots,n$ and $r=3,4,\cdots,k$, we have*

$$\begin{aligned}|a_j^{(r-1)}| &\le 4^rd^{3/s}(h+1)^2t^2(1/2)^{4r}/\sigma_n^3\\ &\quad+c\frac{d^{1/s}}{\sigma_n}(\alpha(h+1))^{(s-2)/s}\Big[r\Big(\frac{1}{2}\Big)^r+r4^r(\alpha(h+1))^{1/s}\Big]\\ &=: a^{(r-1)}\end{aligned} \tag{7.1.14}$$

and

$$|a_j^{(1)}|\le 2d^{2/s}(h+1)|t|/\sigma_n^2. \tag{7.1.15}$$

Proof. Note that $a_j^{(r-1)}=E\Big\{Y_j\prod_{l=1}^{r-1}\xi_j^{(l)}\Big\}$. From

$$\begin{aligned}|a_j^{(1)}| = |EY_j\xi_j^{(1)}| &\le |t|E|X_j|\sum_{|p-j|\le h}|X_p|/\sigma_n^2\\ &\le 2d^{2/s}(h+1)|t|/\sigma_n^2,\end{aligned}$$

(7.1.15) follows.

Now, to prove (7.1.14). Denote

$$\xi_{j1}^{(l)} = \exp\Big\{it \sum_{p=j-lh+1}^{j-(l-1)h} Y_p\Big\} - 1,$$

$$\xi_{j2}^{(l)} = \exp\Big\{it \sum_{p=j+(l-1)h+1}^{j+lh} Y_p\Big\} - 1.$$

Obviously

$$|\xi_j^{(l)}| \le |\xi_{j1}^{(l)}| + |\xi_{j2}^{(l)}|.$$

It follows that

$$\begin{aligned} |a_j^{(r-1)}| &\le E\Big|Y_j \prod_{l=1}^{r-1} \xi_j^{(l)}\Big| \\ &\le \sum_{k=0}^{r-1} \sum{}^* E\Big|Y_j \prod_{\nu=1}^{k} \xi_{j1}^{(l_\nu)} \prod_{\mu=k+1}^{r-1} \xi_{j2}^{(l_\mu)}\Big| \end{aligned} \tag{7.1.16}$$

where the summation $\sum^*$ sums up for $1 \le l_1 < \cdots < l_k \le r-1, \quad 1 \le l_{k+1} < \cdots < l_{r-1} \le r-1, \quad l_\nu \ne l_\mu \ (\nu \ne \mu)$. For each term of the right hand side of (7.1.16) we have

$$\begin{aligned} & E\Big|Y_j \prod_{\nu=1}^{k} \xi_{j1}^{(l_\nu)} \prod_{\mu=k+1}^{r-1} \xi_{j2}^{(l_\mu)}\Big| \\ &\le \Big(E\Big|Y_j {\prod_\nu}' \xi_{j1}^{(l_\nu)} {\prod_\mu}' \xi_{j2}^{(l_\mu)}\Big|^{\frac{2+\delta}{1+\delta}}\Big)^{\frac{1+\delta}{2+\delta}} \cdot \\ &\quad \cdot \Big(E\Big|{\prod_\nu}'' \xi_{j1}^{(l_\nu)} {\prod_\mu}'' \xi_{j2}^{(l_\mu)}\Big|^{2+\delta}\Big)^{\frac{1}{2+\delta}} \end{aligned} \tag{7.1.17}$$

where $\prod'$ is the product for all even l, $\prod''$ is the product for all odd l. It is easy to see that

$$\begin{aligned} & E\Big|Y_j {\prod_\nu}' \xi_{j1}^{(l_\nu)} {\prod_\mu}' \xi_{j2}^{(l_\mu)}\Big|^{\frac{2+\delta}{1+\delta}} \\ &\le (E|Y_j|^{2+\delta})^{\frac{1}{1+\delta}} {\prod_\nu}' E|\xi_{j1}^{(l_\nu)}|^{\frac{2+\delta}{1+\delta}} {\prod_\mu}' E|\xi_{j2}^{(l_\mu)}|^{\frac{2+\delta}{1+\delta}} \\ &\quad + cr2^{r-1}(\alpha(h+1))^{\frac{\delta}{1+\delta}} (E|Y_j|^{2+\delta})^{\frac{1}{1+\delta}}, \end{aligned} \tag{7.1.18}$$

$$\begin{aligned} & E\Big|{\prod_\nu}'' \xi_{j1}^{(l_\nu)} {\prod_\mu}'' \xi_{j2}^{(l_\mu)}\Big|^{2+\delta} \\ &\le cr2^{r-1}\alpha(h+1) + {\prod_\nu}'' E|\xi_{j1}^{(l_\nu)}|^{2+\delta} {\prod_\mu}'' E|\xi_{j2}^{(l_\mu)}|^{2+\delta} \end{aligned} \tag{7.1.19}$$

and

$$\max\Big(E|\xi_{j1}^{(l_\nu)}|^{2+\delta},\quad E|\xi_{j2}^{(l_\mu)}|^{2+\delta}\Big) \le \Big(\frac{|t|b_h}{\sigma_n}\Big)^{2+\delta} \wedge 2^{2+\delta},$$

where $b_h = \max_{1\le i\le h+1}\Big\|\sum_{j=1}^{i} X_j\Big\|_{2+\delta}$. It follows from (7.1.17)–(7.1.19) that we have

$$\begin{aligned}
&E\Big|Y_j \prod_{\nu} \xi_{j1}^{(l_\nu)} \prod_{\mu} \xi_{j2}^{(l_\mu)}\Big| \\
&\quad \le 2^r d^{3/s}(h+1)^2 t^2 (1/2)^{4r}/\sigma_n^3 \\
&\qquad + c\frac{d^{1/s}}{\sigma_n}(\alpha(h+1))^{\frac{s-2}{s}}\Big[r(\frac{1}{2})^{2r} + r2^r(\alpha(h+1))^{1/s}\Big]. \quad (7.1.20)
\end{aligned}$$

Note that the number of terms of the right hand side of (7.1.16) does not exceed 2^r. (7.1.14) is proved.

We shall use the following results: for any finite $p \ge 1$

$$\sum_{r=1}^{\infty} r^p/e^r < \infty. \quad (7.1.21)$$

And for $j = 1,\cdots,n$; $r = 2,\cdots,k$

$$(E|\eta_j^{(r)}|^s)^{1/s} \le |t|d^{1/s}(2rh+1)/\sigma_n. \quad (7.1.22)$$

Assume that $1 \le h \le n-1$, $2 \le k \le n-1$ and

$$k^{3/2}4^k(\alpha(h+1))^{1/s} \le 1. \quad (7.1.23)$$

Lemma 7.1.3. *If (7.1.23) is satisfied and $|t| \le T_1$. Then*

$$\begin{aligned}
&\Big|\sum_{j=1}^{n} a_j^{(1)} E\eta_j^{(2)} + \sum_{r=3}^{k}\sum_{j=1}^{n} a_k^{(r-1)} E(\eta_j^{(r)}+1)\Big| \\
&\quad \le \frac{c}{c_0}\Big[d^{3/s}(h+1)^2t^2/\sigma_n + d^{1/s}\sigma_n(\alpha(h+1))^{(s-2)/s}\Big], \quad (7.1.24)
\end{aligned}$$

$$\begin{aligned}
&\sum_{r=2}^{k}\Big|\sum_{j=1}^{n} a_j^{(r-1)} E[(\eta_j^{(r)} - E\eta_j^{(r)})e^{itZ_n}]\Big| \\
&\quad \le \frac{c}{c_0^{1/2}}[(h+1)^{1/2} + \alpha^{1/2}][d^{3/s}(h+1)^2t^2/\sigma_n^2 \\
&\qquad + d^{1/s}(\alpha(h+1))^{(s-2)/s}]. \quad (7.1.25)
\end{aligned}$$

In addition, if $k \geq (\log n)/8\log 2$, *then*

$$\sum_{j=1}^{n}\Big|E\Big(Y_j\prod_{l=1}^{k}\xi_j^{(l)}e^{itz_j^{(k)}}\Big)\Big|$$
$$\leq cc_0^{-1/2}d^{3/s}(h+1)^2t^2/\sigma_n^2$$
$$+cc_0^{-1}d^{1/s}\sigma_n(\alpha(h+1))^{(s-2)/s}. \tag{7.1.26}$$

Proof.

$$\Big|\sum_{j=1}^{n}a_j^{(1)}E\eta_j^{(2)}+\sum_{r=3}^{k}\sum_{j=1}^{n}a_j^{(r-1)}E(\eta_j^{(r)}+1)\Big|$$
$$\leq\sum_{j=1}^{n}|a_j^{(1)}|E|\eta_j^{(2)}|+\sum_{r=3}^{k}\sum_{j=1}^{n}|a_j^{(r-1)}|,$$

which implies (7.1.24) by using Lemma 7.1.2, (7.1.22), (7.1.23) and (7.1.6).

Further, $\mathrm{Cov}(\xi,\eta)=E(\xi-E\xi)(\overline{\eta-E\eta})$. Applying the Hölder inequality and Lemma 1.2.4, we obtain that for $r=2,3,\cdots,k$

$$\Big|\sum_{j=1}^{n}a_j^{(r-1)}E\Big[(\eta_j^{(r)}-E\eta_j^{(r)})e^{itZ_n}\Big]\Big|$$
$$\leq\Big\{\Big(\sum_{j=1}^{n}\sum_{|p-j|\leq 2rh}+\sum_{j=1}^{n}\sum_{|p-j|>2rh}\Big)a_j^{(r-1)}\overline{a_p^{(r-1)}}\mathrm{Cov}(\eta_j^{(r)},\eta_p^{(r)})\Big\}^{1/2}$$
$$\leq\Big\{\sum_{j=1}^{n}\sum_{|p-j|\leq 2rh}|a_j^{(r-1)}||a_p^{(r-1)}|\|\eta_j^{(r)}\|_2\|\eta_p^{(r)})\|_2\Big\}^{1/2}$$
$$+\Big\{24\sum_{j=1}^{n}\sum_{|p-j|>2rh}|a_j^{(r-1)}||a_p^{(r-1)}|$$
$$\times(\alpha(|p-j|-2rh))^{(s-2)/s}\|\eta_j^{(r)}\|_s\|\eta_p^{(r)})\|_s\Big\}^{1/2}. \tag{7.1.27}$$

Let $r=3,4,\cdots,k$. Since for $j=1,\cdots,n$, $|\eta_j^{(r)}|\leq 2$ and for $|t|\leq T_1$, $|a_j^{(r-1)}|\leq a^{(r-1)}$, and $a^{(r-1)}$ is independent of j, considering (7.1.6) we get that right hand side of (7.1.27) does not exceed

$$2a^{(r-1)}\Big\{\sum_{j=1}^{n}\sum_{|p-j|\leq 2rh}1\Big\}^{1/2}$$
$$+ca^{(r-1)}\Big\{\sum_{j=1}^{n}\sum_{|p-j|>2rh}(\alpha(|p-j|-2rh))^{(s-2)/s}\Big\}^{1/2}$$
$$\leq cc_0^{-1/2}\sigma_n[r^{1/2}(h+1)^{1/2}+\alpha^{1/2}]a^{(r-1)}. \tag{7.1.28}$$

For $r = 2$ we proceed in the same way but we estimate $\eta_j^{(2)}$ for $j = 1, 2 \cdots, n$ according to (7.1.22). One gets that for $r = 2$ the right hand side of (7.1.27) does not exceed

$$cc_0^{-1/2} d^{3/s}(h+1)^{5/2}t^2/\sigma_n^2 + cc_0^{-1/2}\alpha^{1/2}d^{3/s}(h+1)^2t^2/\sigma_n^2. \tag{7.1.29}$$

Adding over r from 2 to k in (7.1.27) and considering (7.1.28), (7.1.29) and (7.1.23), we get the proof of (7.1.25).

Finally, since for $k \geq (\log n)/(8\log 2)$ we have that $(1/2)^{4k} \leq n^{-1/2}$, (7.1.26) follows from the proof of Lemma 7.1.2, (7.1.6) and (7.1.23). Lemma 7.1.3 is proved.

Lemma 7.1.4. *We have*

$$\sum_{j=1}^{n} |EY_j e^{itz_j^{(1)}}| \leq 32c_0^{-1}d^{1/s}\sigma_n(\alpha(h+1))^{(s-1)/s}. \tag{7.1.30}$$

If (7.1.23) is satisfied, then

$$\sum_{r=2}^{k}\sum_{j=1}^{n}\Big|EY_j\prod_{l=1}^{r-1}\xi_j^{(l)}e^{itz_j^{(r)}} - E\Big(Y_j\prod_{l=1}^{r-1}\xi_j^{(l)}\Big)Ee^{itz_j^{(r)}}\Big|$$
$$\leq 48c_0^{-1}d^{1/s}\sigma_n(\alpha(h+1))^{(s-1)/s}.$$

Proof. We only prove the second inequality since the first can be proved analogously. From the definition of $z_j^{(r)}$, we have that for all $r = 2, 3, \cdots, k$, $\widehat{z_j}^{(r)}$ and $\widetilde{z_j}^{(r)}$ cannot be equal to zero simultaneously. Without loss of generality we assume that $\widehat{z_j}^{(r)} \neq 0$ and $\widetilde{z_j}^{(r)} \neq 0$. Then, according to Lemma 1.2.4

$$\Big|E\Big(Y_j\prod_{l=1}^{r-1}\xi_j^{(l)}e^{itz_j^{(r)}}\Big) - E\Big(Y_j\prod_{l=1}^{r-1}\xi_j^{(l)}\Big)Ee^{itz_j^{(r)}}\Big|$$
$$\leq \Big|E\Big(Y_j\prod_{l=1}^{r-1}\xi_j^{(l)}e^{itz_j^{(r)}}\Big) - E\Big(Y_j\prod_{l=1}^{r-1}\xi_j^{(l)}e^{it\widehat{z_j}^{(r)}}\Big)Ee^{it\widetilde{z_j}^{(r)}}\Big|$$
$$+\Big|E\Big(Y_j\prod_{l=1}^{r-1}\xi_j^{(l)}e^{it\widehat{z_j}^{(r)}}\Big) - EY_j\prod_{l=1}^{r-1}\xi_j^{(l)}Ee^{it\widehat{z_j}^{(r)}}\Big|$$
$$+\Big|Ee^{it\widehat{z_j}^{(r)}}Ee^{it\widetilde{z_j}^{(r)}} - Ee^{itz_j^{(r)}}\Big|\cdot\Big|EY_j\prod_{l=1}^{r-1}\xi_j^{(l)}\Big|$$
$$\leq 48d^{1/s}(\alpha(h+1))^{(s-1)/s}2^{r-1}/\sigma_n.$$

Adding the inequalities obtained over all $j = 1, 2, \cdots, n$ and $r = 2, 3, \cdots, k$, we use (7.1.6) and (7.1.23). Lemma 7.1.4 is proved.

Proof of Theorem 7.1.1.

According to the basic method of the proof , we first establish the differential equation (7.1.4). Denote $\alpha = \sum_{k=1}^{n} \alpha(k)^{\delta/(2+\delta)}$, let

$$\begin{aligned}
a_0 &= cd^{1/s}\sigma_n(\alpha(h+1))^{(s-2)/s}/c_0, \\
a_1 &= cd^{2/s}(1+\alpha)^{1/2}\sigma_n(\alpha(h+1))^{(s-2)/2s}/c_0^{3/2}, \\
a_2 &= cd^{3/s}(h+1)^2/c_0\sigma_n, \qquad a_3 = cd(h+1)^{s-1}/c_0\sigma_n^{s-2}, \\
b_0 &= cd^{1/s}((h+1)^{1/2}+\alpha^{1/2})(\alpha(h+1))^{(s-2)/s}/c_0^{1/2} + a_0, \\
b_2 &= cd^{3/s}((h+1)^{1/2}+\alpha^{1/2})(h+1)^2/c_0^{1/2}\sigma_n^2, \\
T_2 &= \min\{1/a_0, 1/(6a_2), (1/6a_3)^{1/(s-2)}\}.
\end{aligned}$$

Suppose that $1 \le h \le n-1$, $\quad 2 \le k \le n-1$ such that

$$k \ge \frac{\log n}{8 \log 2}, \qquad k^{3/2}4^k(\alpha(h+1))^{1/s} \le 1, \quad 2kh+1 < n. \tag{7.1.31}$$

From (7.1.3), Lemmas 7.1.1, 7.1.3 and 7.1.4, it follows that

$$f_n'(t) = (-t + \Theta a(t))f_n(t) + \Theta b(t) \qquad \text{as} \quad |t| \le T_1 \tag{7.1.32}$$

where $|\Theta| \le 1$,

$$a(t) = a_0 + a_1|t| + a_2t^2 + a_3|t|^{s-1}, \qquad b(t) = b_0 + b_2t^2.$$

Next, we solve the linear differential equation (7.1.32) and get

$$\begin{aligned}
&|f_n(t) - e^{-t^2/2}| \\
&\quad \le |x_0|e^{-t^2/2+|x_0|} \\
&\qquad + e^{-t^2/2}\int_0^{|t|} b(u)\exp\Big\{\frac{u^2}{2} + \int_{|u|}^{|t|} a(v)dv\Big\}du,
\end{aligned} \tag{7.1.33}$$

where $x_0 = \int_0^t \Theta a(u)du$. Let $0 \le u \le t$. Then for $a_1 \le 1/6$ and $|t| \le T_2$ we have

$$\int_u^t a(v)dv \le 1 + \frac{t^2 - u^2}{4}, \tag{7.1.34}$$

and it is easy to see that

$$\int_0^{|t|} u^2 \exp(u^2/4)du \le 2|t|\exp(t^2/4), \tag{7.1.35}$$

$$\int_0^{|t|} \exp(u^2/4)du \le \min(4/|t|, |t|)\exp(t^2/4). \tag{7.1.36}$$

From (7.1.33)–(7.1.36) it follows that for $|t| \le \min(T_1, T_2)$ and $a_1 \le 1/6$

$$|f_n(t) - e^{-t^2/2}| \le (a_0|t| + \frac{a_1}{2}t^2 + \frac{a_2}{3}|t|^3 + \frac{a_3}{s}|t|^s)e^{-t^2/4+1} + eb_0\Big\{\frac{4}{|t|} \wedge t\Big\} + 2eb_2|t|. \quad (7.1.37)$$

By Lemma 1.2.4 we have

$$\sigma_n^2 \le (1+12\alpha)d^{2/s}n,$$

and hence $1 \le (1+12\alpha)^{1/2}\mathcal{K}$, $\mathcal{K} = d^{1/s}/c_0^{1/2}$. Thus from (7.1.37) and (7.1.1) we get

$$\begin{aligned}\Delta_n \le & c\Big\{\mathcal{K}^s\frac{(h+1)^{s-1}}{\widetilde{\sigma}_n^{s-2}} + \mathcal{K}^3\frac{(h+1)^2}{\widetilde{\sigma}_n} \\ & + \mathcal{K}^2((h+1)^{1/2}+\alpha^{1/2})\frac{h+1}{\widetilde{\sigma}_n} \\ & + \mathcal{K}^2(1+\alpha)^{1/2}\widetilde{\sigma}_n(\alpha(h+1))^{(s-2)/2s} \\ & + \mathcal{K}((h+1)^{1/2}+\alpha^{1/2})(\alpha(h+1))^{(s-2)/s}\Big\} \quad (7.1.38)\end{aligned}$$

where $\widetilde{\sigma}_n^2 = \sigma_n^2/c_0$.

Finally, we prove (7.1.7). Let $0 < \lambda \le \lambda_2 = \frac{4(2+\delta)}{\delta}\log\widetilde{\sigma}_n$ and $h = \left[\frac{4(2+\delta)}{\lambda\delta}\log\widetilde{\sigma}_n\right]$, $n \ge 1$. Then $\frac{4(2+\delta)}{\lambda\delta} \le \frac{h+1}{\log\widetilde{\sigma}_n}$. Hence, by the condition $\alpha(n) \le Ke^{-\lambda n}$, we obtain

$$\begin{aligned}\alpha^{1/2} &= \Big(\sum_{\tau=1}^{n}(\alpha(\tau))^{\delta/(2+\delta)}\Big)^{1/2} \\ &\le \frac{1}{2}K^{\delta/2(2+\delta)}(h+1)^{1/2}/(\log\widetilde{\sigma}_n)^{1/2} \\ &\le C(K,\delta)(h+1)^{1/2}. \quad (7.1.39)\end{aligned}$$

This means

$$1 \le C(K,\delta)(h+1)^{1/2}\mathcal{K}. \quad (7.1.40)$$

For the chosen h,

$$\begin{aligned}(\alpha(h+1))^{\delta/2(2+\delta)} &\le K^{\delta/2(2+\delta)}e^{-\lambda\delta(h+1)/2(2+\delta)} \\ &\le K^{\delta/2(2+\delta)}\widetilde{\sigma}_n^{-2}, \quad (7.1.41) \\ (\alpha(h+1))^{\delta/(2+\delta)} &\le K^{\delta/(2+\delta)}\widetilde{\sigma}_n^{-4}. \quad (7.1.42)\end{aligned}$$

We get from (7.1.39)-(7.1.42) and (7.1.37) that

$$\Delta_n \le C(K,\delta)\Big[\mathcal{K}^{2+\delta}\frac{(h+1)^{1+\delta}}{\widetilde{\sigma}_n^\delta} + \mathcal{K}^3\frac{(h+1)^2}{\widetilde{\sigma}_n}\Big]$$
$$\le C(k,\delta)\mathcal{K}^{2+\delta}\frac{(h+1)^{1+\delta}}{\widetilde{\sigma}_n^\delta}. \qquad (7.1.43)$$

It remains to check whether there exists a $k, 2 \le k \le n-1$, such that for $h = \left[\frac{4(2+\delta)}{\lambda\delta}\log\widetilde{\sigma}_n\right]$, where $\lambda_1 \le \lambda \le \lambda_2$ we have

$$k \ge \frac{\log n}{8\log 2}, \qquad k^{3/2}4^k(\alpha(h+1))^{1/(2+\delta)} \le 1, \qquad 2kh+1 < n. \quad (7.1.44)$$

It is easy to verify that for all large $n (n > n_0)$ and for

$$k = \Big[\frac{(4/\delta)\log\widetilde{\sigma}_n - (\log K)/(2+\delta)}{(3/2)+\log 4}\Big]$$

the first two inequalities hold and the third inequality does not constrict the interval of change of the parameter λ.

Considering (7.1.40), we have

$$n^{-1} \le C(K,\delta)\mathcal{K}^{2+\delta}(h+1)^{1+\delta}/\widetilde{\sigma}_n^\delta.$$

This also proves Theorem 7.1.1 for all $n \le n_0$. The proof of Theorem 7.1.1 is completed.

The proof of Theorem 7.1.2 is analogous to that of Theorem 7.1.1. We only indicate that for this it is sufficient to set $h = [\widetilde{\sigma}_n^\alpha]$, where $\alpha = 2\delta/(\beta+1)(\delta+1)$, and

$$k = \Big[\frac{(\alpha\mu/(2+\delta))\log\widetilde{\sigma}_n - (\log K)/(2+\delta)}{(3/2)+\log 4}\Big]$$

in (7.1.31)-(7.1.33).

Remark 7.1.2. Tikhomirov (1980) proved the Berry-Esseen inequality for the ρ-mixing sequence with the exponential rate of decay of $\rho(n)$. Zuparov (1991) improved his method to prove the following theorem in which there is a better order of Δ_n when $E|X_1|^s < \infty, \quad 2 < s < s_0(< 1+\sqrt{3})$, and the rate of $\rho(\cdot)$ was weakened.

Theorem 7.1.3. *Let $\{X_n, n \ge 1\}$ be a strictly stationary ρ-mixing sequence with $EX_1 = 0$ and*

$$\rho(n) \le Kn^{-\theta}, \qquad \theta > 0, \quad K > 0, E|X_1|^s < \infty,$$

$$2 < s < s_0(\theta) = \frac{\theta - 1}{\theta} + \sqrt{\left(\frac{\theta - 1}{\theta}\right)^2 + \frac{4 + 2\theta}{\theta}}.$$

If $\sigma_n^2 = ES_n^2 \geq \tau n E X_1^2$. *Then there exists a constant* $C = C(s, \theta, K, \tau)$ *depending only on* s, θ, K *and* τ *such that*

$$\Delta_n \leq C(s, \theta, K, \tau) \beta_s / n^{(s-2)/2},$$

where $\beta_s = E|X_1|^s / (EX_1^2)^{s/2}$, $\quad s_0 < 1 + \sqrt{3}$.

The proof of Theorem 7.1.3 will not be presented here.

7.2 The rate of weak convergence for a φ-mixing sequence

Let $\{X_n, n \geq 1\}$ be a sequence of random variables with $EX_n = 0$, $\sigma_n^2 = ES_n^2$. Define the partial sum process as follows:

$$W_n(t) = \frac{1}{\sigma_n}\Big(S_{[nt]} + \frac{t\sigma_n^2 - \sigma_{[nt]}^2}{\sigma_{[nt]+1}^2 - \sigma_{[nt]}^2} X_{[nt]+1}\Big), \tag{7.2.1}$$

if $\sigma_{[nt]}^2 \leq t\sigma_n^2 \leq \sigma_{[nt]+1}^2$.

Denote by P_n the distribution of W_n in $C[0,1]$, i.e. $P_n = P \circ W_n^{-1}$. The weak invariance principle holds, i.e. $P_n \Longrightarrow W$, that is equivalent to

$$L(P_n, W) \longrightarrow 0, \tag{7.2.2}$$

where $L(P, Q)$ is the *Lévy-Prohorov distance*:

$$L(P, Q) = \inf\{\varepsilon : \forall B \in \mathcal{F}, P(B) \leq Q(B^\varepsilon) + \varepsilon, Q(B) \leq P(B^\varepsilon) + \varepsilon\}, \tag{7.2.3}$$

where $\mathcal{F}$ is the Borel σ-field of $C[0,1]$, B^ε is the ε-neighborhood of Borel set B,

$$B^\varepsilon = \{y : y \in C[0,1], \exists z \in B, \|y - z\| < \varepsilon\}, \qquad \|y - z\| = \sup_{0 \leq t \leq 1} |y(t) - z(t)|.$$

Denote

$$\Lambda_n(\varepsilon) = \sup_{B \in \mathcal{F}} |P(W_n \in B) - P(W \in B^\varepsilon)|,$$

It is obvious that

$$L(P_n, W) = \inf_{\varepsilon}(\varepsilon \vee \Lambda_n(\varepsilon)). \tag{7.2.4}$$

Then without changing the distribution of $\{W_n, n \geq 1\}$, we can redefine process $\{\widetilde{W}_n, n \geq 1\}$ on a richer probability space together with the standard Wiener process $\{W(t), t \geq 0\}$ such that

$$\Lambda_n(\varepsilon) \leq P\{\|\widetilde{W}_n - W\| \geq \varepsilon\}.$$

Therefore, in order to estimate the rate of weak convergence, we need only consider the following inequality:

$$L(P_n, W) \le \inf_{\varepsilon}(\varepsilon \vee P\{\|W_n - W\| \ge \varepsilon\}).$$

For the independent random variables $\{X_n, n \ge 1\}$, Prohorov (1956) gave a precise estimation :

$$L(P_n, W) = O(L_3^{1/4} \log^2 L_3),$$

where $L_3 = \sum_{k=1}^n E|X_k|^3/\sigma_n^3$. For i.i.d.r.v.'s, Borovkov (1973) proved: if $E|X_1|^{2+\delta}$
$< \infty$, $0 < \delta \le 1$, then

$$L(P_n, W) = O(L_{2+\delta}^{1/(3+\delta)}) = O(n^{-\frac{\delta}{2(3+\delta)}}). \tag{7.2.5}$$

Borovkov (1973) also pointed out that the weak convergence rate (7.2.5) can not be improved if we use only the Skorohod method. Utev (1981) generalized this estimation to the case $0 < \delta < 3$ and gave

$$L(P_n, W) = O(n^{-\frac{\delta}{2(3+\delta)}}), \text{when} E|X_1|^{2+\delta} < \infty,\ 0 < \delta < 3. \tag{7.2.6}$$

Utev (1984) gave the same estimation as (7.2.6) for φ-mixing sequence under a stronger condition on the rate of decay of $\varphi(n)$.

Theorem 7.2.1. *Let $\{X_n, n \ge 1\}$ be a strictly stationary φ-mixing sequence of random variables with $EX_1 = 0, EX_1^2 = 1$ and $E|X_1|^{2+\delta} < \infty$ for some $0 < \delta < 3$. Suppose*

$$\varphi(n) \le cn^{-g}$$

where $g > j(u)(j(u) - 1)$, $u = (12 + 5\delta)/(2(3 - \delta))$, $j(u) = \min\{2k : 2k \ge u, k \in \mathbb{N}\}$. Then

$$L(P_n, W) \le cn^{-\frac{\delta}{2(3+\delta)}}.$$

By using Lemma 2.2.10, Lu (1993) weakened the conditions and proved the following theorem.

Theorem 7.2.2. *Let $\{X_n, n \ge 1\}$ be a φ-mixing sequence of random variables with $EX_n = 0$, $A_0 = \sup_n EX_n^2 < \infty$, $A_\delta = \sup_n E|X_n|^{2+\delta} < \infty$, $0 < \delta < 3$. Suppose that*

(i) if $\sigma_n^2 \to \infty$ as $n \to \infty$,

(ii) $\varphi(n) \le cn^{-g}$, $g \ge (3\delta+\varepsilon)/(2(3-\delta)) \vee (2+\varepsilon)$ *for any* $\varepsilon > 0$.

Then we have

$$L(P_n, W) = O(n^{-\frac{\delta}{2(3+\delta)}}). \tag{7.2.7}$$

Proof. By using (i), $A_\delta < \infty$, Lemma 2.2.2 and Lemma 2.2.10, we have

$$\begin{aligned} E \max_{1\le i\le n} |S_k(i)|^{2+\delta} &\le c\Big((n)^{\frac{2+\delta}{2}} + E \max_{k\le i\le k+n} |X_i|^{2+\delta}\Big) \\ &\le cn^{1+\delta/2}. \end{aligned} \tag{7.2.8}$$

Without loss of generality, we can assume that $\sigma_n^2 = c_0 n$, $c_0 = 1$. Then W_n is a random polygonal line with nodes at $(k/n, S_k/\sqrt{n})$, $k = 0, 1, \cdots, n$. Denote

$$X_i^{(1)} = X_i I(|X_i| \le y\sqrt{n}) - EX_i I(|X_i| \le y\sqrt{n}) \qquad i = 1, \cdots, n,$$

where $y = n^{-1/2}(\sum_{i=1}^n E|X_i|^{2+\delta})^{1/(2+\delta)} \le n^{-\frac{\delta}{2(2+\delta)}} A_\delta^{\frac{1}{2+\delta}}$. Note that there exist all finite moments of $X_i^{(1)}$ for any given n.

Let $S_k^{(1)} = \sum_{i=1}^k X_i^{(1)}$ and $W_n^{(1)}$ be a random polygonal line with nodes at $(k/n, S_k^{(1)}/\sqrt{n})$. By using Lemma 2.2.10 , we have

$$\begin{aligned} &P\Big\{\|W_n^{(1)} - W_n\| \ge cn^{-\delta/2(3+\delta)}\Big\} \\ &\le P\Big\{\max_{0\le k\le n}\Big|\frac{1}{\sqrt{n}}(S_k - S_k^{(1)})\Big| \ge cn^{-\delta/2(3+\delta)}\Big\} \\ &\le cn^{-3(2+\delta)/2(3+\delta)} E \max_{o\le k\le n} |S_k - S_k^{(1)}|^{2+\delta} \\ &\le cn^{-3(2+\delta)/2(3+\delta)}\Big[\Big(\max_{1\le k\le n} E|S_k - S_k^{(1)}|^2\Big)^{\frac{2+\delta}{2}} \\ &\quad + \sum_{i=1}^n E|X_i - X_i^{(1)}|^{2+\delta}\Big] \\ &\le cn^{-\delta/2(3+\delta)}. \end{aligned} \tag{7.2.9}$$

Next, put $l = [n^{3/(3+\delta)-\eta}]$, $m = [n^{\delta/(3+\delta)+\eta}]$, $\eta > 0$ (specified later on), write $n = lm + r$, $0 \le r < l$. Let $W_n^{(2)}$ be a random polygonal line with nodes at $(kl/n, S_{kl}^{(1)}/\sqrt{n})$, $k = 0, 1, \cdots, m$ and $(1, S_n^{(1)}/\sqrt{n})$. We have

$$\begin{aligned}
&P\{\|W_n^{(2)} - W_n^{(1)}\| \geq b\} \\
&\leq \sum_{k=0}^{m} P\{\max_{1\leq i\leq l} |S_{kl+i}^{(1)} - S_{kl}^{(1)}| \geq b\sqrt{n}\} \\
&\leq c\sum_{k=0}^{m} b^{-u}n^{-u/2}E\max_{1\leq i\leq l}|S_{kl}^{(1)}(i)|^u \\
&\leq cmb^{-u}n^{-u/2}l^{u/2} \\
&= cb^{-u}(\frac{n}{l})^{1-u/2},
\end{aligned}$$

where we have applied Lemma 2.2.10. If we take $b = cn^{-\delta/2(3+\delta)}$, $u \geq 2 + 3\delta/[(3+\delta)\eta]$, then

$$\left(\frac{n}{l}\right)^{1-u/2} \leq n^{-\frac{\delta(u+1)}{2(3+\delta)}}.$$

Therefore we have

$$P\{\|W_n^{(2)} - W_n^{(1)}\| \geq cn^{-\frac{\delta}{2(3+\delta)}}\} \leq cn^{-\frac{\delta}{2(3+\delta)}}. \tag{7.2.10}$$

Put $h = [n^\theta]$, $\theta > 0$ (specified later on). Let $d = l - h$,

$$\xi_k^{(1)} = \sum_{i=1}^{d} X_{(k-1)l+i}^{(1)},$$

$W_n^{(3)}$ be a random polygonal line with nodes at $(kl/n, \sum_{j=1}^k \xi_j^{(1)}/\sqrt{n})$, we have

$$\begin{aligned}
&P\{\|W_n^{(3)} - W_n^{(2)}\| \geq b\} \\
&= P\Big\{\max_{1\leq k\leq m+1} |S_{kl}^{(1)} - \sum_{j=1}^{k}\sum_{i=1}^{d} X_{(j-1)l+i}^{(1)}| \geq b\sqrt{n}\Big\} \\
&= P\Big\{\max_{1\leq k\leq m+1} |\sum_{j=1}^{k}\sum_{i=1}^{h} X_{(j-1)l+d+i}^{(1)}| \geq b\sqrt{n}\Big\} \\
&\leq cb^{-u}n^{-u/2}\Big\{\Big[\max_{1\leq k\leq m+1} E\Big(\sum_{j=1}^{k}\sum_{i=1}^{h}|X_{(j-1)l+d+i}^{(1)}|\Big)^2\Big]^{u/2} \\
&\qquad + \sum_{j=1}^{m+1} E\Big[\sum_{i=1}^{h}|X_{(j-1)l+d+i}^{(1)}|\Big]^u\Big\}
\end{aligned}$$

where the first term does not exceed $c(mh)^{u/2}$ and the last sum does not exceed $c\,m\,h^{u/2} = o((mh)^{u/2})$ by Lemma 2.2.4, therefore

$$P\{\|W_n^{(3)} - W_n^{(2)}\| \geq b\}$$
$$\leq cb^{-u}n^{-u/2}(mh)^{u/2} \leq cb^{-u}(h/l)^{u/2}.$$

If we take $b = cn^{-\delta/2(3+\delta)}$ and

$$0 < \theta \leq \frac{3-\delta}{3+\delta} - \frac{4\eta}{3}, \tag{7.2.11}$$

then $(h/l)^{u/2} \leq cn^{-\delta(u+1)/2(3+\delta)}$, so that

$$P\{\|W_n^{(3)} - W_n^{(2)}\| \geq cn^{-\delta/2(3+\delta)}\} \leq cn^{-\delta/2(3+\delta)}. \tag{7.2.12}$$

In the remain of the proof, imitating the proof of Utev (1984), i.e. by Berkes and Philipp (1979) Theorem 2, we have

$$P\{\|W_n^{(4)} - W_n^{(3)}\| \geq c\varphi(h)n/l\} \leq c\varphi(h)n/l, \tag{7.2.13}$$

where $W_n^{(4)}$ is the random polygonal line with nodes $(kl/n, \sum_{j=1}^k \xi_j^{(2)}/\sqrt{n})$, $k = 0, 1, \cdots, m+1$, the $\xi_j^{(2)}$ are independent and distributed same as $\xi_j^{(1)}$. And by Sakhanenko (1981), we have

$$P\{\|W_n^{(5)} - W_N^{(4)}\| \geq q_u^{\frac{1}{u+1}}\} \leq q_u^{\frac{1}{u+1}} \qquad u > 2, \tag{7.2.14}$$

where $W_n^{(5)}$ is the random polygonal line with nodes $(kl/n, \sum_{j=1}^k Y_j/\sqrt{n})$, $k = 0, 1, \cdots, m+1$, the Y_j are independent normally distributed random variables with $EY_j = 0$, $\quad \mathrm{Var}Y_j = \mathrm{Var}\xi_j^{(1)}, j = 1, 2, \cdots, m+1$ and

$$q_u = c\Big(\sum_{i=1}^{m+1} E|\xi_i^{(1)}|^u\Big)\Big/\Big(\sum_{i=1}^{m+1} \mathrm{Var}\xi_i^{(1)}\Big)^{u/2}.$$

It is easy to see that by (7.2.8) we have

$$q_u = cn^{-u/2}\sum_{i=1}^{m} E|\xi_i^{(1)}|^u \leq cn^{-u/2}ml^{u/2}$$
$$\leq c(n/l)^{1-u/2}, \tag{7.2.15}$$

so that

$$q_u^{\frac{1}{u+1}} \leq cn^{-\delta/2(3+\delta)}$$

when $u \geq 2 + 3\delta/(3+\delta)\eta$.

At last, if we take

$$\theta \geq \frac{2(3-\delta)}{(3+\varepsilon)\delta}\Big(\frac{3\delta}{2(3+\delta)}+\eta\Big),$$

then

$$\varphi(h)n/l = O(n^{-g\theta+\frac{\delta}{3+\delta}+\eta}) \leq cn^{-\delta/2(3+\delta)}. \tag{7.2.16}$$

Combining with (7.2.11), we have

$$\frac{2(3-\delta)}{(3+\varepsilon)\delta}\Big(\frac{3\delta}{2(3+\delta)}+\eta\Big) \leq \theta < \frac{3-\delta}{3+\delta}-\frac{4\eta}{3},$$

that is to say, we need take η as follows:

$$0<\eta<\frac{3-\delta}{3+\delta}\frac{3\varepsilon\delta}{2(9+3\delta+2\varepsilon\delta)}.$$

Combining (7.2.9)-(7.2.16) together we obtain

$$P\{\|W_n-W\| \geq cn^{-\delta/2(3+\delta)}\} \leq cn^{-\delta/2(3+\delta)}.$$

Since $L(W_n, W) \leq \inf_\varepsilon(\varepsilon + P\{\|W_n - W\| \geq \varepsilon\})$, (7.2.7) holds true. The proof of Theorem 7.2.2 is completed.

Remark 7.2.1. Utev (1984) discussed the rate of weak convergence for an α-mixing sequence and obtained the following result:

Let $\{X_n, n \geq 1\}$ be a strictly stationary α-mixing sequence with $EX_1 = 0$, $EX_1^2 = 1$ and $E|X_1|^{2+\delta} < \infty$, $0 < \delta < 3$. Assume that

$$0<\sigma^2 := EX_1^2 + 2\sum_{n=1}^{\infty} EX_1X_{n+1} < \infty$$

and

$$\alpha(n) \leq cn^{-g}, \qquad n = 1, 2, \cdots$$

where

$$g > \max(\delta_1^{-1}j(u)(j(u)+\delta_1), \delta_1^{-1}j(2(2+\delta))(j(2(2+\delta))+\delta_1)),$$

$$u = \max\Big(2(2+\delta), \frac{12+5\delta}{2(3-\delta)}\Big), \quad \delta_1 = \frac{\delta}{(2+\delta)(3+\delta)},$$

$j(u)$ is defined as in Theorem 7.2.1. Then

$$L(P_n, W) \leq cA_\delta n^{-\delta/(2(3+\delta))},$$

where $c = c(A, g, \delta)$.

If $\delta = 1$, i.e., assume that $E|X_1|^3 < \infty$, $\alpha(n) = O(n^{-(438+\varepsilon)})$, then $L(P_n, W) = O(n^{-1/8})$.

Gorodezkii (1983) has also given a similar result.

Remark 7.2.2. Yoshihara (1979) discussed the rate of the weakly invariance principle for the strictly stationary absolutely regular sequence $\{X_n, n \geq 1\}$, and proved that if $EX_1 = 0$, $E|X_1|^{4+\delta} < \infty$ for some $\delta > 0$ and $\sum_{n=1}^{\infty} n\beta(n)^{\delta/(4+\delta)} < \infty$, then

$$L(P_n, W) = O(n^{-1/8}(\log n)^{1/2}).$$

Part III Almost Sure Convergence and Strong Approximations

In this part, we study the almost sure convergence and the strong approximations of the partial sums of a mixing dependent sequence. Since 1960s, the almost sure convergence has been discussed by some authors, e.g., Iosifescu and Theodorescu (1969) obtained the 0-1 law, the strong law of large numbers and the convergence of random series of a φ-mixing sequence, etc. The complete convergence of various mixing sequences has been studied deeply by Shao et al. in the past ten years, from which the elegant results of the strong laws of large numbers follow. We shall discuss these in Chapter 8.

The strong approximations of the partial sums $S_n = \sum_{k=1}^{n} X_k$ for a mixing sequence $\{X_n, n \geq 1\}$ by a Wiener process were done by Philipp and Stout (1975) et al., and were improved comprehensively by Shao and Lu; the limiting behaviour of the increments of partial sums for a mixing sequence was obtained by Lin et al. Those elegant theorems will be discussed in Chapter 9 and Chapter 10 respectively.

The strong approximation of a mixing sequence with set-indexed will be introduced in Chapter 11.

Chapter 8 Laws of Large Numbers and Complete Convergence

We shall introduce the Borel-Cantelli lemma, the weak law of large numbers and the strong law of large numbers for a φ-mixing sequence in the first two sections. Since the concept of complete convergence was raised by Hsu and Robbins (1947), this subject has attracted the attention of many mathematicians. The complete convergence of weakly dependent sequences has been obtained by some mathematicians. We shall introduce the complete convergence of a φ-mixing sequence, a ρ-mixing sequence and an α-mixing sequence in Sections 8.3-8.5 respectively. At last, we discuss three problems which are posed by Prohorov for the complete convergence of ρ-mixing sequence in Section 8.6.

8.1 Weak law of large numbers

Theorem 10.1.1 of Chow and Teicher (1978) gave a sufficient and necessary condition for the weak law of large numbers for an array of independent random variables $\{X_{nj}, 1 \leq j \leq k_n\}$. Du (1993) generalized this result to the φ-mixing case and proved the following theorem. Denote $S_n = \sum_{j=1}^{k_n} X_{nj}$.

Theorem 8.1.1. *Let $\{X_{nj}, 1 \leq j \leq k_n \to \infty\}$ be a random variable array which is φ-mixing in each row. There exists an integer M such that $\varphi(M) < 1/2$. Then for some real numbers A_n*

$$S_n - A_n \xrightarrow{P} 0, \tag{8.1.1}$$

$$P\Big\{\max_{1\leq j\leq k_n} |X_{nj}| \geq \varepsilon\Big\} \to 0 \qquad \text{for any} \quad \varepsilon > 0, \tag{8.1.2}$$

as $n \to \infty$ *iff*

$$\sum_{j=1}^{k_n} P\{|X_{nj}| \geq \varepsilon\} \to 0 \qquad \text{for any} \quad \varepsilon > 0, \tag{8.1.3}$$

$$\operatorname{Var}\Big(\sum_{j=1}^{k_n} X_{nj} I(|X_{nj}| < 1)\Big) \to 0, \tag{8.1.4}$$

in which we can take

$$A_n = \sum_{j=1}^{k_n} E X_{nj} I(|X_{nj}| < 1) + o(1).$$

The proof of Theorem 8.1.1 will needs the following lemmas.

Lemma 8.1.1. *Let* $\{X_n, n \geq 1\}$ *be a* φ*-mixing sequence. Denote by* $m(Y)$ *the median of a random variable* Y*. Then we have*

$$\begin{aligned} &P\{\max_{1 \leq j \leq n} |S_j - m(S_j - S_{n+k-1} + S_j(k-1))| \geq \varepsilon\Big\} \\ &\leq \frac{2}{1-2C} P\Big\{(k-1) \max_{1 \leq j \leq n+k-1} |X_j| + |S_{n+k-1}| \geq \varepsilon\Big\} \end{aligned} \tag{8.1.5}$$

for large k *and some* C*,* $0 < C < 1/2$*, where* $S_j(k) = \sum_{i=j+1}^{j+k} X_i$*.*

Proof. Take k so large that $C := \varphi(k) < 1/2$. Let

$$T = \begin{cases} \min_{1 \leq j \leq n} \{j : S_j - m(S_j - S_{n+k-1} + S_j(k-1)) \geq \varepsilon\}, \\ n+1, \qquad \text{if the above set is empty.} \end{cases}$$

Denote

$$\begin{aligned} B_j = \{&S_j - S_{n+k-1} + S_j(k-1) \\ &\leq m(S_j - S_{n+k-1} + S_j(k-1))\}, \quad 1 \leq j \leq n. \end{aligned}$$

It is clear that $P\{B_j\} \geq 1/2$. Since $\{T = j\} \in \mathcal{F}_1^j = \sigma\{X_1, \cdots, X_j\}, B_j \in \mathcal{F}_{j+k}^{n+k-1}$, we have

$$\begin{aligned} \bigcup_{j=1}^{n} (B_j \bigcap (T = j)) &\subset \{S_{n+k-1} - S_j(k-1) \geq \varepsilon\} \\ &\subset \Big\{(k-1) \max_{1 \leq j \leq n+k-1} |X_j| + S_{n+k-1} \geq \varepsilon\Big\}. \end{aligned}$$

From the φ-mixing property it follows that

$$\begin{aligned} P\Big\{(k-1)\max_{1\le j\le n+k-1}|X_j| + S_{n+k-1} \ge \varepsilon\Big\} \\ &\ge \sum_{j=1}^{n} P\{B_j, T=j\} \\ &\ge \sum_{j=1}^{n} (P(B_j)-\varphi(k))P\{T=j\} \\ &\ge (\frac{1}{2}-C)P\{1\le T\le n\} \\ &= (\frac{1}{2}-C)P\Big\{\max_{1\le j\le n}(S_j \\ &\quad - m(S_j - S_{n+k-1}+S_j(k-1)) \ge \varepsilon\Big\}. \end{aligned} \tag{8.1.6}$$

By the same way, we have another inequality when X_j is replaced by $-X_j$. (8.1.5) is proved.

Denote

$$B = \Big\{\max_{0\le l\le n}|S_l(n-l)| > \varepsilon\Big\},$$
$$B_j = \Big\{|S_j(n-j)| > \varepsilon,\ \max_{j<k\le n}|S_k(n-k)| \le \varepsilon\Big\}, \quad 0\le j<n,$$
$$B_n = \{|X_n| > \varepsilon\}, \quad F_j = \bigcup_{i=j}^{n} B_i,\ 0\le j\le n \quad F_{n+1} = \emptyset.$$

Lemma 8.1.2. *Let $\{X_n, n\ge 1\}$ be a φ-mixing sequence with $|X_j| \le r < \infty$, $j=1,2,\cdots,n$, and $\varphi(M) < 1/6$ for some $M>0$. Then*

$$P(B) \ge \frac{(1-6\varphi(M))E\{\max_{1\le l\le n} S_l^2\} - 4\varepsilon^2}{3E\{\max_{1\le l\le n} S_l^2\} + 5M^2r^2 + 12(\varepsilon+r)^2 - 2\varepsilon^2}. \tag{8.1.7}$$

Proof. For $1\le i<k\le n$ we have

$$\max_{1\le i\le k}|S_i| \le |S_k| + \max_{0\le i\le k}|S_i(k-i)| \le 2\max_{0\le i\le k}|S_i(k-i)|. \tag{8.1.8}$$

If $0<M<n, 0<i<n-M$, we have

$$\max_{1\le l\le j}|S_l| \le \max_{1\le l\le i}|S_l| + Mr, \quad i<j\le i+M, \tag{8.1.9}$$

and

$$\max_{i+M\le l\le j}|S_l| \le |S_i| + Mr + \max_{i+M<l\le j}|S_{i+M}(l-i-M)| \tag{8.1.10}$$

for $i + M < l < j \le n$. From (8.1.8)-(8.1.10) it follows that

$$\begin{aligned}\max_{1\le l\le j}|S_l| &= \max\Big\{\max_{1\le k\le i+M}|S_k|, \max_{i+M<l\le j}|S_l|\Big\}\\ &\le \max_{1\le k\le i}|S_k| + Mr + \max_{i+M<l\le j}|S_{i+M}(l-i-M)|\\ &\le \max_{1\le k\le i}|S_k| + Mr + 2\max_{i+M\le l\le j}|S_l(j-l)|\\ &\le \max_{1\le k\le i}|S_k| + Mr + 4\max_{i+M\le l\le n}|S_l(n-l)|\end{aligned}$$

for $j = 1, 2, \cdots, n$. Therefore for any given $i, 1 \le i \le n - M$ we have

$$\max_{1\le l\le n}|S_l| \le \max_{1\le k\le i}|S_k| + Mr + 4\max_{i+M\le l\le n}|S_l(n-l)|. \tag{8.1.11}$$

Note that

$$\begin{aligned}&\max_{j\le l\le n}|S_l(n-l)|I_{B_j}\\ &\quad\le |X_{j+1}|I_{B_j} + \max_{j+1\le l\le n}|S_l(n-l)|I_{B_j} \le (r+\varepsilon)I_{B_j}\end{aligned}$$

for $j = 1, 2, \cdots, n$ and $(a+b+c)^2 \le 3(a^2+b^2+c^2)$. By (8.1.11) we obtain

$$\begin{aligned}E\max_{1\le l\le n}S_l^2 I_B &= \sum_{j=0}^{n} E\max_{1\le l\le n}S_l^2 I_{B_j}\\ &\le \sum_{j=0}^{M} E\max_{1\le l\le n}S_l^2 I_{B_j} + \sum_{j=M+1}^{n} E\Big\{\max_{1\le k\le j-M}|S_k|\\ &\quad + Mr + 4\max_{j\le l\le n}|S_l(n-l)|\Big\}^2 I_{B_j}\\ &\le \sum_{j=0}^{M} E\max_{1\le l\le n}S_l^2 I_{B_j} + 3\sum_{j=M+1}^{n} E\max_{1\le k\le j-M}S_k^2 I_{B_j}\\ &\quad + (3M^2r^2 + 12(r+\varepsilon)^2)P(B).\end{aligned} \tag{8.1.12}$$

Put $Y_j = \max_{1\le k\le j-M}|S_k|$, $M < j \le n$. By Lemma 1.2.10 and noting that $P(F_i)$ is non-increasing for i, we get

$$\begin{aligned}
&\sum_{j=M+1}^{n} E \max_{1\le k\le j-M} S_k^2 I_{B_j} \\
&\qquad = \sum_{j=M+1}^{n} EY_j^2 I_{B_j} \\
&\qquad = \sum_{j=M+1}^{n} (EY_j^2 I_{F_j} - EY_j^2 I_{F_{j+1}}) \\
&\qquad = EY_{M+1}^2 I_{F_{M+1}} + \sum_{j=M+2}^{n} E(Y_j^2 - Y_{j-1}^2) I_{F_j} \\
&\qquad \le EY_{M+1}^2 \{P(F_{M+1}) + 2\varphi(M)\} \\
&\qquad\quad + \sum_{j=M+2}^{n} E(Y_j^2 - Y_{j-1}^2)\{P(F_j) + 2\varphi(M)\} \\
&\qquad \le EY_n^2 \{P(F_{M+1}) + 2\varphi(M)\}.
\end{aligned}$$

Moreover

$$\begin{aligned}
\sum_{j=0}^{M} E \max_{1\le l\le n} S_l^2 I_{B_j} &\le (Mr+\varepsilon)^2 P(B) \\
&\le 2(M^2r^2+\varepsilon^2)P(B).
\end{aligned}$$

Inserting these into (8.1.12) we have

$$\begin{aligned}
&E \max_{1\le l\le n} S_l^2 I_B \\
&\qquad \le 3E \max_{1\le l\le n} S_l^2 (P(B) + 2\varphi(M)) \\
&\qquad\quad + (5M^2r^2 + 12(r+\varepsilon)^2 + 2\varepsilon^2)P(B). \qquad (8.1.13)
\end{aligned}$$

On the other hand

$$\begin{aligned}
E \max_{1\le l\le n} S_l^2 I_B &= E \max_{1\le l\le n} S_l^2 - E \max_{1\le l\le n} S_l^2 I_{B^c} \\
&\ge E \max_{1\le l\le n} S_l^2 - 4\varepsilon^2(1-P(B)). \qquad (8.1.14)
\end{aligned}$$

Combining (8.1.14) with (8.1.13) yields (8.1.7).

Lemma 8.1.3. *Let $\{X_n, n \ge 1\}$ be as in Lemma 8.1.2. If S_n almost surely converge to a random variable as $n \to \infty$, then $E\max_{1\le k\le n} S_k^2$ are convergent.*

Proof. Denote $D_{n,n+m} = \bigcup_{i=n}^{n+m}\{|S_i(n+m-i)| > \varepsilon\}$. If $\varphi(M) < 1/6$, by Lemma 8.1.2

$$P\{D_{n,n+m}\} \geq \frac{(1-6\varphi(M))E\{\max_{n\leq l\leq n+m} S_l^2\} - 4\varepsilon^2}{3E\{\max_{n\leq l\leq n+m} S_l^2\} + 5M^2r^2 + 12(\varepsilon+r)^2 - 2\varepsilon^2}.$$

Using the reduction to absurdity, if $E\max_{1\leq k\leq n} S_k^2$ are divergent, then for any given $n \geq 1$

$$E\max_{n\leq l\leq n+m} S_l^2 \geq E\max_{1\leq l\leq n+m} S_l^2 - E\max_{1\leq l\leq n} S_l^2 \geq \delta_0 > 0 \tag{8.1.15}$$

for large m. On the other hand, from $S_n \xrightarrow{\text{a.s.}} S$ we have

$$\begin{aligned} P\{D_{n,n+m}\} &= P\Big\{\max_{n\leq i\leq n+m} |S_i(n+m-i)| > \varepsilon\Big\} \\ &\leq P\Big\{\max_{0<l\leq m} |S_n(l)| > \varepsilon/2\Big\} \\ &\quad + P\{|S_n(m)| > \varepsilon/2\} \to 0, \end{aligned}$$

as $n, m \to \infty$.The contradiction to (8.1.15) implies that Lemma 8.1.3 holds true.

Lemma 8.1.4. *Let* $B = \Big\{\max_{1\leq j\leq n} |X_j| \geq \varepsilon\Big\}, T_n = \sum_{i=1}^n I(|X_i| \geq \varepsilon)$. *If* $\varphi(M) < 1/2$, *we have*

$$P(B) \geq \frac{(1-2\varphi(M))ET_n}{ET_n + M + 1}. \tag{8.1.16}$$

Proof. Denote

$$\begin{aligned} &T_0 = 0,\ T_n(m) = T_{n+m} - T_n, \\ &B_n = \{|X_n| \geq \varepsilon\}, \\ &B_j = \{\max_{j<i\leq n} |X_i| < \varepsilon, |X_j| \geq \varepsilon\}, \\ &F_j = \bigcup_{i=j}^n B_i,\ j = 1, \cdots, n,\ F_{n+1} = \emptyset. \end{aligned}$$

We have

$$\begin{aligned}
ET_nI_B &= \sum_{j=1}^{M} E(T_{j-1} + T_{j-1}(n-j+1))I(B_j) \\
&\quad + \sum_{j=M+1}^{n} E(T_{j-M-1} + T_{j-M-1}(M) \\
&\quad + T_{j-1}(n-j-1))I(B_j) \\
&\le \sum_{j=1}^{M} MP(B_j) + \sum_{j=1}^{M} P(B_j) \\
&\quad + \sum_{j=M+1}^{n} (ET_{j-M+1}I(B_j) + MP(B_j) + P(B_j)) \\
&\le \sum_{j=M+1}^{n} ET_{j-M-1}I(B_j) + (M+1)P(B).
\end{aligned}$$

By Lemma 1.2.10

$$\begin{aligned}
&\sum_{j=M+1}^{n} E(T_{j-M-1}I(B_j)) \\
&\qquad = \sum_{j=M+1}^{n} E(T_{j-M-1}(I(F_j) - I(F_{j+1}))) \\
&\qquad = ET_1I(F_{M+2}) + \sum_{j=M+3}^{n} EI(|X_{j-M-1}| \ge \varepsilon)I(F_j) \\
&\qquad \le \Big\{\sum_{i=1}^{n-M} EI(|X_i| \ge \varepsilon)\Big\}\{P(F_{M+1}) + 2\varphi(M)\} \\
&\qquad \le ET_n\{P(B) + 2\varphi(M)\}.
\end{aligned}$$

Therefore

$$\begin{aligned}
ET_n &= ET_nI(B) \\
&\le ET_n\{P(B) + 2\varphi(M)\} + (M+1)P(B),
\end{aligned}$$

as desired.

Proof of Theorem 8.1.1.

The part of "*if*". It is clear that (8.1.3) implies (8.1.2). In order to show (8.1.1), denote

$$Y_{nj} = X_{nj}I(|X_{nj}| < 1), \quad V_n = \sum_{j=1}^{k_n} Y_{nj}.$$

By (8.1.4) we have $V_n - EV_n \xrightarrow{P} 0$. From (8.1.3) it follows that

$$P(V_n \neq S_n) \leq \sum_{j=1}^{k_n} P\{|X_{nj}| \geq 1\} = o(1), \tag{8.1.17}$$

i.e., $S_n - EV_n \xrightarrow{P} 0$. (8.1.1) holds true for $A_n = EV_n + o(1)$.

The part of "*only if*". From Lemma 8.1.4 we have

$$P\Big\{\max_{1\leq j\leq k_n} |X_{nj}| \geq \varepsilon\Big\} \geq \frac{(1-2\varphi(M))E\sum_{j=1}^{k_n} I(|X_{nj}| \geq \varepsilon)}{E\sum_{j=1}^{k_n} I(|X_{nj}| \geq \varepsilon) + M + 1},$$

so that (8.1.2) implies (8.1.3). In order to show (8.1.4), let Y'_{nj} be an independent copy of Y_{nj} and put

$$Y^*_{nj} = Y_{nj} - Y'_{nj}, \quad V^*_n = \sum_{j=1}^{k_n} Y^*_{nj}.$$

From (8.1.17) and (8.1.1) it follows that $V_n - A_n \xrightarrow{P} 0$. Therefore $V^*_n \xrightarrow{P} 0$. Denote $V^*_{nk} = \sum_{j=1}^{k} Y^*_{nj}$. By Lemma 8.1.1

$$\begin{aligned} &P\Big\{\max_{1\leq k\leq k_n-M} |V^*_{nk}| \geq \varepsilon\Big\} \\ &\qquad \leq \frac{2}{1-2C} P\Big\{(M-1)\max_{1\leq j\leq k_n} |Y^*_{nj}| + |V^*_n| \geq \varepsilon\Big\} \\ &\qquad = o(1). \end{aligned} \tag{8.1.18}$$

Hence by (8.1.2) and (8.1.18)

$$\begin{aligned} &P\Big\{\max_{1\leq k\leq k_n} |V^*_{nk}| \geq \varepsilon\Big\} \\ &\quad \leq P\Big\{\max_{1\leq k\leq k_n-M} |V^*_{nk}| \geq \varepsilon\Big\} \\ &\qquad + \sum_{j=k_n-M+1}^{k_n} P\{|Y^*_{nj}| \geq \varepsilon/M\} = o(1). \end{aligned}$$

Using the same method as in the proof of Lemma 8.1.3 we get that $\mathrm{Var}V^*_n$ tends to 0. Thus $\mathrm{Var}V_n = \mathrm{Var}V^*_n/2$ approaches 0 as well. Theorem 8.1.1 is proved.

8.2 Strong laws of large numbers

First we show the Borel-Cantelli lemma for a φ-mixing sequence.

Theorem 8.2.1. *Let $\{X_n, n \geq 1\}$ be a φ-mixing sequence and let $\{\mathcal{F}_n = \sigma(X_n), n \geq 1\}$ be a sequence of σ-fields. Then for any given $A_n \in \mathcal{F}_n$, $P\{A_n, \text{ i.o.}\} = 0$ iff*

$$\sum_{n=1}^{\infty} P\{A_n\} < \infty. \tag{8.2.1}$$

Furthermore, that $\sum_{n=1}^{\infty} P\{A_n\} = \infty$ implies $P\{A_n, i.o.\} = 1$.

Proof. For the first part, we need only to show that $P\{A_n, \text{ i.o.}\} = 0$ implies $\sum_{n=1}^{\infty} P\{A_n\} < \infty$. Otherwise, if $\sum_{n=1}^{\infty} P\{A_n\} = \infty$. By the assumption, there exists an $l > 0$ such that $\delta := \varphi(l) < 1$, and there is some $j, 0 \leq j \leq l-1$ such that $\sum_{n=1}^{\infty} P\{A_{nl+j}\} = \infty$. We have

$$\begin{aligned} P\Big\{\bigcup_{i=m}^{n} A_{il+j}\Big\} =& P\{A_{nl+j}\} + P\{A_{nl+j}^c \cap A_{(n-1)l+j}\} + \cdots \\ &+ P\{A_{nl+j}^c \cap \cdots \cap A_{(m+1)l+j}^c \cap A_{ml+j}\}. \end{aligned}$$

By the φ-mixing property we get

$$\begin{aligned} &P\Big\{\bigcup_{i=m}^{n} A_{il+j}\Big\} \\ &\quad\geq P\{A_{nl+j}\} + P\{A_{(n-1)l+j}\}(P\{A_{nl+j}^c\} - \delta) + \cdots \\ &\quad\quad + P\{A_{ml+j}\}(P\{A_{(m+1)l+j}^c \cap \cdots \cap A_{nl+j}^c\} - \delta) \\ &\quad\geq \sum_{i=m}^{n} P\{A_{il+j}\}\Big(P\Big\{\Big(\bigcup_{i=m+1}^{n} A_{il+j}\Big)^c\Big\} - \delta\Big). \end{aligned}$$

From $\sum_{k=1}^{\infty} P\{A_{kl+j}\} = \infty$ it follows that $P\Big\{\Big(\bigcup_{i=m}^{\infty} A_{il+j}\Big)^c\Big\} \leq \delta$, i.e., $P\Big\{\bigcup_{i=m}^{\infty} A_{il+j}\Big\} \geq 1 - \delta$, therefore $P\{A_n, \text{ i.o.}\} \geq 1 - \delta$, which contradict with $P\{A_n, \text{ i.o.}\} = 0$.

Secondly, if $\sum_{n=1}^{\infty} P\{A_n\} = \infty$, it follows from the above discussion that

$$P\Big\{\limsup_n A_n\Big\} = P\{A_n, \text{ i.o.}\} \neq 0.$$

It is clear that $\limsup A_n \in \bigcap_{n=1}^{\infty} \bigvee_{i=n}^{\infty} \mathcal{F}_i$. We now show that $\bigcap_{n=1}^{\infty} \bigvee_{i=n}^{\infty} \mathcal{F}_i$ is a trivial σ-field. Otherwise, there exists a set $B \in \bigcap_{n=1}^{\infty} \bigvee_{i=n}^{\infty} \mathcal{F}_i$, $0 < P(B) < 1$. By the φ-mixing property, for every $A \in \bigvee_{i=1}^{n} \mathcal{F}_i$ we have

$$|P(AB) - P(A)P(B)| < \eta P(A), \tag{8.2.2}$$

where $\eta < 1/2$. It is easy to check that the class of sets which satisfies (8.2.2) includes the σ-field $\bigvee_{i=1}^{\infty} \mathcal{F}_i$, so that taking $A = B$ we obtain

$$P(B) - P(B)^2 \leq \eta P(B),$$

which implies $P(B) > 1/2$. On the other hand, we have also $B^c \in \bigcap_{n=1}^{\infty} \bigvee_{i=n}^{\infty} \mathcal{F}_i$. A contradiction leads $P\{A_n, \quad \text{i.o.}\} = 1$ as desired.

Corollary 8.2.1. *Let $\{X_n, n \geq 1\}$ be a φ-mixing sequence of identically distributed random variables. Put $S_n = \sum_{k=1}^{n} X_k$. If $S_n/n \to b$ a.s., where b is a finite constant, then $E|X_1| < \infty$.*

Proof. From $S_n/n \to b$ (a.s.) we have

$$\frac{X_n}{n} = \frac{S_n}{n} - \frac{S_{n-1}}{n-1} \cdot \frac{n-1}{n} \to 0 \qquad \text{a.s.}$$

Therefore $P\{|X_n/n| \geq \varepsilon, \quad \text{i.o.}\} = 0$ for any $\varepsilon > 0$. By Theorem 8.2.1

$$E|X_1| \leq \sum_{n=0}^{\infty} P\{|X_1| \geq n\} < \infty.$$

For the φ-mixing sequence of identically distributed random variables, we have the following Marcinkiewicz strong law of large numbers.

Theorem 8.2.2. *Let $\{X_n, n \geq 1\}$ be a φ-mixing (ρ-mixing) sequence of identically distributed random variables with*

$$\sum_{n=1}^{\infty} \varphi^{1/2}(2^n) < \infty \ (\sum_{n=1}^{\infty} \rho(2^n) < \infty), E|X_1|^r < \infty$$

for some $1 \leq r < 2$. Then

$$\frac{1}{n}\sum_{i=1}^{n}(X_i - EX_i) = o(n^{-(1-1/r)}) \qquad a.s. \tag{8.2.3}$$

Theorem 8.2.2 is an immediate consequence of Corollary 8.3.4 (Corollary 8.4.2), so we omit its proof here.

Corollary 8.2.2. *Let $\{X_n, n \geq 1\}$ be a φ-mixing sequence of identically distributed random variables with $\sum_{n=1}^{\infty} \varphi^{1/2}(2^n) < \infty$. Then $S_n/n \to b$ a.s. iff $E|X_1| < \infty$ and $b = EX_1$.*

Remark 8.2.1. Xue (1994) showed that Theorem 8.2.2 is also true for a φ-mixing sequence $\{X_n, n \geq 1\}$ with $\sum \varphi^{1/2}(2^n) < \infty$, if there exists a random variable X such that for any $x > 0$, $P\{|X_n| \geq x\} \leq P\{|X| \geq x\}$ and $E|X|^r < \infty$ for some $1 \leq r < 2$.

Remark 8.2.2. Iosifescu and Theodorescu (1969) have discussed the convergence of a series of a φ-mixing sequence, for example, they showed that for $\sum X_n$ a.s. convergence is equivalent to convergence in probability under some condition. They have also discussed the following strong law of large numbers: ii $a_n \uparrow \infty$, $\sum_{n=1}^{\infty} \mathrm{Var}X_n/a_n^2 < \infty$ and $\sum_{n=1}^{\infty} \varphi^{1/2}(n) < \infty$ then

$$\frac{1}{a_n}(S_n - ES_n) \to 0 \quad \text{a.s.}$$

Remark 8.2.3. Chen and Wu (1989) have proved a strong law of large numbers for an α-mixing sequence. Suppose that $\sup_n E|X_n|^p < \infty$ for some $p > 1$, and

$$\alpha(n) = \begin{cases} O(n^{-\frac{p}{2p-2}-\varepsilon}) & if \quad 1 < p < 2, \\ O(n^{-\frac{2}{p}-\varepsilon}) & if \quad p \geq 2. \end{cases}$$

Then $(S_n - ES_n)/n = o(1)$ a.s.

8.3 Complete convergence for φ-mixing sequences

The concept of complete convergence was introduced by Hsu and Robbins (1947). They showed: if $\{X_n, n \geq 1\}$ is a sequence of i.i.d. random variables with $EX_1 = 0$, $EX_1^2 < \infty$, then

$$\sum_{n=1}^{\infty} P\{|S_n| \geq \varepsilon n\} < \infty$$

for any $\varepsilon > 0$. Baum and Katz (1965) showed that $EX_1 = 0$, $E|X_1|^{rt} < \infty$ for $r \geq 1$, $1 \leq t < 2$ iff

$$\sum_{n=1}^{\infty} n^{r-2} P\{|S_n| \geq \varepsilon n^{1/t}\} < \infty$$

for any $\varepsilon > 0$. Bai and Su (1985) obtained:

Theorem 8.3.1. *Let $\{X_n, n \ge 1\}$ be a sequence of i.i.d. random variables and $r > 1$, $0 < t < 2$, $h(x)$ be a slowly varying function as $x \to \infty$. Then the following conditions are equivalent:*

(i) $E|X_1|^{rt}h(|X_1|^t) < \infty$,

(ii) $\sum_{n=1}^{\infty} n^{r-2}h(n)P\{|S_n - nb| \ge \varepsilon n^{1/t}\} < \infty$ *for any* $\varepsilon > 0$,

(iii) $\sum_{n=1}^{\infty} n^{r-2}h(n)P\{\sup_{k\ge n} |S_k - kb|/k^{1/t} \ge \varepsilon\} < \infty$ *for any* $\varepsilon > 0$,

where $b = EX_1$ *if* $1 \le t < 2$, *and* 0 *if* $0 < t < 1$.

Since 1970s, the complete convergence of a mixing sequence has been discussed by some research works. The best result corresponding to that for an i.i.d. sequence is due to Shao (1988a).

Let $l(x)$ and $\beta(x)$ be positive even functions such that

$$\left\{\begin{array}{l} l(x),\ \beta(x)/x^{\theta}\ (\text{some}\ \theta > 0)\ \text{and}\ x^2/\beta(x) \\ \text{are monotonically nondecreasing} \end{array}\right. \tag{8.3.1}$$

for $x > 0$. Shao (1988a) proved the following theorems.

Theorem 8.3.2. *Let $\alpha(x) = \operatorname{inv}\beta(x)$, $\{X_n, n \ge 1\}$ be a φ-mixing sequence of identically distributed random variables with $EX_1 = 0$. Suppose that*

$$E\beta(X_1)l(\beta(X_1)) < \infty. \tag{8.3.2}$$

If one of the following conditions is satisfied:

a) $\beta(x)l(\beta(x))/x \uparrow$, $\beta(x)l(\beta(x))/x^2 \downarrow$ *and there exists a* $r \ge 2$ *such that*

$$\sum_{n=1}^{\infty} \frac{1}{nl^r(n)} < \infty \tag{8.3.3}$$

and

$$\sum_{i=1}^{[\log n]} \varphi^{1/2}(2^i) \le \frac{13}{42} \log l(n); \tag{8.3.4}$$

b) $\beta(x)l(\beta(x))/x \uparrow$, $\beta(x)l(\beta(x))/x^2 \downarrow$,

$$\sum_{j=n}^{\infty} \frac{l(j)}{\alpha^2(j)} = O\Big(\frac{nl(n)}{\alpha^2(n)}\Big) \quad \text{as} \quad n \to \infty \tag{8.3.5}$$

$$\sum_{i=1}^{\infty} \varphi^{1/2}(2^i) < \infty; \tag{8.3.6}$$

c) $\beta(x)l(\beta(x))/x^2 \uparrow$ *and there exist* $q_1 > q_2 > 2$ *such that*

$$\beta(x)l(\beta(x))/x^{q_2} \downarrow,$$

$$\sum_{n=1}^{\infty} \frac{l(n)}{n} \Big(\frac{n^{1/2}}{\alpha(n)}\Big)^{q_1} \exp\Big\{3q_1 \sum_{i=1}^{[\log n]} \varphi^{1/2}(2^i)\Big\} < \infty. \tag{8.3.7}$$

Then

$$\sum_{n=1}^{\infty} \frac{l(n)}{n} P\Big\{\max_{1\le i\le n} \beta(S_i) \ge \varepsilon n\Big\} < \infty, \quad \textit{for any} \quad \varepsilon > 0. \tag{8.3.8}$$

Theorem 8.3.3. *Suppose that $l(x)$ is strictly monotone. Let $\{X_n, n \ge 1\}$ be a φ-mixing sequence with a common distribution. If (8.3.8) is satisfied then (8.3.2) holds and*

$$\sum_{n=1}^{\infty} \frac{l(n) - l([n/2])}{n} P\Big\{\sup_{i\ge n} \frac{\beta(S_i)}{i} > \varepsilon\Big\} < \infty, \quad \text{for any} \quad \varepsilon > 0. \tag{8.3.9}$$

Remark 8.3.1. If $l(n) = O(l(n) - l([n/2]))$, (8.3.8) is equivalent to (8.3.9).

Let $l(n) = n^{r-1}, r > 1, \beta(n) = n^t, 1 \le t < 2$ in Theorems 8.3.2 and 8.3.3, we have

Corollary 8.3.1. *Let $\{X_n, n \ge 1\}$ be a φ-mixing sequence with a common distribution. The following results are equivalent:*

(i) $E|X_1|^{rt} < \infty$, $EX_1 = 0$;

(ii) $\sum_{n=1}^{\infty} n^{r-2} P\Big\{\max_{1\le i\le n} |S_i| > \varepsilon n^{1/t}\Big\} < \infty$, for any $\varepsilon > 0$;

(iii) $\sum_{n=1}^{\infty} n^{r-2} P\Big\{\sup_{k\ge n} |S_k|/k^{1/t} > \varepsilon\Big\} < \infty$, for any $\varepsilon > 0$.

Choosing $l(n) = \log n$, $\beta(n) = n^t$, $1 \le t < 2$, we have

Corollary 8.3.2. *Let $\{X_n, n \ge 1\}$ be a φ-mixing sequence with a common distribution and $\varphi(n) \le \frac{1}{49}(\log n)^{-2}$ for large n. Then the following results are equivalent:*

(i) $E|X_1|^t \log(1 + |X_1|) < \infty$, $EX_1 = 0$;

(ii) $\sum_{n=1}^{\infty} \frac{\log n}{n} P\Big\{\max_{1\le i\le n} |S_i| \ge \varepsilon n^{1/t}\Big\} < \infty$, *for any* $\varepsilon > 0$.

Corollary 8.3.3. *Let $\{X_n, n \ge 1\}$ be a φ-mixing sequence with a common distribution and $EX_1 = 0$, $E|X_1|^t < \infty$ $1 \le t < 2$. If $\sum_{n=1}^{\infty} \varphi^{1/2}(2^n) < \infty$ then*

$$\sum_{n=1}^{\infty} \frac{1}{n} P\Big\{\max_{1\le i\le n} |S_i| \ge \varepsilon n^{1/t}\Big\} < \infty$$

for any $\varepsilon > 0$.

Remark 8.3.2. Corollary 8.3.1 can be strengthened as follows: the same result as Theorem 8.3.1 holds true for a φ-mixing sequence with a common distribution, i.e., if $h(x) > 0$ is a slowly varying function, the following are equivalent:

(i)' $E|X_1|^{rt}h(|X_1|^t) < \infty$, $EX_1 = 0$;

(ii)' $\sum_{n=1}^{\infty} n^{r-2}h(n)P\Big\{\max_{1\le i\le n}|S_i| \ge \varepsilon n^{1/t}\Big\} < \infty$, for any $\varepsilon > 0$;

(iii)' $\sum_{n-1}^{\infty} n^{r-2}h(n)P\Big\{\max_{k\ge n}|S_k|/k^{1/t} \ge \varepsilon\Big\} < \infty$, for any $\varepsilon > 0$.

where $r > 1$, $1 < t < 2$.

Corollary 8.3.2 can be strengthened similarly.

Proof of Theorem 8.3.2.

It is easy to see that (8.3.1) implies

$$\alpha(x)/x^{1/2}\uparrow, \quad \alpha(x)/x^{1/\theta}\downarrow. \tag{8.3.10}$$

Put

$$X_{ni} = X_i I\{|X_i| \le \alpha(n)\}, \quad S_{ni} = \sum_{j=1}^{i} X_{nj}.$$

We have

$$\begin{aligned}
&\sum_{n=1}^{\infty}\frac{l(n)}{n}P\Big\{\max_{i\le n}\beta(S_i) > \varepsilon n\Big\}\\
&\quad\le \sum_{n=1}^{\infty}\frac{l(n)}{n}P\Big\{\max_{i\le n}|X_i| > \alpha(n)\Big\}\\
&\qquad + \sum_{n=1}^{\infty}\frac{l(n)}{n}P\Big\{\max_{i\le n}|S_{ni}| > \alpha(\varepsilon n)\Big\}\\
&\quad =: I_1 + I_2.
\end{aligned} \tag{8.3.11}$$

It is clear that

$$\begin{aligned}
I_1 &\le \sum_{n=1}^{\infty} l(n)P\{|X_i| > \alpha(n)\}\\
&= \sum_{n=1}^{\infty} l(n)\sum_{j=n}^{\infty} P\{j \le \beta(X_1) < j+1\}\\
&\le \sum_{j=1}^{\infty} jl(j)P\{j \le \beta(X_1) < j+1\}\\
&\le E\beta(X_1)l(\beta(X_1)) < \infty.
\end{aligned} \tag{8.3.12}$$

Now we estimate I_2. Without loss of generality we can assume that $0 < \varepsilon < 1/2$. By (8.3.10)

$$\begin{aligned} n|EX_1I\{|X_1| \le \alpha(n)\}| &\le nE|X_1|I\{|X_1| > \alpha(n)\} \\ &\le \frac{\alpha(n)}{l(n)} E\beta(X_1)l(\beta(X_1))I\{|X_1| > \alpha(n)\} \\ &\le \frac{\alpha(\varepsilon n)}{\varepsilon^{1/\theta} l(1)} E\beta(X_1)l(\beta(X_1))I\{|X_1| > \alpha(n)\}. \end{aligned}$$

Noting that $l(1) > 0$, $\lim_{n\to\infty} E\beta(X_1)l(\beta(X_1))I\{|X_1| > \alpha(n)\} = 0$, we have

$$n|EX_1I\{|X_1| \le \alpha(n)\}| \le \frac{1}{2}\alpha(\varepsilon n)$$

for large n. Therefore

$$I_2 \le c\sum_{n=1}^{\infty} \frac{l(n)}{n} P\Big\{\max_{i\le n} |S_{ni} - ES_{ni}| > \frac{1}{2}\alpha(\varepsilon n)\Big\}. \tag{8.3.13}$$

By Lemma 2.2.2 and Lemma 2.2.10, for given $q \ge 2$ (specified later on), there exists a $C_q = C(q)$ depending only on q such that

$$\begin{aligned} &P\Big\{\max_{1\le i\le n} |S_{ni} - ES_{ni}| > \frac{1}{2}\alpha(\varepsilon n)\Big\} \\ &\quad \le C_q(\alpha(\varepsilon n))^{-q}\Big(\Big(n\exp\Big(3\sum_{i=1}^{[\log n]} \varphi^{1/2}(2^i)\Big) \\ &\qquad \cdot EX_1^2 I\{|X_1| \le \alpha(n)\}\Big)^{q/2} \\ &\qquad + nE|X_1|^q I\{|X_1| \le \alpha(n)\}\Big). \end{aligned} \tag{8.3.14}$$

If condition a) is satisfied, let $q = 28(r+1)$. From (8.3.10), (8.3.13) and (8.3.14) we have

$$\begin{aligned} I_2 &\le c\sum_{n=1}^{\infty} \frac{l(n)}{n}\Big(\frac{n}{\alpha^2(n)} \exp\Big(3\sum_{i=1}^{[\log n]} \varphi^{1/2}(2^i)\Big) EX_1^2 I\{|X_1| \le \alpha(n)\}\Big)^{q/2} \\ &\quad + c\sum_{n=1}^{\infty} \frac{l(n)}{\alpha^q(n)} E|X_1|^q I\{|X_1| \le \alpha(n)\} \\ &=: I_3 + I_4. \end{aligned} \tag{8.3.15}$$

By (8.3.4)

$$\begin{aligned} I_3 &\le c\sum_{n=1}^{\infty}\frac{\exp\Big((3q/2)\sum_{i=1}^{[\log n]}\varphi^{1/2}(2^i)\Big)}{nl(n)^{\frac{q-2}{2}}}(E\beta(X_1)l(\beta(X_1)))^{q/2} \\ &\le c\sum_{n=1}^{\infty}\frac{1}{nl(n)^{\frac{q}{28}-1}} \\ &\le c\sum_{n=1}^{\infty}\frac{1}{nl(n)^{r}} < \infty. \end{aligned} \tag{8.3.16}$$

Since $\beta(x)l(\beta(x))/x^2 \downarrow$ implies $xl(x)/\alpha^2(x) \downarrow$, it follows from (8.3.10) that

$$\begin{aligned} \sum_{j=n}^{\infty}\frac{l(j)}{\alpha^q(j)} &= \sum_{j=n}^{\infty}\frac{jl(j)}{\alpha^2(j)}\cdot\frac{j^{\frac{q-2}{2}}}{\alpha^{q-2}(j)j^{q/2}} \\ &\le \frac{n^{q/2}l(n)}{\alpha^q(n)}\sum_{j=n}^{\infty}\frac{1}{j^{q/2}} \\ &= O\Big(\frac{nl(n)}{\alpha^q(n)}\Big). \end{aligned} \tag{8.3.17}$$

Therefore

$$\begin{aligned} I_4 &\le c\sum_{n=1}^{\infty}\frac{l(n)}{\alpha^q(n)}\sum_{j=0}^{n}E|X_1|^qI(j\le\beta(X_1)<j+1) \\ &\le c\sum_{j=0}^{\infty}\Big(\sum_{n=j}^{\infty}\frac{l(n)}{\alpha^q(n)}\Big)E|X_1|^qI(j\le\beta(X_1)<j+1) \\ &\le c\sum_{j=1}^{\infty}\frac{jl(j)}{\alpha^q(j)}E|X_1|^qI(j\le\beta(X_1)<j+1) \\ &\le cE\beta(X_1)l(\beta(X_1)) < \infty. \end{aligned} \tag{8.3.18}$$

From (8.3.15), (8.3.16) and (8.3.18) it follows that $I_2 < \infty$, i.e., Theorem 8.3.2 holds true under condition a).

When condition b) is satisfied, let $q = 2$. By the similar discussion we have $I_2 < \infty$. If condition c) is satisfied, let $q = q_1$. Note that from (8.3.7) we have $I_3 < \infty$. Since $\beta(x)l(\beta(x))/x^{q_2} \downarrow$ implies $xl(x)/\alpha(x)^{q_2} \downarrow$, by (8.3.10) we have

$$\sum_{j=n}^{\infty}\frac{l(j)}{\alpha^{q_1}(j)} \le \frac{n^{1+\frac{q_1-q_2}{2}}l(n)}{\alpha^{q_1}(n)}\sum_{j=n}^{\infty}j^{-1-\frac{q_1-q_2}{2}} = O\Big(\frac{nl(n)}{\alpha^{q_1}(n)}\Big).$$

It follows that $I_4 < \infty$. Therefore $I_2 < \infty$. The proof of Theorem 8.3.2 is completed.

Proof of Theorem 8.3.3.

First we show that (8.3.8) implies (8.3.9). Put $d_n = \sum_{j=2^n}^{2^{n+1}-1} l(j)$. We have

$$\begin{aligned}
&\sum_{n=2}^{\infty} \frac{l(n) - l([n/2])}{n} P\Big\{\sup_{i\ge n} \frac{\beta(S_i)}{i} \ge \varepsilon\Big\} \\
&\le \sum_{j=1}^{\infty} \sum_{n=2^j}^{2^{j+1}-1} \frac{l(n) - l([n/2])}{2^j} P\Big\{\sup_{i\ge 2^j} \frac{\beta(S_i)}{i} \ge \varepsilon\Big\} \\
&= \sum_{j=1}^{\infty} 2^{-j} \Big(\sum_{n=2^j}^{2^{j+1}-1} l(n) - 2 \sum_{n=2^{j-1}}^{2^j-1} l(n)\Big) P\Big\{\sup_{i\ge 2^j} \frac{\beta(S_i)}{i} \ge \varepsilon\Big\} \\
&\le \sum_{j=1}^{\infty} 2^{-j}(d_j - 2d_{j-1}) \sum_{k=j}^{\infty} P\Big\{\max_{2^k\le i<2^{k+1}} \beta(S_i) > \varepsilon 2^k\Big\} \\
&\le \sum_{k=1}^{\infty} 2^{-k} d_k P\Big\{\max_{2^k\le i<2^{k+1}} \beta(S_i) \ge \varepsilon 2^k\Big\} \\
&\le \sum_{k=1}^{\infty} 4 \sum_{2^{k+1}\le n<2^{k+2}-1} \frac{l(n)}{n} P\Big\{\max_{i\le 2^{k+1}} \beta(S_i) \ge \varepsilon 2^k\Big\} \\
&\le 4 \sum_{n=1}^{\infty} \frac{l(n)}{n} P\Big\{\max_{i\le n} \beta(S_i) > \frac{\varepsilon n}{2}\Big\} < \infty. \qquad (8.3.19)
\end{aligned}$$

This proves that (8.3.9) holds true.

Now we show that (8.3.2) holds true. From (8.3.19) it follows that

$$\sum_{k=1}^{\infty} l(2^k) P\Big\{\max_{i\le 2^k} |X_i| \ge \alpha(\varepsilon 2^k)\Big\} < \infty \qquad (8.3.20)$$

for any $\varepsilon > 0$, which implies

$$P\Big\{\max_{i\le 2^k} |X_i| \ge \alpha(\varepsilon 2^k)\Big\} \longrightarrow 0, \quad \text{as } k \to \infty. \qquad (8.3.21)$$

Since $\varphi(p) \to 0$ as $p \to \infty$, there exist k_0 and p_0 such that for any $k \ge k_0$

$$\varphi(p_0) + P\Big\{\max_{i\le 2^k} |X_i| \ge \alpha(\varepsilon 2^k)\Big\} < 1/2.$$

Then, as in (3.2) of Lai (1977) we have

$$P\Big\{\max_{1\le i\le 2^k} |X_i| \ge \alpha(\varepsilon 2^k)\Big\} \ge \frac{2^k}{2p_0} P\{|X_1| \ge \alpha(\varepsilon 2^k)\}.$$

Whence

$$\sum_{k=1}^{\infty} l(2^k)2^k P\{|X_1| \geq \alpha(\varepsilon 2^k)\}$$
$$\leq c\sum_{k=1}^{\infty} l(2^k)P\Big\{\max_{1\leq i\leq 2^k} |X_i| \geq \alpha(\varepsilon 2^k)\Big\} < \infty. \qquad (8.3.22)$$

On the other hand

$$E\beta(X_1)l(\beta(X_1))$$
$$\leq l(1) + \sum_{j=1}^{\infty} 2^j l(2^j)P\{2^{j-1} < \beta(X_1) \leq 2^j\}$$
$$\leq l(1) + \sum_{j=1}^{\infty} 2^j l(2^j)P\{|X_1| \geq \alpha(2^{j-1})\}.$$

From (8.3.22) we obtain $E\beta(X_1)l(\beta(X_1)) < \infty$. (8.3.2) is proved. This completes the proof of Theorem 8.3.3.

8.4 Complete convergence for ρ-mixing sequences

The conclusions of complete convergence for a ρ-mixing sequence have not arrived completely as for the φ-mixing case, but some ideal sufficient conditions can be given.

Definition 8.4.1. A function $l(x) > 0\,(x > 0)$ is said to be quasi-monotone non-decreasing, if

$$\limsup_{x\to\infty} \sup_{0<t\leq x} l(t)/l(x) < \infty.$$

A function $l(x)$ is said to be quasi-monotone non-increasing, if

$$\limsup_{x\to\infty} \sup_{t\geq x} l(t)/l(x) < \infty.$$

Throughout this section, we assume that $l(x)$ is a positive, even and quasi-monotone non-decreasing function, $\beta(x)$ is a positive even function and $\beta(x)/x^\theta$, $x^2/\beta(x)$ are monotone non-decreasing for some $\theta > 0$. Denote $\alpha(x) = \mathrm{inv}\beta(x)$.

Shao (1989c) proved the following theorems.

Theorem 8.4.1. *Suppose that $\beta(x)l(\beta(x))$ and $x^{2-\varepsilon_0}/\beta(x)l(\beta(x))$ for some $0 < \varepsilon_0 < 1$ are quasi-monotone non-decreasing functions, $\{X_n, n \geq 1\}$ is a ρ-mixing sequence with a common distribution and $EX_1 = 0$, $E\beta(X_1)l(\beta(X_1)) < \infty$. If*

$$\sum_{n=1}^{\infty} \rho(2^n) < \infty, \tag{8.4.1}$$

then for any $\varepsilon > 0$

$$\sum_{n=1}^{\infty} \frac{l(n)}{n} P\Big\{\max_{1\leq i\leq n} \beta(S_i) \geq \varepsilon n\Big\} < \infty. \tag{8.4.2}$$

Theorem 8.4.2. *Suppose that there exist $q \geq 2$, $q_0 \geq 2$ and $\delta > 0$ such that $x^q/\beta(x)l(\beta(x))$, $\beta(x)l(\beta(x))/x^2$ are quasi-monotone non-decreasing functions, and $l(x) \geq x^\delta (x > 0)$,*

$$\sum_{n=1}^{\infty} \frac{l(n)}{n} \Big(\frac{n^{1/2}}{\alpha(n)}\Big)^{q_0} < \infty. \tag{8.4.3}$$

Let $\{X_n\}$ be a ρ-mixing sequence with a common distribution and $EX_1 = 0$, $E\beta(X_1)l(\beta(X_1)) < \infty$. If

$$\sum_{n=1}^{\infty} \rho^{2/r}(2^n) < \infty \tag{8.4.4}$$

for some $r > q$, then (8.4.2) holds.

Corollary 8.4.1. *Let $\{X_n, n \geq 1\}$ be a ρ-mixing sequence with a common distribution and $EX_1 = 0$, $E|X_1|^p h(|X_1|^{1/\alpha}) < \infty$ for $p \geq 1$, $p\alpha > 1$, $\alpha > 1/2$ and a slowly varying function $h(x) > 0$. If*

$$\sum_{n=1}^{\infty} \rho^{2/r}(n) < \infty \tag{8.4.5}$$

for $r = 2$ when $1 \leq p < 2$; $r > p$, when $p \geq 2$, then

$$\sum_{n=1}^{\infty} n^{p\alpha-2} h(n) P\Big\{\max_{1\leq i\leq n} |S_i| \geq \varepsilon n^\alpha\Big\} < \infty \tag{8.4.6}$$

for any $\varepsilon > 0$.

An immediate consequence of the above complete convergence result is the following Marcinkiewicz-Zygmund law of large numbers.

Corollary 8.4.2. *Let $\{X_n, n \geq 1\}$ be a ρ-mixing sequence with a common distribution and $EX_1 = 0, E|X_1|^p < \infty, 1 \leq p < 2$. Assume that*

$$\sum_{n=1}^{\infty} \rho(2^n) < \infty.$$

Then

$$\lim_{n \to \infty} S_n / n^{1/p} = 0. \qquad a.s.$$

Corollary 8.4.3. *Let $\{X_n, n \geq 1\}$ be a ρ-mixing sequence with a common distribution and $EX_1 = 0$, $E|X_1|^p h(|X_1|^p) < \infty$, $1 \leq p < 2$, and $h(x)$ be a monotone non-decreasing slowly varying function. Suppose that*

$$\sum_{n=1}^{\infty} \rho(2^n) < \infty.$$

Then for any $\varepsilon > 0$

$$\sum_{n=1}^{\infty} \frac{h(n)}{n} P\Big\{\max_{1 \leq i \leq n} |S_i| \geq \varepsilon n^{1/p}\Big\} < \infty.$$

If $E|X_1|^{p+\delta} < \infty$ for some $\delta > 0$ instead of the condition $E|X_1|^p h(|X_1|^{1/\alpha}) < \infty$ in Corollary 8.4.1, condition (8.4.5) can be deleted. Furthermore Kong and Zhang (1994) pointed out that (8.4.6) also holds true for non-identically distributed random variables.

The proofs of Theorems 8.4.1 and 8.4.2 need the following lemma.

Lemma 8.4.1. *Let $\{\xi_n, n \geq 1\}$ be a ρ-mixing sequence with $E\xi_n = 0$, $Eg(|\xi_n|) < \infty$, where $g(x)$ is a function for which there exists a constant $0 < C < \infty$ such that $\sup_{x \geq t} \frac{x}{g(x)} \leq C \frac{t}{g(t)}$ for any $t > 0$. Denote $T_k(i) = \sum_{j=k+1}^{k+i} \xi_j$. Then for any $q_1, q_2 \geq 2$ there exists a constant K depending* *only on q_1, q_2 and $\rho(\cdot)$ such that*

$$
\begin{aligned}
&P\Big\{\max_{i\le n}|T_0(i)|\ge x\Big\}\\
&\quad\le\sum_{i=1}^{n}P\{|\xi_i|\ge A\}+K\Big\{x^{-q_1}\Big(n^{q_1/2}\\
&\qquad\cdot\exp\Big(K\sum_{i=1}^{[\log n]}\rho(2^i)\Big)\\
&\qquad\cdot\max_{i\le n}\|\xi_i I(|\xi_i|\le B)\|_2^{q_1}\\
&\qquad+n\exp\Big(K\sum_{i=1}^{[\log n]}\rho^{2/q_1}(2^i)\Big)\\
&\qquad\cdot\log^{q_1}(2n)\max_{i\le n}E|\xi_i|^{q_1}I(|\xi_i|\le B)\Big)\\
&\qquad+x^{-q_2}\Big(n^{q_2/2}\exp\Big(K\sum_{i=1}^{[\log n]}\rho(2^i)\Big)\\
&\qquad\cdot\max_{i\le n}\|\xi_i I(B<|\xi_i|<A)\|_2^{q_2}\\
&\qquad+\Big(\frac{n}{k}\Big)^{q_2/2}k\exp\Big(K\sum_{i=1}^{[\log n]}\rho^{2/q_2}(2^i)\Big)\\
&\qquad\cdot\max_{i\le n}E|\xi_i|^{q_2}I(B<|\xi_i|<A)\Big)\\
&\qquad+x^{-2}\cdot n\rho^2(k)\cdot\exp\Big(K\sum_{i=1}^{[\log n]}\rho(2^i)\Big)\\
&\qquad\cdot\log^4\Big[\frac{n}{k}\Big]\cdot\max_{i\le n}E\xi_i^2 I(B<|\xi_i|<A)\Big\}
\end{aligned}
$$

for any k ($\le n$), $x>0$ and $A\ge B>0$ which satisfy the following conditions

$$\frac{4nCA}{g(A)}\max_{1\le i\le n}Eg(|\xi_i|)I(|\xi_i|\ge A)\le x, \tag{8.4.7}$$

$$\frac{48kBC}{g(B)}\max_{1\le i\le n}Eg(|\xi_i|)I(|\xi_i|\ge B)\le x. \tag{8.4.8}$$

Proof. For simplicity, we assume that $\{\xi,\xi_i,i\ge 1\}$ have a common distribution. Put

$$\xi_{i1} = \xi_i I(|\xi_i| \le B) - E\xi_i I(|\xi_i| \le B);$$
$$\xi_{i2} = \xi_i I(B < |\xi_i| < A) - E\xi_i I(B < |\xi_i| < A);$$
$$\xi_{i3} = \xi_i I(|\xi_i| \ge A) - E\xi_i I(|\xi_i| \ge A);$$
$$T_{ik} = \sum_{j=1}^{i} \xi_{jk}, \qquad k = 1, 2, 3.$$

It is clear that

$$\begin{aligned} &P\Big\{\max_{i\le n}|T_0(i)| \ge x\Big\} \\ &\quad \le P\Big\{\max_{i\le n}|T_{i1}| \ge \frac{x}{4}\Big\} + P\Big\{\max_{i\le n}|T_{i2}| \ge \frac{x}{4}\Big\} \\ &\qquad + P\Big\{\max_{i\le n}|T_{i3}| \ge \frac{x}{2}\Big\} \\ &\quad =: I_1 + I_2 + I_3. \end{aligned} \tag{8.4.9}$$

From (8.4.7)

$$\begin{aligned} I_3 &\le P\Big\{\sum_{i=1}^{n}|\xi_i|I(|\xi_i| \ge A) \ge \frac{x}{2} \\ &\qquad - \sum_{i=1}^{n} E|\xi_i|I(|\xi_i| \ge A)\Big\} \\ &\le P\Big\{\sum_{i=1}^{n}|\xi_i|I(|\xi_i| \ge A) \ge \frac{x}{2} \\ &\qquad - \sum_{i=1}^{n}\frac{CA}{g(A)}Eg(|\xi_i|)I(|\xi_i| \ge A)\Big\} \\ &\le P\Big\{\sum_{i=1}^{n}|\xi_i|I(|\xi_i| \ge A) \ge \frac{x}{4}\Big\} \\ &\le \sum_{i=1}^{n}P(|\xi_i| \ge A). \end{aligned} \tag{8.4.10}$$

By Lemma 2.2.5 and Lemma 4.1.2, there exists a constant K_1 depending only on q_1 and $\rho(\cdot)$ such that

$$I_1 \le K_1 x^{-q_1}\Big(n^{q_1/2}\|\xi I(|\xi| \le B)\|_2^{q_1} \exp\Big(K_1 \sum_{i=0}^{[\log n]} \rho(2^i)\Big)$$
$$+ n \log^{q_1}(2n)\|\xi I(|\xi| \le B)\|_{q_1}^{q_1}$$
$$\cdot \exp\Big(K_1 \sum_{i=0}^{[\log n]} \rho^{2/q_1}(2^i)\Big)\Big). \tag{8.4.11}$$

In order to estimate I_2, put

$$Y_i = \sum_{j=2ik+1}^{(2i+1)k} \xi_{j2}, \qquad i = 0, 1, \cdots, p_1,$$
$$Z_i = \sum_{j=(2i+1)k+1}^{2(i+1)k} \xi_{j2}, \qquad i = 0, 1, \cdots, p_2 \tag{8.4.12}$$

where

$$p_1 = \Big[(\frac{n}{k} - 1)/2\Big], \; p_2 = \Big[(\frac{n}{k} - 2)/2\Big].$$

Denote

$$W_i = \sum_{j=0}^{i} Y_j, \quad W_i^* = \sum_{j=0}^{i} Z_j.$$

Then

$$I_2 \le P\Big\{\max_{0\le i\le p_1} |W_i| \ge \frac{x}{12}\Big\} + P\Big\{\max_{0\le i\le p_2} |W_i^*| \ge \frac{x}{12}\Big\}$$
$$+ P\Big\{\max_{0\le i\le [n/k]} \max_{ik+1\le j<(i+1)k} \Big|\sum_{l=ik+1}^{j} \xi_{l2}\Big| \ge \frac{x}{12}\Big\}$$
$$=: I_4 + I_5 + I_6. \tag{8.4.13}$$

By Lemma 2.2.5 and (8.4.8) we have

$$
\begin{aligned}
I_6 \le 2\left[\frac{n}{k}\right] \max_{0\le i\le [n/k]} P\Big\{ & \sum_{j=ik+1}^{(i+1)k} |\xi_j| I(B < |\xi_i| < A) \\
& - E|\xi_j| I(B < |\xi_j| < A) \\
& \ge \frac{x}{12} - 2 \sum_{j=ik+1}^{(i+1)k} E|\xi_j| I(B < |\xi_j| < A)\Big\} \\
\le 2\cdot \left[\frac{n}{k}\right] \max_{0\le i\le [n/k]} P\Big\{ & \sum_{j=ik+1}^{(i+1)k} |\xi_j| I(B < |\xi_j| < A) \\
& - E|\xi_j| I(B < |\xi_j| < A) \ge \frac{x}{24}\Big\} \\
\le K_2 \left[\frac{n}{k}\right] x^{-q_2} \Big(& k^{q_2/2} \|\xi I(B < |\xi| < A)\|_2^{q_2} \\
& \cdot \exp\Big(K_2 \sum_{i=0}^{[\log n]} \rho(2^j)\Big) \\
& + k\|\xi I(B < |\xi| < A)\|_{q_2}^{q_2} \exp\Big(K_2 \sum_{i=0}^{[\log n]} \rho^{2/q_2}(2^j)\Big)\Big) \\
\le K_2 x^{-q_2} \Big(& n^{q_2/2} \|\xi I(B < |\xi| < A)\|_2^{q_2} \\
& \cdot \exp\Big(K_2 \sum_{i=0}^{[\log n]} \rho(2^j)\Big) \\
& + n\cdot \|\xi I(B < |\xi| < A)\|_{q_2}^{q_2} \\
& \cdot \exp\Big(K_2 \sum_{i=0}^{[\log n]} \rho^{2/q_2}(2^j)\Big). \qquad (8.4.14)
\end{aligned}
$$

Now we estimate I_4. Denote $\mathcal{F}_i = \sigma(\xi_j, j \le 2(i+1)k)$, $i = 0, 1, \cdots, p_1$, $\mathcal{F}_{-1}$ to be a trivial σ-field. Put

$$
U_i = Y_i - E(Y_i|\mathcal{F}_{i-1}), \quad G_i = \sum_{j=0}^{i} U_j,
$$

$$
H_i = \sum_{j=1}^{i} E(Y_j|\mathcal{F}_{j-1}), \quad i = 0, 1, \cdots, p_1.
$$

It is easy to see that

$$
\begin{aligned}
I_4 &\le P\Big\{\max_{0\le i\le p_1} |G_i| \ge \frac{x}{24}\Big\} + P\Big\{\max_{0\le i\le p_1} |H_i| \ge \frac{x}{24}\Big\} \\
&=: I_7 + I_8. \qquad (8.4.15)
\end{aligned}
$$

Note that $\{U_i, \mathcal{F}_i, i \geq 0\}$ is a martingale difference sequence. Hence by the maximum velue inequality for a martingale difference sequence, the Marcinkiewicz-Zygmund inequality and Lemma 2.2.5, we have

$$\begin{aligned}
I_7 &\leq \left(\frac{x}{24}\right)^{-q_2}\left(\frac{q_2}{q_2-1}\right)^{q_2} E|G_{p_1}|^{q_2} \\
&\leq K_3 x^{-q_2} \cdot p_1^{q_2/2} \max_{0\leq i\leq p_1} E|U_i|^{q_2} \\
&\leq K_3 x^{-q_2} \cdot p_1^{q_2/2}\Big(k^{q_2/2}\|\xi I(B<|\xi|<A)\|_2^{q_2} \\
&\quad \cdot \exp\Big(K_3 \sum_{i=0}^{[\log n]} \rho(2^j)\Big) \\
&\quad + k\|\xi I(B<|\xi|<A)\|_{q_2}^{q_2} \\
&\quad \cdot \exp\Big(K_3 \sum_{i=0}^{[\log n]} \rho^{2/q_2}(2^i)\Big)\Big) \\
&\leq K_3 x^{-q_2}\Big(n^{q_2/2}\|\xi I(B<|\xi|<A)\|_2^{q_2} \\
&\quad \cdot \exp\Big(K_3 \sum_{i=0}^{[\log n]} \rho(2^i)\Big) \\
&\quad + \left(\frac{n}{k}\right)^{q_2/2} k \cdot E|\xi|^{q_2} I(B<|\xi|<A) \\
&\quad \cdot \exp\Big(K_3 \sum_{i=0}^{[\log n]} \rho^{2/q_2}(2^j)\Big). \qquad (8.4.16)
\end{aligned}$$

By imitating the proof of Lemma 2.2.2 and using Lemma 2.2.5, there exists a constant K_4 depending only on $\rho(\cdot)$ such that

$$\begin{aligned}
&E\Big(\sum_{j=i+1}^{i+m} E(Y_j|\mathcal{F}_{j-1})\Big)^2 \\
&\quad \leq K_4 m k \rho^2(k) E\xi^2 I(B<|\xi|<A) \\
&\quad\quad \cdot \log^2(2m)\exp\Big(K_4 \sum_{i=0}^{[\log n]} \rho(2^i)\Big). \qquad (8.4.17)
\end{aligned}$$

From Lemma 4.1.2 it follows that

$$\begin{aligned}
E\max_{1\leq i\leq p_1} H_i^2 \leq &3K_4 p_1 k\rho^2(k)\log^4(2p_1) E\xi^2 I(B<|\xi|<A) \\
&\cdot \exp\Big(K_4 \sum_{i=0}^{[\log n]} \rho(2^i)\Big). \qquad (8.4.18)
\end{aligned}$$

Thus we obtain

$$I_8 \le K_4 x^{-2} n\rho^2(k)\log^4\Big[\frac{n}{k}\Big] E\xi^2 I(B<|\xi|<A) \cdot \exp\Big(K_4 \sum_{i=0}^{[\log n]} \rho(2^i)\Big). \tag{8.4.19}$$

(When $[n/k] \le 2$, we have $p_1 = 0$ and hence $I_8 = 0$, i.e., (8.4.19) holds true; when $[n/k] > 2$, (8.4.18) implies (8.4.19).) From (8.4.15), (8.4.16) and (8.4.19) it follows that

$$\begin{aligned} I_4 \le K_5 x^{-q_2}\Big(& n^{q_2/2}\|\xi_i I(B<|\xi_i|<A)\|_2^{q_2} \\ & \cdot \exp\Big(K_5 \sum_{i=0}^{[\log n]} \rho(2^i)\Big) \\ & + \Big(\frac{n}{k}\Big)^{q_2/2} k \|\xi_i I(B<|\xi_i|<A)\|_{q_2}^{q_2} \\ & \cdot \exp\Big(K_5 \sum_{i=0}^{[\log n]} \rho^{2/q_2}(2^i)\Big) \\ & + K_5 x^{-2} n\rho^2(k)\log^4\Big[\frac{n}{k}\Big] \|\xi_i I(B<|\xi_i|<A)\|_{q_2}^{q_2} \\ & \cdot \|\xi_i I(B<|\xi_i|<A)\|_2^2 \\ & \cdot \exp\Big(K_5 \sum_{i=0}^{[\log n]} \rho(2^i)\Big). \end{aligned} \tag{8.4.20}$$

By the same way, we have also (8.4.20) for I_5. Lemma 8.4.1 is proved.

Proof of Theorem 8.4.1.

From the assumption that $l(x), \beta(x)l(\beta(x))/x, x^{2-\varepsilon_0}/\beta(x)l(\beta(x))$ are quasi-monotone non-decreasing, it follows that there exists a constant $C > 0$ such that for any $t \ge x > 0$

$$l(x) \le Cl(t), \tag{8.4.21}$$
$$\beta(x)l(\beta(x))/x \le C\beta(t)l(\beta(t))/t, \tag{8.4.22}$$
$$x^{2-\varepsilon_0}/\beta(x)l(\beta(x)) \le Ct^{2-\varepsilon_0}/\beta(t)l(\beta(t)). \tag{8.4.23}$$

By (8.4.22), (8.4.23) and the monotonicity of $\beta(x)$ we have

$$\frac{\alpha(x)}{xl(x)} \ge \frac{1}{C}\frac{\alpha(t)}{tl(t)}, \tag{8.4.24}$$

$$\frac{\alpha^{2-\varepsilon_0}(x)}{xl(x)} \le C\frac{\alpha^{2-\varepsilon_0}(t)}{tl(t)}. \tag{8.4.25}$$

Let $C_n = \log^{-8/\varepsilon_0} n$ and take $A = \alpha(n), B = \alpha(nC_n), k = [nC_n], x = \alpha(\varepsilon n), g(x) = \beta(x)l(\beta(x))$ in Lemma 8.4.1. Then (8.4.7) and (8.4.8) are satisfied for large n. In fact, for $n \geq 1/\varepsilon$, from (8.4.24) and (8.4.21) we have

$$\begin{aligned}
&\frac{4Cn\alpha(n)}{nl(n)} E\beta(X_1)l(\beta(X_1))I(|X_1| \geq \alpha(n)) \\
&\quad \leq \frac{4C^2\alpha(\varepsilon n)}{\varepsilon l(\varepsilon n)} E\beta(X_1)l(\beta(X_1))I(|X_1| \geq \alpha(n)) \\
&\quad \leq \frac{4C^3\alpha(\varepsilon n)}{\varepsilon l(1)} E\beta(X_1)l(\beta(X_1))I(|X_1| \geq \alpha(n)).
\end{aligned}$$

Noting that $\lim_{n\to\infty} E\beta(X_1)l(\beta(X_1))I(|X_1| \geq \alpha(n)) = 0$, for large n we obtain

$$\frac{4Cn\alpha(n)}{nl(n)} E\beta(X_1)l(\beta(X_1))I(|X_1| \geq \alpha(n)) \leq \alpha(\varepsilon n).$$

This proves that (8.4.7) is satisfied. Similarly we can verify (8.4.8). By Lemma 8.4.1, letting $q_1 = q_2 = 2$, we have

$$\begin{aligned}
&\sum_{n=1}^{\infty} \frac{l(n)}{n} P\Big\{\max_{i\leq n} \beta(S_i) \geq \varepsilon n\Big\} \\
&\quad = \sum_{n=1}^{\infty} \frac{l(n)}{n} P\Big\{\max_{i\leq n} |S_i| \geq \alpha(\varepsilon n)\Big\} \\
&\quad \leq \sum_{n=1}^{\infty} l(n) P\{|X_1| \geq \alpha(n)\} \\
&\qquad + c\sum_{n=1}^{\infty} \frac{l(n)}{\alpha^2(\varepsilon n)} EX_1^2 I(|X_1| \leq \alpha(nC_n)) \log^2 n \\
&\qquad + c\sum_{n=1}^{\infty} \frac{l(n)}{\alpha^2(\varepsilon n)} EX_1^2 I(|X_1| \leq \alpha(n)) \\
&\qquad + c\sum_{n=1}^{\infty} \frac{l(n)}{\alpha^2(\varepsilon n)} \rho^2(nC_n)(\log\log n)^4 EX_1^2 I(|X_1| \leq \alpha(n)) \\
&\quad =: J_1 + J_2 + J_3 + J_4. \qquad (8.4.26)
\end{aligned}$$

From (8.4.21)

$$
\begin{aligned}
J_1 &\le \sum_{n=1}^{\infty} l(n) \sum_{j=n}^{\infty} P\{j \le \beta(X_1) < j+1\} \\
&= \sum_{j=1}^{\infty} \sum_{n=1}^{j} l(n) P\{j \le \beta(X_1) < j+1\} \\
&\le c \sum_{j=1}^{\infty} j l(j) P\{j \le \beta(X_1) < j+1\} \\
&\le cE\beta(X_1) l(\beta(X_1)) < \infty. \qquad (8.4.27)
\end{aligned}
$$

From (8.4.23), (8.4.25) and the monotonicity of $x/\alpha^{\theta}(x), \alpha^2(x)/x$ we have

$$
\begin{aligned}
J_2 &\le c \sum_{n=1}^{\infty} \frac{l(n) \log^2 n \alpha^2(nC_n)}{\alpha^2(\varepsilon n) n C_n l(nC_n)} E\beta(X_1) l(\beta(X_1)) \\
&\le c\varepsilon^{-2/\theta} \sum_{n=1}^{\infty} \frac{\log^2 n}{n} \frac{\alpha^{\varepsilon_0}(nC_n)}{\alpha^{\varepsilon_0}(n)} E\beta(X_1) l(\beta(X_1)) \\
&\le c \sum_{n=1}^{\infty} \frac{\log^2 n}{n} C_n^{\varepsilon_0/2} \\
&\le c \sum_{n=1}^{\infty} \frac{1}{n \log^2 n} < \infty. \qquad (8.4.28)
\end{aligned}
$$

Since $\sum_{n=1}^{\infty} \rho(2^n) < \infty$ we have $\rho(n) \le c \log^{-1} n$. Hence

$$
\begin{aligned}
J_4 &\le c \sum_{n=1}^{\infty} \frac{l(n)}{\alpha^2(\varepsilon n)} \log^{-3/2} n E X_1^2 I(|X_1| \le \alpha(n)) \\
&\le c \sum_{n=1}^{\infty} \frac{l(n)}{\alpha^2(\varepsilon n)} \frac{\alpha^2(n)}{n l(n)} \log^{-3/2} n E\beta(X_1) l(\beta(X_1)) \\
&\le c \sum_{n=1}^{\infty} \frac{1}{n \log^{3/2} n} < \infty. \qquad (8.4.29)
\end{aligned}
$$

At last we estimate J_3. From (8.4.25) and $x/\alpha^2(x) \downarrow$, we have

$$
\begin{aligned}
\sum_{j=n}^{\infty} \frac{l(j)}{\alpha^2(j)} &= \sum_{j=n}^{\infty} \frac{l(j) j}{\alpha^{2-\varepsilon_0}(j)} \frac{j^{\varepsilon_0/2}}{\alpha^{\varepsilon_0}(j)} \frac{1}{j^{1+\varepsilon_0/2}} \\
&\le \frac{c l(n) n^{1+\varepsilon_0/2}}{\alpha^2(n)} \sum_{j=n}^{\infty} j^{-1-\varepsilon_0/2} \\
&\le \frac{c}{\varepsilon_0} \cdot \frac{n l(n)}{\alpha^2(n)}.
\end{aligned}
$$

Therefore

$$
\begin{aligned}
J_3 &\le c\sum_{n=1}^{\infty} \frac{l(n)}{\alpha^2(n)}\alpha^2(1) \\
&\quad + \sum_{n=1}^{\infty} \frac{l(n)}{\alpha^2(n)} \sum_{j=1}^{n} EX_1^2 I(j < \beta(X_1) \le j+1) \\
&\le c\sum_{j=1}^{\infty}\sum_{n=j}^{\infty} \frac{l(n)}{\alpha^2(n)} EX_1^2 I(j < \beta(X_1) \le j+1) \\
&\le c\sum_{j=1}^{\infty} \frac{jl(j)}{\alpha^2(j)} EX_1^2 I(j < \beta(X_1) \le j+1) \\
&\le cE\beta(X_1) l(\beta(X_1)) < \infty. \qquad (8.4.30)
\end{aligned}
$$

Theorem 8.4.1 follows from (8.4.27)–(8.4.30).

Proof of Theorem 8.4.2.

From the assumption that $l(x), x^q/\beta(x)l(\beta(x))$ and $\beta(x)l(\beta(x))/x^2$ are quasi-monotone non-decreasing it follows that there exists a $C > 0$ such that (8.4.21) is satisfied and for any $t \ge x > 0$ we have

$$
\frac{\alpha^q(x)}{xl(x)} \le C\frac{\alpha^q(t)}{tl(t)}, \quad \frac{\beta(x)l(\beta(x))}{x^2} \le C\frac{\beta(t)l(\beta(t))}{t^2}. \qquad (8.4.31)
$$

Let $C_n = n^{-\varepsilon_1}$, where $\varepsilon_1 = \frac{\delta}{2(1+\delta)}$, let $\beta = \alpha(nC_n), k = n, g(x) = \beta(x)l(\beta(x))$. Take $x = \alpha(\varepsilon n)$ in Lemma 8.4.1. Then (8.4.7) and (8.4.8) are satisfied. In fact, by $l(x) \ge x^\delta$ we have for large n

$$
\begin{aligned}
&\frac{48C\alpha(nC_n)}{C_n l(nC_n)} E\beta(X_1)l(\beta(X_1)) \\
&\qquad \le \frac{48C\alpha(\varepsilon n)}{C_n(nC_n)^\delta} E\beta(X_1)l(\beta(X_1)) \\
&\qquad \le \frac{48C\alpha(\varepsilon n)}{n^{\delta/2}} E\beta(X_1)l(\beta(X_1)).
\end{aligned}
$$

Thus (8.4.8) is satisfied for large n. By the same way (8.4.7) is satisfied as well.

By Lemma 8.4.1 with $q_1 = q_0 + q + 4, q_2 = r$ we have

$$\begin{aligned}
&\sum_{n=1}^{\infty} \frac{l(n)}{n} P\Big\{\max_{i\le n} \beta(S_i) \ge \varepsilon n\Big\} \\
&\quad \le c\sum_{n=1}^{\infty} l(n) P\{|X_1| \ge \alpha(n)\} \\
&\quad + c\sum_{n=1}^{\infty} \frac{l(n)}{n\alpha^{q_1}(n)} \Big(n^{q_1/2} + n\exp\Big(K\sum_{i=0}^{[\log n]} \rho^{2/q_1}(2^i)\Big) \\
&\quad \cdot \log^{q_1} n E|X_1|^{q_1} I(|X_1| \le \alpha(nC_n))\Big) \\
&\quad + c\sum_{n=1}^{\infty} \frac{l(n)}{n\alpha^{r}(n)} \Big(n^{r/2} (EX_1^2 I(|X_1| \ge \alpha(nC_n)))^{r/2} \\
&\quad + nE|X_1|^r I(|X_1| \le \alpha(n))\Big) \\
&\quad =: L_1 + L_2 + L_3. \qquad (8.4.32)
\end{aligned}$$

As in the estimation of J_1 in the proof of Theorem 8.4.1, we have

$$L_1 < \infty. \qquad (8.4.33)$$

Noting that $\exp\Big(K\sum_{i=1}^{[\log n]} \rho^{2/q_1}(2^i)\Big) \log^{q_1} n$ is a slowly varying function, $\frac{\alpha^2(x)}{x} \uparrow$, and using (8.4.31) and condition (8.4.3), we have

$$\begin{aligned}
L_2 &\le c\sum_{n=1}^{\infty} \frac{l(n)}{\alpha^{q_1}(n)} n^{q_1/2} E|X_1|^q I(|X_1| \le \alpha(nC_n)) \\
&\le c\sum_{n=1}^{\infty} \frac{l(n) n^{q_1/2} \alpha^{q_1}(nC_n)}{\alpha^{q_1}(n) l(nC_n) nC_n} \\
&\le c\sum_{n=1}^{\infty} n^{\varepsilon_1/2-1} \frac{\alpha^{q_1-q}(nC_n)}{\alpha^{q_1-q}(n)} \\
&\le c\sum_{n=1}^{\infty} n^{\varepsilon_1/2-1} C_n^2 \\
&\le c\sum_{n=1}^{\infty} n^{-1-\varepsilon_1} < \infty. \qquad (8.4.34)
\end{aligned}$$

Next we estimate L_3. From (8.4.31) it follows that

$$\begin{aligned}
&\sum_{n=1}^{\infty}\frac{l(n)}{n}\Big(\frac{nEX_1^2I(|X_1|\geq\alpha(nC_n))}{\alpha^2(n)}\Big)^{r/2}\\
&\quad\leq c\sum_{n=1}^{\infty}\frac{l(n)}{n}\frac{\alpha^r(nC_n)}{C_n^{r/2}l^{r/2}(nC_n)\alpha^r(n)}\\
&\quad\leq c\sum_{n=1}^{\infty}\frac{l(n)n}{\alpha^r(n)}\frac{\alpha^r(nC_n)}{nC_nl(nC_n)}\frac{1}{nC_n^{r/2-1}l^{r/2-1}(nC_n)}\\
&\quad\leq c\sum_{n=1}^{\infty}\frac{1}{nC_n^{r/2-1}(nC_n)^{(r/2-1)\delta}}\\
&\quad\leq c\sum_{n=1}^{\infty}n^{-1-(\frac{r-1}{4})\delta}<\infty.
\end{aligned}\tag{8.4.35}$$

By (8.4.35) and (8.4.31) we obtain

$$\begin{aligned}
L_3&\leq c\sum_{n=1}^{\infty}\frac{l(n)}{\alpha^r(n)}E|X_1|^rI(|X_1|\leq\alpha(n))\\
&\leq c\sum_{n=1}^{\infty}\frac{l(n)}{\alpha^r(n)}\cdot\alpha^r(1)\\
&\qquad+\sum_{n=1}^{\infty}\frac{l(n)}{\alpha^r(n)}\cdot\sum_{j=1}^{n}E|X_1|^rI(j\leq\beta(X_1)<j+1)\\
&\leq c\sum_{n=1}^{\infty}\frac{nl(n)}{\alpha^q(n)}\cdot\frac{n^{\frac{r-q}{2}}}{\alpha^{r-q}(n)}\frac{\alpha^r(1)}{n^{1+(r-q)/2}}\\
&\qquad+\sum_{j=1}^{\infty}\sum_{n=j}^{\infty}\frac{nl(n)}{\alpha^q(n)}\frac{n^{(r-q)/2}}{\alpha^{r-q}(n)}\frac{1}{n^{1+(r-q)/2}}\\
&\qquad\cdot E|X_1|^rI(j\leq\beta(X_1)<j+1)\\
&\leq 1+c\sum_{j=1}^{\infty}\frac{jl(j)}{\alpha^r(j)}E|X_1|^rI(j\leq\beta(X_1)<j+1)\\
&\leq cE\beta(X_1)l(\beta(X_1))<\infty.
\end{aligned}\tag{8.4.36}$$

Then Theorem 8.4.2 follows from (8.4.32)–(8.4.36).

In order to prove Corollaries 8.4.1 and 8.4.2, we need the following lemma.

Lemma 8.4.2. *Let $h(x)$ be a positive slowly varying function. Then $x^{\varepsilon}h(x)$ is a quasi-monotone non-decreasing function for any $\varepsilon>0$.*

Proof. By the property of a slowly varying function, we have

$$\lim_{N\to\infty} \sup_{2^N\le x\le 2^{N+1}} \frac{h(x)}{h(2^{N+1})} = \lim_{N\to\infty} \inf_{2^N\le x\le 2^{N+1}} \frac{h(x)}{h(2^{N+1})} = 1.$$

Therefore there exists an N such that for $m \ge N$ we have

$$\sup_{2^m\le x\le 2^{m+1}} \frac{h(x)}{h(2^{m+1})} \le 2^{\varepsilon}, \quad 2^{-\varepsilon} \le \inf_{2^m\le x\le 2^{m+1}} \frac{h(x)}{h(2^{m+1})}. \tag{8.4.37}$$

For any $t \ge x \ge 2^N$, let $M_1, M_2 \ge N$ such that

$$2^{M_1} \le x < 2^{M_1+1}, \quad 2^{M_2} \le t < 2^{M_2+1}. \tag{8.4.38}$$

From (8.4.37) and (8.4.38) we have

$$\begin{aligned} x^{\varepsilon}h(x) &\le x^{\varepsilon} \sup_{2^{M_1}\le s<2^{M_1+1}} \frac{h(s)}{h(2^{M_1+1})} \cdot h(2^{M_1+1}) \\ &\le x^{\varepsilon}2^{\varepsilon(M_2-M_1+1)}h(2^{M_2+1}) \\ &\le x^{\varepsilon}2^{\varepsilon(M_2-M_1+2)}h(t) \\ &\le t^{\varepsilon}2^{\varepsilon(M_2-M_1+1)} \cdot 2^{\varepsilon(M_2-M_1+2)}h(t) \\ &= 2^{\varepsilon} \cdot t^{\varepsilon}h(t). \end{aligned}$$

This proves that $x^{\varepsilon}h(x)$ is a quasi-monotone nondecreasing function.

Proof of Corollary 8.4.1. From $p\alpha > 1$ and Lemma 8.4.2 it follows that $n^{p\alpha-1}h(n)$ is quasi-monotone nondecreasing. If $1 \le p < 2$, $x^p h(x), x^{2-p}/h(x)$ are quasi-monotone nondecreasing. Then Corollary 8.4.1 follows from Theorem 8.4.1. If $p > 2$, $x^{p-2}h(x)$ and $x^{r-p}/h(x)$ are quasi-monotone non-decreasing. Then Corollary 8.4.1 follows from Theorem 8.4.2. If $p = 2$, $\lim_{n\to\infty} n^{-\varepsilon}EX_1^2 \cdot I(|X_1| \le n^{\alpha}) = 0$ for any $\varepsilon > 0$, and hence Corollary 8.4.1 can be obtained by repeating the proof of Theorem 8.4.2.

Proof of Corollary 8.4.2. Obviously $x^{p-1}h(x)$ is increasing. By Lemma 8.4.2, $x^{2-p}/h(x)$ is quasi-monotone non-decreasing. Hence Corollary 8.4.2 follows from Theorem 8.4.1.

8.5 Complete convergence for α-mixing sequences

For the complete convergence of α-mixing sequences, Hipp (1979) obtained the following result:

Theorem 8.5.0. *Let*

$$\frac{1}{2} < \alpha \le 1, 2 < r \le \infty, 1/\alpha < p < r, \{X_n, n \ge 1\}$$

be a strictly stationary α-mixing sequence of random variables with $EX_1 = 0, E|X_1|^r < \infty$. Assume that

$$\sum_{n=1}^{\infty} \alpha^{1/\theta}(n) < \infty \quad \text{for some} \quad \theta > \Big[2 + \frac{r}{r-p}\Big] \cdot \frac{p\alpha}{p\alpha - 1}. \tag{8.5.1}$$

Then

$$\sum_{n=1}^{\infty} n^{\alpha-2} P\Big\{\max_{1\le i\le n} |S_i| \ge \varepsilon n^{\alpha}\Big\} < \infty \quad \text{for any} \quad \varepsilon > 0. \tag{8.5.2}$$

However, an example contrary to Hipp's conclusion was given by Berbee (1987) when $r = \infty$, i.e., in the case of X_1 bounded.

Shao (1993c) proved the following theorem.

Theorem 8.5.1. *Let $1/2 < \alpha \le 1$, $1/\alpha \le p < r \le \infty$, $\{X_n, n \ge 1\}$ be an α-mixing sequence of random variables with $EX_n = 0$, $\sup_{n\ge 1} \|X_n\|_r < \infty$. Assume that*

$$\alpha(n) = O(n^{-r(p-1)/(r-p)} \log^{-\beta} n) \quad \text{for some} \quad \beta > rp/(r-p). \tag{8.5.3}$$

Then (8.5.2) holds true.

An immediate consequence of Theorem 8.5.1 with $p = \alpha = 1$ is:

Corollary 8.5.1. *Let $1 < r \le \infty$, $\{X_n, n \ge 1\}$ be an α-mixing sequence of random variables with $EX_n = 0, \sup_{n\ge 1} \|X_n\|_r < \infty$. Assume that*

$$\alpha(n) = O(\log^{-\beta} n) \quad \text{for some} \quad \beta > r/(r-1).$$

Then

$$\sum_{n=1}^{\infty} \frac{1}{n} P\Big\{\max_{1\le i\le n} |S_i| \ge \varepsilon n\Big\} < \infty \quad \text{for any} \quad \varepsilon > 0.$$

Particularly, we have

$$S_n/n \to 0 \quad a.s.$$

The proof of Theorem 8.5.1 will needs the following lemmas.

Lemma 8.5.1. *Let $\{X_n, n \geq 1\}$ be a sequence of random variables with $EX_n = 0$ for every $n \geq 1$. Then*

$$P\Big\{\max_{1\leq i\leq n} |S_i| \geq x\Big\} \leq 4x^{-1}\sum_{i=1}^{n} E|X_i|I(|X_i| > c) + 4x^a + 32^3 ncx^{-1}\alpha(k) \tag{8.5.4}$$

for any $a \geq 1$, $x \geq 1$, $c > 0$ and integer k satisfying

$$1 \leq k \leq x/(64ac\log x) \tag{8.5.5}$$

and for some $s \geq 2$,

$$\Big(\sum_{i=1}^{n} \|X_i I(|X_i| \leq c)\|_s^2\Big)\sum_{i=0}^{k} \alpha^{1-2/s}(i) \leq x^2/(32^3 a\log x). \tag{8.5.6}$$

Proof. Let

$$\overline{X}_i = X_i I(|X_i| \leq c) - EX_i I(|X_i| \leq c),$$

$$Y_{i,1} = \sum_{j=1+2ik}^{(2i+1)k\wedge n} \overline{X}_j, \quad i = 0, 1, \cdots, q_1 := \Big[\frac{n}{2k} - \frac{1}{2}\Big],$$

$$Y_{i,2} = \sum_{j=1+(2i+1)k}^{2(i+1)k\wedge n} \overline{X}_j, \quad i = 0, 1, \cdots, q_2 := \Big[\frac{n}{2k} - 1\Big],$$

$$\overline{S}_i = \sum_{j=1}^{i} \overline{X}_j, \qquad T_{i,1} = \sum_{j=0}^{i} Y_{j,1} \qquad T_{i,2} = \sum_{j=0}^{i} Y_{j,2}.$$

It is easy to see that

$$\max_{1\leq i\leq n} |S_i| \leq \max_{1\leq i\leq n} |\overline{S}_i| + \sum_{i=1}^{n} |X_i|I(|X_i| > c) + \sum_{i=1}^{n} E|X_i|I(|X_i| > c)$$

and

$$P\Big\{\max_{1\leq i\leq n} |S_i| \geq x\Big\} \leq P\Big\{\max_{1\leq i\leq n} |\overline{S}_i| \geq x/2\Big\} + 4x^{-1}\sum_{i=1}^{n} E|X_i|I(|X_i| > c). \tag{8.5.7}$$

Since

$$\max_{1\le i\le n}|\overline{S}_i| \le \max_{0\le i\le q_1}|T_{i,1}| + \max_{0\le i\le q_2}|T_{i,2}| + 2kc$$
$$\le \max_{0\le i\le q_1}|T_{i,1}| + \max_{0\le i\le q_2}|T_{i,2}|x + x/32$$

by (8.5.5), we have

$$P\Big\{\max_{1\le i\le n}|\overline{S}_i| \ge x/2\Big\}$$
$$\le P\Big\{\max_{0\le i\le q_1}|T_{i,1}| \ge x/8\Big\} + P\Big\{\max_{0\le i\le q_2}|T_{i,2}| \ge x/8\Big\}$$
$$=: I_1 + I_2. \tag{8.5.8}$$

We first estimate I_1. The estimation of I_2 is completely similar. Put

$$G_{-1} = \sigma(\Omega), \quad G_i = \sigma(X_j, 1 \le j \le (2i+1)k),$$
$$u_i = Y_{i,1} - E(Y_{i,1}|G_{i-1}), \quad U_i = \sum_{j=0}^{i} u_j, \quad i = 0, 1, \cdots.$$

Then

$$I_1 \le P\Big\{\sum_{i=0}^{q_1}|E(Y_{i,1}|G_{i-1})| \ge x/16\Big\}$$
$$+ P\Big\{\max_{0\le i\le q_1}|U_i| \ge x/16\Big\}$$
$$=: I_3 + I_4 \tag{8.5.9}$$

and $\{U_i, G_i, i \ge 0\}$ is a martingale with $|u_i| \le 2kc$ for every $i \ge 0$. Noting that for each real t and $i \ge 1$,

$$E(e^{tu_i}|G_{i-1}) = 1 + \sum_{l=2}^{\infty} E\Big(\frac{(tu_i)^l}{l!}\Big|G_{i-1}\Big)$$
$$\le 1 + t^2E(u_i^2|G_{i-1})\sum_{l=0}^{\infty}\frac{(|t|2kc)^l}{l!}$$
$$\le \exp(t^2e^{2|t|kc}E(u_i^2|G_{i-1}))$$
$$\le \exp(t^2e^{2|t|kc}E(Y_{i,1}^2|G_{i-1})),$$

we find that $\Big\{\exp\Big(tU_i - t^2e^{2|t|kc}\sum_{j=0}^{i}E(Y_{j,1}^2|G_{j-1})\Big), G_i, i \ge 0\Big\}$ is a non-negative supermartingale for every t and hence

$$P\Big\{\max_{0\le i\le q_1}\exp\Big(tU_i - t^2e^{2|t|kc}\sum_{j=0}^{i}E(Y_{j,1}^2|G_{j-1})\Big) \ge y\Big\} \le \frac{1}{y} \tag{8.5.10}$$

for $y > 0$, by the maximum inequality of the non-negative supermartingale. Take $t = (32a\log x)/x$ in (8.5.10). By (8.5.10) and (8.5.5), we have

$$\begin{aligned}
&P\Big\{\max_{0\le i\le q_1} U_i \ge x/16\Big\}\\
&\le P\Big\{\max_{0\le i\le q_1} \exp\Big(tU_i - t^2e^{2|t|kc}\sum_{j=0}^{i} E(Y_{j,1}^2|G_{j-1})\Big)\\
&\qquad \ge \exp\Big(\frac{xt}{16} - t^2e^{2|t|kc}\sum_{j=0}^{i} E(Y_{j,1}^2|G_{j-1})\Big)\Big\}\\
&\le P\Big\{\sum_{j=0}^{q_1} E(Y_{j,1}^2|G_{j-1}) \ge \frac{x^2}{4(32)^2a\log x}\Big\}\\
&\quad + P\Big\{\max_{0\le i\le q_1} \exp\Big(tU_i - t^2e^{2|t|kc}\sum_{j=0}^{i} E(Y_{j,1}^2|G_{j-1})\Big)\\
&\qquad \ge \exp\Big(\frac{xt}{16} - \frac{t^2e^{2|t|kc}x^2}{4(32)^2a\log x}\Big)\Big\}\\
&\le P\Big\{\sum_{j=0}^{q_1} E(Y_{j,1}^2|G_{j-1}) \ge \frac{x^2}{4(32)^2a\log x}\Big\}\\
&\quad + \exp\Big(-\frac{xt}{16} + \frac{t^2e^{2|t|kc}x^2}{4(32)^2a\log x}\Big)\\
&\le P\Big\{\sum_{j=0}^{q_1} E(Y_{j,1}^2|G_{j-1}) \ge \frac{x^2}{4(32)^2a\log x}\Big\} + x^{-a}. \qquad (8.5.11)
\end{aligned}$$

Using Lemma 1.2.4, one can obtain

$$\begin{aligned}
\sum_{j=0}^{q_1} EY_{j,1}^2 &\le 4\Big(\sum_{i=0}^{k}\alpha^{1-2/s}(i)\Big)\sum_{j=0}^{q_1}\sum_{i=2jk+1}^{(2j+1)k}\|X_iI(|X_i|\le c)\|_s^2\\
&\le 4\Big(\sum_{i=0}^{k}\alpha^{1-2/s}(i)\Big)\sum_{i=0}^{n}\|X_iI(|X_i|\le c)\|_s^2\\
&\le \frac{x^2}{8(32)^2a\log x}
\end{aligned}$$

by (8.5.6). Hence

$$\begin{aligned}
&P\Big\{\sum_{j=0}^{q_1} E(Y_{j,1}^2|G_{j-1}) \ge \frac{x^2}{4(32)^2a\log x}\Big\}\\
&\quad \le \frac{8(32)^2a\log x}{x^2}\sum_{j=0}^{q_1} E|E(Y_{j,1}^2|G_{j-1}) - EY_{j,1}^2|. \qquad (8.5.12)
\end{aligned}$$

Write $\xi_j = E(Y_{j,1}^2|G_{j-1}) - EY_{j,1}^2$. We find that

$$E|\xi_j| = E(Y_{j,1}^2 - EY_{j,1}^2)\text{sgn}\xi_j \leq 4(kc)^2\alpha(k) \tag{8.5.13}$$

by Lemma 1.2.4 again. Inserting (8.5.13) into (8.5.12), we obtain

$$\begin{aligned} P\Big\{\sum_{j=0}^{q_1} E(Y_{j,1}^2|G_{j-1}) \geq \frac{x^2}{4(32)^2 a \log x}\Big\} \\ \leq \frac{(32)^3 anc^2 k\alpha(k)\log x}{x^2} \leq \frac{(32)^2 nc\alpha(k)}{x} \end{aligned} \tag{8.5.14}$$

by (8.5.5). Now a combination of (8.5.11) and (8.5.14) yields

$$P\Big\{\max_{0\leq i\leq q_1} U_i \geq \frac{x}{16}\Big\} \leq x^{-a} + (32)^2 ncx^{-1}\alpha(k).$$

Similarly, we have

$$P\Big\{\max_{0\leq i\leq q_1} U_i \leq -\frac{x}{16}\Big\} \leq x^{-a} + (32)^2 ncx^{-1}\alpha(k).$$

Hence

$$I_4 \leq 2x^{-a} + 2(32)^2 ncx^{-1}\alpha(k). \tag{8.5.15}$$

Also, one can get that

$$E|E(Y_{i,1}|G_{i-1})| = EY_{i,1}\text{sgn}(E(Y_{i,1}|G_{i-1})) \leq 4kc\alpha(k)$$

and hence

$$I_3 \leq 64ncx^{-1}\alpha(k). \tag{8.5.16}$$

It follows from (8.5.9), (8.5.15) and (8.5.16) that

$$I_1 \leq 2x^{-a} + 3(32)^2 ncx^{-1}\alpha(k). \tag{8.5.17}$$

Similarly, we have

$$I_2 \leq 2x^{-a} + 3(32)^2 ncx^{-1}\alpha(k). \tag{8.5.18}$$

This proves (8.5.4) by (8.5.7), (8.5.8), (8.5.17) and (8.5.18).

Lemma 8.5.2. *Let $\{X_n, n \geq 1\}$ be an α-mixing sequence with*

$$EX_i = 0, \qquad \|X_i\|_\nu \leq D \quad \text{and} \quad \alpha(i) \leq C_0 i^{-\tau}\log^{-\lambda} i$$

for $i \geq 1$ and for some $1 < \nu \leq \infty$, $C_0 \geq 1$, $\tau \geq 0$ and real λ. Then there exists a finite positive constant K depending only on ν, τ, λ and C_0 such that

$$P\Big\{\max_{1\leq i\leq n}|S_i| \geq x\Big\} \leq Kn(D/x)^{\nu(\tau+1)/(\nu+\tau)}\log^{(\nu-1)(\tau-\lambda)/(\nu+\tau)}(x/D) \tag{8.5.19}$$

for any $x \geq KDn^{1/2}\log^{1+|\lambda|/2} n$.

Proof. We assume , without loss of generality, that $D = 1$. It suffices to show that there exists a constant K such that

$$P\Big\{\max_{1\leq i\leq n}|S_i| \geq x\Big\} \\ \leq Knx^{-\nu(\tau+1)/(\nu+\tau)}\log^{(\nu-1)(\tau-\lambda)/(\nu+\tau)} x \tag{8.5.20}$$

for any $x \geq Kn^{1/2}\log^{1+|\lambda|/2} n$. Take

$$C = 2x^{\tau/(\nu+\tau)}\log^{(\lambda-\tau)/(\nu+\tau)} x,$$
$$a = \tau+2, \quad k = [x/(64ac\log x)]$$

in Lemma 8.5.1. Assume that

$$Knx^{-\nu(\tau+1)/(\nu+\tau)}\log^{(\nu-1)(\tau-\lambda)/(\nu+\tau)} x \leq 1. \tag{8.5.21}$$

Otherwise, (8.5.20) is trivial. If the conditions of Lemma 8.5.1 are satisfied, then (8.5.20) will follow from (8.5.4) immediately. So we need only to verify (8.5.6). In what follows we denote by K_1 the finite positive constant depending only on ν, τ, λ and C_0, whose value may be different from line to line. If $1 < \nu \leq 2$, then

$$\begin{aligned} &\Big(\sum_{i=1}^{n}\|X_iI(|X_i|\leq c)\|_2^2\Big)\sum_{i=0}^{k}\alpha^{1-2/\nu}(i) \\ &\leq 2nkc^{2-\nu} \leq xnc^{1-\nu}/(32a\log x) \\ &= \frac{x^2}{(32)^3 a\log x}(32)^2 2^{1-\nu}nx^{-\nu(\tau+1)/(\nu+\tau)} \\ &\quad\cdot \log^{(\nu-1)(\tau-\lambda)/(\nu+\tau)} x \\ &\leq \frac{x^2}{(32)^3 a\log x} \end{aligned} \tag{8.5.22}$$

by (8.5.21). When $\nu > 2$, we have

$$\begin{aligned}
&\Big(\sum_{i=1}^{n} \|X_i I(|X_i| \le c)\|_\nu^2\Big) \sum_{i=0}^{k} \alpha^{1-2/\nu}(i) \\
&\le n\Big(1 + C_0 \sum_{i=1}^{k} i^{-\tau(1-2/\nu)} \log^{-\lambda(1-2/\nu)} i\Big) \\
&\le K_1 n(\log^{1+|\lambda|(1-2/\nu)} k + k^{-\tau(1-2/\nu)+1} \log^{-\lambda(1-2/\nu)} k) \\
&\le K_1 n(\log^{1+|\lambda|} x + (x/(c\log x))^{-\tau(1-2/\nu)+1} \log^{-\lambda(1-2/\nu)} x) \\
&\le \frac{x^2}{(32)^3 a \log x}(K_1 n x^{-2} \log^{2+|\lambda|(1-2/\nu)} x \\
&\quad + K_1 n x^{-\nu(\tau+1)/(\nu+\tau)} \log^{(\nu-1)(\tau-\lambda)/(\nu+\tau)} x) \\
&\le \frac{x^2}{(32)^3 a \log x} \qquad (8.5.23)
\end{aligned}$$

by (8.5.21) and $x \ge K n^{1/2} \log^{1+|\lambda|/2} n$. Now we conclude from (8.5.22) and (8.5.23) that (8.5.6) is satisfied. This completes the proof of the lemma.

Proof of Theorem 8.5.1. By Lemma 8.5.2, we have that for any $\varepsilon > 0$, there exists a positive constant K such that

$$\begin{aligned}
&P\Big\{\max_{1\le i\le n} |S_i| \ge \varepsilon n^\alpha\Big) \\
&\le K n^{1-\alpha r(1+r(p-1)/(r-p))/(r+r(p-1)/(r-p))} \\
&\quad \cdot \log^{(1-r)(\beta-r(p-1)/(r-p))/(r+r(p-1)/(r-p))} n \\
&= K n^{1-p\alpha} \log^{-1-(r-p)(\beta-rp/(r-p))/r} n,
\end{aligned}$$

which yields (8.5.5) immediately by (8.5.6), as desired.

The example below shows that Theorem 8.5.1 does not remain valid, if the assumption $\beta > rp/(r-p)$ is replaced by $\beta \ge r/(r-p)$.

Example 8.5.1. Let $1 < p < r$, $1/p \le \alpha \le 1$. Put

$$\begin{aligned}
&a = \frac{\alpha(r-p)}{r(1-\alpha)+p\alpha-1}, \quad b = \frac{a(p-1)}{r-p}, \quad d = \frac{-1}{r-1-\alpha(r-p)}, \\
&g(x) = x^a \log^d x, \quad x \ge 0, \\
&G(0) = 0, \quad G(n) = \sum_{i=1}^{n} [g(i)], \ n = 1, 2, \cdots, \\
&f(x) = (g(x))^{r(p-1)/(r-p)} (\log g(x))^{r/(r-p)} \log\log g(x), \quad x \ge 0.
\end{aligned}$$

Let $\{Y_n, n \geq 1\}$ be an independent sequence of random variables with

$$P\{Y_n = \pm f^{1/r}(n)\} = \frac{1}{2f(n)},$$
$$P\{Y_n = 0\} = 1 - \frac{1}{f(n)}.$$

Define a sequence $\{X_n, n \geq 1\}$ by $X_n = Y_j$ for $G(j-1) < n \leq G(j)$. Then $\{X_n, n \geq 1\}$ has the following properties.

$$EX_n = 0, \qquad E|X_n|^r = 1, \tag{8.5.24}$$
$$\alpha(n) = O(n^{-r(p-1)/(r-p)} \log^{-r/(r-p)} n (\log\log n)^{-1}), \tag{8.5.25}$$
$$\sum_{n=1}^{\infty} n^{p\alpha-2} P\{|S_n| \geq \varepsilon n^{\alpha}\} = \infty \quad \text{for any} \quad \varepsilon > 0. \tag{8.5.26}$$

Example 8.5.2. Let $r > 1$. Put

$$a = (r-1)/r, \quad g(n) = [n^{a-1}\exp(n^a)], \quad G(n) = \sum_{i=1}^{n} g(i).$$

Let $\{Y_n, n \geq 1\}$ be an independent sequence of random variables with

$$P\{Y_n = \pm n^{1/r} \log^{1/r} n\} = \frac{1}{2n\log n}, \quad P\{Y_n = 0\} = 1 - \frac{1}{n\log n}.$$

Define $X_n = Y_j$ for $G(j-1) < n \leq G(j)$. Then $\{X_n, n \geq 1\}$ has the following properties.

$$EX_n = 0, \qquad E|X_n|^r = 1,$$
$$\alpha(n) = O(\log^{-r/(r-1)}(\log\log n)^{-1}),$$
$$\sum_{n=1}^{\infty} \frac{1}{n} P\{|S_n| \geq \varepsilon n\} = \infty, \quad \text{for any} \quad \varepsilon > 0.$$

Remark 8.5.1. We now can discuss whether Theorem 8.5.0 is true or not in the case of $r < \infty$. Assume that $1 < p \leq r/2$. Then (8.5.1) will be satisfied with $\theta > 8$ provided p large enough. But Example 8.5.1 says that the mixing rate $n^{-(p-1)/2}$ is required at least. This means that Theorem 8.5.0 is quite possibly not true. Unfortunately, $\{X_n, n \geq 1\}$ in Example 8.5.1 is not strictly stationary. Shao (1993c) conjectured that there is a strictly stationary α-mixing sequence satisfying (8.5.24), (8.5.26) and $\alpha(n) = O(n^{-r(p-1)/(r-p)} \log^{-1} n)$. He also conjectured that the assumption $\beta > rp/(r-p)$ in Theorem 8.5.1 can be replaced by $\beta > r/(r-p)$.

8.6 A further discussion on the complete convergence for partial sums of a mixing sequence

Let $\{X_n, n \geq 1\}$ be a sequence of i.i.d.r.v.'s, $S_n = \sum_{i=1}^n X_i$. Assume that positive functions $H(t)$ and $\psi(t)$ are defined on $(0, \infty)$ and $H(t) \uparrow \infty (t \to \infty)$, $\widehat{\psi}(t) = \int_0^t \psi(u)du$, $t > 0$. Denote

$$\nu(\varepsilon) = \sum_{n=1}^{\infty} I((|S_n| \geq \varepsilon H(n)),$$

$$\eta(\varepsilon) = \sup_{n \geq 1}\{(|S_n| - \varepsilon H(n))^+/\varepsilon\},$$

$$\chi(\varepsilon) = \sup\{n \geq 1 : |S_n| \geq \varepsilon H(n)\},$$

where ε is an arbitrary positive number. Three problems posed by Prohorov are as follows:

(i) Is

$$\sum_{n=1}^{\infty} \psi(n) P\{|S_n| \geq \varepsilon H(n)\} < \infty, \quad \forall \varepsilon > 0 \tag{8.6.1}$$

equivalent to

$$E\hat{\psi}(\nu(\varepsilon)) < \infty, \quad \forall \varepsilon > 0?$$

(ii) Is

$$\sum_{n=1}^{\infty} \psi(n) P\{\max_{1 \leq k \leq n} |S_k| \geq \varepsilon H(n)\} < \infty, \quad \forall \varepsilon > 0, \tag{8.6.2}$$

equivalent to

$$E\hat{\psi}(\text{inv}H(\eta(\varepsilon))) < \infty, \quad \forall \varepsilon > 0?$$

(iii) Is

$$\sum_{n=1}^{\infty} \psi(n) P\{\sup_{k \geq n} |S_k|/H(k) \geq \varepsilon\} < \infty, \quad \forall \varepsilon > 0, \tag{8.6.3}$$

equivalent to

$$E\hat{\psi}(\chi(\varepsilon)) < \infty, \quad \forall \varepsilon > 0?$$

Let

$$M_1 = \{\psi : \psi(x) > 0, x \in [1, \infty), \exists \delta_1, \delta_2 > 0, \quad \text{such that} \\ x^{1-\delta_1}\psi(x) \uparrow, \quad e^{-\delta_2 x}\psi(x) \downarrow, \quad x \to \infty\}.$$

Sirazgimov and Gafrov (1987) discussed these problems and showed that under the following assumptions:

$$\limsup_{x\to\infty} H(Cx)/H(x) < \infty, \quad \forall C > 1; \tag{8.6.4}$$

$$\liminf_{n\to\infty} P\{S_n \geq -\varepsilon H(n)\} > 0, \quad \forall \varepsilon > 0 \tag{8.6.5}$$

and

$$\psi \in M_1 \quad \text{or} \quad \psi(x), \quad \text{as} \quad x \to \infty,$$

the following three conclusions are equivalent

$$P(\psi,H,\varepsilon) := \sum_{n=1}^{\infty} \psi(n)P\{S_n \geq \varepsilon H(n)\} < \infty, \quad \forall \varepsilon > 0, \tag{8.6.6}$$

$$M(\psi,H,\varepsilon) := \sum_{n=1}^{\infty} \psi(n)P\{\max_{1\leq k\leq n} S_k \geq \varepsilon H(n)\} < \infty, \quad \forall \varepsilon > 0, \tag{8.6.7}$$

$$S(\psi,H,\varepsilon) := \sum_{n=1}^{\infty} \psi(n)P\{\sup_{k\geq n} S_k/H(k) \geq \varepsilon\} < \infty, \quad \forall \varepsilon > 0. \tag{8.6.8}$$

Su (1989) studied these problems and gave the following result. Suppose that the following conditions are satisfied:

(A) There exist constants $C > 0,\ t_0 > 0$ such that $\psi(t) \leq C\hat{\psi}(t)$ for any $t \geq t_0$;

(B) $t\psi(t) \uparrow \infty (t \to \infty)$ and there exist $C' > 0, t_0 > 0$ such that for any $t \geq t_0$

$$\int_0^t u\psi(u)du \geq C't^2\psi(t);$$

(C) $0 < \beta_1 \leq H(t)/t \leq \beta_2 < +\infty$ or $H(t)/t \uparrow \infty (t \to \infty)$ or $H(t)/t \downarrow 0$ and $H^2(t)/t \uparrow \infty (t \to \infty)$;

(D) There exist a positive integer N and $\delta_2 > 0$ such that for every integer $k > 0$

$$\sum_{n=1}^{N^{k+1}} \psi(n) \geq (1+\delta_2)\sum_{n=1}^{N^k} \psi(n).$$

Then (8.6.1) $\Longleftrightarrow$ (8.6.2) $\Longleftrightarrow$ (8.6.3) $\Longleftrightarrow$ $E\hat{\psi}(\chi(\varepsilon)) < \infty$.

Wang (1993) discussed Prohorov's three problems for a strictly stationary ρ-mixing sequence. Denote

$$\varphi^*(1) = \sup_k \sup_{A\in\mathcal{F}_1^k, B\in\mathcal{F}_{k+1}^\infty, P(A)P(B)>0} \cdot \max\{|P(B|A) - P(B)|, |P(A|B) - P(A)|\} \quad (8.6.9)$$

$$\psi^*(x) = \frac{1}{x}\Big(x\psi(x) - \Big[\frac{x}{2}\Big]\psi\Big(\Big[\frac{x}{2}\Big]\Big)\Big), \quad x \geq 1, \quad (8.6.10)$$

$$M_1^* = \{\psi : \psi(x) > 0, x \geq 1, n\psi(n) \to \infty, n\psi(n) > [n/2]\psi([n/2])\}.$$

Wang (1993) proved the following theorem.

Theorem 8.6.1. *Suppose $\{X_n, n \geq 1\}$ is a strictly stationary ρ-mixing sequence satisfying*

$$\liminf_{n\to\infty} P\{S_n \geq -\varepsilon H(n)\} > \varphi^*(1) \quad \textit{for any } \varepsilon > 0. \quad (8.6.11)$$

Suppose that $\psi \in M_1^*$, $H(n)$ satisfies (8.6.4) and

$$\sum_{n=1}^\infty \psi(n)\rho(n) < \infty \quad (8.6.12)$$

and for $\psi \in M_1^* - M_1$

$$P\Big\{\sup_{k\geq n}\frac{S_k}{H(k)} \geq \varepsilon\Big\} \geq CP\Big\{\inf_{k\geq n}\frac{S_k}{H(k)} \leq -C\varepsilon\Big\}, \quad n = 1, 2, \cdots, \quad (8.6.13)$$

for some $C > 0$. Then for any $\varepsilon > 0$ (8.6.6) $\Longleftrightarrow$ (8.6.7) $\Longleftrightarrow$

$$S(\psi^*, H, \varepsilon) < \infty. \quad (8.6.8')$$

The proof of Theorem 8.6.1 will be given by several lemmas.

Lemma 8.6.1. *If for any $x > 0$*

$$\liminf_{n\to\infty} \inf_{k\leq n-1} P\{S_n - S_k \geq -x\} > \varphi^*(1), \quad (8.6.11')$$

then there exists a constant $C > 0$ such that for any y and $n \in \mathbb{N}$

$$P\Big\{\max_{k\leq n} S_k \geq y\Big\} \leq CP\Big\{S_n \geq y - x\Big\}. \quad (8.6.14)$$

Proof. By the definition of $\varphi^*(1)$ and (8.6.11′) we have

$$\begin{aligned}
&P\{S_n \geq y - x\} \\
&\quad \geq P\Big\{S_n \geq y - x, \max_{k\leq n} S_k \geq y\Big\} \\
&\quad = \sum_{k=1}^{n} P\Big\{S_n \geq y - x, S_k \geq y, \max_{i\leq k-1} S_i < y\Big\} \\
&\quad \geq \sum_{k=1}^{n} P\Big\{S_n - S_k \geq -x, S_k \geq y, \max_{i\leq k-1} S_i < y\Big\} \\
&\quad \geq \sum_{k=1}^{n} \Big(P\{S_n - S_k \geq -x\} P\Big\{S_k \geq y, \max_{i\leq k-1} S_i < y\Big\} \\
&\qquad - \varphi^*(1) P\Big\{S_k \geq y, \max_{i\leq k-1} S_i < y\Big\}\Big) \\
&\quad \geq CP\Big\{\max_{k\leq n} S_k \geq y\Big\}, \quad n = 1, 2, \cdots.
\end{aligned}$$

Lemma 8.6.2. *Let $\{X_n, n \geq 1\}$ be a strictly stationary sequence satisfying (8.6.11) and $n\psi(n) \geq C$, $n = 1, 2, \cdots$. Then* (8.6.6) $\Longleftrightarrow$ (8.6.7).

Proof. We need only to prove (8.6.6) $\Longrightarrow$ (8.6.7). Take $y = \varepsilon H(n), x = \varepsilon H(n)/2$ for any $\varepsilon > 0$ in (8.6.14). It is enough to prove (8.6.11′). First we have

$$P\{S_n \geq \varepsilon H(n)\} \longrightarrow 0, \quad \text{as} \quad n \to \infty. \tag{8.6.15}$$

Otherwise there exist $\{n_i\}, r$ and $\varepsilon_0 > 0$, such that

$$P\{S_{n_i} \geq \varepsilon_0 H(n_i)\} \geq r > 0.$$

Without loss of generality, we assume that $n_{i+1} \geq 3n_i$, $i = 1, 2, \cdots$. So by (8.6.11) for any $2n_i \leq n \leq 3n_i$, $i = 1, 2, \cdots$ we have

$$\begin{aligned}
&P\Big\{S_n \geq \frac{\varepsilon_0}{2} H(n)\Big\} \\
&\quad \geq P\Big\{S_{n_i} \geq \varepsilon_0 H(n), S_n - S_{n_i} \geq -\frac{\varepsilon_0}{2} H(n)\Big\} \\
&\quad \geq P\{S_{n_i} \geq \varepsilon_0 H(n)\} \Big(P\Big\{S_n - S_{n_i} \geq -\frac{\varepsilon_0}{2} H(n)\Big\} - \varphi^*(1)\Big) \\
&\quad \geq P\{S_{n_i} \geq \varepsilon_0 H(n)\} \Big(P\Big\{S_{n-n_i} \geq -\frac{\varepsilon_0}{2} H(n - n_i)\Big\} - \varphi^*(1)\Big) \\
&\quad \geq r_1 > 0
\end{aligned}$$

for some $r_1 > 0$. It follows from $n\psi(n) \geq C > 0, \ n = 1, 2, \cdots$ that

$$\begin{aligned} P(\psi, H, \varepsilon_0/2) &\geq \sum_{i=1}^{\infty} \sum_{m=2ni}^{3ni} \psi(m) P\{S_m \geq \varepsilon_0 H(m)/2\} \\ &\geq r_1 \sum_{i=1}^{\infty} \sum_{m=2ni}^{3ni} m\psi(m)/m \\ &\geq r_1 C \sum_{i=1}^{\infty} \sum_{m=2ni}^{3ni} \frac{1}{m} = \infty, \end{aligned}$$

which contradicts with (8.6.6). Therefore from (8.6.15) and (8.6.11) we obtain that for any $\varepsilon > 0$

$$\begin{aligned} &\liminf_{n\to\infty} \inf_{k\leq n-1} P\{S_n - S_k \geq -2\varepsilon H(n)\} \\ &\quad \geq \liminf_{n\to\infty} \inf_{k\leq n-1} P\{S_n \geq -\varepsilon H(n), S_k < \varepsilon H(n)\} \\ &\quad \geq \liminf_{n\to\infty} \inf_{k\leq n-1} (P\{S_n \geq -\varepsilon H(n)\} - P\{S_k \geq \varepsilon H(n)\}) \\ &\quad > \varphi^*(1). \end{aligned}$$

Lemma 8.6.2 is proved.

Lemma 8.6.3. *Let $\{X_n\}$ be a sequence of random variables, $H(x)$ satisfy (8.6.4) and $n\psi(n) > [n/2]\psi([n/2])$, $n = 1, 2, \cdots$. Then (8.6.7) $\Longrightarrow$ (8.6.8′).*

Proof. Noting that $n\psi(n) \uparrow$ and $H(n)/H(2n) \geq \delta$ we have

$$\begin{aligned} &\sum_{n=2}^{\infty} \frac{n\psi(n) - [n/2]\psi([n/2])}{n} P\Big\{\sup_{i\geq n} \frac{S_i}{H(i)} \geq \varepsilon\Big\} \\ &\quad \leq \sum_{j=1}^{\infty} \sum_{n=2^j}^{2^{j+1}-1} \frac{n\psi(n) - [n/2]\psi([n/2])}{n} P\Big\{\sup_{i\geq 2^{j-1}} \frac{S_i}{H(i)} \geq \varepsilon\Big\} \end{aligned}$$

$$
\begin{aligned}
&\le \sum_{j=1}^{\infty} \frac{1}{2^j} \sum_{n=2^j}^{2^{j+1}-1} (n\psi(n) - [n/2]\psi([n/2])) \\
&\quad \cdot \sum_{k=j}^{\infty} P\Big\{ \sup_{2^{k-1}\le i<2^k} \frac{S_i}{H(i)} \ge \varepsilon \Big\} \\
&\le \sum_{j=1}^{\infty} \Big(\frac{1}{2^j} \sum_{n=2^j}^{2^{j+1}-1} n\psi(n) - \frac{1}{2^{j-1}} \sum_{n=2^{j-1}}^{2^j-1} n\psi(n) \Big) \\
&\quad \cdot \sum_{k=j}^{\infty} P\Big\{ \sup_{i\le 2^k} S_i \ge \varepsilon H(2^{k-1}) \Big\} \\
&\le \sum_{k=1}^{\infty} \frac{1}{2^k} \sum_{n=2^k}^{2^{k+1}-1} n\psi(n) P\Big\{ \sup_{i\le 2^k} S_i \ge \varepsilon H(2^{k-1}) \Big\} \\
&\le 2 \sum_{k=1}^{\infty} \sum_{n=2^k}^{2^{k+1}-1} \psi(n) P\Big\{ \sup_{i\le 2^k} S_i \ge \varepsilon\delta^2 H(2^{k+1}) \Big\} \\
&\le 2 \sum_{n=2}^{\infty} \psi(n) P\Big\{ \sup_{i\le n} S_i \ge \varepsilon\delta^2 H(n) \Big\}.
\end{aligned}
$$

By the arbitrariness of ε and (8.6.7) one gets Lemma 8.6.3.

Lemma 8.6.4. *Under the conditions of Theorem 8.6.1, we have (8.6.8′) $\Longrightarrow$ (8.6.7).*

Proof. For any given $\varepsilon > 0$, put $\varepsilon_1 = \max\{\varepsilon, C\varepsilon\}$. From (8.6.13) we have

$$
\begin{aligned}
&(1+C^{-1}) P\Big\{ \sup_{i\ge 2^j} S_i/H(i) \ge \varepsilon \Big\} \\
&\ge P\Big\{ \sup_{i\ge 2^j} S_i/H(i) \ge \varepsilon \Big\} + P\Big\{ \inf_{i\ge 2^j} S_i/H(i) \le -C\varepsilon \Big\} \\
&\ge P\Big\{ \sup_{i\ge 2^j} |S_i|/H(i) \ge \varepsilon_1 \Big\}.
\end{aligned}
\tag{8.6.16}
$$

By (8.6.4), there exists a $\delta, 0 < \delta < 1$, such that for every $x \ge 1$

$$
\delta H(2x) \le H(x). \tag{8.6.17}
$$

Hence, we have

$$
\begin{aligned}
&\Big\{\sup_{i\geq 2^j}|S_i|/H(i)\geq\varepsilon_1\Big\}\\
&\quad=\bigcup_{i=j}^{\infty}\bigcup_{k=2^i}^{2^{i+1}-1}\{|S_k|\geq\varepsilon_1 H(k)\}\\
&\quad\supset\bigcup_{i=j}^{\infty}\bigcup_{k=2^i}^{2^{i+1}-1}\Big\{|S_k|\geq\frac{\varepsilon_1}{\delta}H(2^i)\Big\}\bigcup_{i=j}^{\infty}\Big\{|S_{2^i}|\geq\frac{\varepsilon_1}{\delta}H(2^i)\Big\}\\
&\quad\supset\bigcup_{i=j}^{\infty}\bigcup_{k=2^i}^{2^{i+1}-1}\Big\{|S_k-S_{2^i}|\geq\frac{2\varepsilon_1}{\delta}H(2^i)\Big\}\\
&\quad=\Big\{\sup_{i\geq j}\max_{2^i\leq k<2^{i+1}}|S_k-S_{2^i}|\geq\frac{2\varepsilon_1}{\delta}H(2^i)\Big\},
\end{aligned}
$$

Combining it with (8.6.16), we obtain

$$
\begin{aligned}
&P\Big\{\sup_{i\geq 2^j}S_i/H(i)\geq\varepsilon_1\delta\Big\}\\
&\quad\geq(1+c^{-1})^{-1}P\Big\{\sup_{i\geq j}\max_{2^i\leq k<2^{i+1}}|S_k-S_{2^i}|/H(2^i)\geq 2\varepsilon_1\Big\}.\qquad(8.6.18)
\end{aligned}
$$

Denote

$$
\begin{aligned}
B_i&=\Big\{\max_{2^i\leq k<2^{i+1}}\frac{S_k-S_{2^i}}{H(2^i)}\geq 2\varepsilon_1\Big\},\\
A_i&=\bigcap_{l=i+1}^{\infty}B_l^c,\\
B_i^*&=\Big\{\max_{2^i\leq k<2^{i+1}-2^{i-1}}\frac{S_k-S_{2^i}}{H(2^i)}\geq 2\varepsilon_1\Big\}.
\end{aligned}
$$

It is easy to check that any mixing sequence obeys the 0-1 law. Since $\{B_i, i.o.\}\in\bigcap_{n=1}^{\infty}\sigma(X_k, k\geq n)$, we have $P(B_i, i.o.)=1$ or 0. If $P(B_i, i.o.)=1$, then

$$
P\Big\{\sup_{i\geq j}\max_{2^i\leq k\leq 2^{i+1}}\frac{S_k-S_{2^i}}{H(2^i)}\geq 2\varepsilon_1\Big\}=P\Big\{\bigcup_{i=j}^{\infty}B_i\Big\}=1,\quad j=1,2,\cdots.
$$

By this equality, (8.6.18) and the property of M_1^*, we have

$$\begin{aligned}
S(\psi^*,H,\varepsilon_0) &\geq \sum_{j=2}^{\infty} 2^{-j}\Big(\sum_{n=2^{j-1}}^{2^j-1} n\psi(n) \\
&\quad - 2\sum_{n=2^{j-2}}^{2^{j-1}-1} n\psi(n)\Big)P\Big\{\sup_{i\geq 2^j}\frac{S_i}{H(i)}\geq \varepsilon_0\Big\} \\
&\geq c\sum_{j=2}^{\infty}\Big(2^{-j}\sum_{n=2^{j-1}}^{2^j-1} n\psi(n) - 2^{-(j-1)}\sum_{n=2^{j-2}}^{2^{j-1}-1} n\psi(n)\Big) \\
&= c\lim_{N\to\infty}\Big(2^{-N}\sum_{n=2^{N-1}}^{2^N-1} n\psi(n) - \frac{1}{2}\psi(1)\Big) = \infty. \qquad (8.6.19)
\end{aligned}$$

In fact, for any given $M > 0$ there exists an N_0, when $N \geq N_0, n \geq 2^{N_0-1}$, we have $n\psi(n) \geq 2M$, so that $2^{-N}\sum_{n=2^{N-1}}^{2^N-1} n\psi(n) \geq M$. (8.6.19) contradicts with (8.6.8′). Therefore we have

$$P(B_i,\ i.o.) = 0,$$

i.e. $P\Big(\bigcup_{l=i+1}^{\infty} B_l\Big) \to 0$, as $i \to \infty$, so that $P(A_i) \to 1$ as $i \to \infty$.

Combining it with (8.6.17), stationarity and the definition of ρ-mixing, for large j we have

$$\begin{aligned}
&P\Big\{\sup_{i\geq j}\max_{2^i\leq k\leq 2^{i+1}}\frac{S_k - S_{2^i}}{H(2^i)}\geq 2\varepsilon_1\Big\} \\
&= P\Big\{\bigcup_{i=j}^{\infty} B_i\Big\} \\
&\geq P\Big\{\bigcup_{i=j}^{\infty} B_iA_i\Big\} = \sum_{i=j}^{\infty} P(B_iA_i) \geq \sum_{i=j}^{\infty} P(B_i^*A_i) \\
&\geq \sum_{i=j}^{\infty}\{P(B_i^*)P(A_i) - \rho(2^{i-1})\} \\
&\geq c\sum_{i=j}^{\infty}\{P(B_i^*) - c\rho(2^{i-1})\} \\
&\geq c\sum_{i=j}^{\infty}\Big\{P\Big\{\max_{k\leq 2^{i-1}} S_k \geq \frac{2\varepsilon_1}{\delta}H(2^{i-1})\Big\} - c\rho(2^{i-1})\Big\}. \qquad (8.6.20)
\end{aligned}$$

Since $n\psi(n) \geq [n/2]\psi([n/2])$, $\sum_{2^{j-1}}^{2^j-1} n\psi(n) \geq \sum_{2^{j-2}}^{2^{j-1}-1} n\psi(n)$, combining it with (8.6.18), (8.6.20) and (8.6.12) for any given $\varepsilon > 0$, we have

$$
\begin{aligned}
\infty &> S(\psi^*, H, \varepsilon\delta) \\
&\geq c\sum_{j=2}^{\infty} 2^{-j}\Big(\sum_{n=2^{j-1}}^{2^j-1} n\psi(n) - 2\sum_{n=2^{j-2}}^{2^{j-1}-1} n\psi(n)\Big) \\
&\quad \cdot P\Big\{\sup_{i\geq j}\max_{2^i\leq k<2^{i+1}} \frac{S_k}{H(k)} \geq \varepsilon_1\delta\Big\} \\
&\geq c\sum_{j=2}^{\infty} 2^{-j}\Big(\sum_{n=2^{j-1}}^{2^j-1} n\psi(n) - 2\sum_{n=2^{j-2}}^{2^{j-1}-1} n\psi(n)\Big) \\
&\quad \cdot P\Big\{\sup_{i\geq j}\max_{2^i\leq k<2^{i+1}} \frac{S_k - S_{2^i}}{H(2^i)} \geq 2\varepsilon_1\Big\} \\
&\geq c\sum_{i=2}^{\infty}\sum_{j=2}^{i}\Big(2^{-j}\sum_{n=2^{j-1}}^{2^j-1} n\psi(n) \\
&\quad - 2^{-(j-1)}\sum_{n=2^{j-2}}^{2^{j-1}-1} n\psi(n)\Big)P(B_i^* A_i) \\
&\geq c\sum_{i=2}^{\infty} 2^{-i}\sum_{n=2^{i-2}}^{2^{i-1}-1} n\psi(n)P(B_i^* A_i) \\
&\geq c\sum_{i=2}^{\infty}\sum_{n=2^{i-2}}^{2^{i-1}-1}\psi(n)\Big(P\Big\{\max_{k\leq 2^{i-1}} S_k \geq \frac{2\varepsilon_1}{\delta}H(2^{i-1})\Big\} - c\rho(2^{i-1})\Big) \\
&\geq c\sum_{n=1}^{\infty}\psi(n)P\Big\{\max_{k\leq n} S_k \geq \frac{2\varepsilon_1}{\delta}H(n)\Big\} - c\sum_{n=1}^{\infty}\psi(n)\rho(n) \\
&= cM(\psi, H, 2\varepsilon_1/\delta) - c\sum_{n=1}^{\infty}\psi(n)\rho(n).
\end{aligned}
$$

Lemma 8.6.4 is proved.

Theorem 8.6.1 follows from Lemmas 8.6.1 – 8.6.4.

Remark 8.6.1. The condition (8.6.13) is only used in the proof of Lemma 8.6.4. It can be removed when one considers the convergence of two-sided tail probability series.

Further discussion of Prohorov's problem for a sequence of independent random variables were given by Su (1989) and Shao (1991), etc.

Chapter 9 Strong Approximations

The law of the iterated logarithm and the strong invariance principle for a sequence of mixing random variables have been discussed by some mathematicians in the sixties. Applying Strassen's martingale embedding method to strong approximations for partial sums of dependent random variables by a Wiener process, Philipp and Stout (1975) established the almost sure invariance principle for α-mixing sequences, φ-mixing sequences, lacunary trigonometric series, a class of Gaussian sequences and additive functional of Markov chains. The results for strictly stationary φ-mixing and ρ-mixing sequences have been improved by Berkes and Philipp (1979), Dabrowski (1982) and Bradley (1985) respectively. These results are improved essentially by Lu and Shao. In Shao and Lu (1987), they got a better approximation order for a φ-mixing sequence. This will be introduced in Section 9.1. For a sequence of stationary φ-mixing random variables $\{X_n,\ n \geq 1\}$ with $EX_1 = 0$, $EX_1^2 < \infty$ and $\sum_{n=1}^{\infty} \varphi^{1/2}(n) < \infty$, Heyde and Scott (1973) first gave that the order of strong approximation is $o((n \log\log n)^{1/2})$. Shao (1989a, 1993b) improved this result and gave that the same order of strong approximations for φ-mixing and ρ-mixing sequences when the rate of mixing is $O((\log n)^{-1-\varepsilon})$. These results imply the law of the iterated logarithm. When the $(2+\delta)$-th moment is finite, he gives the further results which imply the Chung law of the iterated logarithm. We shall discuss these results for the φ-mixing and ρ-mixing cases in Section 9.2. Shao and Lu(1987) and Shao (1989a) have also studied the strong approximations for an α-mixing sequence. These will be given in Section 9.3.

9.1 Strong approximations for a φ-mixing sequence

Let $\{X_n,\ n \geq 1\}$ be a sequence of φ-mixing random variables with $EX_n = 0$. Put $S_n = \sum_{k=1}^{n} X_k$, $S(t) = S_{[t]}$ $(t > 0)$. In this section, we first discuss how fast is the rate of strong approximations for a sequence of φ-mixing random variables when applying Strassen's martingale embedding method. The rate of strong approximations gotten by Shao and Lu (1987) approaches the ideal order $O((n \log\log n)^{1/4}(\log n)^{1/2})$ for a sequence of i.i.d.r.v.'s. The following theorem is proved.

Theorem 9.1.1. (Shao, Lu 1987) *Let $\{X_n,\ n \geq 1\}$ be a sequence of φ-mixing random variables with $EX_n = 0$. Suppose that*
(i) $\sigma_n^2 = ES_n^2 \geq Cn$ for some $C > 0$,
(ii) $\sup_n EX_n^4 < \infty$,
(iii) $\varphi(n) = O(1/n)$.
Then without changing the distribution of $\{S(t),\ t \geq 0\}$, we can redefine the process $\{S(t),\ t \geq 0\}$ on a richer probability space together with a standard Wiener process $\{W(t),\ t \geq 0\}$ such that for any $\varepsilon > 0$

$$S(t) - W(\sigma_t^2) = O(\sigma_t^{1/2}(\log \sigma_t)^{9/4+\varepsilon}) \quad a.s. \tag{9.1.1}$$

as $t \to \infty$, where $\sigma_t^2 = \sigma_{[t]}^2$.

From Theorem 9.1.1 we have the following corollary immediately.

Corollary 9.1.1. *Let $\{X_n,\ n \geq 1\}$ be a sequence of stationary φ-mixing random variables with $EX_1 = 0$. If*

$$\sigma^2 = EX_1^2 + 2\sum_{k=2}^{\infty} EX_1X_k \tag{9.1.2}$$

is absolutely convergent, by assuming $\sigma = 1$ without loss of generality, then under the conditions (ii) and (iii), we have

$$S(t) - W(t) = O(t^{1/4}(\log t)^{9/4+\varepsilon}) \quad a.s. \tag{9.1.3}$$

In order to prove Theorem 9.1.1, we first give a fundamental proposition. The proof of the proposition points out how to use the Strassen's martingale embedding method for a mixing sequence to get the results of strong approximations.

Let $\mathcal{F}_0 = \{\emptyset, \Omega\}$ and $\{\mathcal{F}_n, n \geq 0\}$ be a sequence of non-decreasing σ-fields. Assume that X_n is $\mathcal{F}_n$-measurable, $n = 1, 2, \cdots$. Define

$$\begin{aligned} Y_n &= \sum_{k=0}^{2^n}\{E(X_{n+k}|\,\mathcal{F}_n) - E(X_{n+k}|\mathcal{F}_{n-1})\} \\ &= X_n + u_n - u_{n-1} - v_n \end{aligned} \tag{9.1.4}$$

for every $n \geq 1$, where

$$u_n = \sum_{k=1}^{2^n} E(X_{n+k}|\mathcal{F}_n), \quad v_n = \sum_{k=2^{n-1}}^{2^n} E(X_{n+k}|\mathcal{F}_{n-1}). \tag{9.1.5}$$

$\{Y_n, \mathcal{F}_n, n \geq 1\}$ is a martingale difference sequence.

Proposition 9.1.1. *Let $\{X_n, n \geq 1\}$ be a sequence of random variables with $EX_n = 0$, $\sup_n E|X_n|^{2+\delta} < \infty\,(0 < \delta \leq 2)$. Let $\mathcal{F}_n = \sigma\{X_k, 1 \leq k \leq n\}$ be a natural σ-field sequence. If the following conditions are satisfied:*

(a) $\sigma_n^2 = ES_n^2 \geq C\,n$ *for some* $C > 0$,

(b) $\|u_n\|_{2+\delta} = O(1)$, $\sum_{n=1}^{\infty} \|v_n\|_{2+\delta} < \infty$,

(c) for some $\lambda \geq 0$

$$E|\sum_{m<k\leq m+n} (Y_k^2 - EY_k^2)|^{(2+\delta)/2} = O(n(\log n)^{\lambda}), \tag{9.1.6}$$

then without changing the distribution of $\{S(t), t \geq 0\}$, we can redefine the process $\{S(t), t \geq 0\}$ on a richer probability space together with a standard Wiener process $\{W(t), t \geq 0\}$ such that for any $\varepsilon > 0$

$$S(t) - W(\sigma_t^2) = O\Big(\sigma_t^{2/(2+\delta)}(\log \sigma_t)^{1+\varepsilon+(1+\lambda)/(2+\delta)}\Big) \quad a.s. \tag{9.1.7}$$

as $t \to \infty$. Particularly, for $\delta = 2$ we have

$$S(t) - W(\sigma_t^2) = O\Big(\sigma_t^{1/2}(\log \sigma_t)^{(5+\lambda)/4+\varepsilon}\Big) \quad a.s. \tag{9.1.8}$$

Proof. 1) We first prove that for each $\varepsilon > 0$,

$$S(t) - \sum_{k\leq t} Y_k = O(t^{1/(2+\delta)}(\log t)^{(1+\varepsilon)/(2+\delta)}\Big) \quad \text{a.s.} \tag{9.1.9}$$

In fact, noting that

$$S(t) - \sum_{k\leq t} Y_k = \sum_{k\leq t} v_k - u_{[t]},$$

using condition (b) and the Borel-Cantelli lemma, we conclude that (9.1.9) holds.

We now apply a martingale version of the Skorohod-Strassen representation theorem. There exists a probability space, on which a standard

Wiener process and a sequence of non-negative stopping time T_n are defined such that

$$\Big\{W\Big(\sum_{j\le n} T_j\Big),\, n\ge 1\Big\} \quad \text{and} \quad \Big\{\sum_{j\le n} Y_j,\, n\ge 1\Big\}$$

have the same distribution. Hence without loss of generality, on this new probability space we can redefine Y_n by

$$Y_n = W\Big(\sum_{j\le n} T_j\Big) - W\Big(\sum_{j<n} T_j\Big),$$

and keep the same notation. Now, write $\mathcal{F}_n = \sigma\{W(\sum_{j\le k} T_j);\ 1\le k\le n\}$, and $\mathcal{G}_n = \sigma\{W(t);\ 0\le t\le \sum_{j\le n} T_j\}$. It is easy to see that $\mathcal{F}_n \subseteq \mathcal{G}_n$, T_n is $\mathcal{G}_n$-measurable and for every $n\ge 1$,

$$E(T_n|\mathcal{G}_{n-1}) = E(Y_n^2|\mathcal{G}_{n-1}) = E(Y_n^2|\mathcal{F}_{n-1}) \quad \text{a.s.} \tag{9.1.10}$$

Moreover, for each $1<p\le 2$,

$$E|T_n|^p \le E|Y_n|^{2p}.$$

2) Write

$$S(t) - W(\sigma_t^2) = S(t) - \sum_{k\le t} Y_k + W\Big(\sum_{k\le t} T_k\Big) - W(\sigma_t^2). \tag{9.1.11}$$

In order to estimate the second difference of the right hand side of (9.1.11), we have to estimate $\sum_{k\le t} T_k - \sigma_t^2$. Write

$$\begin{aligned}\sum_{k\le t} T_k - \sigma_t^2 &\\ &= \sum_{k\le t}\{T_k - E(T_k|\mathcal{G}_{k-1})\}\\ &\quad + \sum_{k\le t}\{E(Y_k^2|\mathcal{F}_{k-1}) - Y_k^2\} + \Big\{\sum_{k\le t} Y_k^2 - \sigma_t^2\Big\}.\end{aligned} \tag{9.1.12}$$

We shall show that

$$\sum_{k\le t} T_k - \sigma_t^2 = O(\sigma_t^{4/(2+\delta)}(\log\sigma_t)^{1+\varepsilon+2(1+\lambda)/(2+\delta)}) \quad \text{a.s.} \tag{9.1.13}$$

Put $R_j = Y_j^2 - E(Y_j^2|\mathcal{F}_{j-1})$. Then $\{R_j, \mathcal{F}_j\}$ is a martingale difference sequence. We obtain from condition (b) that

$$E|R_j|^{(2+\delta)/2} \le 16E|Y_j|^{2+\delta} = O(1).$$

Hence by the fundamental theorem on the martingale (cf. Chow 1965), we have

$$\sum_{k\le t} R_k = O(t^{2/(2+\delta)}(\log t)^{(2+\varepsilon)/(2+\delta)}) \quad \text{a.s.} \tag{9.1.14}$$

Similarly, we have

$$\sum_{k\le t}(T_k - E(T_k|\mathcal{G}_{k-1})) = O(t^{2/(2+\delta)}(\log t)^{(2+\varepsilon)/(2+\delta)}) \quad \text{a.s.} \tag{9.1.15}$$

For the third term of the right hand side of (9.1.12), by the condition (c), Móricz's theorem and the Borel-Cantelli lemma, it is easy to show

$$\sum_{k\le t}(Y_k^2 - EY_k^2) = O(t^{2/(2+\delta)}(\log t)^{1+\varepsilon+2(1+\lambda)/(2+\delta)}) \quad \text{a.s.} \tag{9.1.16}$$

Noting that $\{Y_k, \mathcal{F}_k\}$ is a martingale difference sequence, from the conditions $(a), (b)$ and the Schwarz inequality, we obtain

$$\begin{aligned}\sum_{k\le t} EY_k^2 - \sigma_t^2 &= E\Big(\sum_{k\le t} Y_k\Big)^2 - \sigma_t^2 \\ &= 2E\Big(\sum_{k\le t} X_k\Big)\Big(u_t - \sum_{k\le t} v_k\Big) + E\Big(u_t - \sum_{k\le t} v_k\Big)^2 \\ &= O(\sigma_t).\end{aligned} \tag{9.1.17}$$

Hence by $(9.1.16), (9.1.17)$, and condition (a), we have

$$\sum_{k\le t} Y_k^2 - \sigma_t^2 = O(\sigma_t^{4/(2+\delta)}(\log \sigma_t)^{1+\varepsilon+2(1+\lambda)/(2+\delta)}) \quad \text{a.s.} \tag{9.1.18}$$

Equality (9.1.13) follows from $(9.1.14), (9.1.15)$ and $(9.1.18)$.

We now conclude

$$W\Big(\sum_{k\le t} T_k\Big) - W(\sigma_t^2) = O(\sigma_t^{2/(2+\delta)}(\log \sigma_t)^{1+\varepsilon+(1+\lambda)/(2+\delta)}) \quad \text{a.s.}$$

by (9.1.13) and the proof of Theorem 3.2B in Hanson and Russo (1983). Combining (9.1.11) with (9.1.9) yields the proposition.

Proof of Theorem 9.1.1. We now verify that the conditions (b) *and* (c) of proposition 9.1.1 are satisfied. By Lemma 1.2.8 and Lemma

2.2.8, we have

$$\begin{aligned}\|u_n\|_{2+\delta}^{2+\delta} &= E|u_n|^{2+\delta} = E(u_n(\mathrm{sgn}u_n)|u_n|^{1+\delta})\\ &= E(X_{n+1}|u_n|^{1+\delta}\mathrm{sgn}u_n) + \sum_{i=0}^{n-1} E\Big(\sum_{k=2^i+1}^{2^{i+1}} X_{n+k}|u_n|^{1+\delta}\mathrm{sgn}u_n\Big)\\ &= O\Big(\|u_n\|_{2+\delta}^{1+\delta} + \sum_{i=0}^{n-1}\varphi^{(1+\delta)/(2+\delta)}(2^i)2^{i/2}\|u_n\|_{2+\delta}^{1+\delta}\Big)\\ &= O(\|u_n\|_{2+\delta}^{1+\delta}),\end{aligned}$$

which implies that

$$\|u_n\|_{2+\delta} = O(1). \tag{9.1.19}$$

Similarly, we have

$$\|v_n\|_{2+\delta} = O(\varphi^{(1+\delta)/(2+\delta)}(2^{n-1})2^{(n-1)/2}).$$

Hence

$$\sum_{n=1}^{\infty} \|v_n\|_{2+\delta} < \infty. \tag{9.1.20}$$

Therefore condition (b) is satisfied.

For condition (c), put

$$T_m(n) = \sum_{m<k\le m+n} (Y_k^2 - EY_k^2), \quad \tau_n = \sup_m \|T_m(n)\|_2. \tag{9.1.21}$$

It is easy to obtain

$$\|T_m(n)\|_2 \le \|T_m([n/2]) + T_{m+[n/2]+l}([n/2])\|_2 + 2\tau_l + 2\tau_1, \tag{9.1.22}$$

where $l = [2n(\log(2n))^{-2-\varepsilon}]$. Note that

$$\begin{aligned}&E(T_m([n/2]) + T_{m+[n/2]+l}([n/2]))^2\\ &\quad\le 2\tau_{[n/2]}^2 + 2E\{T_m([n/2])T_{m+[n/2]+l}([n/2])\}.\end{aligned} \tag{9.1.23}$$

By an elementary calculation, from (9.1.4) and Lemma 2.2.8 we obtain

$$\begin{aligned}&ET_m([n/2])\,T_{m+[n/2]+l}([n/2])\\ &\quad\le 2\varphi^{1/2}(l)\,\|T_m([n/2])\|_2\Big\|\sum_k X_k\Big\|_4^2\\ &\qquad+ \|T_n([n/2])\|_2\Big(\|u_{N_1}\|_4 + \|u_{N_2}\|_4 + \sum_k \|v_k\|_4\Big)\Big\|\sum_k X_k\Big\|_4\\ &\qquad+ \|T_m([n/2])\|_2\Big(\|u_{N_1}\|_4^2 + \|u_{N_2}\|_4^2 + \sum_k \|v_k\|_4^2\Big),\end{aligned}$$

where $\sum_k$ is extended over $m + [n/2] + l < k \le m + 2[n/2] + l$ and

$$N_2 = m + [n/2] + l, \quad N_1 = m + 2[n/2] + l.$$

Using Lemma 2.2.8, (9.1.12), (9.1.13) and conditions (i) , (ii), we conclude that there exists a constant C such that for each $m, n \ge 1$,

$$ET_m([n/2])T_{m+[n/2]+l}([n/2]) \le C\tau([n/2])\, n^{1/2}\, (\log n)^{1+\varepsilon}. \qquad (9.1.24)$$

From (9.1.22), (9.1.23) and (9.1.24), we obtain

$$\|T_m(n)\|_2 \le 2^{1/2}\tau([n/2]) + C\, n^{1/2}\, (\log n)^{1+\varepsilon}$$
$$+\tau([2n(\log n)^{-2-\varepsilon}]). \qquad (9.1.25)$$

Finally, from (9.1.25) and by induction we have

$$\tau_n \le C_0 n^{1/2} (\log n)^{2+\varepsilon},$$

with $C_0 = \max(\exp(2^{2/\varepsilon}), 2C)$. This shows that condition (c) holds for $\lambda = 4 + \varepsilon$, and Theorem 9.1.1 follows.

Remark 9.1.1. Let $\{X_n;\, n \ge 1\}$ be a φ-mixing sequence with $EX_n = 0$. Suppose that condition (i) of Theorem 9.1.1 is satisfied and for some $0 < \delta \le 2$,

$$\sup_n E|X_n|^{2+\delta} < \infty$$

and $\varphi(n)$ goes to zero with the polynomial rate. Shao and Lu (1987) have also given the following results:

1) If $0 < \delta < 2$ and $\varphi(n) = O(n^{-\alpha})$ for some $\alpha > 1$, then

$$S(t) - W(\sigma_t^2) = O(\sigma_t^{2/(2+\delta)} (\log \sigma_t)^{1+\varepsilon+(1+\lambda)/(2+\delta)}) \quad \text{a.s.}$$

for any $\varepsilon > 0$, where $\lambda = 2\log 3/\log \theta^{-1}, \theta = 1 - 2(\alpha - 1)/(\alpha(2+\delta)) > 0$.

2) If $0 < \delta \le 2$ and $\varphi(n) = O(n^{-\alpha}), (2+\delta)/(2(1+\delta)) < \alpha \le 1$, then

$$S(t) - W(\sigma_t^2) = O(\sigma_t^{1-\alpha\delta/(2+\delta)+\varepsilon}) \quad \text{a.s.}$$

for any $\varepsilon > 0$.

9.2 Strong approximation for a ρ-mixing sequence

In the introduction of this chapter, we have mentioned an interesting problem: what the rate of strong approximation of partial sum process $S(t)$

for a mixing sequence by a Wiener process is when we only assume that $\sup_n EX_n^2 < \infty$ and $\varphi(n) = O((\log n)^{-a})$, $a > 1$ (or $\rho(n) = O((\log n)^{-a})$, $a > 1$) ? In this section, we deal with a ρ-mixing sequence, and the strong approximation order $O((n \log\log n)^{1/2})$ is given. With this result the law of the iterated logarithm holds true for this ρ-mixing sequence. In order to get the Chung law of iterated logarithm for a ρ-mixing sequence, we need the assumption $\sup_n E|X_n|^{2+\delta} < \infty$.

Theorem 9.2.1.(Shao 1989a, 1993b) *Let $\{X_n, n \geq 1\}$ be a strictly stationary ρ-mixing sequence with $EX_1 = 0$, $EX_1^2 < \infty$. Assume that*

(i) $\sigma_n^2 = ES_n^2 \to \infty$ as $n \to \infty$,

(ii) $\rho(n) = O((\log n)^{-1-\varepsilon'})$ for some $\varepsilon' > 0$.

Then without changing the distribution of $\{S(t), t \geq 0\}$, we can redefine the process $\{S(t), t \geq 0\}$ on a richer probability space together with the standard Wiener process $\{W(t), t \geq 0\}$ such that

$$S(t) - W(\sigma_t^2) = o(\sigma_t(\log\log t)^{1/2}) \quad a.s. \tag{9.2.1}$$

as $t \to \infty$.

Theorem 9.2.2.(Shao 1989a, 1993b) *Let $\{X_n, n \geq 1\}$ be a strictly stationary φ-mixing sequence with $EX_1 = 0, EX_1^2 < \infty$. Assume that*

(i)' $\sigma_n^2 \to \infty$ *as $n \to \infty$,*

(ii) $\sum_{n=1}^{\infty} \varphi^{1/2}(2^n) < \infty$. *Then (9.2.1) is also true.*

With the help of the law of the iterated logarithm for a Wiener process, from the previous theorems we have the following corollary.

Corollary 9.2.1. *Under the conditions of Theorem 9.2.1 or 9.2.2, we have*

$$\limsup_{n\to\infty} |S_n| / \sqrt{2\sigma_n^2 \log\log \sigma_n^2} = 1 \quad a.s. \tag{9.2.2}$$

The following strong approximation theorem implies the Chung law of the iterated logarithm.

Theorem 9.2.3.(Shao 1989a, 1993b) *Let $\{X_n, n \geq 1\}$ be a strictly stationary ρ-mixing sequence with $EX_1 = 0$, $E|X_1|^{2+\delta} < \infty$ for some $\delta > 0$. Assume that*

(i) $\sigma_n^2 \to \infty$, as $n \to \infty$,

(ii) $\rho(n) = O((\log n)^{-r})$ *for some $r > 1/2$.*

Then for any $0 < \theta < r/2 - 1/4$, *we have*

$$S(t) - W(\sigma_t^2) = o(\sigma_t (\log t)^{-\theta}) \quad a.s. \tag{9.2.3}$$

Corollary 9.2.2. *Under the conditons of Theorem 9.2.3, we have (9.2.2) and*

$$\liminf_{N\to\infty} \Big(\frac{8 \log\log N}{\pi^2 \sigma_N^2} \Big)^{1/2} \max_{1\le n\le N} |S_n| = 1 \quad a.s. \tag{9.2.4}$$

Remark 9.2.1. The results of Theorem 9.2.3 and Corollary 9.2.2 hold true for a stationary φ-mixing sequence with $EX_1 = 0, E|X_1|^{2+\delta} < \infty$, $\sigma_n^2 \to \infty$ and $\varphi(n) = O((\log n)^{-r})$ for some $r > (2+\delta)/(2(1+\delta))$.

Remark 9.2.2. The stationarity condition in these theorems and corollaries can been replaced by the identical distribution condition.

We point out in passing that the results in this section improve those of Bradley(1985), Dabrowski(1982) and Theorem 4 of Berkes and Philipp(1979)

In the proof of these theorems , we shall use both the Bernstein divided-section method and the Strassen martingale embedding method. Put

$$\overline{X}_k = X_k I(X_k^2 \ge k) - EX_k I(X_k^2 \ge k), \tag{9.2.5}$$

$$\widehat{X}_k = X_k I(X_k^2 < k) - EX_k I(X_k^2 < k). \tag{9.2.6}$$

Let

$$\overline{S}(n) = \sum_{k=1}^{n} \overline{X}_k, \qquad \widehat{S}(n) = \sum_{k=1}^{n} \widehat{X}_k,$$

$$\overline{S}_k(n) = \sum_{i=k+1}^{k+n} \overline{X}_i \qquad \widehat{S}_k(n) = \sum_{i=k+1}^{k+n} \widehat{X}_i.$$

We shall first prove

$$\overline{S}(n) = o(n^{1/2}) \qquad \text{a.s.} \tag{9.2.7}$$

Define blocks of integers $H_1, I_1, H_2, I_2, \cdots$ by requiring that H_k contains h_k and I_k contains i_k consecutive integers and that there are no gaps between consecutive blocks, where

$$\begin{aligned} h_k &= \text{Card } H_k = [ak^{a-1}\exp(k^a)], \\ i_k &= \text{Card } I_k = [ak^{a-1}\exp(k^a/2)] \end{aligned} \tag{9.2.8}$$

and $0 < a < 1$ will be chosen later on. Put

$$\begin{aligned}
N_k &:= \sum_{j\le k} \mathrm{Card}(H_j \cup I_j) \sim \exp(k^a),\\
u_k &:= \sum_{j\in H_k} \widehat{X}_j, \qquad v_k := \sum_{j\in I_k} \widehat{X}_j,\\
\xi_k &:= u_k - E(u_k|\,\mathcal{F}_{k-1}) \quad \text{where } \mathcal{F}_k = \sigma(X_i, i \le N_{k-1} + h_k),\\
m_n &:= \{k : n \in H_k \cup I_k\} - 1 \sim (\log n)^{1/a}.
\end{aligned}$$

It is clear that

$$S_n = \widehat{S}(n) + \overline{S}(n). \tag{9.2.9}$$

$$\begin{aligned}
\widehat{S}(n) &= \sum_{i=1}^{m_n} \xi_i + \sum_{i=1}^{m_n} E(u_i|\,\mathcal{F}_{i-1})\\
&\quad + \sum_{i=1}^{m_n} v_i + \sum_{i=N_{m_n}+1}^{n} \widehat{X}_i.
\end{aligned} \tag{9.2.10}$$

In order to prove Theorem 9.2.1, by this two representations , we shall show at first that one needs only to check (9.2.7) and to show that the following relations hold true:

$$\sum_{i=1}^{n} v_i = o(\exp(n^a/2)) \quad \text{a.s.} \tag{9.2.11}$$

$$\max_{N_n < j \le N_{n+1}} \Big| \sum_{i=N_n+1}^{j} \widehat{X}_i \Big| = o(\exp(n^a/2)) \quad \text{a.s.} \tag{9.2.12}$$

$$\sum_{i=1}^{n} E(u_i|\,\mathcal{F}_{i-1}) = o(\exp(n^a/2)) \quad \text{a.s.} \tag{9.2.13}$$

$$\sum_{i=1}^{\infty} e^{-(1+\delta_1)i^a} E|\xi_i|^{2(1+\delta_1)} < \infty \quad \text{for some } \delta_1 \in (0,1) \tag{9.2.14}$$

$$\sum_{i=1}^{n} (E(\xi_i^2|\,\mathcal{F}_{i-1}) - E\xi_i^2) = o(\exp(n^a)) \quad \text{a.s.} \tag{9.2.15}$$

$$\sum_{i=1}^{m_n} E\xi_i^2 - ES_n^2 = o(n) \quad \text{a.s.} \tag{9.2.16}$$

Proof of Theorem 9.2.1.

Assume that (9.2.7) and (9.2.11)–(9.2.13) hold. Denote $\tilde{\sigma}_n^2 = \sum_{i=1}^{m_n} E\xi_i^2$ Then

$$S(n) - \sum_{i=1}^{m_n} \xi_i = o(n^{1/2}) \quad \text{a.s.} \tag{9.2.17}$$

By the Strassen martingale embedding method, for the martingale difference sequence $\{\xi_k, \mathcal{F}_k, k \geq 1\}$ there exist the stopping times $\{T_n, n \geq 1\}$ such that

$$\Big\{W(\sum_{i=1}^{n} T_i),\, n \geq 1\Big\} \stackrel{\mathrm{d}}{=} \Big\{\sum_{i=1}^{n} \xi_i,\, n \geq 1\Big\}.$$

Redefine

$$\xi_1 = W(T_1), \quad \xi_n = W\Big(\sum_{i=1}^{n} T_i\Big) - W\Big(\sum_{i=1}^{n-1} T_i\Big), \quad n \geq 2.$$

As in (9.1.10), we have

$$E(T_n | \mathcal{G}_{n-1}) = E(\xi_n^2 | \mathcal{G}_{n-1}) = E(\xi_n^2 | \mathcal{F}_{n-1}).$$

Write

$$\sum_{i=1}^{m_n} \xi_i - W(\tilde{\sigma}_n^2) = W\Big(\sum_{i \leq m_n} T_i\Big) - W\Big(\sum_{i \leq m_n} ET_i\Big) \tag{9.2.18}$$

and

$$\begin{aligned}\sum_{i=1}^{n}&(T_i - ET_i)\\ &= \sum_{i=1}^{n}(T_i - E(T_i | \mathcal{G}_{i-1}))\\ &\quad + \sum_{i=1}^{n}(E(\xi_i^2 | \mathcal{F}_{i-1}) - \xi_i^2) + \sum_{i=1}^{n}(\xi_i^2 - E\xi_i^2).\end{aligned} \tag{9.2.19}$$

By a martingale result of Chow (1965) and (9.2.14), the series

$$\sum_{i=1}^{\infty} e^{-n^a} \xi_n^2 < \infty \quad \text{a.s.}$$

implies that the series

$$\sum_{i=1}^{\infty} e^{-n^a} (\xi_n^2 - E\xi_n^2)$$

is also almost surely convergent. It follows from the Kronecker lemma that

$$\sum_{i\le n}(\xi_i^2 - E\xi_i^2) = o(\exp(n^a)) \quad \text{a.s.}$$

By the same way, we have $\sum_{i\le n}(T_i - E(T_i|\mathcal{G}_{i-1})) = o(\exp(n^a))$. Hence, in combination with (9.2.15), we have

$$\sum_{i=1}^{n}(T_i - ET_i) = o(\exp(n^a)) \quad \text{a.s.}$$

Using the Hanson-Russo Theorem for the lag increments of a Wiener process (see Hanson and Russo 1983 Theorem 3.2.B) and (9.2.18), we get

$$\sum_{i=1}^{m_n}\xi_i - W(\tilde{\sigma}_n^2) = o((n\log\log n)^{1/2}) \quad \text{a.s.} \tag{9.2.20}$$

Similarly, from (9.2.16), we have $W(\tilde{\sigma}_n^2) - W(\sigma_n^2) = o((n\log\log n)^{1/2})$ a.s. Combining it with (9.2.20) and (9.2.17), the conclusion of Theorem 9.2.1 follows.

Now in order to complete the proof of Theorem 9.2.1, it is sufficient to prove (9.2.7), (9.2.11)–(9.2.16). They will be given by a series of lemmas in the following. At first, we give three lemmas for any sequence of random variables. Lemma 9.2.1 implies (9.2.7).

Lemma 9.2.1. *Let $\{X_n\}$ be a sequence of strictly stationary random variables, $EX_1^2 < \infty$. Then*

$$\sum_{i=1}^{n}\{|X_i|I(|X_i| \ge i^{1/2}) + E|X_i|I(|X_i| \ge i^{1/2})\} = o(n^{1/2}) \quad a.s. \tag{9.2.21}$$

Proof. We have

$$\begin{aligned}
&\sum_{i=1}^{\infty} i^{-1/2}E|X_i|I(X_i^2 \ge i)\\
&\le \sum_{j=1}^{\infty}\sum_{i=1}^{j} i^{-1/2}E|X_1|I(j-1 < X_1^2 \le j)\\
&\le 4\sum_{j=1}^{\infty}(j-1)^{1/2}E|X_1|I(j-1 < X_1^2 \le j)\\
&\le 4\sum_{j=1}^{\infty}EX_1^2 I(j-1 < X_1^2 \le j)\\
&= 4EX_1^2 < \infty.
\end{aligned}$$

Therefore $\sum_{i=1}^{\infty} i^{-1/2}|X_i|I(X_i^2 \geq i) < \infty$ a.s., then (9.2.21) follows from the Kronecker lemma.

Lemma 9.2.2. *If $E|X| < \infty$, then for any $0 < b \leq 1, \varepsilon > 0$*

$$\sum_{k=1}^{\infty} k^{b-1} e^{-\varepsilon k^b} E|X|^{1+\varepsilon} I\Big(|X| < e^{k^b}\Big) < \infty.$$

Proof. It is easy to see that the left hand side of the above inequality equals to

$$\begin{aligned}
&\sum_{k=1}^{\infty} k^{b-1} e^{-\varepsilon k^b} E|X|^{1+\varepsilon} I(|X| < 1) \\
&\qquad + \sum_{k=1}^{\infty} k^{b-1} e^{-\varepsilon k^b} \sum_{l=1}^{k} E|X|^{1+\varepsilon} I\Big(e^{(l-1)^b} \leq |X| < e^{l^b}\Big) \\
&\qquad \leq c + c \sum_{l=1}^{\infty} E|X|^{1+\varepsilon} I\Big(e^{(l-1)^b} \leq |X| < e^{l^b}\Big) e^{-\varepsilon l^b} \\
&\qquad \leq c E|X| < \infty.
\end{aligned}$$

Lemma 9.2.3. *Let $\{c_n,\ n \geq 1\}$ be a sequence of nonincreasing positive numbers. Then, for any real sequence $\{\eta_n,\ n \geq 1\}$, we have*

$$\max_{1 \leq i \leq n} \Big|\sum_{j=1}^{i} \eta_j\Big| c_i \leq 2 \max_{1 \leq i \leq n} \Big|\sum_{j=1}^{i} c_j \eta_j\Big|. \tag{9.2.22}$$

Proof. Denote $D_i = \sum_{j=1}^{i} c_j \eta_j$. We have

$$\eta_j = (D_j - D_{j-1})/c_j = \sum_{i=1}^{j} \Big(\frac{1}{c_i} - \frac{1}{c_{i-1}}\Big)(D_j - D_{j-1}),$$

where $1/c_0 = 0$. Therefore

$$\begin{aligned}
\sum_{j=1}^{k} \eta_j &= \sum_{j=1}^{k} \sum_{i=1}^{j} \Big(\frac{1}{c_i} - \frac{1}{c_{i-1}}\Big)(D_j - D_{j-1}) \\
&= \sum_{i=1}^{k} \Big(\frac{1}{c_i} - \frac{1}{c_{i-1}}\Big)(D_k - D_{i-1}).
\end{aligned}$$

It follows that

$$\max_{1\le k\le n} c_k\Big|\sum_{j=1}^{k}\eta_j\Big| \le \max_{1\le k\le n}\max_{1\le i\le k}|D_k - D_{i-1}|$$
$$\le 2\max_{1\le k\le n}\Big|\sum_{j=1}^{k} c_j\eta_j\Big|. \tag{9.2.23}$$

The Lemma is proved.

Lemma 9.2.4. *Let $\{X_n,\ n\ge 1\}$ be a ρ-mixing sequence with $EX_n = 0$, $EX_n^2<\infty$. Let u_k, v_k be as above. Denote $u_k(n)=\sum_{i=k+1}^{k+n}u_i$, $v_k(n)=\sum_{i=k+1}^{k+n}v_i$. Suppose that*

$$\sum_{n=1}^{\infty}\rho(2^n)<\infty.$$

Then there exists a constant $c=c(\rho(\cdot))$ such that for any $k\ge 0, n\ge 1$

$$Eu_k^2(n)\le c\Big(\sum_{i=k+1}^{k+n}Eu_i^2\Big), \tag{9.2.24}$$

$$Ev_k^2(n)\le c\Big(\sum_{i=k+1}^{k+n}Ev_i^2\Big). \tag{9.2.25}$$

Proof. We only give the proof of (9.2.24). It is obvious for $n=1$. For $n\ge 2$, denote $n_1=n-[n/2]$, $n_2=[n/2]$. Then by the definition of $\rho(\cdot)$ we have

$$Eu_k^2(n) = Eu_k^2(n_1)+Eu_{k+n_1}^2(n_2)+2Eu_k(n_1)u_{k+n_1}(n_2)$$
$$\le (Eu_k^2(n_1)+Eu_{k+n_1}^2(n_2))(1+\rho(i_{n_1})). \tag{9.2.26}$$

Let $c_1=1, c_n=c_{n_1}(1+\rho(i_{n_1}))\,(n\ge 2)$. It is easy to see that c_n is non-decreasing. It follows from (9.2.26) that for any $k\ge 0$, $n\ge 1$

$$Eu_k^2(n)\le c_n\Big(\sum_{i=k+1}^{k+n}Eu_i^2\Big).$$

Now, we prove that $\{c_n\}$ is a bounded sequence. Note that

$$c_{2^m} = c_{2^{m-1}}(1+\rho(i_{2^{m-1}}))$$
$$\le c_{2^{m-1}}\exp(\rho(i_{2^{m-1}}))\le \exp\Big\{\sum_{j=0}^{m-1}\rho(i_{2^j})\Big\}.$$

From the definition (9.2.8) of i_k, for any large j, we have

$$\rho(i_{2^j}) \le \rho(2^{2^{ja}}),$$

and further

$$\sum_{j=0}^{m-1} \rho(i_{2^j}) \le \sum_{j=0}^{m-1} \rho(2^{2^{ja}}) \le \sum_{j=0}^{m-1} \rho(2^j) < \infty$$

by monotoncity of $\rho(\cdot)$. This proves $c_{2^m} \le c < \infty$. It follows from the monotoncity of c_n that $\{c_n\}$ is bounded. The lemma is proved.

The following lemma gives (9.2.11).

Lemma 9.2.5. *Let $\{X_n, n \ge 1\}$ be as in Theorem 9.2.1, we have*

$$\sum_{j=1}^{n} v_j = o(\exp(n^a/3)) \quad a.s. \tag{9.2.27}$$

Proof. Using Lemma 2.2.5 with $q = 2$ and condition (ii), we have

$$\begin{aligned} E(\sum_{j=1}^{n} v_j)^2 &\le cn \max_{1\le j\le n} Ev_j^2 \\ &\le cn^a \exp(n^a/2). \end{aligned}$$

It follows from the Borel-Cantelli lemma that (9.2.27) holds true.

The following lemma gives (9.2.12).

Lemma 9.2.6. *Assume that the conditions of Theorem 9.2.1 are satisfied and*

$$0 < a < \varepsilon'/(8+\varepsilon'). \tag{9.2.28}$$

Then

$$\max_{N_k<j\le N_{k+1}} \Big|\sum_{i=N_k}^{j} \widehat{X}_i\Big| = o(\exp(k^a/2)) \quad a.s. \tag{9.2.29}$$

Proof. By the Borel-Cantelli lemma, we need only to prove that for any $\varepsilon > 0$

$$\sum_{k=1}^{\infty} P\Big\{\max_{N_k<j\le N_{k+1}} \Big|\sum_{i=N_k}^{j} \widehat{X}_i\Big| \ge \varepsilon \exp(k^a/2)\Big\} < \infty. \tag{9.2.30}$$

Put $d_k = N_{k+1} - N_k \sim ak^{a-1}\exp(k^a)$, and let

$$B = k^{-a(16+\varepsilon')/\varepsilon'} \exp(k^a/2), \qquad m = [k^{-a(16+\varepsilon')/\varepsilon'} \exp(k^a)].$$

For any $\varepsilon > 0$, we have

$$\frac{48m}{B} E X_1^2 I(|X_1| \geq B) \leq \varepsilon \exp(k^a/2) \tag{9.2.31}$$

for large k. It follows from Lemma 4.3.2 that

$$\begin{aligned}
&P\Big\{\max_{N_k<j\leq N_{k+1}} \Big|\sum_{i=N_k}^{j} \widehat{X}_i\Big| \geq \varepsilon \exp(k^a/2)\Big\} \\
&\leq c\Big\{\exp\Big(-\frac{8+\varepsilon'}{8} k^a\Big)(d_k^{(8+\varepsilon')/8} \\
&\qquad + d_k \log^{2+\varepsilon'/4}(2d_k) E|X_1|^{2+\varepsilon'/4} I(|X_1| \leq B) \\
&\qquad + d_k E|X_1|^{2+\varepsilon'/4} I(|X_1| \leq N_{k+1})) \\
&\qquad + \exp(-k^a) d_k (1 + \rho^2(m) \log^4[d_k/m])\Big\} \\
&\leq c\Big\{k^{-(1-a)(8+\varepsilon')/8} + k^{-1-a} \\
&\qquad + k^{a-1} \exp(-\varepsilon' k^a/8) E|X_1|^{2+\varepsilon'/4} I(|X_1|^2 \leq 2e^{k^a})\Big\} \\
&\qquad + c\Big\{k^{a-1-2a(1+\varepsilon')} \log^4 k\Big\}.
\end{aligned}$$

By Lemma 9.2.2 and (9.2.28), (9.2.30) holds true.

The following lemma gives (9.2.13).

Lemma 9.2.7. *Suppose that condition (ii) of Theorem 9.2.1 is satisfied. Then we have*

$$\sum_{k=1}^{n} E(u_k | \mathcal{F}_{k-1}) = o(\exp(n^a/2)) \quad a.s. \tag{9.2.32}$$

Proof. We first prove that for any $k \geq 0, n \geq 1$ and a sequence of real numbers $\{c_n\}$

$$EG_k^2(n) \leq c\Big\{\sum_{j=k+1}^{k+n} \rho^2(i(j/2))\, c_j^2\, Eu_j^2\Big\} \log^2(2n) \tag{9.2.33}$$

where $G_k(n) = \sum_{j=k+1}^{k+n} c_j\, E(u_j | \mathcal{F}_{j-1})$ and $i(x)$ is the linear interpolation function of i_k. From (9.2.24) , there exists a constant c' such that for any $k \geq 0, n \geq 1$

$$E\Big(\sum_{i=k+1}^{k+n} c_i\, u_i\Big)^2 \leq c'\Big(\sum_{i=k+1}^{k+n} c_i^2\, Eu_i^2\Big). \tag{9.2.34}$$

Put $c = 100c' \log^{-2}(3/2)$. Apply the induction on n. When $n = 1$

$$\begin{aligned} EG_k^2(1) &= c_{k+1}^2 \, E(E(u_{k+1} | \mathcal{F}_k))^2 \\ &\le c_{k+1}^2 \, \rho(i_k) \|u_{k+1}\|_2 \|E(u_{k+1} | \mathcal{F}_k)\|_2 \end{aligned}$$

i.e.

$$EG_k^2(1) \le c_{k+1} \rho^2(i_k) E u_{k+1}^2.$$

This proves (9.2.33) for $n = 1$. Suppose that (9.2.33) holds true for any integer less than n. We prove that (9.2.33) remains true for n. Denote $n_1 = n - [n/2]$, $n_2 = [n/2]$. We have

$$\begin{aligned} EG_k^2(n) &= EG_k^2(n_1) + EG_{k+n_1}^2(n_2) + 2EG_k(n_1)G_{k+n_1}(n_2) \\ &= EG_k^2(n_1) + EG_{k+n_1}^2(n_2) + 2EG_k(n_1)\Big(\sum_{j=k+n_1+1}^{k+n} c_j u_j \Big) \\ &\le EG_k^2(n_1) + EG_{k+n_1}^2(n_2) \\ &\quad + 2\rho(i_{k+n_1}) \| G_k(n_1)\|_2 \Big\| \sum_{j=k+n_1+1}^{k+n} c_j u_j \Big\|_2. \end{aligned}$$

By (9.2.34) and the assumption of induction, we have

$$\begin{aligned} EG_k^2(n) &\le c \Big\{ \sum_{j=k+1}^{k+n} \rho^2(i(j/2)) c_j^2 E u_j^2 \Big\} \log^2(2n_1) \\ &\quad + 2\sqrt{cc'} \rho(i_{k+n_1}) \Big\{ \sum_{j=k+n_1+1}^{k+n} c_j^2 E u_j^2 \Big\}^{1/2} \\ &\quad \cdot \Big\{ \sum_{j=k+1}^{k+n_1} \rho^2(i(j/2)) c_j^2 E u_j^2 \Big\}^{1/2} \log(2n_1) \\ &\le (c(\log 2n_1)^2 + \sqrt{cc'} \log(2n_1)) \Big\{ \sum_{j=k+1}^{k+n} \rho^2(i(j/2)) c_j^2 E u_j^2 \Big\} \\ &\le c \Big\{ \sum_{j=k+1}^{k+n} \rho^2(i(j/2)) c_j^2 E u_j^2 \Big\} \log^2(2n). \end{aligned}$$

(9.2.33) is proved.

It follows from (9.2.33), $\sum_{i=1}^{\infty}\rho(2^i)<\infty$ and Lemma 2.2.2 that

$$\begin{aligned}
&P\Big\{\big|\sum_{i=1}^{n}E(u_i|\mathcal{F}_{i-1})\big|\ge\varepsilon\exp(n^a/2)\Big\}\\
&\qquad\le ce^{-n^a}\Big\{\sum_{i=1}^{n}i^{-2a}Eu_i^2\Big\}\log^2(2n)\\
&\qquad\le ce^{-n^a}\Big\{\sum_{i=1}^{n}i^{-a-1}e^{i^a}\Big\}\log^2(2n)\\
&\qquad\le cn^{-2a}\log^2(2n).
\end{aligned}$$

Let $n_k=[k^{1/a}]$. By the Borel-Cantelli lemma, we have

$$\sum_{i=1}^{n_k}E(u_i|\,\mathcal{F}_{i-1})=o\left(\exp(n_k^a/2)\right)\quad\text{a.s.}\tag{9.2.35}$$

Moreover from Lemma 9.2.3, Lemma 2.2.2, Lemma 4.1.2 and (9.2.33), we have

$$\begin{aligned}
&P\Big\{\max_{n_k<j\le n_{k+1}}\Big|\sum_{l=n_k}^{j}e^{-j^a/2}E(u_l|\,\mathcal{F}_{l-1})\Big|\ge\varepsilon\Big\}\\
&\le P\Big\{\max_{n_k<j\le n_{k+1}}\Big|\sum_{l=n_k}^{j}e^{-l^a/2}E(u_l|\mathcal{F}_{l-1})\Big|\ge\varepsilon/2\Big\}\\
&\le c\Big\{\sum_{j=n_k+1}^{n_{k+1}}j^{-2a}e^{-j^a}Eu_j^2\Big\}\log^4(n_{k+1}-n_k)\\
&\le ck^{-2}(\log k)^4,
\end{aligned}$$

which implies that

$$\max_{n_k<j\le n_{k+1}}\Big|\sum_{l=n_k}^{j}e^{-j^a/2}E(u_l|\,\mathcal{F}_{l-1})\Big|\to 0\quad\text{a.s.}\tag{9.2.36}$$

Thus (9.2.32) follows from (9.2.35) and (9.2.36).

The following lemma gives (9.2.14).

Lemma 9.2.8. *Suppose that a satisfies (9.2.28), and condition (ii) of Theorem 9.2.1 is satisfied. Then for $\delta_1=\varepsilon'/4$, we have*

$$\sum_{k=1}^{\infty}e^{-(1+\delta_1)k^a}\,E|\xi_k|^{2+2\delta_1}<\infty.\tag{9.2.37}$$

Proof. From Lemma 2.2.5, Lemma 9.2.2 and condition (ii), we have

$$\begin{aligned} E|\xi_k|^{2+2\delta_1} &\le cE\,|u_k|^{2+2\delta_1} \\ &\le c\Big\{(k^{a-1}\exp(k^a))^{1+\delta_1} \\ &\quad + k^{a-1}\exp(k^a)\,E|X_1|^{2+2\delta_1} I(X_1^2 \le 2\exp(k^a))\Big\}. \end{aligned}$$

Then, by $0 < a < \varepsilon'/(8+\varepsilon')$ and Lemma 9.2.2, (9.2.37) is proved.

The following lemma gives (9.2.15).

Lemma 9.2.9. *Suppose that a satisfies (9.2.28) and condition (ii) of Theorem 9.2.1 is satisfied. Then*

$$\sum_{j=1}^{n}(E(\xi_j^2|\,\mathcal{F}_{j-1}) - E\xi_j^2) = o(\exp(n^a)) \quad a.s. \tag{9.2.38}$$

Proof.

$$\begin{aligned} &\Big|\sum_{j=1}^{n}(E(\xi_j^2|\,\mathcal{F}_{j-1}) - E\xi_j^2)\Big| \\ &\quad \le \Big|\sum_{j=1}^{n}(E(u_j^2|\,\mathcal{F}_{j-1}) - Eu_j^2)\Big| \\ &\quad + \sum_{j=1}^{n}(E^2(u_j|\,\mathcal{F}_{j-1}) + E(E(u_j|\,\mathcal{F}_{j-1}))^2). \end{aligned} \tag{9.2.39}$$

We first prove that

$$\sum_{j=1}^{n}(E^2(u_j|\,\mathcal{F}_{j-1}) + E(E(u_j|\,\mathcal{F}_{j-1}))^2) = o(\exp(n^a)) \quad \text{a.s.} \tag{9.2.40}$$

In fact, by the definition of $\rho(\cdot)$, we have

$$\begin{aligned} E(E(u_j|\,\mathcal{F}_{j-1}))^2 &= E(u_jE(u_j|\,\mathcal{F}_{j-1})) \\ &\le \rho(i_{j-1})\|u_j\|_2\,\|E(u_j|\,\mathcal{F}_{j-1})\|_2. \end{aligned}$$

Therefore

$$E(E(u_j|\,\mathcal{F}_{j-1}))^2 \le \rho^2(i_{j-1})Eu_j^2 \le c\,j^{-2a(1+\varepsilon)+a-1}\,e^{j^a},$$

and further

$$\sum_{j=1}^{\infty} E(E(u_j|\mathcal{F}_{j-1}))^2/e^{j^a} \le c\sum_{j=1}^{\infty} j^{-1-a(1+2\varepsilon)} < \infty.$$

By the Kronecker lemma, (9.2.40) holds true.

Secondly, we prove

$$\sum_{j=1}^{n}(E(u_j^2|\mathcal{F}_{j-1}) - Eu_j^2) = o(\exp(n^a)) \quad \text{a.s.} \tag{9.2.41}$$

Denote $\overline{u}_i = u_i^2 I(|u_i| \le e^{i^a/2})$. It is easy to see that

$$\begin{aligned}
&|\sum_{i=1}^{n}(E(u_i^2|\mathcal{F}_{j-1}) - Eu_i^2)| \\
&\quad \le |\sum_{i=1}^{n}(E(\overline{u}_i|\mathcal{F}_{j-1}) - E\overline{u}_i)| \\
&\qquad + \sum_{i=1}^{n}(E(u_i^2 I(|u_i| \ge e^{i^a/2})|\mathcal{F}_{i-1})) \\
&\qquad + Eu_i^2 I(|u_i| \ge e^{i^a/2})).
\end{aligned} \tag{9.2.42}$$

It follows from Lemma 2.2.5 and Lemma 9.2.2 that

$$\begin{aligned}
&\sum_{j=1}^{\infty} E(E(u_j^2 I(|u_j| \ge e^{j^a/2})|\mathcal{F}_{j-1}))/e^{j^a} \\
&\quad = \sum_{j=1}^{\infty} e^{-j^a} Eu_j^2 I(|u_j| \ge e^{j^a/2}) \\
&\quad \le c\sum_{j=1}^{\infty} e^{-(1+\delta_1/2)j^a} E|u_j|^{2+\delta_1} \\
&\quad \le c\sum_{j=1}^{\infty} e^{-(1+\delta_1/2)j^a}\Big\{(j^{a-1}e^{j^a})^{1+\delta_1/2} \\
&\qquad + e^{j^a} j^{a-1} E|X_1|^{2+\delta_1} I(X_1^2 \le 2e^{j^a})\Big\} \\
&\quad < \infty.
\end{aligned}$$

By the Kronecker lemma

$$\begin{aligned}
&\sum_{j=1}^{n}\Big(E(u_j^2 I(|u_j| \ge e^{j^a/2})|\mathcal{F}_{j-1}) + Eu_j^2 I(|u_j| \ge e^{j^a/2})\Big) \\
&\qquad = o(\exp(n^a)) \quad \text{a.s.}
\end{aligned} \tag{9.2.43}$$

On the other hand, by the same way as in the proof of (9.2.33) in Lemma 9.2.7, there exists a constant C such that

$$E\Big(\sum_{j=1}^{n} E(\overline{u}_j - E(\overline{u}_j | \mathcal{F}_{j-1}))\Big)^2 \le C\Big(\sum_{j=1}^{n} \rho(i(j/2))\, E\overline{u}_j^2\Big) \log^2(2n).$$

Then by Lemma 2.2.2 and condition (ii), we have

$$\begin{aligned} &P\{|\sum_{j=1}^{n} E(\overline{u}_j - E(\overline{u}_j | \mathcal{F}_{j-1}))| \ge \varepsilon e^{n^a}\} \\ &\le c e^{-2n^a}\Big(\sum_{j=1}^{n} j^{-2a(1+\varepsilon')} E\overline{u}_j^2\Big) (\log n)^2 \\ &\le c e^{-2n^a}\Big(\sum_{j=1}^{n} j^{-2a(1+\varepsilon')} e^{j^a}\Big) (\log n)^2 \\ &\le c n^{-2a(1+\varepsilon')} (\log n)^2. \end{aligned}$$

By the same discussion as in the proof of second half part of Lemma 9.2.7, it follows that

$$\sum_{j=1}^{n} E(\overline{u}_j - E\overline{u}_j | \mathcal{F}_{j-1}) = o(\exp(n^a)) \quad \text{a.s.}$$

Combining it with (9.2.42), (9.2.43) yields (9.2.41). Lemma 9.2.9 is proved.

The following lemma gives (9.2.16).

Lemma 9.2.10. *If the conditions of Theorem 9.2.1 are satisfied, we have*

$$\sum_{i=1}^{m_n} E\xi_i^2 - ES_n^2 = o(n). \tag{9.2.44}$$

Proof. Denote $\widehat{u}_j = \sum_{l\in H_j} X_l$, $\widehat{v}_j = \sum_{l\in I_j} X_l$. We have

$$\begin{aligned} ES_n^2 &= E\Big(\sum_{j=1}^{m_n} \widehat{u}_j + \sum_{j=1}^{m_n} \widehat{v}_j + \sum_{j=N(m_n)+1}^{n} X_j\Big)^2 \\ &= E\Big(\sum_{j=1}^{m_n} \widehat{u}_j\Big)^2 + E\Big(\sum_{j=1}^{m_n} \widehat{v}_j + \sum_{j=N(m_n)+1}^{n} X_j\Big)^2 \\ &\quad + 2E\Big(\sum_{j=1}^{m_n} \widehat{u}_j\Big)\Big(\sum_{j=1}^{m_n} \widehat{v}_j + \sum_{j=N(m_n)+1}^{n} X_j\Big). \end{aligned} \tag{9.2.45}$$

By Lemma 9.2.4 and the definitions of N_k and m_n, it follows that

$$E\Big(\sum_{j=1}^{m_n}\widehat{v}_j+\sum_{j=N(m_n)+1}^{n}X_j\Big)^2=O\Big(\sum_{j=1}^{m_n}E\widehat{v}_j^2+(n-N_{m_n})\Big)$$
$$=O\big(n(\log n)^{\frac{a-1}{a}}\big). \tag{9.2.46}$$

Write

$$E\Big(\sum_{j=1}^{m_n}\widehat{u}_j\Big)^2=E\Big(\sum_{j=1}^{m_n}u_j+\sum_{j=1}^{m_n}(\widehat{u}_j-u_j)\Big)^2.$$

From Lemma 9.2.4 and Lemma 2.2.2 we obtain

$$E\Big(\sum_{j=1}^{m_n}(\widehat{u}_j-u_j)\Big)^2\le c\sum_{j=1}^{m_n}E(\widehat{u}_j-u_j)^2$$
$$\le c\sum_{j=1}^{m_n}j^{a-1}e^{j^a}EX_1^2I(|X_1|^2\ge N_{j-1})$$
$$=o(n).$$

On the other hand,

$$\sum_{j=1}^{m_n}E\xi_j^2-E\Big(\sum_{j=1}^{m_n}u_j\Big)^2$$
$$=E\Big(\sum_{j=1}^{m_n}\xi_j\Big)^2-E\Big(\sum_{j=1}^{m_n}u_j\Big)^2$$
$$=E\Big(\sum_{j=1}^{m_n}E(u_j|\mathcal{F}_{j-1})\Big)^2+2E\Big(\sum_{j=1}^{m_n}u_j\Big)\sum_{j=1}^{m_n}E(u_j|\mathcal{F}_{j-1}).$$

It follows from (9.2.33) that

$$E\Big(\sum_{j=1}^{m_n}E(u_j|\mathcal{F}_{j-1})\Big)^2$$
$$\le c\Big(\sum_{j=1}^{m_n}j^{-2(1+\varepsilon)a}Eu_j^2\Big)(\log m_n)^2$$
$$\le cn(\log n)^{-2(1+\varepsilon)}(\log\log n)^2.$$

Combining it with the above relations yields (9.2.44) .

Now the proof of Theorem 9.2.1 is completed.

The proof of Theorem 9.2.2 is the exactly same as that of Theorem 9.2.1, provided Lemma 2.2.10 is used instead of Lemma 2.2.5 and 4.3.2. The proof of Theorem 9.2.3 is similar to that of Theorem 9.2.1 as well. We need only to apply the Bernstein divided-section method for $\{X_n\}$ immediately. The details are omitted.

9.3 Strong approximations for an α-mixing sequence

Shao and Lu (1987) improved, the strong approximation result of Philipp and Stout (1975) for an α-mixing sequence, and obtained a better rate of strong approximation when using the Strassen martingale embedding method.

Theorem 9.3.1. *Let $\{X_n, n \geq 1\}$ be an α-mixing sequence with $EX_n = 0$ and $g(x)$ a function such that $g(x)/x^{2+\delta}$ for some $0 < \delta \leq 2$ is increasing to infinity. Denote*

$$\|X\|_g = \inf\{t > 0,\ Eg(|X|/t) \leq 1\}.$$

If $\sup_n \|X_n\|_g < \infty$ and the following conditions are satisfied

(i) $\sigma_n^2 = ES_n^2 \geq C\,n$, *for some* $C > 0$,

(ii) $\sum\limits_{n=1}^{\infty} \alpha(n)^{\frac{1}{2+\delta}} \operatorname{invg}\left(\frac{1}{\alpha(n)}\right) < \infty$,

then

$$S(t) - W(\sigma_t^2) = O(\sigma_t^{2/(2+\delta)}(\log \sigma_t^2)^{1+(1+\lambda)/(2+\delta)}) \quad a.s. \tag{9.3.1}$$

where $\lambda = (\log 2)/\log((2+\delta)/\delta) \leq 1$.

Proof. Write

$$\begin{aligned} Y_n &= \sum_{k=0}^{\infty}\{E(X_{k+n}|\mathcal{F}_n) - E(X_{k+n}|\mathcal{F}_{n-1})\} \\ &= X_n + u_n - u_{n-1}, \end{aligned} \tag{9.3.2}$$

where $u_n = \sum\limits_{k=0}^{\infty} E(X_{k+n}|\mathcal{F}_{n-1})$. Let us verify the conditions of Proposition 9.1.1. Condition (a) is assumed. To check condition (b), denote

$$u_{kn} = E(X_{k+n}|\mathcal{F}_{n-1}).$$

Take $f(x) = x^{(2+\delta)/(1+\delta)}$ in Lemma 1.2.3, we have

$$\begin{aligned} E|u_{kn}|^{2+\delta} &= E(X_{k+n}u_{kn}|u_{kn}|^{\delta}) \\ &\leq c \operatorname{invg}\Big(\frac{1}{\alpha(k)}\Big)\alpha(k)^{1/(2+\delta)}\|X_{k+n}\|_g\|u_{kn}\|_{2+\delta}^{1+\delta}. \end{aligned}$$

Therefore

$$\|u_{kn}\|_{2+\delta} \le c \operatorname{inv} g(1/\alpha(k))\alpha(k)^{1/(2+\delta)}.$$

It follows from (ii) that condition (b) of Proposition 9.1.1 is satisfied.

Denote

$$T_m(n) = \sum_{m<k\le m+n} (Y_k^2 - EY_k^2),$$
$$\tau_n = \sup_m E|T_m(n)|^{(2+\delta)/2}.$$

We shall show

$$\tau_n \le cn(\log n)^\lambda \tag{9.3.3}$$

by induction on n, where

$$\lambda = (\log \frac{2+M_\delta}{2})/\log \frac{2+\delta}{\delta} \le 1,$$
$$1 \le M_\delta \le 2,\ 0 < \delta \le 2.$$

Put

$$\theta = \delta/(2+\delta),\ n_1 = n - [n^\theta],\ n_2 = [n^\theta].$$

From an lementary inequality

$$|1+x|^p \le 1 + px + C_p|x|^p, \quad \text{for } 1 < p \le 2,\ 1 \le C_p \le 2,$$

we have

$$\begin{aligned} E|T_m(n)|^{\frac{2+\delta}{2}} &\le E|T_m(n_1)|^{\frac{2+\delta}{2}} + M_\delta E|T_{m+n_1}(n_2)|^{\frac{2+\delta}{2}} \\ &\quad + \Big(\frac{2+\delta}{2}\Big) E|T_m(n_1)|^{\frac{\delta}{2}} |T_{m+n_1}(n_2)| \\ &\le \tau_{n_1} + M_\delta \tau_{n_2} + \Big(\frac{2+\delta}{2}\Big) E\Big(|T_m(n_1)|^{\frac{\delta}{2}} \\ &\quad \cdot \Big\{\Big(\sum_{j=N_2}^{m+n} X_j + u_{N_1} - u_{N_2}\Big)^2 \\ &\quad - E\Big(\sum_{j=N_2}^{m+n} X_j + u_{N_1} - u_{N_2}\Big)^2\Big\}\Big) \\ &=: \tau_{n_1} + M_\delta \tau_{n_2} + \frac{2+\delta}{2} I, \end{aligned} \tag{9.3.4}$$

where

$$N_1 = m+n+1,\ N_2 = m+n_1+1.$$

Note that

$$\begin{aligned}
I = {} & E|T_m(n_1)|^{\frac{\delta}{2}} \sum_{j=N_2}^{m+n} (X_j^2 - EX_j^2) \\
& + E|T_m(n_1)|^{\frac{\delta}{2}}((u_{N_1}-u_{N_2})^2 - E(u_{N_1}-u_{N_2})^2) \\
& + 2E|T_m(n_1)|^{\frac{\delta}{2}}\Big\{\Big(\sum_{j=N_2}^{m+n} X_j\Big)(u_{N_1}-u_{N_2}) \\
& - E\Big(\sum_{j=N_2}^{m+n} X_j\Big)(u_{N_1}-u_{N_2})\Big\} \\
& + 2E|T_m(n_1)|^{\frac{\delta}{2}} \sum_{1\le i,j,i+j\le n_2} (X_{j+N_2-1}X_{i+j+N_2-1} \\
& - EX_{j+N_2-1}X_{i+j+N_2-1}) \\
=: {} & I_1 + I_2 + 2I_3 + 2I_4. \qquad (9.3.5)
\end{aligned}$$

From Lemma 1.2.3 (take $g_1(x^2) = g(x), f(x) = x^{(2+\delta)/\delta}$) and condition (ii), we have

$$\begin{aligned}
I_1 & \le c(E|T_m(n_1)|^{\frac{2+\delta}{2}})^{\frac{\delta}{2+\delta}} \sum_{k=0}^{\infty} \mathrm{inv} g_1\Big(\frac{1}{\alpha(k)}\Big)\mathrm{inv} f\Big(\frac{1}{\alpha(k)}\Big)\alpha(k) \\
& \le c(E|T_m(n_1)|^{\frac{2+\delta}{2}})^{\frac{\delta}{2+\delta}} \sum_{k=0}^{\infty}\Big(\mathrm{inv} g\Big(\frac{1}{\alpha(k)}\Big)\alpha(k)^{\frac{1}{2+\delta}}\Big)^2 \\
& \le c(E|T_m(n_1)|^{\frac{2+\delta}{2}})^{\frac{\delta}{2+\delta}}.
\end{aligned}$$

By the Hölder inequality and condition (b), we have

$$\begin{aligned}
I_2 & \le c(E|T_m(n_1)|^{\frac{2+\delta}{2}})^{\frac{\delta}{2+\delta}}(\|u_{N_1}\|_{2+\delta}^2 + \|u_{N_2}\|_{2+\delta}^2) \\
& \le c(E|T_m(n_1)|^{\frac{2+\delta}{2}})^{\frac{\delta}{2+\delta}}.
\end{aligned}$$

By the Hölder inequality and Lemma 1.2.3, we have

$$\begin{aligned}
I_3 &= E|T_m(n_1)|^{\frac{\delta}{2}}\Big\{\sum_{j=N_2+1}^{N_1-1} X_j \sum_{k=1}^{\infty} X_{N_1+k} - \sum_{j=N_2+1}^{N_1-1} u_{N_2} X_j \\
&\quad - E\sum_{j=N_2+1}^{N_1-1} X_j \sum_{k=1}^{\infty} X_{N_1+k} + E\sum_{j=N_2+1}^{N_1-1} u_{N_2} X_j\Big\} \\
&\le c\,(E|T_m(n_1)|^{\frac{2+\delta}{2}})^{\frac{\delta}{2+\delta}} \\
&\quad \cdot \sum_j \sum_k \operatorname{inv} f\Big(\frac{1}{\alpha(N_1+k-j)}\Big)\alpha(N_1+k-j)^{\frac{2}{2+\delta}} \\
&\le c\,(E|T_m(n_1)|^{\frac{2+\delta}{2}})^{\frac{\delta}{2+\delta}} \\
&\quad \cdot \sum_j \sum_k \Big(\operatorname{inv} g\Big(\frac{1}{\alpha(N_1+k-j)}\Big)\alpha(N_1+k-j)^{\frac{1}{2+\delta}}\Big)^2 \\
&\le c\,(E|T_m(n_1)|^{\frac{2+\delta}{2}})^{\frac{\delta}{2+\delta}},
\end{aligned}$$

where $f(x^{(2+\delta)/2}) = g(x)$, $\operatorname{inv} f(x) = (\operatorname{inv} g(x))^{(2+\delta)/2}$. For I_4, we have

$$\begin{aligned}
I_4 &= E|T_m(n_1)|^{\frac{\delta}{2}}\Big(\sum_{j=1}^{n_2}\sum_{1\le i<j} + \sum_{i=1}^{n_2}\sum_{1\le j<i}\Big) \\
&\quad \cdot \Big(X_{j+N_2-1}X_{i+j+N_2-1} - EX_{j+N_2-1}X_{i+j+N_2-1}\Big) \\
&\le c\Big(\sum_{j=1}^{n_2}\sum_{1\le i<j} + \sum_{i=1}^{n_2}\sum_{1\le j<i}\Big)\alpha(i)^{\frac{2}{2+\delta}} \operatorname{inv} f\Big(\frac{1}{\alpha(i)}\Big) \\
&\quad \cdot (E|T_m(n_1)|^{\frac{2+\delta}{2}})^{\frac{\delta}{2+\delta}},
\end{aligned}$$

where $f(x^2) = g(x)$, $\operatorname{inv} f(x) = (\operatorname{inv} g(x))^2$. It is easy to see that

$$\begin{aligned}
I_4 &\le c\Big(\sum_{j=1}^{n_2}\sum_{1\le i<j} + \sum_{i=1}^{n_2}\sum_{1\le j<i}\Big) \\
&\quad \cdot \alpha(i)^{\frac{1}{2+\delta}} \operatorname{inv} g\Big(\frac{1}{\alpha(i)}\Big)^2 (E|T_m(n_1)|^{\frac{2+\delta}{2}})^{\frac{\delta}{2+\delta}} \\
&\le c\,(E|T_m(n_1)|^{\frac{2+\delta}{2}})^{\frac{\delta}{2+\delta}}.
\end{aligned}$$

Thus combining these with (9.3.4) and (9.3.5) we have

$$\tau_n \le \tau_{n_1} + M_\delta \tau_{n_2} + c\,\tau_{n_1}^{\delta/(2+\delta)}.$$

Hence condition (c) of Proposition 9.1.1 is satisfied and the proof of Theorem 9.3.1 is completed.

Corollary 9.3.1. *Let $\{X_n, n \geq 1\}$ be an α-mixing sequence, $g(x) = x^r$, $r > 2+\delta$, $0 < \delta \leq 2$. If*

(i) $\sigma_n^2 = ES_n^2 \geq Cn$, *for some $C > 0$,*

(ii) $\sum_{n=1}^{\infty} \alpha(n)^{\frac{1}{2+\delta}-\frac{1}{r}} < \infty$,

then

$$S_n - W(\sigma_n^2) = O\left(\sigma_n^{\frac{2}{2+\delta}} (\log \sigma_n)^{1+(1+\lambda)/(2+\delta)}\right) \quad a.s.$$

Particularly, if $\delta = 2$ and $\lim_n \sigma_n^2/n = \sigma^2$ (without loss of generality, assume that $\sigma^2 = 1$), we have

$$S_n - W(n) = O(n^{1/4}(\log n)^{3/2}) \quad a.s.$$

Furthermore, if $\sup_n |X_n| < \infty$, $\sum \alpha(n)^{1/4} < \infty$ and $\sigma_n^2 = n\sigma^2$, then for any $\varepsilon > 0$

$$S_n - W(n) = O(n^{1/4}(\log n)^{5/4+\varepsilon}). \quad a.s.$$

Shao (1989a) gave the following theorem and corollary, which improve the results in Bradley (1983) and Dehling (1983) under the weaker conditions.

Theorem 9.3.2. *Let $\{X_n, n \geq 1\}$ be an α-mixing sequence with $EX_n = 0$, $\sup_n E|X_n|^{2+\delta} < \infty$ for some $\delta > 0$. Assume that*

$$\sup_{n\geq 1} \sup_{k\geq 0} E|S_k(n)|^{2+\delta}/n^{1+\delta/2} < \infty, \tag{9.3.6}$$

$$\alpha(n) = O((\log n)^{-r}), \quad r > 1 + 2/\delta. \tag{9.3.7}$$

Then for any $0 < \theta < \frac{1}{4}(\frac{r\delta}{2+\delta} - 1)$, we have

$$S(t) - W(\sigma_t^2) = O(t^{1/2}(\log t)^{-\theta}) \quad a.s.$$

Corollary 9.3.2. *Let $\{X_n, n \geq 1\}$ be an α-mixing sequence with $EX_n = 0$, $\sup_n E|X_n|^{2+\delta} < \infty (\delta > 0)$. Assume that $\alpha(n) = O(n^{-r})$ for $r > 1 + 2/\delta$ and*

$$\lim_{n\to\infty} \sigma_n^2/n = \sigma^2 > 0.$$

Then $\{X_n\}$ Obeys the law of the iterated logarithm and the Chung law of the iterated logarithm.

The proofs of Theorem 9.3.2 and Corollary 9.3.2 are omitted.

Chapter 10 The Increments of Partial Sums

In Chapter 9, we studied strong approximations of partial sums of a mixing sequence by a Wiener process. But, with the help of these theorems, it is not enough to obtain the ideal increment results, similar to that in the case of i.i.d. sequence (see M.Csörgő and P.Révész 1981). In this chapter, we intend to study the increments of partial sums of a sequence of φ-mixing random variables, which may be non-stationary, even non-identically distributed, by a direct approach. Under the restriction on the rate of convergence to zero for mixing coefficients, the following results are close to those corresponding to independent random variables. Our method can be used to deal with other kinds of mixing sequences as well.

10.1 Some lemmas

In order to show increment results, we need to establish some exponential inequalities. To this end, we first give the next result (cf. Stout 1974, Lemma 5.4.1 and its corollary).

Lemma 10.1.1. *Let $\{Z_n, \mathcal{F}_n, n \geq 1\}$ be a supermartingale with $EZ_n = 0$. Let $Z_0 = 0$ and $U_i = Z_i - Z_{i-1}$ for $i \geq 1$. Suppose $U_i \leq C$ a.s. for some $0 \leq C < \infty$ and all $i \geq 1$. Fix $\lambda > 0$ such that $\lambda C \leq 1$. Put*

$$M_n = \exp(\lambda Z_n) \exp\Big\{-(\lambda^2/2)(1+\lambda C/2)\sum_{i=1}^{n} E(U_i^2 \,|\, \mathcal{F}_{i-1})\Big\}$$

for $n \geq 1$ and $M_0 = 1$ a.s. Then $\{M_n, \mathcal{F}_n, n \geq 0\}$ is a nonnegative supermartingale, and further,

$$P\Big\{\sup_{n\geq 0} M_n > \alpha\Big\} \leq \alpha^{-1}$$

for any $\alpha > 0$.

Let $\{Y_n\}$ be a sequence of φ-mixing random variables with mixing coefficients $\varphi_n = \varphi(n) \downarrow 0$. Without loss of generality, assume that $EY_n =$

0 for every n. Put $T_n = \sum_{i=1}^{n} Y_i$, $\tau_n^2 = ET_n^2$. In this section, we always assume that the following conditions are satisfied:

(a) $|Y_n| \le b_n < \infty$;

(b) there exist $0 < \sigma^2 \le \sigma'^2 < \infty$, such that $\sigma^2 n \le E(Y_{m+1} + \cdots + Y_{m+n})^2 \le \sigma'^2 n$ for every $m \ge 0$ and all large n.

Let p, q, k be integers satisfying that $p = p_n \le n, q = q_n = o(p_n)$, $q_n \uparrow \infty$, $k = k_n = [n/(p_n + q_n)]$. And put $b = \max_{1\le i\le n} b_i$.

Lemma 10.1.2. *Suppose that*

$$\varphi_q p b^2 = o(1), \quad \sum_{j=1}^{k} \varphi^{1/2}(jp) = O(1). \tag{10.1.1}$$

Then for $x = x_n$ and small $\varepsilon > 0$ satisfying

$$\frac{4}{\varepsilon} b n \varphi_q \le x \le \frac{\varepsilon \tau_n^2}{pb} \tag{10.1.2}$$

and

$$x^2/n \to \infty, \tag{10.1.3}$$

we have

$$P\Big\{\max_{1\le i\le n} |T_i| \ge x\Big\} \le 3 \exp\Big\{-\frac{(1-6\varepsilon)x^2}{2\tau_n^2}\Big\} \tag{10.1.4}$$

for all large n. If (10.1.2) is replaced by

$$x > \frac{\varepsilon \tau_n^2}{pb}, \tag{10.1.5}$$

then

$$P\Big\{\max_{1\le i\le n} |T_i| \ge x\Big\} \le 3 \exp\Big\{-\frac{\varepsilon(1-5\varepsilon)x}{2pb}\Big\}. \tag{10.1.6}$$

Proof. We always assume that n is large enough. First, we prove (10.1.4). Define

$$\xi_i = \sum_{j=i(p+q)+1}^{(i+1)p+iq} Y_j,$$

$$\eta_i = \sum_{j=(i+1)p+iq+1}^{(i+1)(p+q)} Y_j, \ i = 0, 1, \cdots, k-1,$$

$$\eta_k = \sum_{j=k(p+q)+1}^{n} Y_j.$$

Put σ-fields $\mathcal{F}_{-1} = \{\emptyset, \Omega\}, \mathcal{F}_i = \sigma(Y_j, j \le (i+1)p+iq), i = 0, 1, \cdots, k-1$. Define martingale differences $\gamma_i = \xi_i - E(\xi_i|\mathcal{F}_{i-1}), i = 0, 1, \cdots, k-1$. Write

$$\begin{aligned} P\{|T_n| \ge x\} &\le P\Big\{|\sum_{i=0}^{k-1} \xi_i| \ge (1-\frac{\varepsilon}{2})x\Big\} + P\Big\{|\sum_{i=0}^{k} \eta_i| \ge \frac{1}{2}\varepsilon x\Big\} \\ &=: J_1 + J_2. \end{aligned} \tag{10.1.7}$$

Consider J_1. Write

$$\begin{aligned} J_1 &\le P\Big\{|\sum_{i=0}^{k-1} \gamma_i| \ge (1-\varepsilon)x\Big\} + P\Big\{|\sum_{i=0}^{k-1} E(\xi_i|\mathcal{F}_{i-1})| \ge \frac{1}{2}\varepsilon x\Big\} \\ &=: J_{11} + J_{12}. \end{aligned} \tag{10.1.8}$$

By Lemma 2.2.8 (note $|\xi_i| \le pb$), for any $B_{i-1} \in \mathcal{F}_{i-1}$,

$$|E\,\xi_i I_{B_{i-1}}| \le 2\varphi_q\, p\, b\, P(B_{i-1}) \qquad (i = 0, 1, \cdots, k-1),$$

which implies

$$|E(\xi_i|\mathcal{F}_{i-1})| \le 2\varphi_q pb \quad \text{a.s.} \quad (i = 0, 1, \cdots, k-1). \tag{10.1.9}$$

Using condition (10.1.2), we obtain

$$J_{12} = 0. \tag{10.1.10}$$

Next, we estimate J_{11}. Since $\{\gamma_i, \mathcal{F}_i, i = 0, 1, \cdots, k-1\}$ possesses martingale difference property, by using Lemma 10.1.1 and noting $|\gamma_i| \le (1+\varepsilon)\,p\,b$ a.s. for large n, for $0 < \lambda \le ((1+\varepsilon)\,p\,b)^{-1}$

$$\zeta_j = \exp\Big(\lambda \sum_{i=0}^{j} \gamma_i\Big) \exp\Big\{-\frac{\lambda^2}{2}\Big(1 + \frac{1}{2}(1+\varepsilon)pb\lambda\Big) \sum_{i=0}^{j} E(\gamma_i^2|\mathcal{F}_{i-1})\Big\}$$

$$(j = 0, 1, \cdots, k-1)$$

possess nonnegative supermartingale property. Write

$$\begin{aligned} &P\Big\{\sum_{i=0}^{k-1} \gamma_i \ge (1-\varepsilon)x\Big\} \\ &\quad = P\Big\{\zeta_{k-1} \ge \exp(\lambda(1-\varepsilon)x) \\ &\qquad \cdot \exp\Big\{-\frac{\lambda^2}{2}\Big(1 + \frac{1}{2}(1+\varepsilon)pb\lambda\Big) \sum_{i=0}^{k-1} E(\gamma_i^2|\mathcal{F}_{i-1})\Big\}\Big\}. \end{aligned} \tag{10.1.11}$$

We are going to estimate $\sum_{i=0}^{k-1} E(\gamma_i^2|\mathcal{F}_{i-1})$. Write

$$\sum_{i=0}^{k-1} E(\gamma_i^2|\mathcal{F}_{i-1}) = \sum_{i=0}^{k-1} E(\xi_i^2|\mathcal{F}_{i-1}) - \sum_{i=0}^{k-1}(E(\xi_i|\mathcal{F}_{i-1}))^2.$$

We have $|E({\xi_i}^2|\mathcal{F}_{i-1}) - E{\xi_i}^2| \le 2\varphi_q(pb)^2$ a.s. by a similar proof to (10.1.9). Hence, by conditions (b) and (10.1.1), we get

$$\sum_{i=0}^{k-1} E({\xi_i}^2|\mathcal{F}_{i-1}) = (1+o(1))\sum_{i=0}^{k-1} E{\xi_i}^2 \quad \text{a.s.}$$

Moreover inequality (10.1.9) implies that

$$\sum_{i=0}^{k-1}(E(\xi_i|\mathcal{F}_{i-1}))^2 \le 4\varphi_q^2 pb^2 n = o(n) \quad \text{a.s.}$$

Thus

$$\sum_{i=0}^{k-1} E({\gamma_i}^2|\mathcal{F}_{i-1}) = (1+o(1))\sum_{i=0}^{k-1} E{\xi_i}^2 \quad \text{a.s.} \tag{10.1.12}$$

Now we estimate $\sum\limits_{i=0}^{k-1} E\xi_i^2$. Write

$$\sum_{i=0}^{k-1} E\xi_i^2 = E\Big(\sum_{i=0}^{k-1}\xi_i\Big)^2 - 2\sum_{0\le i<j\le k-1} E\xi_i\xi_j, \tag{10.1.13}$$

where $|E\xi_i\xi_j| \le 2\varphi((j-i-1)p)pb\sigma' p^{1/2}$ by Lemma 1.2.10 and so for fixed i

$$\sum_{j\ge i+2} |E\xi_i\xi_j| \le 2\sigma'\varphi_p^{1/2} p^{3/2} b \sum_{j=1}^{\infty} \varphi^{1/2}(jp) = o(p)$$

by assumptions in (10.1.1). Summing on i leads to $o(n)$. Also

$$|E\xi_i\xi_{i+1}| \le 2\varphi_q pb\sigma' p^{1/2} = o(p)$$

by (10.1.1) and summing leads to $o(n)$ again. Thus

$$\sum_{i=0}^{k-1} E\xi_i^2 = E\Big(\sum_{i=0}^{k-1}\xi_i\Big)^2 + o(n). \tag{10.1.14}$$

Furthermore

$$E\Big(\sum_{i=0}^{k-1}\xi_i\Big)^2 = ET_n^2 - 2ET_n\Big(\sum_{i=0}^{k}\eta_i\Big) + E\Big(\sum_{i=0}^{k}\eta_i\Big)^2. \tag{10.1.15}$$

By condition (b) and the second equality in (10.1.1), we can prove that

$$E\Big(\sum_{i=0}^{k}\eta_i\Big)^2 = O(kq+p).$$

In fact

$$E\Big(\sum_{i=0}^{k}\eta_i\Big)^2 \le 2\,E\Big(\sum_{i=0}^{k-1}\eta_i\Big)^2 + 2\,E\eta_k^2,$$

where $E\eta_k^2 = O(p)$ and

$$\begin{aligned} E\Big(\sum_{i=0}^{k-1}\eta_i\Big)^2 &= \sum_{i=0}^{k-1} E\eta_i^2 + 2\sum_{0\le i<j\le k-1} E\eta_i\eta_j \\ &\le \sigma'^2 kq + 2\sigma'^2 q\sum_{j=1}^{k-1}(k-j)\varphi^{1/2}(jp) = O(kq). \end{aligned}$$

Hence (10.1.15) implies that $E(\sum_{i=0}^{k-1}\xi_i)^2 = (1+o(1))\,\tau_n^2$. Combining it with (10.1.14) and (10.1.12) implies that

$$\sum_{i=0}^{k-1} E(\gamma_i^2\,|\,\mathcal{F}_{i-1}) = (1+o(1))\tau_n^2 \quad \text{a.s.} \tag{10.1.16}$$

Inserting it into (10.1.11) and using Lemma 10.1.1, we obtain

$$\begin{aligned} &P\Big\{\sum_{i=0}^{k-1}\gamma_i \ge (1-\varepsilon)x\Big\} \\ &\quad\le \exp\{-\lambda(1-\varepsilon)x\} \\ &\quad\quad\cdot\exp\Big\{\frac{\lambda^2}{2}\Big(1+\frac{1}{2}(1+\varepsilon)pb\lambda\Big)(1+\varepsilon)\tau_n^2\Big\} \end{aligned} \tag{10.1.17}$$

for all large n. Choosing $\lambda = x/((1+\varepsilon)\tau_n^2)$, we have $\lambda \le \varepsilon((1+\varepsilon)pb)^{-1}$ by (10.1.2). Inserting it into (10.1.17) implies that

$$P\Big\{\sum_{i=0}^{k-1}\gamma_i \ge (1-\varepsilon)x\Big\} \le \exp\Big\{-\frac{(1-4\varepsilon)x^2}{2\tau_n^2}\Big\}. \tag{10.1.18}$$

Replacing Y_j with $-Y_j$, we obtain

$$J_{11} = P\Big\{\Big|\sum_{i=0}^{k-1}\gamma_i\Big| \ge (1-\varepsilon)x\Big\} \le 2\exp\Big\{-\frac{(1-4\varepsilon)x^2}{2\tau_n^2}\Big\}.$$

Combining it with (10.1.10), we have

$$J_1 \le 2\exp\Big\{-\frac{(1-4\varepsilon)x^2}{2\tau_n^2}\Big\}.$$

For J_2, it is clear that $3^{-1}\exp\{-(1-4\varepsilon)x^2/(2\tau_n^2)\}$ is one of its upper bounds in view of $q_n = o(p_n)$. Then we have proved

$$P\{|T_n| \ge x\} \ge \frac{7}{3}\exp\Big\{-\frac{(1-4\varepsilon)x^2}{2\tau_n^2}\Big\}. \qquad (10.1.19)$$

In order to obtain (10.1.4) from (10.1.19), we use Lemma 5.1.2. First of all, since $\varphi_n \downarrow 0$, there exists an interger n_0 such that $\varphi_{n_0} \le 10^{-1}$. Using condition (b) and (10.1.3), we obtain

$$\begin{aligned}\max_{1\le i\le n} P\Big\{|T_n - T_i| \ge \frac{\varepsilon x}{2(1+\varepsilon)}\Big\} \\ &\le \max_{1\le i\le n}\frac{4(1+\varepsilon)^2E(T_n-T_i)^2}{\varepsilon^2x^2} \\ &\le \frac{4(1+\varepsilon)^2\sigma'^2 n}{\varepsilon^2x^2} \le \frac{1}{10}\end{aligned}$$

for large n. So η in Lemma 5.1.2 can be chosen as 5^{-1}. Furthermore

$$\max_{1\le i\le n}|Y_i| \le b \le \frac{\varepsilon\sigma'^2 n}{px} \le \frac{\varepsilon\sigma'^2 k}{x^2}x = o(x) \qquad (10.1.20)$$

by (10.1.2) and (10.1.3). Hence

$$P\Big\{\max_{1\le i\le n}|Y_i| \ge \varepsilon x/(2(1+\varepsilon)(n_0-1))\Big\} = 0$$

for large n. Therefore, from the above results and Lemma 5.1.2, we obtain

$$\begin{aligned}&P\Big\{\max_{1\le i\le n}|T_i| \ge x\Big\} \\ &\le \frac{5}{4}P\Big\{|T_n| \ge \frac{x}{1+\varepsilon}\Big\} + \frac{5}{4}P\Big\{\max_{1\le i\le n}|Y_i| \ge \frac{\varepsilon x}{2(1+\varepsilon)(n_0-1)}\Big\} \\ &< 3\exp\Big\{-\frac{(1-6\varepsilon)}{2\tau_n^2}x^2\Big\}.\end{aligned}$$

This is namely (10.1.4).

Now we prove (10.1.6). It is clear that we can assume $x/pb > \delta$ for some $\delta > 0$. If condition (10.1.5) is satisfied, choosing $\lambda = \varepsilon((1+\varepsilon)pb)^{-1}$ in (10.1.17), we have

$$P\Big\{\sum_{i=0}^{k-1}\gamma_i \ge (1-\varepsilon)x\Big\} \le \exp\Big\{-\frac{\varepsilon(1-4\varepsilon)x}{2pb}\Big\}. \qquad (10.1.21)$$

By imitating the procedure from (10.1.18) to (10.1.4) and replacing (10.1.20) by $\max_{1\le i\le n}|Y_i| \le b < x/(p\delta) = o(x)$, we obtain (10.1.16) from (10.1.21). The lemma is proved.

Lemma 10.1.3. *Suppose that condition (10.1.1) in Lemma 10.1.2 is replaced by*

$$\varphi_p q\, b^2 = o(1), \quad \sum_{j=1}^{k} \varphi^{1/2}(jp) = O(1). \tag{10.1.1'}$$

Then for any $\nu > 0$, there exists an $\alpha(\nu)$, such that for all large n and α satisfying $\alpha \ge \alpha(\nu)$, we have

$$P\{T_n \ge \alpha\tau_n\} \ge \frac{1}{2}\exp\{-(1+\nu)\alpha^2/2\}$$

provided that

$$\varphi_q = o(\alpha/b) \tag{10.1.22}$$

and

$$p\, b\, \alpha\, \tau_n = o(n). \tag{10.1.23}$$

Proof. For $0 < \delta < 1$ write

$$P\{T_n \ge \alpha\tau_n\} \ge P\Big\{\sum_{i=0}^{k-1}\xi_i \ge (1+\delta)\alpha\tau_n\Big\} - P\Big\{\sum_{i=0}^{k}\eta_i \le -\delta\alpha\tau_n\Big\}$$
$$=: I_1 \;-\; I_2 \tag{10.1.24}$$

and

$$I_1 \ge P\Big\{\sum_{i=0}^{k-1}\gamma_i \ge (1+2\delta)\alpha\tau_n\Big\} - P\Big\{\sum_{i=0}^{k-1}E(\xi_i|\mathcal{F}_{i-1}) \le -\delta\alpha\tau_n\Big\}$$
$$=: I_{11} \;-\; I_{12}. \tag{10.1.25}$$

Using condition (10.1.22) and recalling (10.1.9), we also have for large n

$$I_{12} = 0. \tag{10.1.26}$$

We are going to estimate I_{11}. Let ε be a small positive number indicated later. Put $v = 3\sqrt{\varepsilon}/(1-3\sqrt{\varepsilon}), u = (1+2\delta)\alpha/(1-v)$. Write

$$E\Big\{\exp\Big[u\Big(\sum_{i=0}^{k-1}\gamma_i\Big)/\tau_n\Big]\Big|\,\mathcal{F}_{k-2}\Big\}$$
$$= \exp\Big\{u\Big(\sum_{i=0}^{k-2}\gamma_i\Big)\Big/\tau_n\Big\}E\{\exp(u\gamma_{k-1}/\tau_n)|\,\mathcal{F}_{k-2}\}. \tag{10.1.27}$$

Condition (10.1.23) implies

$$upb/\tau_n \to 0. \tag{10.1.28}$$

Hence

$$\begin{aligned} E\{\exp(u\gamma_{k-1}/\tau_n)|\,\mathcal{F}_{k-2}\} & \\ &\geq 1+\frac{u^2}{2\tau_n^2}\Big(1-\frac{upb}{\tau_n}\Big)E(\gamma_{k-1}^2|\,\mathcal{F}_{k-2}) \\ &\geq \exp\Big\{\frac{u^2}{2\tau_n^2}\Big(1-\frac{2upb}{\tau_n}\Big)E(\gamma_{k-1}^2|\,\mathcal{F}_{k-2})\Big\}. \end{aligned}$$

Inserting it into (10.1.27) we have

$$\begin{aligned} E\Big\{\Big[\exp\Big(u\Big(\sum_{i=0}^{k-1}\gamma_i\Big)/\tau_n\Big)\exp\Big(-\frac{u^2}{2\tau_n^2}\Big(1-\frac{2upb}{\tau_n}\Big) \\ \cdot E(\gamma_{k-1}^2|\,\mathcal{F}_{k-2})\Big)\Big]|\mathcal{F}_{k-2}\Big\} \geq \exp\Big\{u\Big(\sum_{i=0}^{k-2}\gamma_i\Big)/\tau_n\Big\}. \end{aligned}$$

Thus , by induction,

$$E\Big\{\exp\Big[u\Big(\sum_{i=0}^{k-1}\gamma_i\Big)\Big/\tau_n\Big]\exp\Big[-\frac{u^2}{2\tau_n^2}\Big(1-\frac{2upb}{\tau_n}\Big)\sum_{i=0}^{k-1}E(\gamma_i^2|\,\mathcal{F}_{i-1})\Big]\Big\}\geq 1.$$

Since we also have (10.1.16), the above inequality can be rewritten as

$$\begin{aligned} E\Big\{\exp\Big[u\Big(\sum_{i=0}^{k-1}\gamma_i\Big)/\tau_n\Big]\Big\} &\geq \exp\Big\{\frac{u^2}{2}\Big(1-\frac{2upb}{\tau_n}-\frac{\varepsilon}{4}\Big)\Big\} \\ &\geq \exp\Big\{\frac{u^2}{2}\Big(1-\frac{\varepsilon}{2}\Big)\Big\} \end{aligned} \tag{10.1.29}$$

for given $\varepsilon > 0$, provided that n is large enough.

Then the following proof is similar to the corresponding one of Theorem 5.2.2 in Stout (1974) except that some details are different (noting the choice of v and u). It is omitted.

In addition to Lemmas 10.1.2 and 10.1.3, the following generalization (see Erdös 1959) of the Borel-Cantelli lemma will also be utilized for the proof of our theorem.

Lemma 10.1.4. *If $C_1, C_2, \cdots$ are arbitrary events satisfying the conditions*

$$\sum_{n=1}^{\infty} P(C_n) = \infty$$

$$\liminf_{n\to\infty} \sum_{k=1}^{n}\sum_{l=1}^{n} P(C_k C_l) / (\sum_{k=1}^{n} P(C_k))^2 = 1,$$

then

$$P\{C_n, \text{i.o.}\} = 1.$$

10.2 How big are the increments when the moment generation functions exist?

Let $\{X_n\}$ be a sequence of φ-mixing random variables with $EX_n = 0$ $(n \geq 1)$. Put $S_n = \sum_{k=1}^{n} X_k$, ${\sigma_k}^2 = EX_k^2$.

Theorem 10.2.1. (Lin *1991*) *Suppose that $\{X_n\}$ defined above satisfies the following conditions:*

(i) $\underline{\lim}_{n\to\infty} \inf_{m\geq 0} E(X_{m+1} + \cdots + X_{m+n})^2/n > 0$;

(ii) there exist $t_0, M > 0$, such that $Ee^{tX_k} \leq M$ for every k and any $|t| \leq t_0$;

(iii) there exists an $l > 1$, such that $\varphi_n = O(n^{-l})$.

Suppose that $\{a_n\}$ is a non-decreasing sequence of positive integers satisfying

(a) there exists an $a > 0$ such that $a \log^d n \leq a_n \leq n$ where $d > (3l+1)/(l-1)$. Then, putting

$$\sigma_{nN}^2 = E(X_{n+1} + \cdots + X_{n+a_N})^2,$$
$$\beta_{nN} = \sigma_{nN}\{2[\log(N/\sigma_{nN}^2) + \log\log N]\}^{1/2},$$
$$S(n,k) = S_{n+k} - S_n,$$

we have

$$\limsup_{N\to\infty} \beta_{NN}^{-1} |S(N, a_N)| = 1 \quad a.s. \tag{10.2.1}$$

$$\limsup_{N\to\infty} \max_{1\leq n\leq N} \beta_{nN}^{-1} |S(n, a_N)| = 1 \quad a.s. \tag{10.2.2}$$

$$\limsup_{N\to\infty} \max_{1\leq k\leq a_N} \beta_{NN}^{-1} |S(N, k)| = 1 \quad a.s. \tag{10.2.3}$$

$$\limsup_{N\to\infty} \max_{1\le n\le N} \max_{1\le k\le a_N} \beta_{nN}^{-1}|S(n,k)| = 1 \quad a.s. \tag{10.2.4}$$

Furthermore, if condition (a) is replaced by

(b) there exists an $a > 0$, such that $an^{1/(l+1)} \le a_n \le n$ and

$$\lim_{n\to\infty} \log(n/a_n)/\log\log n = \infty,$$

then

$$\lim_{N\to\infty} \max_{1\le n\le N} \beta_{nN}^{-1}|S(n,a_N)| = 1 \quad a.s. \tag{10.2.5}$$

$$\lim_{N\to\infty} \max_{1\le n\le N} \max_{1\le k\le a_N} \beta_{nN}^{-1}|S(n,k)| = 1 \quad a.s. \tag{10.2.6}$$

Proof. From conditions (i)–(iii), there exist constants w_1 and w_2, $0 < w_1 \le w_2 < \infty$ such that

$$w_1 n \le E(X_{m+1} + \cdots + X_{m+n})^2 \le w_2 n \tag{10.2.7}$$

for every integer $m \ge 0$ and n large enough. The right inequality sign is due to that from Lemma 1.2.8

$$\begin{aligned} |EX_iX_j| &\le 2\varphi_{j-i}^{1/l'} E^{1/l'}|X_i|^{l'} E^{1-l/l'}|X_j|^{l'/(l'-1)} \\ &= O((j-i)^{-l/l'}) \end{aligned}$$

for $1 < l' < l, \quad i < j$.

First, we prove (10.2.1)–(10.2.4). Obviously, it is sufficient to verify

$$\limsup_{N\to\infty} \beta_{NN}^{-1}|S(N,a_N)| \ge 1 \quad \text{a.s.} \tag{10.2.8}$$

$$\limsup_{N\to\infty} \max_{1\le n\le N} \max_{1\le k\le a_N} \beta_{nN}^{-1}|S(n,k)| \le 1 \quad \text{a.s.} \tag{10.2.9}$$

For $B > 1/t_0$, define

$$Y_n = X_n I(|X_n| < B\log n), \quad Y_n' = Y_n - EY_n, \quad T_n = \sum_{k=1}^n Y_k',$$

$$\lambda_{nN}^2 = E(Y_{n+1} + \cdots + Y_{n+a_N})^2, \quad T(n,k) = T_{n+k} - T_n,$$

$$\alpha_{nN} = \lambda_{nN}\{2[\log(N/\lambda_{nN}^2) + \log\log N]\}^{1/2}.$$

From condition (ii), $P\{X_n \ne Y_n,\ i.o.\} = 0$ and for t' such that $t' < t_0$ and $t'B > 1$

$$|EY_n| \le cn^{-t'B},$$

Hence, as $N \to \infty$

$$\max_{1\le n\le N} \max_{1\le k\le a_N} \beta_{nN}^{-1}|S(n,k) - T(n,k)| \to 0 \quad \text{a.s.} \tag{10.2.10}$$

Moreover, it is not difficult to show by (10.2.7) and the definition of Y_n that

$$\sigma^2_{nN}/\lambda^2_{nN} \to 1 \quad (N \to \infty) \tag{10.2.11}$$

uniformly in n.

We are now going to prove (10.2.8), which is equivalent to

$$\limsup_{N\to\infty} \alpha^{-1}_{NN}|T(N, a_N)| \geq 1 \quad \text{a.s.} \tag{10.2.12}$$

Put $h = 2/(3l+1)$. Define $p_n = [n^h]$, $q_n = [\alpha_n p_n]$, where $\{\alpha_n\}$ is a sequence of positive numbers tending to zero slowly enough. By conditions (iii), (a) and the definitions of p_n and q_n , it can be seen that

$$\begin{aligned}\varphi_{q_{a_N}} p_{a_N} (\log N)^2 &\leq c\alpha^{-l}_{a_N} p^{-l+1}_{a_N} (\log N)^2 \\ &\leq c\alpha^{-l}_{a_N} a_N^{-2(l-1)/(3l+1)} (\log N)^2 \to 0, \quad \text{as } N \to \infty\end{aligned}$$

provided that α_n tend to zero slowly enough. We also have

$$\sum_{j=1}^{k} \varphi^{1/2}(jp) \leq c\sum_{j=1}^{k}(jp)^{-l/2} \leq ckn^{-l/2} \leq cn^{(3l-1)/(3l+1)-l/2} = o(1)$$

with the help of condition (iii). Hence condition (10.1.1), and further condition (10.1.1'), are satisfied. Similarly, we have

$$\varphi_{q_{a_N}} = o((\log(N/\lambda^2_{NN}) + \log\log N)^{1/2}/\log N)$$

and $p_{a_N}(\log N)\alpha_{nN}/a_N = o(1)$. Hence, we can use Lemma 10.1.3 (choosing $\nu = \varepsilon > 0$), which implies that for large N

$$\begin{aligned}P(C_N) &\geq \frac{1}{2}\exp\{-(1+\varepsilon)(1-\varepsilon)^2\log(N/\lambda^2_{NN}) + \log\log N]\} \\ &\geq c\Big(\frac{a_N}{N\log N}\Big)^{1-\varepsilon},\end{aligned} \tag{10.2.13}$$

where $C_N = \{\alpha^{-1}_{NN} T(N, a_N) \geq 1 - \varepsilon\}$. Put $n_1 = 1$ and define $n_{k+1} = n_k + 2a_{n_k}$. Noting that the sequence is mixing, we have

$$\Big|\sum_{j=1}^{m}\sum_{k=1}^{m} P(C_{n_j}C_{n_k})\Big/\Big(\sum_{k=1}^{m} P(C_{n_k})\Big)^2 - 1\Big|$$

$$\leq \frac{\sum_{j=1}^m P(c_{n_j})(1-P(C_{n_j})) + 4\sum_{j=1}^m\sum_{k=j+1}^m P(C_{n_j})\varphi(n_k - n_j - a_{n_j})}{\left(\sum_{k=1}^m P(C_{n_k})\right)^2}$$

$$\le \Big(1+4\sum_{k=2}^{m}\varphi(n_k-n_1-a_{n_1})\Big)\Big/\sum_{k=1}^{m}P(C_{n_k}).$$

Using condition (iii) and recalling the definition of n_k, one can verify that $\sum\limits_{k=2}^{\infty}\varphi(n_k-n_1-a_{n_1})<\infty$. Moreover $\sum_{k=1}^{\infty}P(C_{n_k})=\infty$ since

$$\sum_{n=n_k+1}^{n_{k+1}}(n\log n)^{-1}\le (n_k\log n_k)^{-1}(n_{k+1}-n_k)$$
$$=(n_k\log n_k)^{-1}2a_{n_k}\le cP(C_{n_k}).$$

Hence, from Lemma 10.1.4, $P\{C_{n_k}, i.o.\}=1$, which yields (10.2.12). The proof of (10.2.8) is completed.

Furthermore, we want to prove (10.2.9). From (10.2.10) and (10.2.11), it is sufficient to show that

$$\limsup_{N\to\infty}\max_{1\le n\le N}\max_{1\le k\le a_N}\alpha_{nN}|T(n,k)|\le 1 \quad \text{a.s.} \tag{10.2.14}$$

Let $r=r(\varepsilon)>1$ be a positive integer indicated later. Put $R=[a_N/r], n_r=R[n/R]$. We have

$$|\lambda_{nN}^2/E(Y_{n_r+1}+\cdots+Y_{(n+a_N)_r})^2-1|\le\frac{2w_2(2r^{1/2}+1)}{w_1(r-1)} \tag{10.2.15}$$

for large N by (10.2.7) and (10.2.11). Write

$$|T_{n+k}-T_n|\le|T_{n+k}-T_{(n+k)_r}|+|T_{(n+k)_r}-T_{n_r}|+|T_{n_r}-T_n|. \tag{10.2.16}$$

Consider the second term of the right hand side of (10.2.16). We have

$$P\Big\{\max_{1\le n\le N}\max_{1\le k\le a_N}\alpha_{nN}^{-1}|T_{(n+k)_r}-T_{n_r}|\ge 1+\varepsilon/3\Big\}$$
$$\le cr\frac{N}{a_N}\max_{1\le n\le N}P\Big\{\max_{1\le k\le a_N}|T_{(n+k)_r}$$
$$-T_{n_r}|\ge(1+\varepsilon/3)\alpha_{nN}\Big\}. \tag{10.2.17}$$

Obviously, $|(n+a_N)_r-n_r|\le a_N(1+1/r)$. We can use (10.1.4) of Lemma 10.1.2, since

$$(\log N)a_N\varphi_{q_{a_N}}=o(\alpha_{nN}),$$
$$\alpha_{nN}=o(E(T_{(n+a_N)_r}-T_{n_r})^2/p_{a_N}\log N).$$

Noting (10.2.15) and choosing r to be large enough, we know the right hand side of (10.2.17) does not exceed

$$\begin{aligned} cr\frac{N}{a_N}\exp\Big\{-(1-\varepsilon/3)(1+\varepsilon/3)^2\alpha_{nN}^2/E(T_{(n+a_N)_r}-T_{n_r})^2\Big\} \\ \le cr\frac{N}{a_N}\Big(\frac{a_N}{N\log N}\Big)^{1+\varepsilon/4} \\ = cr\Big(\frac{a_N}{N}\Big)^{\varepsilon/4}(\log N)^{-(1+\varepsilon/4)} \end{aligned}$$

for large N. Let $N_1 = 1$ and define N_{j+1} by $a_{N_{j+1}} = \min\{a_n : a_n \ge [\theta^j]\}(\theta > 1)$. Obviously, $N_{j+1} \ge [\theta^j]$ since $a_N \le N$. Thus

$$\sum_{j=1}^{\infty} P\Big\{\max_{1\le n\le N_j}\max_{1\le k\le a_{N_j}} \alpha_{nN_j}^{-1}|T_{(n+k)_r} - T_{n_r}| \ge 1+\frac{\varepsilon}{3}\Big\} < \infty$$

which implies

$$\limsup_{j\to\infty}\max_{1\le n\le N_j}\max_{1<k\le a_{N_j}} \alpha_{nN_j}^{-1}|T_{(n+k)_r} - T_{n_r}| \le 1+\frac{\varepsilon}{3} \quad \text{a.s.} \tag{10.2.18}$$

From conditions (i)-(iii) and (10.2.11) and the definition of N_j, it is easy to see that there exists a constant $G > 0$, such that

$$\Big|\lim_{j\to\infty}\frac{\lambda_{nN_{j+1}}^2}{\lambda_{nN_j}^2} - 1\Big| \le G(\theta-1)^{1/2}. \tag{10.2.19}$$

Now, choosing θ near enough to 1 and noting that the ranges in the two max's in (10.2.18) enlarge as j increases we have from (10.2.18)

$$\limsup_{N\to\infty}\max_{1\le n\le N}\max_{1\le k\le a_N} \alpha_{nN}^{-1}|T_{(n+k)_r} - T_{n_r}| \le 1+\frac{\varepsilon}{2} \quad \text{a.s.} \tag{10.2.20}$$

We turn to the first term in the right hand side of (10.2.16). Write

$$\begin{aligned} P\Big\{\max_{1\le n\le N}\max_{1\le k\le a_N} \alpha_{nN}^{-1}|T_{n+k} - T_{(n+k)_r}| \ge \frac{\varepsilon}{6}\Big\} \\ \le \frac{rN}{a_N}\max_n\max_k P\Big\{\max_{(n+k)_r<j\le(n+k)_r+R}|T_j \\ - T_{(n+k)_r}| \ge \frac{\varepsilon}{6}\alpha_{nN}\Big\}. \end{aligned} \tag{10.2.21}$$

Using a similar proof to (10.2.18), one can also employ (10.1.4) of Lemma 10.1.2. By recalling (10.2.7) and (10.2.11) and choosing $r = r(\varepsilon)$ to be large enough, the right hand side of (10.2.21) does not exceed

$$3\frac{rN}{a_N}\exp\Big\{-\frac{(1-6\varepsilon)\varepsilon^2\lambda_{nN}^2}{36E(T_{(n+k)_r+R}-T_{(n+k)_r})^2}\log\frac{N\log N}{\lambda_{nN}^2}\Big\}$$
$$\le\frac{3rN}{a_N}\exp\Big\{-cr\varepsilon^2\log\frac{N\log N}{a_N}\Big\}$$
$$\le\frac{3rN}{a_N}\Big(\frac{a_N}{N\log N}\Big)^{cr\varepsilon^2}\le c\log^{-2}N$$

for all large N. Thus

$$\limsup_{j\to\infty}\max_{1\le n\le N_j}\max_{1\le k\le a_{N_j}}\alpha_{nN_j}^{-1}|T_{n+k}-T_{(n+k)_r}|\le\frac{\varepsilon}{6}\quad\text{a.s.}$$

By imitating the procedure from (10.2.18) to (10.2.20), we have

$$\limsup_{N\to\infty}\max_{1\le n\le N}\max_{1\le k\le a_N}\alpha_{nN}^{-1}|T_{n+k}-T_{(n+k)_r}|\le\frac{\varepsilon}{4}\quad\text{a.s.}\tag{10.2.22}$$

Obviously, we also have

$$\limsup_{N\to\infty}\max_{1\le n\le N}\alpha_{nN}^{-1}|T_n-T_{n_r}|\le\frac{\varepsilon}{4}\quad\text{a.s.}$$

Therefore, (10.2.14), and further (10.2.9), are proved.

Finally, we prove (10.2.5) and (10.2.6). For this purpose, from (10.2.2) and (10.2.4) proved, it is sufficient to verify that

$$\liminf_{N\to\infty}\max_{1\le n\le N}\beta_{NN}^{-1}|S(n,a_N)|\ge 1\quad\text{a.s.}\tag{10.2.23}$$

By (10.2.10) and (10.2.11), (10.2.23) is equivalent to

$$\liminf_{N\to\infty}\max_{1\le n\le N}\alpha_{nN}^{-1}|T(n,a_N)|\ge 1\quad\text{a.s.}\tag{10.2.24}$$

We have for large N

$$P\Big\{\max_{1\le n\le N}\alpha_{nN}^{-1}|T(n,a_N)|\le 1-\varepsilon\Big\}$$
$$\le P\Big\{\max_{1\le j\le[N/a_N]^{1-\varepsilon/2}}\alpha_{2ja_N,N}^{-1}|T(2ja_N,a_N)|\le 1-\varepsilon\Big\}$$
$$\le\prod_{j=1}^{[N/a_N]^{1-\varepsilon/2}}P\{\alpha_{2ja_N,N}^{-1}|T(2ja_N,a_N)|\le 1-\varepsilon\Big\}$$
$$+\Big(\frac{N}{a_N}\Big)^{1-\varepsilon/2}\varphi(a_N)$$

(by (10.2.13) and condition (b))

$$
\begin{aligned}
&\leq \Big\{1 - c\Big(\frac{a_N}{N \log N}\Big)^{1-\varepsilon}\Big\}^{[N/a_N]^{1-\varepsilon/2}} + c\Big(\frac{N}{a_N}\Big)^{-\varepsilon/2} \\
&\leq 2\exp\Big\{-c\Big(\frac{N}{a_N}\Big)^{-\varepsilon/2}(\log N)^{-(1-\varepsilon)}\Big\} + c\Big(\frac{N}{a_N}\Big)^{-\varepsilon/2} \\
&\leq c(\log N)^{-2}.
\end{aligned}
$$

The last inequality is due to condition (b). Thus, if N_j is defined as above, we have

$$\liminf_{j\to\infty} \max_{1\leq n\leq N_j} \alpha_{nN_j}^{-1}|T(n, a_{N_j})| \geq 1 \quad \text{a.s.} \tag{10.2.25}$$

Considering $N_j < N \leq N_{j+1}$, we get

$$
\begin{aligned}
&\liminf_{N\to\infty} \max_{1\leq n\leq N} \alpha_{nN}^{-1}|T(n, a_N)| \\
&\quad \geq \liminf_{j\to\infty} \max_{1\leq n\leq N_j} (\alpha_{nN_j}^{-1}|T(n, a_{N_j})|)(\alpha_{nN_j}\alpha_{nN}^{-1}) \\
&\qquad - \limsup_{N\to\infty} \max_{1\leq n\leq N} \max_{1\leq k\leq a_N - a_{N_j}} \alpha_{nN}^{-1}|T(n, k)|.
\end{aligned}
$$

The first term in the right hand side is a.s. $\geq 1 - G'(\theta-1)^{1/2}$ for some $G' > 0$ by (10.2.25) and (10.2.19). The latter is a.s.$\leq G''(1-1/\theta)^{1/2}$ for some $G'' > 0$ by (10.2.14). Letting $\theta \downarrow 1$, we obtain (10.2.24). The theorem is proved.

Remark 10.2.1. When l in condition (iii) is large enough, in other words, φ_n tends to zero at a great rate, a_n can be $O(\log^{3+\varepsilon} n)$, for any given $\varepsilon > 0$. For an independent sequence, it is required that $a_n/\log n \to \infty$ as $n \to \infty$ (see Lin 1988).

Remark 10.2.2. In the theorem, we don't require that a_n/n is non-increasing. But it is assumed even if a sequence is independent (cf. Csörgő and Révész 1981). In fact, this condition is not realistic, since either $a_n = n$ for all n or $a_n = m \wedge n$ for some fixed m when the condition is added.

10.3 How big are the increments when the moment generating functions do not exist?

Let $\{X_n\}$ be a φ-mixing sequence mentioned in the beginning of the above section.

Theorem 10.3.1.(Lin 1989) *Suppose that $\{X_n\}$ satisfies condition (i) in Theorem 10.2.1 and*

(ii)′ there exists a non-decreasing continuous function $H(x), x > 0$, such that

$$\sum_{n=1}^{\infty} P\{H(|X_n|) > \delta n\} < \infty \quad for\ any\ \delta > 0, \tag{10.3.1}$$

$$\overline{\lim_{k\to\infty}} \frac{1}{k} \sum_{j=n+1}^{n+k} E(H(|X_j|))^{\beta} \le M < \infty \tag{10.3.2}$$

uniformly in n for any $\beta < 1$,

$$x^{-(2+\gamma)} H(x) \text{ is non-decreasing for some } \gamma > 0, \tag{10.3.3}$$

$$\underline{\lim_{x\to\infty}}\ H(x/2)/H(x) > 0, \tag{10.3.4}$$

(iii)′ there exists an $l \ge 1 + \frac{4}{\gamma}$ such that $\varphi_n = O(n^{-l})$.
Let $\{a_n\}$ be a non-decreasing sequence of positive integers satisfying

(a)′ there exists such an $a > 0$ that $a(\mathrm{inv}H(n))^d \le a_n \le n$ where $d = 2(l+1)/(l-1)$.

Then the conclusions of Theorem 10.2.1 remain true.

Proof. The proof of the theorem is similar to that of Theorem 10.2.1. We outline the main differences. It is easy to verify that condition (10.3.1) can be replaced by

$$\sum_{n=1}^{\infty} P\{H(|X_n|) > \delta_n n\} < \infty \tag{10.3.5}$$

for some $\delta_n \downarrow 0$. Define $Y_n = X_n I(|X_n| \le \mathrm{inv}H(\delta_n n)), Y_n' = Y_n - EY_n, T_n = \sum_{k=1}^{n} Y_k'$. By (10.3.5) we have

$$P\{X_n \ne Y_n, i.o.\} = 0. \tag{10.3.6}$$

Without loss of generality we can choose δ_n such that $n^{-\varepsilon} = o(\delta_n)$ for any $\varepsilon > 0$. From this and using the conditions $EX_j = 0$, (10.3.2) and (10.3.3), we have for all large n,

$$\begin{aligned}\sum_{j=n+1}^{n+k} |EY_j| &\le c \sum_{j=n+1}^{n+k} \int_{\{H(|X_j|)>\delta_j j\}} (H(|X_j|))^{\frac{1}{2+\gamma}} dP \\ &\le c \sum_{j=n+1}^{n+k} (\delta_j j)^{-\frac{2+\gamma}{4+\gamma}} E(H(|X_j|))^{\frac{1}{2+\gamma}+\frac{2+\gamma}{4+\gamma}} \\ &\le c \sum_{j=n+1}^{n+k} j^{-\frac{3+\gamma}{6+\gamma}} E(H(|X_j|))^{\frac{\gamma^2+5\gamma+8}{\gamma^2+6\gamma+8}} \\ &\le ck^{-\frac{3+\gamma}{6+\gamma}} \Big\{ \sum_{j=n+1}^{n+k} E(H(|X_j|))^{\frac{\gamma^2+5.5\gamma+8}{\gamma^2+6\gamma+8}} \Big\}^{\frac{\gamma^2+5\gamma+8}{\gamma^2+5.5\gamma+8}} \\ &\le ck^{\frac{3}{6+\gamma}}. \end{aligned} \tag{10.3.7}$$

Consequently (10.2.8) is also equivalent to (10.2.12). Put $h = 1/(l+1)$, $p_n = [n^h], q_n = [\alpha_n p_n]$, where α_n are positive numbers tending to zero slowly enough. Using conditions (iii)$'$ and (a)$'$, we obtain

$$\begin{aligned}&\varphi_{q_{a_N}} p_{a_N} (\text{inv}H(\delta_N N))^2 \\ &\quad \le c\alpha_{a_N}^{-l} a_N^{-h(l-1)} (\text{inv}H(\delta_N N))^2 \\ &\quad \le c\alpha_{a_N}^{-l} (\text{inv}H(\delta_N N)/\text{inv}H(N))^2.\end{aligned}$$

By condition (10.3.4) one can get $\text{inv}H(\delta_N N)/\text{inv}H(N) \to 0$ (see Lin and Lu, 1992, (2.3.12)). Hence ,we have

$$\varphi_{q_{a_N}} p_{a_N} (H(\delta_N N))^2 \to 0, \quad \text{as } N \to \infty.$$

provided that α_N tends to zero slowly enough. Moreover the other conditions in Lemma 10.1.2 are also satisfied. Then we have (10.2.13) as well. The following proof is similar to that of Theorom 10.2.1, hence, is omitted.

Remark 10.3.1. When l in condition (iii)' is large enough, i.e., φ_n tends to zero at a great rate, a_n can be $O((\text{inv}H(n))^{2+\varepsilon})$ for any given $\varepsilon > 0$. For an independent sequence, it is required that $a_n \ge c(\text{inv}H(n))^2/\log n$ (see Lin 1987).

Chapter 11 Strong Approximations for Mixing Random Fields

Let $\{X_{\mathbf{j}}, \mathbf{j} \in \mathbb{N}^d\}$ be a stationary mixing random field with $EX_{\mathbf{j}} = 0$. The various definitions of mixing have been given in Chapter 6. We shall also employ the notations $\mathcal{A} \subset \mathcal{E} \subset \mathcal{B}^d, C_{\mathbf{j}}, \mathcal{A}_\delta, A(\delta)$, etc, which have been posed there. We define the partial-sum process by

$$S_n(A) = \sum_{\mathbf{j}} |nA \cap C_{\mathbf{j}}| X_{\mathbf{j}} \quad \text{for any } A \in \mathcal{A} \tag{11.0.1}$$

where $\mathbf{n} = (n_1, \cdots, n_d) \in \mathbb{N}^d, \mathbf{n}A = \{(n_1 x_1, \cdots, n_d x_d),\ \mathbf{x} = (x_1, \cdots, x_d) \in A\}$.

The main subject of strong approximations of set-indexed processes is to establish a new probability space, on which there exists an independent identically distributed centered Gaussian field $\{Y_{\mathbf{j}}, \mathbf{j} \in \mathbb{N}^d\}$ with covariance $\sigma^2 = EX_{\mathbf{1}}^2 + 2\sum_{\mathbf{j}} EX_{\mathbf{j}+\mathbf{1}} X_{\mathbf{1}}$, and without changing its joint distribution, the field $\{X_{\mathbf{j}}, \mathbf{j} \in \mathbb{N}^d\}$ can be redefined on this new probability space, such that

$$\begin{aligned} D_n &= \sup_{A \in \mathcal{A}} \Big\{ \sum_{\mathbf{j} \in \mathbb{N}^d} |nA \cap C_{\mathbf{j}}| (X_{\mathbf{j}} - Y_{\mathbf{j}}) \Big\} \\ &= O(n^{d/2-\varepsilon}) \quad \text{or} \quad O(n^{d/2} (\log n)^{-a}) \qquad (a, \varepsilon > 0), \end{aligned} \tag{11.0.2}$$

under some conditions on $\mathcal{A}$ and $\{X_{\mathbf{j}}\}$.

Denote

$$\begin{aligned} |\mathbf{n}| &= n_1 n_2 \cdots n_d, \qquad \text{if} \quad \mathbf{n} = (n_1, \cdots, n_d), \\ G_\beta &= \Big\{ \mathbf{n} = (n_1, \cdots, n_d),\ n_i \geq \prod_{k \neq i} n_k^\beta,\ i = 1, 2, \cdots, d \Big\}, \end{aligned} \tag{11.0.3}$$

for $0 < \beta \leq 1$.

Berkes and Morrow (1981) first discussed the strong approximation for a weakly stationary α-mixing random field $\{X_{\mathbf{j}},\ \mathbf{j} \in \mathbb{N}^d\}$. Suppose that

$EX_1 = 0, E|X_1|^{2+\delta} < \infty$ for some $\delta > 0$ and $\alpha(n) = O(n^{-d(1+\varepsilon)(1+2/\delta)})$. They proved that for any $\mathbf{n} \in G_\beta$ and some $\lambda > 0$

$$\sup_{1 \le \mathbf{j} \le \mathbf{n}} \Big| \sum_{\mathbf{i} \le \mathbf{j}} (X_{\mathbf{i}} - Y_{\mathbf{i}}) \Big| = O(|\mathbf{n}|^{1/2-\lambda}) \quad \text{a.s.} \tag{11.0.4}$$

This result was improved by Strittmatter (1990), in which the condition "$\mathbf{n} \in G_\beta$" is left out and the following result is obtained: If

$$N(\varepsilon) := \mathrm{Card}(\mathcal{A}(\varepsilon)) \le c\varepsilon^{-u}, \qquad u < \frac{\delta s - 4 - 2\delta}{d(4+\delta)} - 2, \tag{11.0.5}$$

$$\begin{aligned} b(\varepsilon) := \sup\Big\{ & n^{-d} |((\mathbf{n}A)^{n\varepsilon} \cap (\mathbf{n}A^c)^{n\varepsilon})| : \\ & A \in \bigcup_{\eta > 0} A_\eta,\ n \ge 1/\varepsilon \Big\} \\ \le c\, \varepsilon^h \quad & \text{for some } 0 < h \le 1, \end{aligned} \tag{11.0.6}$$

$$\alpha(n) = O(n^{-s}) \qquad \text{for some } s > 1 + (17/2)^{d-1}/(1 \wedge \delta), \tag{11.0.7}$$

then for some $\gamma > 0$

$$\sup_{A \in \mathcal{A}} \sum_{\mathbf{j} \in \mathbb{N}^d} |nA \cap C_{\mathbf{j}}| (X_{\mathbf{j}} - Y_{\mathbf{j}}) = O(n^{d/2-\gamma}) \quad \text{a.s.} \tag{11.0.8}$$

The class $\mathcal{A}$ of subsets in the above results which satisfies $N(\varepsilon) \le c\varepsilon^{-u}$ is the smaller subset class. Su (1992) proved a strong approximation for a φ-mixing strictly stationary random field with $H(\varepsilon) \le c\varepsilon^{-r}$ for some $r > 0$, and $\mathbf{n} \in G_\beta$. We shall introduce Su's results in Section 11.1 and Strittmatter's results in Section 11.2.

11.1 Strong approximations of a φ-mixing random field

For simplicity, we only give the results for the case of $d = 2$.

Theorem 11.1.1. (Su 1992) *Let $\{X_{\mathbf{j}}, \mathbf{j} \in \mathbb{N}^2\}$ be a strictly stationary φ-mixing random field with $EX_{\mathbf{j}} = 0$ and $E|X_{\mathbf{j}}|^{2+\delta} < \infty$ for some $\delta > 0$. Suppose that the class $\mathcal{A}$ of subsets of $\mathcal{B}^2$ satisfies (11.0.6) and the entropy condition*

$$H(\varepsilon) \le c\varepsilon^{-r} \quad \textit{for some} \quad 0 < r < \delta/(4(1+\delta)), \tag{11.1.1}$$

and suppose that

$$\varphi(x) = O(x^{-q}) \quad \textit{for some} \quad q > \frac{4(2+\delta)}{\delta - r(4+\delta)} - 2. \tag{11.1.2}$$

Then

$$\sigma^2 := EX_1^2 + 2\sum_{\mathbf{v}\neq \mathbf{1}} EX_1X_{\mathbf{v}} = O(1), \tag{11.1.3}$$

further, without changing the distribution of $\{X_{\mathbf{j}},\ \mathbf{j}\in\mathbb{N}^2\}$, *we can redefine the field* $\{X_{\mathbf{j}},\ \mathbf{j}\in\mathbb{N}^2\}$ *on a richer probability space together with an independent normal random field* $\{Y_{\mathbf{j}},\ \mathbf{j}\in\mathbb{N}^2\}$ *with* $EY_{\mathbf{j}}=0, EY_{\mathbf{j}}^2=\sigma^2$, *such that*

$$\sup_{A\in\mathcal{A}}\sum_{\mathbf{j}\in\mathbb{N}^2}|nA\cap C_{\mathbf{j}}|(X_{\mathbf{j}}-Y_{\mathbf{j}}) = O(|n|(\log|n|)^{-\sigma_1}) \quad a.s. \tag{11.1.4}$$

for every $n\in G_\beta, \delta/(2(2+\delta))<\beta\le 1$, *where* $\sigma_1=\sigma_1(r,q,\delta)>0$.

First, we prove the *Bernstein inequality* for a φ-mixing random field.

Lemma 11.1.1. *Let* $\{X_{\mathbf{j}},\ \mathbf{j}\in\mathbb{N}^d\}$ *be a* φ*-mixing random field with* $EX_{\mathbf{j}}=0, |X_{\mathbf{j}}|\le\Delta_{\mathbf{n}}$ *a.s.* $\mathbf{1}\le\mathbf{j}\le\mathbf{n}$. *Denote* $\sigma_{\mathbf{n}}=\max_{\mathbf{1}\le\mathbf{j}\le\mathbf{n}}\|X_{\mathbf{j}}\|_2$. *If* $\sum_{n=1}^{\infty}\varphi^{1/2}(2^n)<\infty$, *then for any* $A\in\mathcal{B}^d$ *we have*

$$P\Big\{\Big|\sum_{\mathbf{j}}|\mathbf{n}A\cap C_{\mathbf{j}}|X_{\mathbf{j}}\Big|\ge x_n\Big\}$$
$$\le 2\exp\Big\{\frac{c|\mathbf{n}|\varphi(m)}{m^d}-\alpha x_n+c\alpha^2|\mathbf{n}A|\sigma_{\mathbf{n}}^2\Big\}, \tag{11.1.5}$$

where $x_n>0$, $m\le\min_{1\le i\le d}n_i$ *and* $2^d\alpha m^d\Delta_{\mathbf{n}}\le 1/2$.

Particularly, if $\alpha=1/(2^{d+1}m^d\Delta_{\mathbf{n}})$, *then*

$$P\Big\{\Big|\sum_{\mathbf{j}}|\mathbf{n}A\cap C_{\mathbf{j}}|X_{\mathbf{j}}\Big|\ge x_n\Big\}$$
$$\le 2\exp\Big\{\frac{c|\mathbf{n}|\varphi(m)}{m^d}-\frac{x_n}{2^{d+1}m^d\Delta_{\mathbf{n}}}+\frac{c|\mathbf{n}A|\sigma_{\mathbf{n}}^2}{(2^{d+1}m^d\Delta_{\mathbf{n}})^2}\Big\}.$$

Proof. Put $\mathbf{N}=(N_1,\cdots,N_d)$ such that $2m(N_i-1)\le n_i\le 2mN_i$, $i=1,\cdots,d$. Denote $V=\{(v_1,\cdots,v_d): v_i=1 \text{ or } 2,\ i=1,\cdots,d\}$. Define

$$I_{\mathbf{v},\mathbf{k}}=[l_1(1),l_2(1)]\times\cdots\times[l_1(d),l_2(d)],\quad \mathbf{v}\in V,\ \mathbf{1}\le\mathbf{k}\le\mathbf{N},$$

where

$$l_1(i)=\{(2k_i+v_i-3)m+1\}\wedge n_i,$$
$$l_2(i)=\{l_1(i)+m-1\}\wedge n_i,\ i=1,2,\cdots,d.$$

Moreover, denote

$$I_{\mathbf{v}} = \sum_{\mathbf{1}\le\mathbf{k}\le\mathbf{N}} I_{\mathbf{v},\mathbf{k}},$$

$$A_{\mathbf{v},\mathbf{k}} = \sum_{\mathbf{j}} |nA \cap I_{\mathbf{v},\mathbf{k}} \cap C_{\mathbf{j}}| X_{\mathbf{j}},$$

$$B_{\mathbf{v},\mathbf{N}} = \sum_{\mathbf{1}\le\mathbf{k}\le\mathbf{N}} A_{\mathbf{v},\mathbf{k}},$$

$$S = \sum_{\mathbf{v}\in V} B_{\mathbf{v},\,\mathbf{N}}.$$

Note that $e^x \le 1 + x + x^2$ if $|x| \le 1/2$ and $1 + x \le e^x$ for any x. Therefore if $2^d \alpha\, m^d \Delta_{\mathbf{n}} \le 1/2$, we have

$$E\exp(2^d \alpha A_{\mathbf{v},\mathbf{k}}) \le \exp((2^d\alpha)^2 E A^2_{\mathbf{v},\mathbf{k}}).$$

By Lemmas 1.2.10 and 6.2.3 we have

$$\begin{aligned}
&E\exp(2^d\alpha B_{\mathbf{v},\mathbf{N}}) \\
&\quad= E\exp\Big(2^d\alpha \sum_{\mathbf{k}\ne\mathbf{N}} A_{\mathbf{v},\mathbf{k}}\Big)\exp(2^d\alpha A_{\mathbf{v},\mathbf{N}}) \\
&\quad\le E\exp\Big(2^d\alpha \sum_{\mathbf{k}\ne\mathbf{N}} A_{\mathbf{v},\mathbf{k}}\Big) E\exp(2^d\alpha A_{\mathbf{v},\mathbf{N}}) \\
&\qquad + 2\varphi(m)\|\exp(2^d\alpha A_{\mathbf{v},\mathbf{N}})\|_\infty E\exp\Big(2^d\alpha \sum_{\mathbf{k}\ne\mathbf{N}} A_{\mathbf{v},\mathbf{k}}\Big) \\
&\quad\le (1 + 2e^{1/2}\varphi(m))\exp(c(2^d\alpha)^2|\,\mathbf{n}A\cap I_{\mathbf{v},\mathbf{N}}|\sigma_n^2) E\exp\Big(2^d\alpha \sum_{\mathbf{k}\ne\mathbf{N}} A_{\mathbf{v},\mathbf{k}}\Big) \\
&\quad\le (1 + 2e^{1/2}\varphi(m))^{|\mathbf{N}|}\exp(c(2^d\alpha)^2|\mathbf{n}A\cap I_{\mathbf{v}}|\sigma_n^2) \qquad (11.1.6)
\end{aligned}$$

where $\|f(x)\|_\infty$ means the superior norm of $f(x)$. From the convexity of an exponential function and (11.1.6), it follows that

$$\begin{aligned}
E\exp(\alpha S) &\le \frac{1}{2^d}\sum_{\mathbf{v}\in V} E\exp(2^d\alpha B_{\mathbf{v},\,\mathbf{N}}) \\
&\le (1 + 2e^{1/2}\varphi(m))^{|\mathbf{N}|}\exp(c(2^d\alpha)^2|\mathbf{n}A|\sigma_n^2). \qquad (11.1.7)
\end{aligned}$$

Since $|\mathbf{N}| \le (3/2)^d|\mathbf{n}|/m^d$ and (11.1.7) holds also true for $E\exp(-\alpha S)$, we obtain (11.1.5) by the Markov inequality.

Proof of Theorem 11.1.1.

First, under the conditions of Theorem 11.1.1, it is easy to see that (11.1.3) holds true and

$$\lim_{\substack{|\mathbf{n}|\to\infty \\ \mathbf{n}\in G_\beta}} E\Big(\sum_{1\le \mathbf{j}\le \mathbf{n}} X_{\mathbf{j}}\Big)^2/|\mathbf{n}| = \sigma^2 < \infty. \tag{11.1.8}$$

In order to prove (11.1.4) for any $\mathbf{n} \in G_\beta, \delta/(2+\delta) < \beta \le 1$, we shall use the truncation, nesting and blocking techniques in the following lemmas.

1. Truncation Take τ small enough (specified later on), let

$$X'_{\mathbf{j}} = X_{\mathbf{j}} I(|X_{\mathbf{j}}| \le |\mathbf{j}|^{(1+\tau)/(2+\delta)}),$$
$$S'_{\mathbf{n}}(A) = \sum_{\mathbf{j}} |\mathbf{n}A \cap C_{\mathbf{j}}| X'_{\mathbf{j}}.$$

Lemma 11.1.2. *We have*

$$\sup_{A\in\mathcal{A}} |S_{\mathbf{n}}(A) - S'_{\mathbf{n}}(A)| = O(1) \quad a.s. \tag{11.1.9}$$

$$\sum_{\mathbf{j}\le n\mathbf{1}} E|X_{\mathbf{j}} - X'_{\mathbf{j}}| = O(n^{d/(2+\delta)}) \quad a.s. \tag{11.1.10}$$

Proof. Since

$$P\{X_{\mathbf{j}} \ne X'_{\mathbf{j}}\} = P\{|X_{\mathbf{j}}| \ge |\mathbf{j}|^{(1+\tau)/(2+\delta)}\} \le c|\mathbf{j}|^{-(1+\tau)},$$

we have $P\{X_{\mathbf{j}} - X'_{\mathbf{j}} \ne 0, \text{ i.o.}\} = 0$ by the Borel-Cantelli lemma. Thus for every $n \in \mathbb{N}$

$$\sup_A |S_{\mathbf{n}}(A) - S'_{\mathbf{n}}(A)| \le \sum_{\mathbf{j}\in n\mathbf{1}} |X_{\mathbf{j}} - X'_{\mathbf{j}}| \le c < \infty \quad \text{a.s.}$$

and

$$\sum_{\mathbf{j}\le n\mathbf{1}} E|X_{\mathbf{j}} - X'_{\mathbf{j}}| \le \sum_{\mathbf{j}\le n\mathbf{1}} \Big(|\mathbf{j}|^{-(1+\tau)(1+\delta)/(2+\delta)}\Big) = O(n^{d/(2+\delta)}).$$

2. Nesting For any given $\mathbf{n} = (n_1, n_2) \in G_\beta, \delta/(2(2+\delta)) < \beta \le 1$, let a, b be least positive integers such that

$$2^a \ge (\log|\mathbf{n}|)^{1/r'}, \quad 2^b \ge |\mathbf{n}|^{\frac{1}{2}(\tau + \frac{4+\delta}{2+\delta})}, \tag{11.1.11}$$

where $r' > r$ is specified later on.

Lemma 11.1.3. *We have*

$$\sup_A |S'_{\mathbf{n}}(A) - S'_{\mathbf{n}}(A(2^{-b}))| = O(|n|^{\frac{1}{2}(1-\frac{\tau\delta}{2+\delta})}) \quad a.s.$$

Proof. For any $A \in \mathcal{A}$, by conditions (11.0.6), (11.1.1) and (11.1.11), we have

$$\begin{aligned}
&|S'_{\mathbf{n}}(A) - S'_{\mathbf{n}}(A(2^{-b}))| \\
&\quad \le \sum_{\mathbf{j}} \Big| |\mathbf{n}A \cap C_{\mathbf{j}}| - |\mathbf{n}A(2^{-b}) \cap C_{\mathbf{j}}| \Big| |X'_{\mathbf{j}}| \\
&\quad \le c|n||A \triangle A(2^{-b})| \cdot |n|^{\frac{1+\tau}{2+\delta}} = O(|n|^{\frac{1}{2}(1-\frac{\tau\delta}{2+\delta})}).
\end{aligned}$$

Lemma 11.1.4. *For every* $\mathbf{n} \in G_\beta$

$$\sup_{A\in\mathcal{A}} |S'_{\mathbf{n}}(A(2^{-b})) - S'_{\mathbf{n}}(A(2^{-a}))| = O(|\mathbf{n}|^{\frac{1}{2}}(\log|\mathbf{n}|)^{-\sigma_1}) \quad a.s.$$

Proof. Take τ, r' and τ' small enough such that

$$\Big(\frac{\delta - 2\tau}{2(2+\delta)} - \frac{1}{2}\tau'\Big(\tau + \frac{4+\delta}{2+\delta}\Big)\Big)(q+1) > 1, \quad \tau' - \tau > r' > r. \tag{11.1.12}$$

Denote $m_i = [|n|^{\frac{\delta-2\tau}{4(2+\delta)}} 2^{-\frac{i\tau'}{2}}]$. Since $n \in G_\beta$, it is clear that $m_i \le n_1 \wedge n_2, i = 1, 2$. On the other hand we have

$$|X'_{\mathbf{j}} - EX'_{\mathbf{j}}| \le 2|n|^{(1+\tau)/(2+\delta)}, \quad \mathbf{j} \le \mathbf{n}.$$

Take $\Delta_{\mathbf{n}} = 2|\mathbf{n}|^{(1+\tau)/(2+\delta)}$ in Lemma 11.1.1, we obtain

$$\begin{aligned}
&P\{|S'_{\mathbf{n}}(A(2^{-(i+1)})) - S'_{\mathbf{n}}(A(2^{-i}))| \ge |n|^{1/2}2^{-i\tau}\} \\
&\quad \le P\Big\{\Big|\sum_{\mathbf{j}} |n(A(2^{-(i+1)})\backslash A(2^{-i}))|(X'_{\mathbf{j}} - EX'_{\mathbf{j}})\Big| > \frac{1}{4}|n|^{1/2}2^{-i\tau}\Big\} \\
&\qquad + P\Big\{\Big|\sum_{\mathbf{j}} |n(A(2^{-i})\backslash A(2^{-(i+1)}))|(X'_{\mathbf{j}} - EX'_{\mathbf{j}})\Big| > \frac{1}{4}|n|^{1/2}2^{-i\tau}\Big\} \\
&\quad \le c\exp(-c2^{i(\tau'-\tau)}).
\end{aligned} \tag{11.1.13}$$

Therefore from (11.1.1) it follows that

$$\sum_{\mathbf{n}\in G_\beta} P\Big\{\sup_{A\in\mathcal{A}} |S'_{\mathbf{n}}(A(2^{-b})) - S'_{\mathbf{n}}(A(2^{-a}))| \ge |n|^{1/2}(\log|n|)^{-\tau/\tau'}\Big\}$$

$$\le \sum_{\mathbf{n}\in G_\beta}\sum_{i=a}^{b}\sum_{A\in\mathcal{A}} P\{|S'_{\mathbf{n}}(A(2^{-(i+1)})) - S'_{\mathbf{n}}(A(2^{-i}))| \ge |n|^{1/2}2^{-i\tau}\}$$

$$\le c\sum_{\mathbf{n}\in G_\beta}\sum_{i=a}^{b}\exp\Big(H(2^{-i}) + H(2^{-(i+1)}) - c2^{i(\tau'-\tau)}\Big) < \infty,$$

which implies Lemma 11.1.4 by the Borel-Cantelli lemma.

Let $\mathbf{k} = (k(1), k(2)) \ge \mathbf{0}$, define $\mathbf{t_k} = (t_{\mathbf{k}}(1), t_{\mathbf{k}}(2))$ as follows:

$$t_{\mathbf{k}}(1) = [\exp(k(1))^{s/4}], \quad t_{\mathbf{k}}(2) = [\exp(k(2))^{s/4}], \quad 0 < s < 1/2.$$

For large $\mathbf{n}$, $\mathbf{n} \in G_\beta$ implies $\mathbf{t_k} \in G_\beta$ if $\mathbf{t_k} + \mathbf{1} \le \mathbf{n} \le \mathbf{t_{k+1}}$.

Lemma 11.1.5. *For $\mathbf{t_k} \in G_\beta$, we have*

$$\sup_{A\in\mathcal{A}}\max_{\mathbf{t_k}<\mathbf{n}\le\mathbf{t_{k+1}}} |S'_{\mathbf{n}}(A(2^{-a})) - S'_{\mathbf{t_k}}(A(2^{-a}))|$$

$$= O(|\mathbf{t_k}|^{1/2}|\log|\mathbf{t_k}|)^{-1/4}) \quad a.s.$$

as $|\mathbf{k}| \to \infty$.

Proof. Note that if $\mathbf{t_k} < \mathbf{n} \le \mathbf{t_{k+1}}, A \subset [0,1]^2$, we have

$$|\mathbf{n}A \,\triangle\, \mathbf{t_k}A| \le c|\mathbf{n}|(\varepsilon_1 + \varepsilon_2),$$

where

$$\varepsilon_i = (t_{\mathbf{k+1}}(i) - t_{\mathbf{k}}(i))/n_i, \; i = 1, 2. \tag{11.1.14}$$

Put $m_k = \Big[|\mathbf{t_k}|^{\frac{\delta-2\tau}{4(2+\delta)}}|\mathbf{k}|^{-\frac{s}{4}}\Big]$. If $\mathbf{t_k} \in G_\beta$, we have $m_k \le t_{\mathbf{k}}(1) \wedge t_{\mathbf{k}}(2)$. Let $\Delta_n = 2|\mathbf{t_k}|^{(1+\tau)/(2+\delta)}$ in Lemma 11.1.1, we obtain

$$P\{|S'_{\mathbf{n}}(A(2^{-a})) - S'_{t_k}(A(2^{-a}))| \ge |\mathbf{t_k}|^{1/2}(k(1)^{-s/16} + k(2)^{-s/16})\}$$

$$\le P\Big\{\Big|\sum_{\mathbf{j}}|(\mathbf{n}A(2^{-a})\backslash\mathbf{t_k}A(2^{-a}))\cap$$

$$\cdot C_{\mathbf{j}}|(X'_{\mathbf{j}} - EX'_{\mathbf{j}})\Big| \ge \frac{1}{4}|\mathbf{t_k}|^{\frac{1}{2}}(k(1)^{-\frac{s}{16}} + k(2)^{-\frac{s}{16}})\Big\}$$

$$+ P\Big\{\Big|\sum_{\mathbf{j}}|(\mathbf{t_k}A(2^{-a})\backslash\mathbf{n}A(2^{-a}))\cap$$

$$\cdot C_{\mathbf{j}}|(X'_{\mathbf{j}} - EX'_{\mathbf{j}})\Big| \ge \frac{1}{4}|\mathbf{t_k}|^{\frac{1}{2}}(k(1)^{-\frac{s}{16}} + k(2)^{-\frac{s}{16}})\Big\}$$

$$\le c\exp(-c(k(1)^{s/2} + k(2)^{s/2})). \tag{11.1.15}$$

For $\mathbf{t_k} \in G_\beta$, we have

$$k(1)^{-s/16} + k(2)^{-s/16} \leq c(\log |\mathbf{t_k}|)^{-1/4}.$$

It follows that

$$\begin{aligned}\sum_{\mathbf{t_k}\in G_\beta} P\Big\{\sup_{A\in\mathcal{A}} \sup_{\mathbf{t_k}<\mathbf{n}\leq\mathbf{t_{k+1}}} &|S'_{\mathbf{n}}(A(2^{-a})) \\ &- S'_{\mathbf{t_k}}(A(2^{-a}))| \geq |\mathbf{t_k}|^{\frac{1}{2}}(\log|\mathbf{t_k}|)^{-\frac{1}{4}}\Big\} \\ &\leq c \sum_{\mathbf{t_k}\in G_\beta} \exp\Big(2^{(\log|\mathbf{t_k}|)^{r/r'}}\Big)|\mathbf{t_{k+1}} - \mathbf{t_k}| \\ &\quad \cdot \exp(-c(k(1)^{s/2} + k(2)^{s/2})) < \infty.\end{aligned}$$

Lemma 11.1.5 is proved by the Borel-Cantelli lemma.

Let $R_{\mathbf{l}} = [t_{\mathbf{l}}(1), t_{\mathbf{l+1}}(1)] \times [t_{\mathbf{l}}(2), t_{\mathbf{l+1}}(2)]$. For any $A \in \mathbf{n}\bigcup_{\delta>0}\mathcal{A}_\delta$, define

$$A^* = \bigcup\{R_{\mathbf{l}} : R_{\mathbf{l}} \cap A \neq \emptyset\}, \quad A_* = \bigcup\{R_{\mathbf{l}} : R_{\mathbf{l}} \subset A\}.$$

It is clear that $A_* \subset A \subset A^*$ and if $\mathbf{t_k} < \mathbf{n} \leq \mathbf{t_{k+1}}$ we have $|A^*\backslash A_*| \leq c|\mathbf{n}|(\varepsilon_1 + \varepsilon_2)$, where ε_i are defined by (11.1.14). Similarly, we have

Lemma 11.1.6. *For $t_k \in G_\beta$, we have*

$$\begin{aligned}\sup_{A\in\mathcal{A}} \max_{\mathbf{t_k}<\mathbf{n}\leq\mathbf{t_{k+1}}} &\Big|\sum_{\mathbf{j}} |(\mathbf{t_k}A(2^{-a})\backslash(\mathbf{t_k}A(2^{-a}))_*) \cap C_{\mathbf{j}}|X'_{\mathbf{j}}\Big| \\ &= O(|\mathbf{t_k}|^{1/2}(\log|\mathbf{t_k}|)^{-1/4}) \quad a.s. \quad as\ |\mathbf{k}| \to \infty.\end{aligned}$$

3. Blocking To estimate

$$\sup_{A\in\mathcal{A}} \max_{\mathbf{t_k}<\mathbf{n}\leq\mathbf{t_{k+1}}} \Big|\sum_{\mathbf{j}} |(\mathbf{t_k}A(2^{-a}))_* \cap C_{\mathbf{j}}|X_{\mathbf{j}}\Big|$$

is the key to the remainder of the proof. Let $\mathbf{u} = (u(1), u(2)) \geq \mathbf{0}$, $H_{\mathbf{u}} = \{\mathbf{v} \in \mathbb{N}^2, \mathbf{t_u} + 1 \leq \mathbf{v} \leq \mathbf{t_{u+1}}\}, 0 < \rho < \beta$. Denote

$$\begin{aligned}L &= \{\mathbf{u} : H_{\mathbf{u}} \subset G_\rho\}, \quad H = \bigcup_{\mathbf{u}\in L} H_{\mathbf{u}}, \\ \Delta_{\mathbf{u}} &= \bigcup_{i=1,2}\Big\{\mathbf{v} \in H_{\mathbf{u}} : t_{\mathbf{u+1}}(i) \\ &\quad - [\exp(\mathbf{u}(i)^{s/4}/4)] \leq v_i \leq t_{\mathbf{u+1}}(i)\Big\}.\end{aligned}$$

For $\mathbf{t_k} \in G_\beta$, define $\mathbf{t}_{\mathbf{k}}^{(p)} = (\mathbf{t}_{\mathbf{k}}^{(p)}(1), \mathbf{t}_{\mathbf{k}}^{(p)}(2)), p = 1, 2$, as follows:

$$\mathbf{t}_{\mathbf{k}}^{(p)}(m) = \delta_{m,p} \min_{\substack{\mathbf{v} \in H, \\ v_l = n_l, \text{if } l \neq m}} v_m + (1 - \delta_{m,p}) n_m, \quad m = 1, 2, \tag{11.1.16}$$

where $\delta_{m,p} = 1$ if $m = p$; $\delta_{m,p} = 0$ if $m \neq p$. Put

$$\begin{aligned} I_1 &= [\mathbf{0}, \mathbf{t}_{\mathbf{k}}^{(1)}] \cup [\mathbf{0}, \mathbf{t}_{\mathbf{k}}^{(2)}], \\ I_2 &= \bigcup_{(t_k^{(1)}(1), t_k^{(2)}(2)) \leq \mathbf{u} \leq \mathbf{k}} H_{\mathbf{u}} \backslash \Delta_{\mathbf{u}}, \\ I_3 &= \bigcup_{\mathbf{u} \in L, \mathbf{u} < \mathbf{k}} \Delta_{\mathbf{u}}. \end{aligned}$$

We have

$$\begin{aligned} &\Big|\sum_{\mathbf{j}} |(\mathbf{t_k} A(2^{-a}))_* \cap C_{\mathbf{j}}| (X_{\mathbf{j}} - Y_{\mathbf{j}})\Big| \\ &\quad \leq \Big|\sum_{\mathbf{j}} |(\mathbf{t_k} A(2^{-a}))_* \cap I_1 \cap C_{\mathbf{j}}| X_{\mathbf{j}}\Big| \\ &\qquad + \Big|\sum_{\mathbf{j}} |(\mathbf{t_k} A(2^{-a}))_* \cap I_1 \cap C_{\mathbf{j}}| Y_{\mathbf{j}}\Big| \\ &\qquad + \Big|\sum_{\mathbf{j}} |(\mathbf{t_k} A(2^{-a}))_* \cap I_2 \cap C_{\mathbf{j}}| (X_{\mathbf{j}} - Y_{\mathbf{j}})\Big| \\ &\qquad + \Big|\sum_{\mathbf{j}} |(\mathbf{t_k} A(2^{-a}))_* \cap I_3 \cap C_{\mathbf{j}}| X_{\mathbf{j}}\Big| \\ &\qquad + \Big|\sum_{\mathbf{j}} |(\mathbf{t_k} A(2^{-a}))_* \cap I_3 \cap C_{\mathbf{j}}| Y_{\mathbf{j}}\Big| \end{aligned}$$

where $Y_{\mathbf{j}}$ will be specified below.

Lemma 11.1.7. *For $t_k \in G_\beta$, we have*

$$\begin{aligned} &\sup_{A \in \mathcal{A}} \max_{\mathbf{t_k} < \mathbf{n} \leq \mathbf{t_{k+1}}} \Big|\sum_{\mathbf{j}} |(\mathbf{t_k} A(2^{-a}))_* \cap I_1 \cap C_{\mathbf{j}}| X_{\mathbf{j}}| \\ &\quad = O(|\mathbf{t_k}|^{\frac{1}{2}} (\log |\mathbf{t_k}|)^{-\sigma_1}) \quad a.s. \quad as\ |\mathbf{k}| \to \infty. \end{aligned}$$

Proof. It is easy to see that

$$\begin{aligned} &\sup_{A \in \mathcal{A}} \max_{\mathbf{t_k} < \mathbf{n} \leq \mathbf{t_{k+1}}} \Big|\sum_{\mathbf{j}} |(\mathbf{t_k} A(2^{-a}))_* \cap I_1 \cap C_{\mathbf{j}}| X_{\mathbf{j}}\Big| \\ &\quad \leq \sum_{\mathbf{l} < \mathbf{k}} \Big|\sum_{\mathbf{j}} |R_{\mathbf{l}} \cap I_1 \cap C_{\mathbf{j}}| X_{\mathbf{j}}\Big|. \end{aligned}$$

Since $t_{\mathbf{k}}^{(1)}(1) \le ct_{\mathbf{k}}(2)^{\rho}$, $t_{\mathbf{k}}^{(2)}(2) \le ct_{\mathbf{k}}(1)^{\rho}$, we have

$$|t_{\mathbf{k}}^{(1)}|/|t_{\mathbf{k}}| \le c|t_{\mathbf{k}}|^{-\beta(\beta-\rho)/2} \quad \text{and} \quad |t_{\mathbf{k}}^{(2)}|/|t_{\mathbf{k}}| \le c|t_{\mathbf{k}}|^{-\beta(\beta-\rho)/2}$$

for $t_{\mathbf{k}} \in G_{\beta}$. By the Markov inequality and Lemma 6.2.3 it follows that

$$\begin{aligned}
P\Big\{\sum_{\mathbf{l}<\mathbf{k}}\Big|\sum_{\mathbf{j}}|R_{\mathbf{l}}\cap I_1\cap C_{\mathbf{j}}|X_{\mathbf{j}}\Big| \ge |\mathbf{t_k}|^{1/2}(\log|\mathbf{t_k}|)^{-\sigma_1}\Big\}&\\
\le \sum_{\mathbf{l}<\mathbf{k}} P\Big\{\Big|\sum_{\mathbf{j}}|R_{\mathbf{l}}\cap I_1\cap C_{\mathbf{j}}|X_{\mathbf{j}}\Big| \ge |\mathbf{t_k}|^{1/2}(\log|\mathbf{t_k}|)^{-\sigma_1}|\mathbf{k}|^{-1}\Big\}&\\
\le c\sum_{\mathbf{l}<\mathbf{k}}|R_{\mathbf{l}}\cap I_1|^{1+\delta/2}\Big/\Big(|\mathbf{t_k}|^{1+\delta/2}((\log|\mathbf{t_k}|)^{-\sigma_1}|\mathbf{k}|^{-1})^{2+\delta}\Big)&\\
\le c|\mathbf{t_k}|^{-\frac{\beta(\beta-\rho)}{2}(1+\frac{\delta}{2})}((\log|\mathbf{t_k}|)^{\sigma_1}|\mathbf{k}|)^{2+\delta}.& \qquad (11.1.17)
\end{aligned}$$

Thus Lemma 11.1.7 holds true by the Borel-Cantelli lemma.

Lemma 11.1.8. *For $t_k \in G_{\beta}$, we have*

$$\begin{aligned}
\sup_{A\in\mathcal{A}}\max_{\mathbf{t_k}<\mathbf{n}\le\mathbf{t_{k+1}}}&\Big|\sum_{\mathbf{j}}|(\mathbf{t_k}A(2^{-a}))_*\cap I_3\cap C_{\mathbf{j}}|X_{\mathbf{j}}\Big|\\
&= O(|\mathbf{t_k}|^{1/2}(\log|\mathbf{t_k}|)^{-\sigma_1}) \quad a.s. \quad as\ |\mathbf{k}|\to\infty. \qquad (11.1.18)
\end{aligned}$$

Proof. As in the proof of Lemma 11.1.7, we have

$$\begin{aligned}
\sup_{A\in\mathcal{A}}\max_{\mathbf{t_k}<\mathbf{n}\le\mathbf{t_{k+1}}}&\Big|\sum_{\mathbf{j}}|(\mathbf{t_k}A(2^{-a}))_*\cap I_3\cap C_{\mathbf{j}}|X_{\mathbf{j}}\Big|\\
&\le \sum_{\mathbf{l}<\mathbf{k}}\Big|\sum_{\mathbf{j}}|R_{\mathbf{l}}\cap I_3\cap C_{\mathbf{j}}|X_{\mathbf{j}}\Big|.
\end{aligned}$$

Denote $\Delta_{\mathbf{u}} = \Delta_{\mathbf{u}}(1)\bigcup\Delta_{\mathbf{u}}(2)$, where $\Delta_{\mathbf{u}}(1)$ and $\Delta_{\mathbf{u}}(2)$ are disjoint rectangles, such that

$$\begin{aligned}
|\Delta_{\mathbf{u}}(1)| =&[\exp(u(1)^{s/4}/4)]\Big([\exp((u(2)+1)^{s/4})]\\
&-[\exp((u(2))^{s/4})]\Big),\\
|\Delta_{\mathbf{u}}(2)| =&[\exp(u(2)^{s/4}/4)]\Big([\exp((u(1)+1)^{s/4})]\\
&-[\exp((u(1))^{s/4})]\Big).
\end{aligned}$$

We have

$$\sum_{\substack{\mathbf{u}\in L\\ \mathbf{u}<\mathbf{k}}} |\Delta_{\mathbf{u}}(1)| \le c\mathbf{k}(1)^{1-s/4}\mathbf{t}_{\mathbf{k}}(1)^{1/4}\mathbf{t}_{\mathbf{k}}(2),$$

$$\sum_{\substack{\mathbf{u}\in L\\ \mathbf{u}<\mathbf{k}}} |\Delta_{\mathbf{u}}(2)| \le c\mathbf{k}(2)^{1-s/4}\mathbf{t}_{\mathbf{k}}(2)^{1/4}\mathbf{t}_{\mathbf{k}}(1).$$

Then, similarly to (11.1.17) in Lemma 11.1.7 we have

$$\begin{aligned} P\Big\{\sum_{\mathbf{l}<\mathbf{k}}\Big|\sum_{\mathbf{j}} |R_{\mathbf{l}}\cap I_3\cap C_{\mathbf{j}}|X_{\mathbf{j}}\Big| \ge |\mathbf{t}_{\mathbf{k}}|^{\frac{1}{2}}(\log|\mathbf{t}_{\mathbf{k}}|)^{-\sigma_1}\Big\} \\ \le c\sum_{\mathbf{l}<\mathbf{k}} |R_{\mathbf{l}}\cap I_3|^{1+\delta/2}\Big/\Big(|\mathbf{t}_{\mathbf{k}}|^{1+\delta/2}((\log|\mathbf{t}_{\mathbf{k}}|)^{-\sigma_1}|\mathbf{k}|^{-1})^{2+\delta}\Big) \\ \le c|\mathbf{t}_{\mathbf{k}}|^{-(3\beta/8)(1+\delta/2)}((\log|\mathbf{t}_{\mathbf{k}}|)^{\sigma_1}|\mathbf{k}|)^{2+\delta} \\ \cdot\Big(\mathbf{k}(1)^{1-s/4}+\mathbf{k}(2)^{1-s/4}\Big)^{1+\delta/2}. \end{aligned} \tag{11.1.19}$$

Lemma 11.1.8 holds true by the Borel-Cantelli lemma .

4. Constructions of $\{X_{\mathbf{j}}\}$ and $\{Y_{\mathbf{j}}\}$.

In order to construct random fields $\{X_{\mathbf{j}}\}$ and $\{Y_{\mathbf{j}}\}$, we quote the following results without any proofs. Denote

$$h_u = \mathrm{Card}((H_{\mathbf{u}}\backslash\Delta_{\mathbf{u}})\cap \mathbb{N}^2), \quad \overline{X}_{\mathbf{u}} = \sum_{\mathbf{j}\in H_{\mathbf{u}}\backslash\Delta_{\mathbf{u}}} X_{\mathbf{j}}/h_{\mathbf{u}}^{1/2}.$$

Proposition 11.1.1. (Berkes, Morrow 1981) Under the conditions of Theorem 11.1.1, there exists a t, $0<t<1$ such that for any $|x|\le h_{\mathbf{u}}^t$ we have

$$|E\exp(ix\overline{X}_{\mathbf{u}}) - \exp(\frac{1}{2}\sigma^2x^2)| \le ch_{\mathbf{u}}^{-t}. \tag{11.1.20}$$

Proposition 11.1.2. (Berkes, Philipp 1979) Let $\{X_k, k\ge 1\}$ be a sequence of random variables with values in R^{d_k}, and let $\{\mathcal{F}_k, k\ge 1\}$ be a non-decreasing sequence of σ-fields such that X_k is $\mathcal{F}_k$-measurable . Finally, let $\{G_k, k\ge 1\}$ be a sequence of probability distributions on R^{d_k} with characteristic functions $g_k(\mathbf{u}), \mathbf{u}\in R^{d_k}$, respectively. Suppose that for some non-negative numbers λ_k, δ_k and $T_k\ge 10^8 d_k$

$$E|E\{\exp(i\langle\mathbf{u}, X_k\rangle)|\mathcal{F}_k\} - g_k(\mathbf{u})| \le \lambda_k \tag{11.1.21}$$

for all $\mathbf{u}$ with $|\mathbf{u}| \leq T_k$ and

$$G_k\{\mathbf{u} : |\mathbf{u}| > \frac{1}{4}T_k\} \leq \delta_k. \tag{11.1.22}$$

Then without changing its distribution we can redefine the sequence $\{X_k, k \geq 1\}$ on a richer probability space together with a sequence $\{Y_k, k \geq 1\}$ of independent random variables such that Y_k has distribution G_k and

$$P\{|X_k - Y_k| \geq \alpha_k\} \leq \alpha_k \qquad k = 1, 2, \cdots \tag{11.1.23}$$

where $\alpha_1 = 1$ and

$$\alpha_k = 16 d_k T_k^{-1} \log T_k + 4\lambda_k^{1/2} T_k^{d_k} + \delta_k. \tag{11.1.24}$$

Proposition 11.1.3. (Berkes, Philipp 1979) Let $S_i, i = 1, 2, 3$ be separable Banach spaces. Let F be a distribution on $S_1 \times S_2$ and let G be a distribution on $S_2 \times S_3$ such that the second marginal of F equals to the first marginal of G. Then there exist a probability space and three random variables $Z_i, i = 1, 2, 3$ defined on it such that the joint distribution of Z_1 and Z_2 is F and the joint distribution of Z_2 and Z_3 is G.

Let $(t_{k_0}^{(1)}(1), t_{k_0}^{(2)}(2))$ be large enough such that for any $\mathbf{u} \in L_0 := \{\mathbf{u} \in L, \mathbf{u} \geq (t_{k_0}^{(1)}(1), t_{k_0}^{(2)}(2))\}$, there exists a $\rho' > 0$ satisfying

$$\begin{aligned} &t_{\mathbf{u+1}}(1) - t_{\mathbf{u}}(1) - [\exp(\mathbf{u}(1)^{s/4}/4)] \\ &\qquad \geq \left(t_{\mathbf{u+1}}(2) - t_{\mathbf{u}}(2) - [\exp(\mathbf{u}(2)^{s/4}/4)]\right)^{\rho'}, \\ &t_{\mathbf{u+1}}(2) - t_{\mathbf{u}}(2) - [\exp(\mathbf{u}(2)^{s/4}/4)] \\ &\qquad \geq \left(t_{\mathbf{u+1}}(1) - t_{\mathbf{u}}(1) - [\exp(\mathbf{u}(1)^{s/4}/4)]\right)^{\rho'}. \end{aligned}$$

Put ψ is a one to one correspondence from $\{1, 2, \cdots\}$ to L_0, denote $\psi(l) = (\psi_1(l), \psi_2(l))$. By Proposition 11.1.1, there exists a t, $0 < t < 1$ such that for any $|x| \leq h_{\psi(l)}^t$ we have

$$|E \exp(ix\overline{X}_{\psi(l)}) - \exp(-\sigma^2 x^2/2)| \leq c h_{\psi(l)}^{-t}. \tag{11.1.25}$$

On the other hand, since $h_{\psi(l)} = O(|t_{\psi(l)}|)$, for any $|x| \leq h_{\psi(l)}^t$ we have

$$\begin{aligned} \lambda_l(x) &:= E|E\{\exp(ix\overline{X}_{\psi(l)})|\overline{X}_{\psi(1)}, \cdots, \overline{X}_{\psi(l-1)}\} - \exp(\sigma^2 x^2/2)| \\ &\leq E|E\{\exp(ix\overline{X}_{\psi(l)})|\overline{X}_{\psi(1)}, \cdots, \overline{X}_{\psi(l-1)}\} - E\exp(ix\overline{X}_{\psi(l)})| \\ &\quad + |E\exp(ix\overline{X}_{\psi(l)}) - \exp(\sigma^2 x^2/2)| \\ &\leq c h_{\psi(l)}^{-t} + 2\pi\varphi([\exp(\psi_1(l)^{s/4}/4)] \wedge [\exp(\psi_2(l)^{s/4}/4)]) \\ &\leq c|t_{\psi(l)}|^{-t_0} \end{aligned}$$

by the property of φ-mixing, where $t_0 = \min(t, \rho q/8)$.

Put $t' < t_0/2$, $T_l = |t_{\psi(l)}|^{t'}$, $\lambda_l = |t_{\psi(l)}|^{-t_0}$, $\delta_l = N(0,\sigma^2)\{x : |x| > T_l/4\}$, $\alpha_l = 16T_l^{-1}\log T_l + 4\lambda_l^{1/2}T_l + \delta_l$. By Proposition 11.1.2, without changing its distribution we can redefine the sequence $\{\overline{X}_{\psi(l)}\}$ on a richer probability space together with a sequence $\{\overline{Y}_{\psi(l)}\}$ of independent normal random variables such that $E\overline{Y}^2_{\psi(l)} = \sigma^2$ and

$$P\{|\overline{X}_{\psi(l)} - \overline{Y}_{\psi(l)}\}| \geq \alpha_l\} \leq \alpha_l. \tag{11.1.26}$$

Moreover, by Proposition 11.1.3, there exist two random fields $\{X_{\mathbf{j}}, \mathbf{j} \in \mathbb{N}^2\}$ and $\{Y_{\mathbf{j}}, \mathbf{j} \in \mathbb{N}^2\}$ on a richer probability space such that the distribution of $\{X_{\mathbf{j}}, \mathbf{j} \in \mathbb{N}^2\}$ is not changed and $\{Y_{\mathbf{j}}, \mathbf{j} \in \mathbb{N}^2\}$ are independent centered normal random variables with $EY_{\mathbf{j}}^2 = \sigma^2$, and further

$$P\Big\{h_{\psi(l)}^{-1/2}\Big|\sum_{\mathbf{j}\in H_{\psi(l)}\backslash\Delta_{\psi(l)}}(X_{\mathbf{j}} - Y_{\mathbf{j}})\Big| \geq \alpha_l\Big\} \leq \alpha_l. \tag{11.1.27}$$

Obviously we have $\sum_l \alpha_l < \infty$, which implies

$$\Big|\sum_{\mathbf{j}\in H_{\psi(l)}\backslash\Delta_{\psi(l)}}(X_{\mathbf{j}} - Y_{\mathbf{j}})\Big| = O(\alpha_l h_{\psi(l)}^{1/2}) \quad \text{a.s.} \tag{11.1.28}$$

Note that $I_2 = \bigcup_{(t_k^{(1)}(1), t_k^{(2)}(2)) \leq \mathbf{u} < \mathbf{k}} H_{\mathbf{u}}\backslash\Delta_{\mathbf{u}}$. Take $\mathbf{t_k} \in G_\beta$ such that $t_{\mathbf{k}}^{(1)}(1) \geq t_{\mathbf{k}_0}^{(1)}(1), t_{\mathbf{k}}^{(2)}(2) \geq t_{\mathbf{k}_0}^{(2)}(2)$. Then for every $A \in \mathcal{A}$ and some $t'' > 0$ we have

$$\begin{aligned}
&\Big|\sum_{\mathbf{j}}|(\mathbf{t_k}A(2^{-a}))_* \cap I_2 \cap C_{\mathbf{j}}|(X_{\mathbf{j}} - Y_{\mathbf{j}})\Big| \\
&\quad \leq \sum_{(t_k^{(1)}(1), t_k^{(2)}(2)) \leq \mathbf{u} < \mathbf{k}} \Big|\sum_{\mathbf{j}\in H_{\mathbf{u}}\backslash\Delta_{\mathbf{u}}}(X_{\mathbf{j}} - Y_{\mathbf{j}})\Big| \\
&\quad \leq \sum_{\psi(l) < \mathbf{k}, \psi_i(l) \geq t_{\mathbf{k}}^{(i)}(i), i=1,2} \alpha_l h_{\psi(l)}^{1/2} \\
&\quad \leq c|t_{\mathbf{k}}|^{1/2 - t''}. \quad \text{a.s.}
\end{aligned} \tag{11.1.29}$$

For the random field $\{Y_{\mathbf{j}}\}$ which defined in (11.1.27), we define $Y'_{\mathbf{j}}$ like

$X'_{\mathbf{j}}$, then for $t_{\mathbf{k}} < n \le t_{\mathbf{k+1}}$, $A \in \mathcal{A}$ we have

$$\Big|\sum_{\mathbf{j}} |\mathbf{n}A \cap C_{\mathbf{j}}|(X_{\mathbf{j}} - Y_{\mathbf{j}})|$$
$$\le \Big|\sum_{\mathbf{j}} |\mathbf{n}A \cap C_{\mathbf{j}}|(X_{\mathbf{j}} - X'_{\mathbf{j}})\Big|$$
$$+ \Big|\sum_{\mathbf{j}} |\mathbf{n}A \cap C_{\mathbf{j}}|(X'_{\mathbf{j}} - Y'_{\mathbf{j}})\Big| + \Big|\sum_{\mathbf{j}} |\mathbf{n}A \cap C_{\mathbf{j}}|(Y'_{\mathbf{j}} - Y_{\mathbf{j}})\Big|,$$

where

$$\Big|\sum_{\mathbf{j}} |\mathbf{n}A \cap C_{\mathbf{j}}|(X'_{\mathbf{j}} - Y'_{\mathbf{j}})\Big|$$
$$\le \Big|\sum_{\mathbf{j}} \big|\,|\mathbf{n}A \cap C_{\mathbf{j}}| - |\mathbf{n}A(2^{-b}) \cap C_{\mathbf{j}}|\,\big|(X'_{\mathbf{j}} - Y'_{\mathbf{j}})\Big|$$
$$+ \Big|\sum_{\mathbf{j}} \big|\,|\mathbf{n}A(2^{-b}) \cap C_{\mathbf{j}}| - |\mathbf{n}A(2^{-a}) \cap C_{\mathbf{j}}|\,\big|(X'_{\mathbf{j}} - Y'_{\mathbf{j}})\Big|$$
$$+ \Big|\sum_{\mathbf{j}} \big|\,|\mathbf{n}A(2^{-a}) \cap C_{\mathbf{j}}| - |t_{\mathbf{k}}A(2^{-a}) \cap C_{\mathbf{j}}|\,\big|(X'_{\mathbf{j}} - Y'_{\mathbf{j}})\Big|$$
$$+ \Big|\sum_{\mathbf{j}} |(t_{\mathbf{k}}A(2^{-a}) \backslash (t_{\mathbf{k}}A(2^{-a}))_* \cap C_{\mathbf{j}}|(X'_{\mathbf{j}} - Y'_{\mathbf{j}})\Big|$$
$$+ \Big|\sum_{\mathbf{j}} |(t_{\mathbf{k}}A(2^{-a}))_* \cap C_{\mathbf{j}}|(X_{\mathbf{j}} - Y_{\mathbf{j}})\Big|$$
$$+ \sum_{\mathbf{j}} |X_{\mathbf{j}} - X'_{\mathbf{j}}| + \sum_{\mathbf{j}} |Y_{\mathbf{j}} - Y'_{\mathbf{j}}| \tag{11.1.30}$$

and

$$\Big|\sum_{\mathbf{j}} |(t_{\mathbf{k}}A(2^{-a}))_* \cap C_{\mathbf{j}}|(X_{\mathbf{j}} - Y_{\mathbf{j}})\Big|$$
$$\le \Big|\sum_{\mathbf{j}} |(t_{\mathbf{k}}A(2^{-a}))_* \cap I_1 \cap C_{\mathbf{j}}|(X_{\mathbf{j}} - Y_{\mathbf{j}})\Big|$$
$$+ \Big|\sum_{\mathbf{j}} |(t_{\mathbf{k}}A(2^{-a}))_* \cap I_2 \cap C_{\mathbf{j}}|(X_{\mathbf{j}} - Y_{\mathbf{j}})\Big|$$
$$+ \Big|\sum_{\mathbf{j}} |(t_{\mathbf{k}}A(2^{-a}))_* \cap I_3 \cap C_{\mathbf{j}}|(X_{\mathbf{j}} - Y_{\mathbf{j}})\Big|. \tag{11.1.31}$$

It is clear that Lemmas 11.1.2-11.1.8 also hold true for the independent normal random field $\{Y_{\mathbf{j}}, \mathbf{j} \in \mathbb{N}^2\}$. Then Theorem 11.1.1 is proved from these lemmas and (11.1.29)–(11.1.31).

11.2 Strong approximations for α-mixing random fields

The strong approximation results for α-mixing random fields were given by Strittmatter (1990) for a smaller class of sets with $N(\varepsilon) \le c\varepsilon^{-u}$ for some $u > 0$ and the condition "$\mathbf{n} \in G_\beta$" was left out.

Theorem 11.2.1. *Let $\{X_{\mathbf{j}}, \mathbf{j} \in \mathbb{N}^d\}$ be a weakly stationary α-mixing random field of R^q-valued random vectors with $EX_{\mathbf{j}} = 0$. Put $\|X_{\mathbf{j}}\|^2 = (X_{\mathbf{j}1}^2 + \cdots + X_{\mathbf{j}q}^2)$ and suppose that there are positive constants $C_i > 0, i = 1, \cdots, 4$, such that*

$$E\|X_{\mathbf{j}}\|^{2+\delta} \le C_1, \quad \text{for every } \mathbf{j} \in \mathbb{N}^d \text{ and some } \delta > 0, \tag{11.2.1}$$

$$\alpha(t) \le C_2 t^{-s} \quad \text{for some } s > 1 + 2(17/2)^{d-1}/(1 \wedge \delta), \tag{11.2.2}$$

$$N(\varepsilon) \le C_3 \varepsilon^{-u} \quad \text{for some } u < \frac{\delta s - 4 - 2\delta}{d(4+\delta)} - 2, \tag{11.2.3}$$

$$\begin{aligned} b(\varepsilon) &:= \sup\{n^{-d}|((\mathbf{n}A)^{n\varepsilon} \cap (\mathbf{n}A^c)^{n\varepsilon})| : A \in \bigcup_{\eta>0} \mathcal{A}_\eta,\ n \ge 1/\varepsilon\} \\ &\le C_4\, \varepsilon^h \qquad \text{for some } 0 < h \le 1. \end{aligned} \tag{11.2.4}$$

Put

$$r(\mathbf{j}) = \operatorname{Cov}(X_{\mathbf{j}}, X_{\mathbf{j}}).$$

Then the series

$$T = \sum_{\mathbf{j} \in \mathbb{Z}^d} r(\mathbf{j})$$

converges absolutely and T is a non-negative definite matrix. Furthermore, without changing its joint distribution, we can redefine the field $\{X_{\mathbf{k}}, \mathbf{k} \in \mathbb{N}^d\}$ on a richer probability space together with a field $\{Y_{\mathbf{k}}, \mathbf{k} \in \mathbb{N}^d\}$ of i.i.d. centered Gaussian random vectors with covariance matrix T such that for some $\gamma = \gamma(u, s, d, h, \delta, N(\varepsilon))$ we have

$$\sup_{A \in \mathcal{A}} \Big\| \sum_{\mathbf{j} \in \mathbb{N}^d} |\mathbf{n}A \cap C_{\mathbf{j}}|(X_{\mathbf{j}} - Y_{\mathbf{j}}) \Big\| = O(n^{d/2-\gamma}) \quad a.s. \tag{11.2.5}$$

The proof of Theorem 11.2.1 will need the following lemmas.

Lemma 11.2.1. *Let $\{X_{\mathbf{j}}, \mathbf{j} \in \mathbb{N}^d\}$ be an α-mixing random field with $EX_{\mathbf{j}} = 0$, $E\|X_{\mathbf{j}}\|^{2+\delta} \le C_1$ and*

$$C_0 = \sum_{r=1}^{\infty} r^{d-1}\alpha(r)^{\delta/(2+\delta)} < \infty. \tag{11.2.6}$$

Then for $0 \le d_{\mathbf{j}} \le 1$, $\mathbf{j} \in \mathbb{N}^d$, where only finitely many $\mathbf{j}$ such that $d_{\mathbf{j}} \ne 0$, we have

$$E\Big\|\sum_{\mathbf{j}\in\mathbb{N}^d} d_{\mathbf{j}}X_{\mathbf{j}}\Big\|^2 \le C\sum_{\mathbf{j}\in\mathbb{N}^d} d_{\mathbf{j}},$$

where $C = (1 + 15d3^dC_0)C_1^{2/(2+\delta)}$.

Proof. By Lemma 1.2.4 we have

$$\begin{aligned}\|EX_{\mathbf{j}}X_{\mathbf{k}}\| &\le 10\alpha(d(\mathbf{j},\mathbf{k}))^{\delta/(2+\delta)}\|X_{\mathbf{j}}\|_{2+\delta}\|X_{\mathbf{k}}\|_{2+\delta}\\ &= 10\alpha(d(\mathbf{j},\mathbf{k}))^{\delta/(2+\delta)}C_1^{2/(2+\delta)},\end{aligned}$$

where $d(\mathbf{j},\mathbf{k}) = \max_{1\le i\le d}|j_i - k_i|$. Lemma 11.2.1 is proved.

Lemma 11.2.2. *Let $\{X_{\mathbf{j}},\mathbf{j} \in \mathbb{N}^d\}$ be a weakly stationary α-mixing random field with $EX_{\mathbf{j}} = 0$, $\|X_{\mathbf{j}}\| \le M < \infty$ and satisfying (11.2.6). Let $D_1, D_2, \cdots, D_q, D_k = (\mathbf{n}_k - 2p\mathbf{1}, \mathbf{n}_k]$, be mutually disjoint cubes with $\mathbf{n}_k \in 2p\mathbb{N}^d$ for some fixed $p \in \mathbb{N}$. Let $0 \le d_{\mathbf{j}} \le 1$, $\mathbf{j} \in \mathbb{N}^d$. Put*

$$D = \bigcup_{k=1}^{q} D_k \quad F_k = \sum_{\mathbf{j}\in D_k\cap\mathbb{N}^d} d_{\mathbf{j}}, \quad F = \sum_{\mathbf{j}\in D} d_{\mathbf{j}}.$$

Then for every $K > 0$ we have

$$\begin{aligned}&P\Big\{\Big\|\sum_{\mathbf{j}\in D} d_{\mathbf{j}}X_{\mathbf{j}}\Big\| > 2^{d+1}K\Big\}\\ &\le c(K^{-2}M^2F^2\alpha(p) + p^{2d}C^{-2}M^4\alpha(p)\\ &\quad + \exp(-K^2/(8CF)) + \exp(-2^{-d-2}K/Mp^d),\end{aligned}$$

where C is defined as in Lemma 11.2.1.

This lemma is a generalization of Theorem 4 in Philipp (1984). The proof follows closely by his lines, and hence is not given here.

Proof of Theorem 11.2.1.

From (11.2.3) it follows that there exist positive numbers τ and ζ such that $\zeta < \frac{1}{2} - \frac{1+\tau}{2+\delta}$ and

$$u < \frac{\zeta s - 1}{d(\frac{1}{2} + \frac{1+\tau}{2+\delta})} - 2. \tag{11.2.7}$$

Define for $\mathbf{j} \in \mathbb{N}^d$

$$X'_{\mathbf{j}} = X_{\mathbf{j}} I(\|X_{\mathbf{j}}\| \le |\mathbf{j}|^{(1+\tau)/(2+\delta)}),$$
$$\overline{X}_{\mathbf{j}} = X'_{\mathbf{j}} - EX'_{\mathbf{j}}, \qquad X''_{\mathbf{j}} = X_{\mathbf{j}} - X'_{\mathbf{j}}.$$

For a constant β specified later define

$$\psi(m) = \sum_{k=1}^{m} k^{\beta} \quad m \in \mathbb{N}. \tag{11.2.8}$$

Assume that m and n are linked by

$$\psi(m) < n \le \psi(m+1). \tag{11.2.9}$$

For $\mathbf{r} = (r_1, \cdots, r_d) \in \mathbb{N}^d$, write

$$R_{\mathbf{r}} = \{(v_1, \cdots, v_d) \in R_+^d : \psi(r_i) < v_i \le \psi(r_i + 1), i \le d\}.$$

For $A \subset R_+^d \cap \mathcal{B}^d$ let

$$A_* = \bigcup_{R_{\mathbf{r}} \subset A} R_{\mathbf{r}}.$$

and for $A \in \mathcal{B}^d \cap [0,1]^d$

$$S_{\mathbf{n}}(A) = \sum_{\mathbf{j}} |\mathbf{n}A \cap C_{\mathbf{j}}| X_{\mathbf{j}},$$
$$\overline{S}_{\mathbf{n}}(A) = \sum_{\mathbf{j}} |\mathbf{n}A \cap C_{\mathbf{j}}| \overline{X}_{\mathbf{j}},$$
$$\overline{V}_n(A) = \sum_{\mathbf{j}} |(\mathbf{n}A \backslash (\mathbf{n}A)_*) \cap C_{\mathbf{j}}| \overline{X}_{\mathbf{j}}.$$

We introduce the following result of Berkes and Morrow (1981) without proof.

Proposition 11.2.1. Let $\{\xi_{\mathbf{j}}, \mathbf{j} \in \mathbb{N}^d\}$ be a weakly stationary α-mixing random field with $E\xi_{\mathbf{j}} = 0$, $E\|\xi_{\mathbf{j}}\|^{2+\delta} < \infty$, $\alpha(t) = O(t^{-d(1+\varepsilon)(1+2/\delta)})$. Denote

$$G_\theta = \bigcap_{k=1}^{d} \{\mathbf{j} \in \mathbb{N}^d : j_k \ge \prod_{l \ne k} j_l^{\theta}\}, \quad 0 < \theta < 1.$$

Then the series

$$\sigma^2 = E\xi_1^2 + 2\sum_{\mathbf{j}} E\xi_1\xi_{\mathbf{j}}$$

converges absolutely. Without loss of generality, assume that $\sigma^2 = 1$. Furthermore without changing the distribution of $\{\xi_{\mathbf{j}}\}$, we can redefine the sequence $\{\xi_{\mathbf{j}}, \mathbf{j} \in \mathbb{N}^d\}$ on a richer probability space together with a random field $\{\eta_{\mathbf{j}}, \mathbf{j} \in \mathbb{N}^d\}$ of i.i.d. centered Gaussian random variables with $E\eta_{\mathbf{j}}^2 = 1$ such that for any $\mathbf{n} \in G_\theta$ we have

$$\sup_{1 \le \mathbf{m} \le \mathbf{n}} \Big\| \sum_{\mathbf{k} \le \mathbf{m}} \xi_{\mathbf{k}} - \sum_{\mathbf{k} \le \mathbf{m}} \eta_{\mathbf{k}} \Big\| = O(|\mathbf{n}|^{1/2-\lambda}) \quad \text{a.s.} \tag{11.2.10}$$

where $\lambda > 0$ depends only on the field $\{\xi_{\mathbf{j}}\}$.

Now we take $\theta = 1/(8d-1)$ and denote

$$G = G_{1/(8d-1)}, \quad L = \{\mathbf{r} \in \mathbb{N}^d : (\psi(r_1), \cdots, \psi(r_d)) \in G\}.$$

By Proposition 11.2.1, we have a random field $\{Y_{\mathbf{j}}, \mathbf{j} \in \mathbb{N}^d\}$ of i.i.d. centered Gaussian random variables with $EY_{\mathbf{j}}^2 = 1$ such that

$$\sum_{\substack{\mathbf{r} \in L \\ \forall i, \psi(r_i) \le n}} \Big\| \sum_{\mathbf{j} \in R_{\mathbf{r}}} (X_{\mathbf{j}} - Y_{\mathbf{j}}) \Big\| = O(n^{d/2-\gamma_1}) \quad \text{a.s.} \tag{11.2.11}$$

for some $\gamma_1 > 0$. Denote

$$\begin{aligned}
Y'_{\mathbf{j}} &= Y_{\mathbf{j}} I(\|Y_{\mathbf{j}}\| \le |\mathbf{j}|^{(1+\tau)/(2+\delta)}), \\
T_{\mathbf{n}}(A) &= \sum_{\mathbf{j}} |\mathbf{n}A \cap C_{\mathbf{j}}| Y_{\mathbf{j}}, \\
T'_{\mathbf{n}}(A) &= \sum_{\mathbf{j}} |\mathbf{n}A \cap C_{\mathbf{j}}| Y'_{\mathbf{j}}, \\
U'_n(A) &= \sum_{\mathbf{j}} |(\mathbf{n}A \backslash (\mathbf{n}A)_*) \cap C_{\mathbf{j}}| Y'_{\mathbf{j}}.
\end{aligned}$$

Then for $\kappa > 0$(specified later on) we have

$$\begin{aligned}
&\Big\| \sum_{\mathbf{j}} |\mathbf{n}A \cap C_{\mathbf{j}}| (X_{\mathbf{j}} - Y_{\mathbf{j}}) \Big\| \\
&\quad = \|S_{\mathbf{n}}(A) - T_{\mathbf{n}}(A)\| \\
&\quad \le \|S_{\mathbf{n}}(A) - \overline{S}_{\mathbf{n}}(A)\| + \|\overline{S}_{\mathbf{n}}(A) - \overline{S}_{\mathbf{n}}(A(n^{-\kappa}))\| \\
&\qquad + \|\overline{V}_n(A(n^{-\kappa}))\| + \|T_{\mathbf{n}}(A) - T'_{\mathbf{n}}(A)\| \\
&\qquad + \|T'_{\mathbf{n}}(A) - T'_{\mathbf{n}}(A(n^{-\kappa}))\| + \|U'_n(A(n^{-\kappa}))\| \\
&\qquad + \sum_{1 \le \mathbf{j} \le n\mathbf{1}} (\|X_{\mathbf{j}} - \overline{X}_{\mathbf{j}}\| + \|Y_{\mathbf{j}} - Y'_{\mathbf{j}}\|) \\
&\qquad + \Big\| \sum_{\mathbf{j}} |(\mathbf{n}A(n^{-\kappa}))_* \cap C_{\mathbf{j}}| (X_{\mathbf{j}} - Y_{\mathbf{j}}) \Big\| \\
&\quad =: I_{11} + I_{12} + I_{13} + I_{21} + I_{22} + I_{23} + I_3 + I_4.
\end{aligned} \tag{11.2.12}$$

From Lemma 11.1.2 it follows that

$$I_{11} \vee I_{21} \vee I_3 = O(n^{d/(2+\delta)}). \tag{11.2.13}$$

From Lemma 11.1.3 it follows that

$$I_{12} \vee I_{22} = O(n^{d/2-\gamma_2}) \tag{11.2.14}$$

where $\gamma_2 = \kappa - d(1+\tau)/(2+\delta) - d/2 > 0$.

In order to estimate I_{13}, I_{23} and I_4, we need the following lemmas.

Lemma 11.2.3. *There exist $d\left(\frac{1}{2} + \frac{1+\tau}{2+\delta}\right) < \kappa \le d$, γ', $\gamma'' > 0$ such that for any $n \in \mathbb{N}$ we have*

$$P\{\sup(\|\overline{V}_n(A(n^{-\kappa}))\| : A(n^{-\kappa}) \in \mathcal{A}(n^{-\kappa})) > cn^{d/2-\gamma'}\} \le cn^{-1-\gamma''}.$$

Proof. Put $p = [n^{\zeta}]$, for fixed $A \in \mathcal{B}^d \cap [0,1]^d$ cover $(\mathbf{n}A)\backslash(\mathbf{n}A)_*$ with cubes in R_+^d with length of edges equal to $2p$ and of the form explained in Lemma 11.2.2; as there let D be the union of these cubes. For $\mathbf{j} \in \mathbb{N}^d$ let $d_{\mathbf{j}} = |((\mathbf{n}A)\backslash(\mathbf{n}A)_*) \cap C_{\mathbf{j}}|$. Then by (11.2.4) we have

$$\begin{aligned} F = \sum_{\mathbf{j}\in D} d_{\mathbf{j}} &= |(\mathbf{n}A)\backslash(\mathbf{n}A)_*| \\ &\le n^d b((\psi(m+1)-\psi(m))/n) \le cn^{d-h}(\psi(m+1)-\psi(m))^h \\ &\le cn^{d-h}n^{h\beta/(\beta+1)} = cn^{d-\gamma}, \end{aligned} \tag{11.2.15}$$

where $\gamma = h/(\beta+1)$. There is a $\gamma' > 0$ such that $\gamma' < \gamma/2$ and

$$\gamma' < d\left(\frac{1}{2} - \frac{1+\tau}{2+\delta} - \zeta\right). \tag{11.2.16}$$

Then by Lemma 11.2.1, (11.2.15) and Lemma 11.2.2, we have

$$\begin{aligned} &P\{\|\overline{V_n}(A)\| > cn^{d/2-\gamma'}\} \\ &\quad \le c\Big(n^{-d+2\gamma'}M^2F^2\alpha(p) + p^{2d}M^4\alpha(p) \\ &\qquad + \exp\Big(-\frac{n^{d-2\gamma'}}{cF}\Big) + \exp\Big(-\frac{cn^{d/2-\gamma'}}{Mp^d}\Big)\Big). \end{aligned}$$

Since $M = 2n^{d(1+\tau)/(2+\delta)}$, $p \sim n^{\zeta}$, we have

$$\begin{aligned}
&P\{\|\overline{V}_n(A)\| > cn^{d/2-\gamma'}\}\\
&\le c\Big(n^{-d+2\gamma'+2d\dfrac{1+\tau}{2+\delta}+2d-2\gamma-\zeta s}\\
&\quad + n^{2d\zeta+4d\dfrac{1+\tau}{2+\delta}-\zeta s}\\
&\quad + \exp\Big(-cn^{-2\gamma'+\gamma}\Big) + \exp\Big(-cn^{d/2-\gamma'-d\dfrac{1+\tau}{2+\delta}-d\zeta}\Big)\Big)\\
&\le cn^{-\zeta s+2\left(1+d\dfrac{1+\tau}{2+\delta}\right)}.
\end{aligned} \tag{11.2.17}$$

There exist θ and κ with $0 < \theta < \zeta s - 1 - d(1 + \dfrac{2(1+\tau)}{2+\delta})$, $d(\frac{1}{2} + \frac{1+\tau}{2+\delta}) < \kappa \le d$ and $\theta/\kappa = u$. We multiply both sides of (11.2.17) by $N(n^{-\kappa}) \le cn^{\kappa u} = cn^{\theta}$ and obtain

$$\begin{aligned}
&P\Big\{\sup_{A(n^{-\kappa})} \|\overline{V_n}(A(n^{-\kappa}))\| > cn^{d/2-\gamma'}\Big\}\\
&\le cn^{\theta-\zeta s+2(1+d\dfrac{1+\tau}{2+\delta})}\\
&= cn^{-1-\gamma''}.
\end{aligned}$$

The proof of Lemma 11.2.3 is completed.

Lemma 11.2.4. *If in (11.2.8)* $\beta \ge 6d$ *then*

$$P\Big\{\sum_{\mathbf{r}\notin L, \psi(r_i)\le n} \Big\|\sum_{\mathbf{j}\in R_{\mathbf{r}}} X_{\mathbf{j}}\Big\| > n^{d/2-1/16}\Big\} \le cm^{-2}, \tag{11.2.18}$$

where n *and* m *are linked by (11.2.9), hence by the Borel-Cantelli lemma*

$$\sum_{\mathbf{r}\notin L, \psi(r_i)\le n} \Big|\sum_{\mathbf{j}\in R_{\mathbf{r}}} X_{\mathbf{j}}\Big| = O(n^{d/2-1/16}) \quad a.s. \tag{11.2.19}$$

Proof. For $r \notin L$ with $\psi(r_i) \le n$, $i = 1, \cdots, d$ we have by definition of L for some $i \le d$, $\psi(r_i)^{8d} < \prod_{i=1}^{d} \psi(r_i) \le n^d$, hence $\psi(r_i) \le n^{1/8}$. And therefore

$$\mathrm{Card}(R_{\mathbf{r}} \cap \mathbb{N}^d) \le n^{1/8} \cdot n^{d-1} = n^{d-7/8}.$$

Furthermore we have

$$\begin{aligned}
&\mathrm{Card}\{t : t \in \mathbb{N}, \psi(t) \le n\} \le \mathrm{Card}\{t : t \in \mathbb{N}, ct^{\beta+1} \le n\} \le cn^{1/(\beta+1)},\\
&\mathrm{Card}\{\mathbf{r} : \mathbf{r} \in \mathbb{N}^d, \psi(r_i) \le n, i = 1, \cdots, d\} \le cn^{d/(\beta+1)}.
\end{aligned}$$

By Lemma 11.2.1 and the Chebyshev inequality this implies

$$\begin{aligned}
&P\Big\{ \sum_{\mathbf{r}\notin L, \psi(r_i)\le n} \Big\| \sum_{\mathbf{j}\in R_{\mathbf{r}}} X_{\mathbf{j}} \Big\| > n^{d/2-1/16} \Big\} \\
&\le cn^{d/(\beta+1)} \max_{\mathbf{r}\notin L, \psi(r_i)\le n} P\Big\{ \Big\| \sum_{\mathbf{j}\in R_{\mathbf{r}}} X_{\mathbf{j}} \Big\| \ge cn^{\frac{d}{2}-\frac{1}{16}-\frac{d}{\beta+1}} \Big\} \\
&\le cn^{\frac{d}{\beta+1}} \cdot n^{d-\frac{7}{8}-d+\frac{1}{8}+\frac{2d}{\beta+1}} \\
&= cn^{-\frac{3}{4}+\frac{3d}{\beta+1}} \\
&\le cm^{-2}.
\end{aligned}$$

Lemma 11.2.4 is proved.

From Lemma 11.2.3 it follows that $I_{13} \vee I_{23} = O(n^{d/2-\gamma'})$. Finally since

$$\begin{aligned}
I_4 \le &\sum_{\mathbf{r}\in L, \psi(r_i)\le n} \Big\| \sum_{\mathbf{j}\in R_{\mathbf{r}}} (X_{\mathbf{j}} - Y_{\mathbf{j}}) \Big\| \\
&+ \sum_{\mathbf{r}\notin L, \psi(r_i)\le n} \Big\| \sum_{\mathbf{j}\in R_{\mathbf{r}}} (X_{\mathbf{j}} - Y_{\mathbf{j}}) \Big\|,
\end{aligned}$$

we have $I_4 = O(n^{d/2-\gamma'})$ a.s. by Lemma 11.2.4 and (11.2.11). Therefore the proof of Theorem 11.2.1 is completed.

Part IV Statistics of a Dependent Sample

The limit behavior of various kinds of statistics with a dependent random sample have been studied by many authors since the sixties. We shall introduce some of them in this part. We first introduce weak convergence and strong approximations of an empirical process with a mixing dependent sample in Chapter 12. The limit behavior of U-statistics, estimations of error variance in a linear model and estimations of density function with a mixing dependent sample are discussed in Chapter 13. We investigate in Chapter 14 the asymptotic fluctuation behavior of sums of other kinds of dependent random variables, such as lacunary trigonometric series, a Gaussian sequence and the additive functional of a Markov process.

Chapter 12 Empirical Processes

Let $\{X_n, n \geq 1\}$ be a sequence of random variables with a common distribution function F. Then the empirical distribution function F_n of $X_1, \cdots, X_n$ is defined by $F_n(t) = n^{-1}\sum_{i=1}^{n} I(X_i \leq t)$, $-\infty < x < \infty$. The nth empirical process β_n is defined by

$$\beta_n(t) = \sqrt{n}(F_n(t) - F(t)), \quad -\infty < t < \infty. \tag{12.0.1}$$

If F is continuous, then $U_n := F(X_n)$ for all $n \geq 1$ are uniformly distributed over [0,1]. The corresponding empirical process is

$$\alpha_n(t) = \sqrt{n}(E_n(t) - t), \quad 0 \leq t \leq 1, \tag{12.0.2}$$

where the empirical distribution function

$$E_n(t) = F_n(\text{inv}F(t)) = \frac{1}{n}\sum_{i=1}^{n} I(U_i \leq t), \quad 0 \leq t \leq 1. \tag{12.0.3}$$

Thus, in term of $U_i = F(X_i)$, $i = 1, \cdots, n$, we have for any continuous F

$$\{\beta_n(\text{inv}F(t)), 0 \leq t \leq 1\} = \{\alpha_n(t), 0 \leq t \leq 1\}, \quad n = 1, 2, \cdots. \tag{12.0.4}$$

This implies that all theorems proved for α_n will hold automatically for β_n as well, simply by letting $y = F(x)$ in (12.0.4). So we shall mainly concern with uniform empirical process α_n in this Chapter.

In the independent case, it is well-known that

$$\alpha_n \Longrightarrow B \quad \text{as } n \to \infty \tag{12.0.5}$$

where B is a Brownian bridge. The strong approximations of $\{\alpha_n(\cdot)\}$ by a sequence of Brownian bridges were discussed by Komlós-Major-Tusnády (1975). They showed that without changing the distribution of $\{\alpha_n(t), n \geq 1\}$, one can redefine the $\{\alpha_n(t)\}$ on a richer probability space together with a sequence $\{B_n\}$ of independent Brownian bridges such that

$$\sup_{0 \leq t \leq 1} |\alpha_n(t) - B_n(t)| = O(n^{-1/2}\log n) \quad \text{a.s..} \tag{12.0.6}$$

For a strictly stationary sequence $\{U_n, n \geq 1\}$, the conditions, under which (12 0.5) and (12.0.6) also hold true, have been given by many mathematicians. This chapter is organized as follows. We shall introduce the best results of weak convergence of empirical processes when the sample is mixing dependent in Section 12.1. The weighted weak convergence for empirical processes of α-mixing and ρ-mixing sequences will be established in Section 12.2. The strong approximations for empirical processes with a mixing sample by Gaussian processes will be introduced in Section 12.3 and the moduli of continuity of empirical processes when the sample is mixing dependent will be studied in Section 12.4.

12.1 Weak convergence

Let $\{U_n, n \geq 1\}$ be a sequence of strictly stationary random variables uniformly distributed over $[0,1]$, denote

$$R(s,t) = s \wedge t - st + \sum_{k=2}^{\infty} E\{[I(U_k \leq s) - s][I(U_1 \leq t) - t] + [I(U_1 \leq s) - s][I(U_k \leq t) - t]\}. \tag{12.1.1}$$

When $\{U_n\}$ is α-mixing, the series in (12.1.1) converges absolutely if $\sum_n \alpha(n) < \infty$ by Lemma 1.2.1.

The weak convergence of empirical processes $\{\alpha_n(t), 0 \leq t \leq 1\}$ of $\{U_n\}$ has been discussed by some scientists. For an insightful view of this subject we refer to Billingsley (1968), Sen (1971, 1974), Yoshihara (1975, 1978) and Shao (1986), etc, until now the best results are due to Shao (1986).

12.1.1 Weak convergence for a φ-mixing sample

Theorem 12.1.1. *Let $\{U_n, n \geq 1\}$ be a sequence of strictly stationary φ-mixing random variables uniformly distributed over [0,1], and let*

$$\alpha_n(t) = \sqrt{n}(E_n(t) - t). \tag{12.1.2}$$

If the series of (12.1.1) converges absolutely and

$$\sum_{n=1}^{\infty} \varphi^{1/2}(2^n) < \infty, \tag{12.1.3}$$

then we have

$$\alpha_n \Longrightarrow Y \quad \textit{in } D[0,1],$$

where $Y = \{Y(t), 0 \le t \le 1\}$ *is a Gaussian process with* $EY(t) = 0$, $EY(s)Y(t) = R(s,t)$.

Proof. For any given $t \in [0,1]$,

$$\alpha_n(t) = \frac{1}{\sqrt{n}} \sum_{k=1}^{n} (I(X_k \le t) - t).$$

By Corollary 4.1.1, $\alpha_n(t)$ converge to $Y(t)$ in distribution and it is easy to see that the finite dimensional distributions of α_n converge to the corresponding finite dimensional distributions of Y.

In order to prove the tightness of $\{\alpha_n\}$, we need only to show that for any $\varepsilon > 0$, $\eta > 0$ there exists a $\delta, 0 < \delta < 1$, such that for any $s, 0 \le s \le 1 - \delta$, and large n

$$P\{\sup_{s \le t \le s+\delta} |\alpha_n(t) - \alpha_n(s)| \ge 4\varepsilon\} < \eta\delta. \tag{12.1.4}$$

For $0 \le s < t \le 1$ denote $\xi_i = (I(U_i \le t) - t) - (I(U_i \le s) - s) = I(s < U_i \le t) - (t - s)$. We have

$$E\xi_i = 0,\ |\xi_i| \le 1,\ E\xi_i^2 = t - s - (t-s)^2 \le t - s.$$

From Lemma 2.2.2 and Lemma 2.2.8 it follows that

$$\begin{aligned} E|\alpha_n(t) - \alpha_n(s)|^4 &\le \frac{1}{n^2} E\Big(\sum_{j=1}^{n} \xi_j\Big)^4 \\ &\le C(t-s)^2 \end{aligned} \tag{12.1.5}$$

for some $C > 0$. Let p be a positive number such that $\varepsilon/n \le p$. Consider the random variables $\alpha_n(s + ip) - \alpha_n(s + (i-1)p)$, $i = 1, 2, \cdots, m$. From Theorem 12.2 of Billingsley (1968) we have

$$P\Big\{\max_{0 \le i \le m} |\alpha_n(s + ip) - \alpha_n(s)| \ge \lambda\Big\} \le \frac{K}{\varepsilon\lambda^4} m^2 p^2 \tag{12.1.6}$$

where constant K depends only on $\varphi(\cdot)$.

For $s \le t \le s + p$, we have

$$|\alpha_n(t) - \alpha_n(s)| \le |\alpha_n(s+p) - \alpha_n(s)| + p\sqrt{n}, \tag{12.1.7}$$

which implies

$$\begin{aligned} &\sup_{s \le t \le s+pm} |\alpha_n(t) - \alpha_n(s)| \\ &\quad \le \sup_{s \le t \le s+pm} \{|\alpha_n(t) - \alpha_n(s+ip)| + |\alpha_n(s+ip) - \alpha_n(s)|\} \\ &\quad \le 3 \max_{i \le m} |\alpha_n(s+ip) - \alpha_n(s)| + p\sqrt{n}. \end{aligned} \tag{12.1.8}$$

If $\varepsilon/n \le p < \varepsilon/\sqrt{n}$, from (12.1.6) and (12.1.8) it follows that

$$P\Big\{\sup_{s\le t\le s+pm} |\alpha_n(t)-\alpha_n(s)| \ge 4\varepsilon\Big\} \le \frac{K}{\varepsilon^5} m^2 p^2. \tag{12.1.9}$$

Take δ such that $K\delta/\varepsilon^5 < \eta$. For large n there exists an m such that $(\delta/\varepsilon)\sqrt{n} < m \le (\delta/\varepsilon)n$ and $mp = \delta$. It follows from (12.1.8) and (12.1.9) that (12.1.4) holds true. The proof of Theorem 12.1.1 is completed.

Corollary 12.1.1. *Let $\{U_n, n \ge 1\}$ be a sequence of strictly stationary ρ-mixing random variables uniformly distributed over $[0,1]$. If the series in (12.1.1) converges absolutely and $\sum_{n=1}^{\infty} \rho(2^n) < \infty$, then $\alpha_n \Longrightarrow Y$.*

Remark 12.1.1. In the case of a p-dimensional sample ($p \ge 2$), the result of Theorem 12.1.1 is also true.

Remark 12.1.2. The Glivenko-Cantelli theorem, i.e. almost sure convergence of empirical processes of mixing sequences, is an immediate consequence of the results in Sections 8.3–8.5. For example, from Corollary 8.3.1 we have

Proposition 12.1.1. Let $\{X_n, \ge 1\}$ be a sequence of φ-mixing random variables with a common continuous distribution $F(x)$. Then for any $\theta > 0$, $\varepsilon > 0$

$$\sum_{n=1}^{\infty} P\{|F_n(x) - F(x)| \ge \varepsilon\, n^{-1/2+\theta}\} < \infty,$$

that is to say the rate of convergence in the Glivenko-Cantelli theorem is given as follows:

$$|F_n(x) - F(x)| = o(n^{-1/2+\theta}) \qquad \text{a.s.}$$

When the sample is ρ-mixing or α-mixng, there is the similar results under certain conditions.

12.1.2 Weak convergence for an α-mixing sample

Theorem 12.1.2. *Let $\{U_n, n \ge 1\}$ be a sequence of strictly stationary α-mixing random variables uniformly distributed over $[0,1]$ and*

$$\alpha(n) = O(n^{-r}) \quad r > 2. \tag{12.1.10}$$

Then $\alpha_n \Longrightarrow Y$.

By the following lemmas, from the proof of Theoem 12.1.1 it follows that Theorem 12.1.2 holds true.

Lemma 12.1.1. *Let $\{\xi_n, n \geq 1\}$ be a sequence of strictly stationary α-mixing random variables with $|\xi_1| \leq 1$ a.s., $E\xi_1 = 0$, $E\xi_1^2 = \tau$, $E|\xi_1| = 2\tau$ and satisfying (12.1.10). Put $S_n = \sum_{j=1}^n \xi_j$. Suppose that $r > 2$ is a non-integer and $m = 2k$, where k is an integer, satisfies $r - 1 < m < r + 1$. Then we have*

$$E|S_n|^m \leq c \sum_{i=0}^{(m-2)/2} \left(n^{i+1} \tau^{i\theta_1 + \theta_{m-1-2i}} \right) \tag{12.1.11}$$

where $[r] - 1 < r\theta_1 < r - 1$, $\theta_k = \theta_1 - (k-1)/r$, $k = 1, \cdots, [r]$.

Proof. By (12.1.10) and the definition of θ_k, we have

$$\sum_{i=0}^{\infty} (i+1)^{k-1} \alpha^{1-\theta_k}(i) < \infty, \tag{12.1.12}$$

for $k = 1, \cdots, [r]$. Let $\sum_{n(k)}$ be a summation for $i_1, \cdots, i_k \geq 0$, $i_1 + \cdots + i_k \leq n$, and $\sum_{n(k)}^{(l)}$ be a summation for $i_1, \cdots, i_k \geq 0$, $i_1 + \cdots + i_k \leq n$ and $i_l = \max_{1 \leq j \leq k} i_j$, where $l = 1, \cdots, k$; $k = 1, \cdots, [r]$. Denote

$$\sum\nolimits_{n(k)}^{(l)} = \sum\nolimits_{n(k)}^{(l)} |E\xi_0 \xi_{i_1} \xi_{i_1+i_2} \cdots \xi_{i_1+\cdots+i_k}|.$$

Note that for any even d, by the stationarity, we have

$$E|S_n|^d \leq d! n \sum\nolimits_{n(d-1)} |E\xi_0 \xi_{i_1} \cdots \xi_{i_1+\cdots+i_{d-1}}|. \tag{12.1.13}$$

From Lemma 1.2.5 and (12.1.12) we have

$$\sum\nolimits_{n(1)} |E\xi_0 \xi_{i_1}| \leq 6 \sum_{i_1=0}^{n} \alpha^{1-\theta_1}(i_1) \tau^{\theta_1} \leq c\, \tau^{\theta_1}, \tag{12.1.14}$$

$$\begin{aligned}
\sum\nolimits_{n(2)} |E\xi_0 \xi_{i_1} \xi_{i_1+i_2}| &\leq \left\{ \sum\nolimits_{n(2)}^{(1)} + \sum\nolimits_{n(2)}^{(2)} \right\} |E\xi_0 \xi_{i_1} \xi_{i_1+i_2}| \\
&\leq 6 \sum\nolimits_{n(2)}^{(1)} \alpha^{1-\theta_2}(i_1) \tau^{\theta_2} + 6 \sum\nolimits_{n(2)}^{(2)} \alpha^{1-\theta_2}(i_2) \tau^{\theta_2} \\
&\leq 6 \sum_{i_1=0}^{n} (i_1+1) \alpha^{1-\theta_2}(i_1) \tau^{\theta_2} + 6 \sum_{i_2=0}^{n} (i_2+1) \alpha^{1-\theta_2}(i_2) \tau^{\theta_2} \\
&\leq c\tau^{\theta_2}.
\end{aligned} \tag{12.1.15}$$

Next we prove

$$\sum\nolimits_{n(k)} \leq c \sum_{i=0}^{[(k-1)/2]} n^i \tau^{i\theta_1 + \theta_{k-2i}} \tag{12.1.16}$$

for $k = 1, 2, \cdots, [r]$ by the induction. From (12.1.14) and (12.1.15) it follows that (12.1.16) holds true for $k = 1, 2$. Assume that (12.1.16) holds true for $k-1$, $2 \le k-1 < [r]$, we show that (12.1.16) also holds true for k. Note that

$$\sum\nolimits_{n(k)} \le \sum\nolimits_{n(k)}^{(1)} + \cdots + \sum\nolimits_{n(k)}^{(k)},$$
$$\sum\nolimits_{n(k)}^{(1)} \le 6\sum\nolimits_{n(k)}^{(1)} \alpha^{1-\theta_k}(i_1)\tau^{\theta_k}$$
$$\le 6\sum_{i_1=0}^{n}(i_1+1)^{k-1}\alpha^{1-\theta_k}(i_1)\tau^{\theta_k} \le c\tau^{\theta_k}.$$

Similarly, $\sum_{n(k)}^{(k)} \le c\tau^{\theta_k}$. For $2 \le l \le k-1$,

$$\sum\nolimits_{n(k)}^{(l)} \le 6\sum\nolimits_{n(k)}^{(l)} \alpha^{1-\theta_k}(i_l)\tau^{\theta_k}$$
$$+ \sum\nolimits_{n(k)}^{(l)} |E\xi_0\xi_{i_1}\cdots\xi_{i_1+\cdots+i_{l-1}}| \cdot |E\xi_{i_1+\cdots+i_l}\cdots\xi_{i_1+\cdots+i_k}|$$
$$=: I_1 + I_2.$$

We also have $I_1 \le c\tau^{\theta_k}$. By the induction and stationarity we have

$$I_2 \le cn\Big(\sum_{i=0}^{[(l-2)/2]} n^i\tau^{i\theta_1+\theta_{l-1-2i}}\Big)\Big(\sum_{j=0}^{[(k-l-1)/2]} n^j\tau^{j\theta_1+\theta_{k-l-2j}}\Big)$$
$$\le c \sum_{0\le i\le[\frac{l-2}{2}],\ 0\le j\le[\frac{k-l-1}{2}]} n^{i+j+1}\tau^{(i+j+1)\theta_1+\theta_{k-2(i+j+1)}}$$
$$\le c\sum_{i=1}^{[(k-1)/2]} n^i\tau^{i\theta_1+\theta_{k-2i}}.$$

Thus (12.1.16) holds true for $k = 1, 2, \cdots, [r]$. From (12.1.13), (12.1.16) and noting $m-1 \le [r]$ we complete the proof of Lemma 12.1.1.

Lemma 12.1.2. *If the conditions of Lemma 12.1.1 are satisfied with* $2 < r < 3$, *then*

$$ES_n^4 \le c(n^{4-r} + n^2\tau^{2\theta_1}). \tag{12.1.17}$$

Proof. From Lemma 1.2.5 and (12.1.14) we have

$$\sum\nolimits_{n(3)}^{(1)} \le 6\sum\nolimits_{n(3)}^{(1)}\alpha(i_1) \le 6\sum_{i_1=0}^{n}(i_1+1)^2\alpha(i_1) \le cn^{3-r}.$$

Similarly $\sum_{n(3)}^{(3)} \le cn^{3-r}$. And by (12.1.16) we have

$$\begin{aligned}\sum\nolimits_{n(3)}^{(2)} &\le 6\sum\nolimits_{n(3)}^{(2)} \alpha(i_2) + \sum\nolimits_{n(3)}^{(2)} |E\xi_0\xi_{i_1}| \cdot |E\xi_{i_1+i_2}\xi_{i_1+i_2+i_3}| \\ &\le c(n^{3-r} + n\tau^{2\theta_1}).\end{aligned}$$

Combining these inequalities with (12.1.13) yields (12.1.17).

Now Theorem 12.1.2 can be proved along the similar lines to that of Theorem 12.1.1 by applying Lemma 12.1.2 instead of Lemma 2.2.2 and Lemma 2.2.8.

Remark 12.1.3. In the case of a p-dimensional sample ($p \ge 2$), by using Lemma 12.1.1, the result of Theorem 12.1.2 is also true if

$r > p + 1$ when p is even;

$r > p^2/(p-1)$ when p is odd.

Remark 12.1.4. The weak convergence of partial-sum processes and empirical processes with random indexes has been discussed by some scientist. For this subject we refer to Rényi (1958, 1960), Billingsley (1968 §17), Aldous (1978) and Lu (1984), etc. We enumerate only a result without proof for the weak convergence of empirical processes with random indexes as follows:

Let $\{\alpha_n, n \ge 1\}$ be a sequence of empirical processes as in Theorem 12.1.2 and let $\{\tau_n, n \ge 1\}$ be a sequence of positive integer-valued random variables on the same probability space. Suppose that $\alpha_n \Longrightarrow Y$, where Y is a Gaussian process and $\{\tau_n\}$ satisfies

$$\tau_n/n \xrightarrow{P} \tau,$$

where τ is a positive random variable. Then we have

$$\alpha_{\tau_n} \Longrightarrow Y.$$

12.2 Weighted weak convergence

Let q be a positive weight function on $(0,1)$, i.e. $\inf_{\delta \le t \le 1-\delta} q(t) > 0$ for all $0 < \delta < 1/2$, and define the weighted uniform empirical processes as $\{\alpha_n(t)/q(t), 0 < t < 1\}$. When $\{U_n, n \ge 1\}$ is a sequence of independent random variables uniformly distributed on $[0,1]$, the weighted weak convergence of empirical processes has been intensively studied in recent years. For an insightful view of this subject we refer to M. Csörgő, S. Csörgő, Horváth and Mason (1986a), Shorack and Wellner (1986) and

Csörgő and Horváth (1993), etc. We restate a theorem of M. Csörgő, S. Csörgő, Horváth and Mason (1986a) as follows. A shorter and more direct proof was given by Csörgő and Horváth (1993 Chapter 4).

Theorem 12.2.1. *Assume that q is positive and continuous on $(0,1)$, and is nondecreasing in a neighborhood of 0 and nonincreasing in a neighborhood of 1. Then*

$$I(q,\lambda) := \int_0^1 \frac{1}{t(1-t)} \exp(-\lambda q^2(t)/(t(1-t)))dt < \infty \tag{12.2.1}$$

for all $\lambda > 0$ if and only if, as $n \to \infty$

$$\alpha_n/q \Longrightarrow B/q \qquad in\ D[0,1]. \tag{12.2.2}$$

There are only a few studies concerned with the weighted weak convergence for empirical processes of a dependent sequence. In the latter case the limit process is changed from being a Brownian bridge, due to the appearence of covariances among observations (cf. (12.1.1)).

Shao and Yu (1995) studied the weighted weak convergence for empirical processes of strictly stationary observations under mixing and associated dependence assumptions. We only introduce the case of mixing dependence here. First we give the following basic theorem.

Theorem 12.2.2. *Let $\{U_n, n \geq 1\}$ be a strictly stationary sequence of uniform-[0, 1] random variables. Assume that for all $0 \leq s,t \leq 1$ and $n \geq 1$ we have*

(A1) $E|\alpha_n(t) - \alpha_n(s)|^p \leq C_1(|t-s|^{p_1} + n^{-p_2/2}|t-s|^{r_1})$ for some $C_1 > 0, p > 2, p_1 > 1, 0 \leq r_1 \leq 1$ and $p_2 > 1 - r_1$;

(A2) $E(\alpha_n(t) - \alpha_n(s))^2 \leq C_2|t-s|^{r_2}$ for some $C_2 > 0$ and $0 < r_2 \leq 1$.

If we have

$$\alpha_n \Longrightarrow Y \qquad in\ D[0,1] \tag{12.2.3}$$

with the Gaussian process $Y(\cdot)$ defined as in § 12.1, then

$$\alpha_n/q \Longrightarrow Y/q \qquad in\ D[0,1], \tag{12.2.4}$$

where q is a weight function such that for some $C > 0$ and $\beta > 1/2$

$$q(t) \geq C(t(1-t))^{\mu}(\log 1/(t(1-t)))^{\beta} \qquad for\ all\ 0 < t < 1 \tag{12.2.5}$$

and

$$\mu = \min\Big(\frac{p_1}{p}, \frac{r_1+p_2}{p+p_2}, \frac{r_2}{2}\Big). \tag{12.2.6}$$

Remark 12.2.1. By using a standard argument (cf. Theorem 12.2 and (22.18) in Billingsley 1968), one can easily verify that $\{\alpha_n(t), 0 \leq t \leq 1\}$ is tight by condition (A1). Hence to show (12.2.3), one needs only to prove that any finite dimensional distribution of $\{\alpha_n(t)\}$ converges to that of $\{Y(t)\}$ and the series in (12.1.1) converges absolutely.

Remark 12.2.2. The weight function q used in Theorem 12.2.1 is usually called a Chibisov-O'Reilly weight function. If we write $q(t) = (t(1-t)\log\log(1/(t(1-t))))^{1/2}g(t)$, then, necessarily, $g(t) \to \infty$ as $t \to 0$ or $t \to 1$. Thus the weight function q in (12.2.5) can be compared to a Chibisov-O'Reilly weight function by taking μ in (12.2.6) close to $1/2$ or exactly $1/2$ for properly chosen p, p_1, p_2, r_1, and r_2. In fact, Theorems 12.2.3 and 12.2.4 below show this possibility of taking $\mu = 1/(2+\varepsilon)$ for some $\varepsilon > 0$ in the case of mixing sequences. In particular our sharpest rate with $\mu = 1/2$ is obtained for ρ-mixing under a stronger mixing decay rate. In most cases, however, $\mu < 1/2$. We note in passing that if for a general weight function q we have $\int_0^1 1/q^2(t)dt < \infty$, then we have (12.2.1) as well for all $\lambda > 0$, i.e., q is then necessarily a Chibisov-O'Reilly weight function.

Remark 12.2.3. If $\mu = (r_1 + p_2)/(p + p_2) < \min(p_1/p, r_2/2)$ in (12.2.6), then from the proof of Theorem 12.2.2, one can relax the restriction on β from $\beta > 1/2$ to $\beta > 1/(p+p_2) = (1-\mu)/(p-r_1)$. Moreover, in the case of $\mu \geq 1/(p+1-r_1)$, one can use a simple sufficient condition $\int_0^1 1/q^{1/\mu}(t)dt < \infty$ to replace (12.2.5).

A direct application of Theorem 12.2.2 is to obtain weak convergence for integral functionals of α_n. For example, we consider the integral functional

$$\Delta_n(t) = \int_0^t \alpha_n(s)dQ(s) = \int_0^t \beta_n(\mathrm{inv}F(s))dQ(s), \quad 0 \leq t \leq 1$$

and its approximating Gaussian counterpart

$$\Delta(t) = \int_0^t Y(s)dQ(s), \quad 0 \leq t \leq 1, \tag{12.2.7}$$

where $Q(s) = \mathrm{inv}F(s)$ is the quantile function of distribution function F of X (recalling (12.0.4)). The function $\Delta(t)$ plays a central role in weak approximation theory for empirical total time on test, mean residual life, empirical Lorenz and Goldie concentration processes which are of interest in reliability and economic concentration theories. (cf. e.g., M. Csörgő, S. Csörgő, Horváth and Mason 1986b).

Corollary 12.2.1. *Under the conditions of Theorem 12.2.2, if*

$$\int_0^1 (t(1-t))^{\mu}(\log 1/(t(1-t)))^{\beta} dQ(t) < \infty, \tag{12.2.8}$$

then

$$\Delta_n \Longrightarrow \Delta \qquad \textit{in } D[0,1].$$

Remark 12.2.4. Let F be the distribution function of a random variable X. Then condition (12.2.8) is sightly stronger than the existence of the $(1/\mu)$-th moment of X. This is not necessarily true conversely, but $E|X|^{1/\mu}(\log(1+|X|))^{(1+\beta)/\mu+\delta} < \infty$, with any $\delta > 0$, implies (12.2.8).

Theorem 12.2.2 enables us to establish weighted weak convergence for empirical processes of a stationary mixing sequence.

Theorem 12.2.3. *Let $\{U_n, n \geq 1\}$ be a strictly stationary α-mixing sequence of uniform-[0, 1] random variables. If*

$$\alpha(n) = O(n^{-\theta-\varepsilon}) \tag{12.2.9}$$

for some $\theta \geq 1+\sqrt{2}$ and $\varepsilon > 0$, then we have

$$\alpha_n/q \Longrightarrow Y/q \quad \text{in } D[0,1]$$

for q satisfying $q(t) \geq C(t(1-t))^{(1-1/\theta)/2}$ for some $C > 0$.

Theorem 12.2.4. *Let $\{U_n, n \geq 1\}$ be a strictly stationary ρ-mixing sequence of uniform-[0, 1] random variables. Suppose that the series in (12.1.1) converges absolutely. If*

$$\sum_{n=1}^{\infty} \rho(2^n) < \infty, \tag{12.2.10}$$

then for any $\varepsilon > 0$ we have

$$\alpha_n/q \Longrightarrow Y/q \quad \textit{in } D[0,1]$$

for q satisfying $q(t) \geq C(t(1-t))^{1/(2+\varepsilon)}$ for some $C > 0$.

If, in addition,

$$\sum_{n=1}^{\infty} \rho^{2/p}(2^n) < \infty \tag{12.2.11}$$

for some $p > 2$, then we have

$$\alpha_n/q \Longrightarrow Y/q \quad \textit{in } D[0,1]$$

for q satisfying $q(t) \geq C(t(1-t))^{1/2}(\log 1/(t(1-t)))^{\beta}$ for some $C > 0$ and $\beta > 1/2$.

Corollary 12.2.2. *Under the conditions of Theorem 12.2.3, if*

$$\int_{-\infty}^{\infty} |x|^{2/(1-1/\theta)} dF(x) < \infty,$$

then

$$\Delta_n \Longrightarrow \Delta \quad \text{in } D[0,1].$$

Corollary 12.2.3. *Let $\{U_n, n \geq 1\}$ be a strictly stationary ρ-mixing sequences of uniform-[0, 1] random variables. If (12.2.10) holds and for any $\varepsilon > 0$ we have*

$$\int_{-\infty}^{\infty} |x|^{2+\varepsilon} dF(x) < \infty,$$

then

$$\Delta_n \Longrightarrow \Delta.$$

If, in addition, (12.2.11) holds and

$$\int_0^1 (t(1-t))^{1/2}(\log 1/(t(1-t)))^{\beta} dQ(t) < \infty,$$

then we have

$$\Delta_n \Longrightarrow \Delta \quad \text{in } D[0,1].$$

In order to prove Theorems, we need the following lemmas.

Lemma 12.2.1. *Let $\{\xi_i, i \geq 1\}$ be a sequence of random variables and let $\mathcal{F}_i = \sigma(\xi_j, j \leq i)$. Then, for any $p \geq 2$, there exists a constant $D = D(p)$ such that*

$$\begin{aligned} E\Big|\sum_{i=1}^n \xi_i\Big|^p \leq & D\Big(\Big(\sum_{i=1}^n E\xi_i^2\Big)^{p/2} + \sum_{i=1}^n E|\xi_i|^p + n^{p-1}\sum_{i=1}^n E|E(\xi_i|\mathcal{F}_{i-1})|^p \\ & + n^{p/2-1}\sum_{i=1}^n E|E(\xi_i^2|\mathcal{F}_{i-1}) - E\xi_i^2|^{p/2}\Big). \end{aligned} \tag{12.2.12}$$

Proof. Let $\eta_i = \xi_i - E(\xi_i|\mathcal{F}_{i-1})$ for $1 \leq i \leq n$. Then, $\{\eta_i, \mathcal{F}_{i-1}, 1 \leq i \leq n\}$ is a martingale difference sequence. By the well-known Burkholder

(1973) inequality, there is a $D = D(p) < \infty$ such that

$$\begin{aligned} E\Big|\sum_{i=1}^n \eta_i\Big|^p &\le D\Big(E\Big(\sum_{i=1}^n E(\eta_i^2|\mathcal{F}_{i-1})\Big)^{p/2} + \sum_{i=1}^n E|\eta_i|^p\Big) \\ &\le 2^p D\Big(\Big(\sum_{i=1}^n E\xi_i^2\Big)^{p/2} + \sum_{i=1}^n E|\xi_i|^p \\ &\quad + E\Big(\sum_{i=1}^n |E(\xi_i^2|\mathcal{F}_{i-1}) - E\xi_i^2|\Big)^{p/2}\Big) \\ &\le 2^{2p} D\Big(\Big(\sum_{i=1}^n E\xi_i^2\Big)^{p/2} + \sum_{i=1}^n E|\xi_i|^p \\ &\quad + n^{p/2-1}\sum_{i=1}^n E|E(\xi_i^2|\mathcal{F}_{i-1}) - E\xi_i^2|^{p/2}\Big). \qquad (12.2.13) \end{aligned}$$

On the other hand, it is easy to see that

$$E\Big|\sum_{i=1}^n \xi_i\Big|^p \le 2^p\Big(E\Big|\sum_{i=1}^n \eta_i\Big|^p + n^{p-1}\sum_{i=1}^n E|E(\xi_i|\mathcal{F}_{i-1})|^p\Big).$$

This proves (12.2.12) by the inequalities above.

We now develop a Rosenthal-type inequality for an α-mixing sequence which is of its own interest.

Lemma 12.2.2. *Let $2 < p < r \le \infty$, $2 < v \le r$ and $\{X_n, n \ge 1\}$ be an α-mixing sequence of random variables with $EX_n = 0$ and $\|X_n\|_r < \infty$. Assume that*

$$\alpha(n) \le Cn^{-\theta} \qquad (12.2.14)$$

for some $C > 0$ and $\theta > 0$. Then, for any $\varepsilon > 0$, there exists a $K = K(\varepsilon, r, p, v, \theta, C) < \infty$ such that

$$E|S_n|^p \le K\Big((nC_n)^{p/2}\max_{i\le n}\|X_i\|_v^p + n^{(p-(r-p)\theta/r)\vee(1+\varepsilon)}\max_{i\le n}\|X_i\|_r^p\Big), \qquad (12.2.15)$$

where $C_n = \Big(\sum_{i=o}^n (i+1)^{2/(v-2)}\alpha(i)\Big)^{(v-2)/v}$. In particular, for any $\varepsilon > 0$,

$$E|S_n|^p \le K\Big(n^{p/2}\max_{i\le n}\|X_i\|_v^p + n^{1+\varepsilon}\max_{i\le n}\|X_i\|_r^p\Big), \qquad (12.2.16)$$

if $\theta > v/(v-2)$ and $\theta \ge (p-1)r/(r-p)$, and

$$E|S_n|^p \le Kn^{p/2}\max_{i\le n}\|X_i\|_r^p, \qquad (12.2.17)$$

if $\theta \geq pr/(2(r-p))$.

Proof. For the sake of convenience of statement, we assume that $\{X, X_n, n \geq 1\}$ is a strictly stationary α-mixing sequence. By a result of Rio (1993), there is $D_1 = D_1(v)$ such that

$$ES_n^2 \leq D_1 n C_n \|X\|_v^2. \tag{12.2.18}$$

We shall prove (12.2.15) by induction on n. Suppose that for each $1 \leq k < n$,

$$E|S_k|^p \leq K\Big((kC_k)^{p/2}\|X\|_v^p + k^{(p-(r-p)\theta/r)\vee(1+\varepsilon)}\|X\|_r^p\Big). \tag{12.2.19}$$

We now prove that (12.2.19) is still true for $k = n$. Let $0 < a < 1/2$ that will be specified later and let $m = [an] + 1$. Define

$$\xi_i = \sum_{j=2(i-1)m+1}^{n\wedge(2i-1)m} X_j \quad \text{and} \quad \eta_i = \sum_{j=(2i-1)m+1}^{n\wedge 2im} X_j$$

for $1 \leq i \leq k_n := [n/(2m)] + 1$. Clearly, we have

$$E|S_n|^p \leq 2^{p-1}\Big(E\Big|\sum_{i=1}^{k_n}\xi_i\Big|^p + E\Big|\sum_{i=1}^{k_n}\eta_i\Big|^p\Big) =: 2^{p-1}(I_1 + I_2).$$

Let $\mathcal{F}_i = \sigma(\xi_j, j \leq i)$. It follows from Lemma 12.2.1 that there is D_2 such that $D_2 \geq (2D_1)^{p/2}$ and

$$\begin{aligned} I_1 \leq\ & D_2\Big(\sum_{i=1}^{k_n} E|\xi_i|^p + \Big(\sum_{i=1}^{k_n} E\xi_i^2\Big)^{p/2} + k_n^{p-1}\sum_{i=1}^{k_n} E|E(\xi_i|\mathcal{F}_{i-1})|^p \\ & + k_n^{p/2-1}\sum_{i=1}^{k_n} E|E(\xi_i^2|\mathcal{F}_{i-1}) - E\xi_i^2|^{p/2}\Big) \\ =:\ & D_2\Big(\sum_{i=1}^{k_n} E|\xi_i|^p + I_{11} + I_{12} + I_{13}\Big). \end{aligned} \tag{12.2.20}$$

In terms of (12.2.18), we have

$$\begin{aligned} I_{11} &\leq (D_1 k_n m C_m \|X\|_v^2)^{p/2} \\ &\leq (2D_1 n C_n)^{p/2}\|X\|_v^p \leq D_2(nC_n)^{p/2}\|X\|_v^p. \end{aligned} \tag{12.2.21}$$

To estimate I_{13}, we write

$$Y_i = E(\xi_i^2|\mathcal{F}_{i-1}) - E\xi_i^2.$$

Then, by Lemma 1.2.4

$$\begin{aligned}
E|Y_i|^{p/2} &= E|Y_i|^{p/2-1}\mathrm{sgn}(Y_i)Y_i = E(|Y_i|^{p/2-1}\mathrm{sgn}(Y_i)(\xi_i^2 - E\xi_i^2)) \\
&= \sum_{2(i-1)m<j,l\le n\wedge(2i-1)m} E|Y_i|^{p/2-1}\mathrm{sgn}(Y_i)(X_jX_l - EX_jX_l) \\
&\le 12 \sum_{2(i-1)m<j,l\le n\wedge(2i-1)m} \alpha^{1-\frac{p-2}{p}-\frac{2}{r}}(m) \\
&\quad \cdot (E|Y_i|^{p/2})^{(p-2)/p}\|X_jX_l\|_{r/2} \\
&\le 12m^2\alpha^{\frac{2}{p}-\frac{2}{r}}(m)(E|Y_i|^{p/2})^{(p-2)/p}\|X\|_r^2,
\end{aligned}$$

and hence

$$E|Y_i|^{p/2} \le 12^{p/2}m^p\alpha^{1-p/r}(m)\|X\|_r^p, \tag{12.2.22}$$

which, together with (12.2.14), yields

$$\begin{aligned}
I_{13} &\le k_n^{p/2}12^p m^p\alpha^{1-p/r}(m)\|X\|_r^p \\
&\le C24^p n^p m^{(p-r)\theta/r}\|X\|_r^p \\
&\le C24^p a^{(p-r)\theta/r} n^{p+(p-r)\theta/r}\|X\|_r^p \\
&\le C24^p a^{(p-r)\theta/r} n^{(p+(p-r)\theta/r)\vee(1+\varepsilon)}\|X\|_r^p.
\end{aligned} \tag{12.2.23}$$

Similarly to (12.2.22), we have

$$E|E(\xi_i|\mathcal{F}_{i-1})|^p \le 12^p m^p\alpha^{1-p/r}(m)\|X\|_r^p. \tag{12.2.24}$$

Therefore

$$\begin{aligned}
I_{12} &\le k_n^p 12^p m^p\alpha^{1-p/r}(m)\|X\|_r^p \\
&\le C24^p a^{(p-r)\theta/r} n^{(p+(p-r)\theta/r)\vee(1+\varepsilon)}\|X\|_r^p.
\end{aligned} \tag{12.2.25}$$

Putting the inequalities above together yields

$$\begin{aligned}
I_1 \le D_2\Big(&\sum_{i=1}^{k_n} E|\xi_i|^p + D_2(nC_n)^{p/2}\|X\|_v^p \\
&+ 2C24^p a^{(p-r)\theta/r} n^{(p+(p-r)\theta/r)\vee(1+\varepsilon)}\|X\|_r^p\Big).
\end{aligned}$$

Similarly,

$$\begin{aligned}
I_2 \le D_2\Big(&\sum_{i=1}^{k_n} E|\eta_i|^p + D_2(nC_n)^{p/2}\|X\|_v^p \\
&+ 2C24^p a^{(p-r)\theta/r} n^{(p+(p-r)\theta/r)\vee(1+\varepsilon)}\|X\|_r^p\Big).
\end{aligned}$$

Consequently, we have

$$E|S_n|^p \le 2^{p-1}D_2\Big(\sum_{i=1}^{k_n}(E|\xi_i|^p + E|\eta_i|^p) + 2D_2(nC_n)^{p/2}\|X\|_v^p + 4C24^p a^{(p-r)\theta/r} n^{(p+(p-r)\theta/r)\vee(1+\varepsilon)}\|X\|_r^p\Big). \qquad (12.2.26)$$

Now we let

$$a = (2^{p+4}D_2)^{-1/\varepsilon - p/(p-2)},$$
$$K = 2^{p+1}D_2(D_2 + 2C24^p a^{(p-r)\theta/r}).$$

By (12.2.26) and induction hypothesis (12.2.19), we get

$$\begin{aligned}
E|S_n|^p &\le 2^p D_2\Big(k_n K\Big((mC_m)^{p/2}\|X\|_v^p + m^{(p+(p-r)\theta/r)\vee(1+\varepsilon)}\|X\|_r^p\Big) \\
&\quad + D_2(nC_n)^{p/2}\|X\|_v^p + 2C24^p a^{(p-r)\theta/r} n^{(p+(p-r)\theta/r)\vee(1+\varepsilon)}\|X\|_r^p\Big) \\
&\le 2^p D_2(n/m)K\Big((mC_n)^{p/2}\|X\|_v^p + m^{(p+(p-r)\theta/r)\vee(1+\varepsilon)}\|X\|_r^p\Big) \\
&\quad + 2^p D_2(D_2 + 2C24^p a^{(p-r)\theta/r}) \\
&\quad \cdot \Big((nC_n)^{p/2}\|X\|_v^p + n^{(p+(p-r)\theta/r)\vee(1+\varepsilon)}\|X\|_r^p\Big) \\
&\le 2^p D_2 K\Big(a^{(p-2)/2}(nC_n)^{p/2}\|X\|_v^p + a^{\varepsilon} n^{(p+(p-r)\theta/r)\vee(1+\varepsilon)}\|X\|_r^p\Big) \\
&\quad + (K/2)\Big((nC_n)^{p/2}\|X\|_v^p + n^{(p+(p-r)\theta/r)\vee(1+\varepsilon)}\|X\|_r^p\Big) \\
&\le (K/2)\Big((nC_n)^{p/2}\|X\|_v^p + n^{(p+(p-r)\theta/r)\vee(1+\varepsilon)}\|X\|_r^p\Big) \\
&\quad + (K/2)\Big((nC_n)^{p/2}\|X\|_v^p + n^{(p+(p-r)\theta/r)\vee(1+\varepsilon)}\|X\|_r^p\Big) \\
&= K\Big((nC_n)^{p/2}\|X\|_v^p + n^{(p+(p-r)\theta/r)\vee(1+\varepsilon)}\|X\|_r^p\Big).
\end{aligned}$$

This proves that (12.2.19) remains valid for $k = n$, as desired.

Proof of Theorem 12.2.2. By Theorem 4.2 in Billingsley (1968), it is sufficient to prove that for any $\varepsilon > 0$

$$\lim_{\theta\to 0}\limsup_{n\to\infty} P\Big\{\sup_{0<t\le\theta}|\alpha_n(t)/q(t)| \ge \varepsilon\Big\} = 0, \qquad (12.2.27)$$

$$\lim_{\theta\to 0}\limsup_{n\to\infty} P\Big\{\sup_{1-\theta\le t<1}|\alpha_n(t)/q(t)| \ge \varepsilon\Big\} = 0, \qquad (12.2.28)$$

$$\lim_{\theta\to 0} P\Big\{\sup_{0<t\le\theta}|Y(t)/q(t)| \ge \varepsilon\Big\} = 0, \qquad (12.2.29)$$

$$\lim_{\theta\to 0} P\Big\{\sup_{1-\theta\le t<1}|Y(t)/q(t)| \ge \varepsilon\Big\} = 0. \qquad (12.2.30)$$

Note that

$$
\begin{aligned}
P\Big\{\sup_{0<t\le\theta}|\alpha_n(t)/q(t)|\ge\varepsilon\Big\}
&\le\sum_{j=1}^{\infty}P\Big\{\sup_{\theta2^{-j}<t\le\theta2^{-j+1}}|\alpha_n(t)/q(t)|\ge\varepsilon\Big\}\\
&\le\sum_{j=1}^{\infty}P\Big\{\sup_{\theta2^{-j}<t\le\theta2^{-j+1}}|\alpha_n(t)|\ge\varepsilon q(\theta2^{-j})\Big\}.
\end{aligned}
$$

Hence, (12.2.27) can be rewritten as

$$
\begin{aligned}
&\lim_{\theta\to0}\limsup_{n\to\infty}P\Big\{\sup_{0<t\le\theta}|\alpha_n(t)/q(t)|\ge\varepsilon\Big\}\\
&\qquad\le\limsup_{\theta\to0}\limsup_{n\to\infty}\sum_{j=1}^{\infty}B_{j,n},
\end{aligned}
\tag{12.2.31}
$$

where $B_{j,n}=P\{\sup_{0\le t\le\theta2^{-j+1}}|\alpha_n(t)|\ge\varepsilon q(\theta2^{-j})\}$.

Put

$$
\varepsilon_j=\varepsilon q(\theta2^{-j}),\qquad G_n=\{j:n^{1/2}\theta2^{-j+1}\le\varepsilon_j/2\},
$$
$$
H_n=\{j:n^{1/2}\theta2^{-j+1}>\varepsilon_j/2\}.
$$

It is easy to check that for any $0\le s\le t\le s+h\le1$

$$
|\alpha_n(t)-\alpha_n(s)|\le|\alpha_n(s+h)-\alpha_n(s)|+n^{1/2}h. \tag{12.2.32}
$$

Hence we have for $j\in G_n$

$$
\begin{aligned}
B_{j,n}&\le P\{|\alpha_n(\theta2^{-j+1})|+n^{1/2}\theta2^{-j+1}\ge\varepsilon_j\}\\
&\le P\{|\alpha_n(\theta2^{-j+1})|\ge\varepsilon_j/2\}\\
&\le c\varepsilon_j^{-2}(\theta2^{-j})^{r_2}\\
&\le c\varepsilon^{-2}\Big(\log(2^j/\theta)\Big)^{-2\beta}
\end{aligned}
\tag{12.2.33}
$$

by (A2), (12.2.5) and (12.2.6). Hence (12.2.33) implies that

$$
\limsup_{\theta\to0}\limsup_{n\to\infty}\sum_{j\in G_n}B_{j,n}=0. \tag{12.2.34}
$$

Note that in the case of $\mu<r_2/2$, (12.2.34) holds also for $\beta=0$.

Write

$$
\Delta:=\Delta_{j,n}=\frac{1}{4}\frac{\varepsilon_j}{n^{1/2}}=\frac{\varepsilon}{4}\frac{q(\theta2^{-j})}{n^{1/2}}. \tag{12.2.35}
$$

When $j \in H_n$, using (12.2.32) again, we obtain

$$
\begin{aligned}
B_{j,n} &\le P\Big\{\max_{1\le i\le \theta 2^{-j+1}/\Delta} |\alpha_n(i\Delta)| \ge \varepsilon_j/2\Big\} \\
&\quad + P\Big\{\max_{0\le i\le \theta 2^{-j+1}/\Delta}\ \sup_{i\Delta<t\le (i+1)\Delta} |\alpha_n(t)-\alpha_n(i\Delta)| \ge \varepsilon_j/2\Big\} \\
&\le P\Big\{\max_{1\le i\le \theta 2^{-j+1}/\Delta} |\alpha_n(i\Delta)| \ge \varepsilon_j/2\Big\} \\
&\quad + P\Big\{\max_{0\le i\le \theta 2^{-j+1}/\Delta} |\alpha_n((i+1)\Delta)-\alpha_n(i\Delta)| + \Delta n^{1/2} \ge \varepsilon_j/2\Big\} \\
&\le P\Big\{\max_{1\le i\le \theta 2^{-j+1}/\Delta} |\alpha_n(i\Delta)| \ge \varepsilon_j/2\Big\} \\
&\quad + P\Big\{\max_{0\le i\le \theta 2^{-j+1}/\Delta} |\alpha_n((i+1)\Delta)-\alpha_n(i\Delta)| \ge \varepsilon_j/4\Big\} \\
&\le 3P\Big\{\max_{1\le i\le \theta 2^{-j+2}/\Delta} |\alpha_n(i\Delta)| \ge \varepsilon_j/8\Big\}. \qquad (12.2.36)
\end{aligned}
$$

From (A1) it follows that for all $0 \le i < k \le \theta 2^{-j+2}/\Delta$

$$
\begin{aligned}
E|\alpha_n(k\Delta)-\alpha_n(i\Delta)|^p &\le C_1\Big(((k-i)\Delta)^{p_1} + n^{-p_2/2}((k-i)\Delta)^{r_1}\Big) \\
&\le C_1\Big(((k-i)\Delta)^{p_1} + n^{-p_2/2}(k-i)\Delta^{r_1}\Big).
\end{aligned}
$$

Thus, by Móricz's theorem (Lemma 4.1.2), there is a constant C, depending only on C_1 and p_1, such that

$$
\begin{aligned}
&E \max_{0\le i\le \theta 2^{-j+2}/\Delta} |\alpha_n(i\Delta)|^p \\
&\le C\Big((\theta 2^{-j+2}/\Delta)^{p_1}\Delta^{p_1} + n^{-p_2/2}(\theta 2^{-j+2}/\Delta)\Delta^{r_1}\log^p(\theta 2^{-j+2}/\Delta)\Big) \\
&\le C4^p\Big((\theta 2^{-j})^{p_1} + n^{-p_2/2}\theta 2^{-j}\Delta^{r_1-1}\log^p(\theta 2^{-j+2}/\Delta)\Big). \qquad (12.2.37)
\end{aligned}
$$

Since $p_2 > 1 - r_1$ in (A1), $l(x) = (\log x)^p/x^{-1+r_1+p_2} \le c$ for all $x \ge 1$. Thus from (12.2.5), (12.2.6), (12.2.35), (12.2.37) and the fact that $\theta 2^{-j+4}n^{1/2}/\varepsilon_j \ge 8$, we conclude that for $j \in H_n$

$$
\begin{aligned}
&P\Big\{\max_{0\le i\le \theta 2^{-j+2}/\Delta} |\alpha_n(i\Delta)| \ge \varepsilon_j/8\Big\} \\
&\le c\varepsilon_j^{-p}\Big((\theta 2^{-j})^{p_1} + n^{-p_2/2}\theta 2^{-j}\Delta^{r_1-1}\log^p(\theta 2^{-j+2}/\Delta)\Big) \\
&\le c\varepsilon_j^{-p}\Big((\theta 2^{-j})^{p_1} + \varepsilon_j^{r_1-1}(\theta 2^{-j})n^{(1-r_1-p_2)/2}\log^p(\theta 2^{-j+4}n^{1/2}/\varepsilon_j)\Big) \\
&\le c\varepsilon_j^{-p}\Big((\theta 2^{-j})^{p_1} + \varepsilon_j^{-p_2}(\theta 2^{-j})^{r_1+p_2} l(\theta 2^{-j+4}n^{1/2}/\varepsilon_j)\Big) \\
&\le c\Big(\varepsilon_j^{-p}(\theta 2^{-j})^{p_1} + \varepsilon_j^{-p-p_2}(\theta 2^{-j})^{r_1+p_2}\Big) \\
&\le c\varepsilon^{-p-p_2}\Big(\log(2^j/\theta)\Big)^{-p\beta}.
\end{aligned}
$$

This, together with (12.2.36), proves that

$$\limsup_{\theta\to 0}\limsup_{n\to\infty}\sum_{j\in H_n} B_{j,n}=0. \tag{12.2.38}$$

Note that in the case of $\mu < p_1/p$, (12.2.38) is true for $\beta > 1/(p+p_2)$. The proof of (12.2.27) is now completed by (12.2.31), (12.2.34) and (12.2.38). Similarly one can prove (12.2.28).

By (12.2.3) we have for any $0 \le s,t \le 1$

$$\alpha_n(t)-\alpha_n(s)\Longrightarrow Y(t)-Y(s).$$

Then by (A2) and Theorem 5.3 in Billingsley (1968)

$$E(Y(t)-Y(s))^2 \le \liminf_{n\to\infty} E(\alpha_n(t)-\alpha_n(s))^2 \le C_2|t-s|^{r_2}. \tag{12.2.39}$$

Thus, based on the fact that $\{Y(t), 0\le t\le 1\}$ is a Gaussian process, we have

$$E(Y(t)-Y(s))^4 \le cE^2(Y(t)-Y(s))^2 \le c|t-s|^{2r_2}$$

for all $0 \le s,t \le 1$. Applying Theorem 12.2 in Billingsley (1968), we can immediately get (12.2.29) and (12.2.30). This completes the proof of Theorem 12.2.2.

Proof of Corollary 12.2.1. First we verify that $\{\Delta_n(t), n\ge 1\}$ and $\Delta(t)$ are well defined on [0, 1]. By Schwarz's inequality, (A2), (12.2.6) and (12.2.8), for $0\le t\le 1$,

$$\begin{aligned} E\Delta_n^2(t) &= \int_0^t\int_0^t E\alpha_n(u)\alpha_n(v)dQ(u)dQ(v) \\ &\le \Big(\int_0^1 E^{1/2}(\alpha_n(t))^2 dQ(t)\Big)^2 \\ &\le 2^{r_2}C_2\Big(\int_0^1 (t(1-t))^{r_2/2}dQ(t)\Big)^2 < \infty, \end{aligned}$$

where the last inequality follows from $\alpha_n(0)=\alpha_n(1)=0$, $E(\alpha_n(t))^2 \le C_2t^{r_2}$ for $0\le t\le 1/2$, and $E(\alpha_n(t))^2 \le C_2(1-t)^{r_2}$ for $1/2\le t\le 1$. Similarly, using (12.2.39) in conjunction with (A2), we have

$$\begin{aligned} E\Delta^2(t) &= \int_0^1\int_0^1 EY(u)Y(v)dQ(u)dQ(v) \\ &\le \Big(\int_0^1 E^{1/2}(Y(t))^2 dQ(t)\Big)^2 \\ &\le 2^{r_2}C_2\Big(\int_0^1 (t(1-t))^{r_2/2}dQ(t)\Big)^2 < \infty. \end{aligned}$$

This shows that $\{\Delta_n(t), 0 \le t \le 1; n \ge 1\}$ and $\{\Delta(t), 0 \le t \le 1\}$ are square integrable processes. Now we have for any $\theta > 0$

$$\sup_{0<t\le\theta} |\Delta_n(t)| \le \sup_{0<t\le\theta} |\alpha_n(t)/q^*(t)| \int_0^1 q^*(t)dQ(t)$$

and

$$\sup_{1-\theta\le t<1} |\Delta_n(1) - \Delta_n(t)| \le \sup_{1-\theta\le t<1} |\alpha_n(t)/q^*(t)| \int_0^1 q^*(t)dQ(t),$$

where $q^*(t) = (t(1-t))^\mu(\log 1/(t(1-t)))^\beta$. Thus (12.2.27), (12.2.28) and (12.2.8) imply that for any $\varepsilon > 0$

$$\lim_{\theta\to 0} \limsup_{n\to\infty} P\Big\{\sup_{0<t\le\theta} |\Delta_n(t)| \ge \varepsilon\Big\}$$
$$= \lim_{\theta\to 0} \limsup_{n\to\infty} P\Big\{\sup_{1-0\le t<1} |\Delta_n(1) - \Delta_n(t)| \ge \varepsilon\Big\} = 0. \quad (12.2.40)$$

Similarly, (12.2.29), (12.2.30) and (12.2.8) imply that for any $\varepsilon > 0$

$$\lim_{\theta\to 0} P\Big\{\sup_{0<t\le\theta} |\Delta(t)| \ge \varepsilon\Big\}$$
$$= \lim_{\theta\to 0} P\Big\{\sup_{1-\theta\le t<1} |\Delta(1) - \Delta(t)| \ge \varepsilon\Big\} = 0. \quad (12.2.41)$$

Hence Corollary 12.2.1 follows from Theorem 12.2.2 and Theorem 4.2 in Billingsley (1968).

Proof of Theorem 12.2.3. Let θ and ε be as in (12.2.9). Since $\theta \ge 1 + \sqrt{2}$, we can take $r = \infty, v$ and p in Lemma 12.2.2 such that

$$\frac{2(\theta+\varepsilon)}{\theta+\varepsilon-1} < v < \frac{2\theta}{\theta-1} \quad \text{and} \quad v < \theta + 1 < p < \theta + 1 + \varepsilon. \qquad (12.2.42)$$

Therefore, by (12.2.16) and (12.2.18), for any $0 < \eta < (p-1-\theta)/\theta$, there is a $K < \infty$ such that for any $0 \le s, t \le 1$

$$E\Big|\sum_{i=1}^n (I(U_i \le t) - I(U_i \le s) - (t-s))\Big|^p$$
$$\le K(n^{p/2}|t-s|^{p/v} + n^{1+\eta/2})$$

and

$$E\Big|\sum_{i=1}^n (I(U_i \le t) - I(U_i \le s) - (t-s))\Big|^2 \le Kn|t-s|^{2/v},$$

which imply that

$$E|\alpha_n(t) - \alpha_n(s)|^p \le K(|t-s|^{p/v} + n^{-(p-2-\eta)/2})$$

and

$$E|\alpha_n(t) - \alpha_n(s)|^2 \le K|t-s|^{2/v}.$$

Hence (A1) and (A2) hold for $p_1 = p/v > 1$, $p_2 = p-2-\eta > 1$, $r_1 = 0$ and $r_2 = 2/v$. Note that $0 < \eta < (p-1-\theta)/\theta$, it is easy to see from (12.2.42) that

$$\begin{aligned}\min\Big(\frac{p_1}{p}, \frac{r_1+p_2}{p+p_2}, \frac{r_2}{2}\Big) &\\ &\ge \min\Big(\frac{1}{v}, \frac{p-2-\eta}{2p-2}\Big)\\ &> \frac{1}{2}\Big(1-\frac{1}{\theta}\Big). \qquad (12.2.43)\end{aligned}$$

By Theorem 12.1.2, (12.2.3) holds. This proves Theorem 12.2.3 by Theorem 12.2.2.

Proof of Theorem 12.2.4. From Corollary 12.1.1 it follows that (12.2.3) holds. By Theorem 1.1 of Shao (1995), we have for $p \ge 2$

$$\begin{aligned}E\Big|\sum_{i=1}^{n}(I(U_i \le t) - I(U_i \le s) - (t-s))\Big|^p &\\ &\le c\Big(n^{p/2}\exp\Big(c\sum_{i=0}^{[\log n]}\rho(2^i)\Big)(E(I(U_1 \le t)\\ &\quad - I(U_1 \le s) - (t-s))^2)^{p/2}\\ &\quad + n\exp\Big(K\sum_{i=0}^{[\log n]}\rho^{2/p}(2^i)\Big)E|I(U_1 \le t)\\ &\quad - I(U_1 \le s) - (t-s)|^p\Big)\\ &\le c\Big(n^{p/2}\exp\Big(c\sum_{i=0}^{[\log n]}\rho(2^i)\Big)|t-s|^{p/2}\\ &\quad + n\exp\Big(c\sum_{i=0}^{[\log n]}\rho^{2/p}(2^i)\Big)|t-s|\Big).\end{aligned}$$

Clearly, under condition (12.2.10), we have for $p \ge 2$ and any $0 < \delta <$

$\min\{\varepsilon(p-1)/(2(2+\varepsilon)), (p-2)/2\}$,

$$E\Big|\sum_{i=1}^{n}(I(U_i \le t) - I(U_i \le s) - (t-s))\Big|^p$$
$$\le c\Big(n^{p/2}|t-s|^{p/2} + n\exp\Big(c\sum_{i=0}^{[\log n]}\rho^{2/p}(2^i)\Big)|t-s|\Big)$$
$$\le c(n^{p/2}|t-s|^{p/2} + n^{1+\delta}|t-s|),$$

since $\exp(c\sum_{i=0}^{[\log n]}\rho^{2/p}(2^i))$ is a slowly varying function. This, in turns, gives us for $p > 2$

$$E|\alpha_n(t) - \alpha_n(s)|^p \le c(|t-s|^{p/2} + n^{-(p-2-2\delta)/2}|t-s|),$$

and for $p = 2$

$$E|\alpha_n(t) - \alpha_n(s)|^2 \le c|t-s|.$$

Thus (A1) and (A2) are satisfied for $p_1 = p/2$, $p_2 = p - 2 - 2\delta > 0$ and $r_1 = r_2 = 1$. Hence by (12.2.6)

$$\mu = (1 + p - 2 - 2\delta)/(p + p - 2 - 2\delta) > 1/(2+\varepsilon). \tag{12.2.44}$$

On the other hand, under condition (12.2.11), we have for $p \ge 2$

$$E\Big|\sum_{i=1}^{n}(I(U_i \le t) - I(U_i \le s) - (t-s))|^p$$
$$\le c(n^{p/2}|t-s|^{p/2} + n|t-s|),$$

which implies that for $p \ge 2$

$$E|\alpha_n(t) - \alpha_n(s)|^p \le c(|t-s|^{p/2} + n^{-(p-2)/2}|t-s|).$$

Thus (A1) and (A2) are satisfied for $p_1 = p/2$, $p_2 = p-2$ and $r_1 = r_2 = 1$. Obviously $\mu = 1/2$. Now the proof of Theorem 12.2.4 is complete.

12.3 Strong approximations

In this section, we introduce the strong approximations for empirical processes with a dependent sample. For simplicity, we only give the results for an α-mixing sample due to Philipp (1982). Strittmatter (1990) has given a further result for an absolutely regular sample.

In order to discuss the strong approximations of empirical processes with a dependent sample, we first introduce a result of strong approximations for random elements with values in a Banach space. Let $\{x_j, j \geq 1\}$ be a sequence of strictly stationary α-mixing random elements on probability space (Ω, Σ, P) and with values in a measurable space $(A, \mathcal{A})$. Let $(S, \|\cdot\|)$ be a Banach space and let $h: A \to S$ be a mapping. We call $\{X_j := h(x_j), j \geq 1\}$ a sequence of strictly stationary α-mixing S-valued random elements.

Theorem 12.3.1. *Let $\{X_j, j \geq 1\}$ be a sequence of strictly stationary α-mixing S-valued random elements as above, such that $\|X_1\| \leq \xi$ for some ξ with $E\xi^{2+\delta} < \infty$, $0 < \delta \leq 1$ and*

$$\alpha(n) \leq cn^{-(2+\rho)(1+2/\delta)} \tag{12.3.1}$$

for some $\rho > 0$. Suppose that for each $m \geq 1$ there is a linear mapping $\Lambda_m : S \to S$ with the following properties:

$$\sup_{m\geq 1} \|\Lambda_m\| < \infty, \tag{12.3.2}$$

$$\dim \Lambda_m S \leq C\, m, \qquad \textit{for some } C > 0. \tag{12.3.3}$$

And suppose that there is a θ with the following property: for each $m \geq 1$ there is an $n_0(m) \leq c\exp(m^{1-\theta})$ and a non-increasing function $g(m) \geq m^{-1/2}$ such that for all $n \geq n_0$

$$P^*\Big\{n^{-1/2}\|\sum_{j\leq n}(X_j - \Lambda_m X_j)\| \geq g(m)\Big\} \leq m^{-1-\theta}, \tag{12.3.4}$$

where P^* is the outer measure of P. Moreover assume that for each $m \geq 1$ the mapping $\Lambda_m \circ h$ is a measurable function from $(A, \mathcal{A})$ into the linear span $L_m S$ of $\Lambda_m S$, and that

$$E\Lambda_m X_1 = 0, \quad E\|\Lambda_m X_1\|^{2+\delta} < \infty. \tag{12.3.5}$$

Let T be the completion of the linear span of $\cup_{m\geq 1}\Lambda_m S$, so T is a separable Banach space. Then there exists a sequence $\{Y_j, j \geq 1\}$ of i.i.d. T-valued Gaussian random variables defined on (Ω, Σ, P) such that $EY_1 = 0$. Moreover, for each $s, t \in T'$, the space of bounded linear functional on T, the following limits exist and the series converge absolutely and the covariance *structure of Y_1 is given by*

$$\begin{aligned}Es(Y_1)t(Y_1) = &\lim_{m\to\infty} E\{s(\Lambda_m X_1)t(\Lambda_m X_1)\}\\&+ \lim_{m\to\infty} \sum_{j\geq 2} E\{s(\Lambda_m X_1)t(\Lambda_m X_j)\}\\&+ \lim_{m\to\infty} \sum_{j\geq 2} E\{s(\Lambda_m X_j)t(\Lambda_m X_1)\} \qquad (12.3.6)\end{aligned}$$

and

$$\begin{aligned}\|\sum_{j\leq n}(X_j - Y_j)\| &\leq V_n\\&= O\Big(n^{1/2}\Big\{(\log n)^{-(1-\beta)/4}\\&\quad + (\log n)^{(1-\beta)/\beta} g\Big(a(\log n)^{1-\theta/2}\Big)\Big\}\Big) \quad a.s. (12.3.7)\end{aligned}$$

for some measurable V_n and some positive constant a and any β with $1 - \theta/4 \leq \beta < 1$.

The proof of Theorem 12.3.1 will not be presented here.

Theorem 12.3.2. *Let $\mathcal{C} \subset \mathcal{A}$ be a class of measurable sets with*

$$N(\delta, \mathcal{C}) \leq c\,\delta^{-\tau}, \qquad \text{for some } \tau > 1/6. \qquad (12.3.8)$$

Let $\{x_n, n \geq 1\}$ be a sequence of strictly stationary A-valued α-mixing random elements with

$$\alpha(n) \leq c\,n^{-2\tau-4}. \qquad (12.3.9)$$

Put $g_n(C) = I(x_n \in C) - P(C), C \in \mathcal{C}$. Then there exists a sequence $\{Y_j, j \geq 1\}$ of i.i.d. Gaussian processes, defined on the same probability space (Ω, Σ, P), indexed by $C \in \mathcal{C}$ with $EY_1(C) = 0$,

$$\begin{aligned}&E(Y_1(C)Y_1(D))\\&\quad = Eg_1(C)g_1(D) + \sum_{n\geq 2} Eg_1(C)g_n(D)\\&\quad\quad + \sum_{n\geq 2} Eg_n(C)g_1(D), \quad C, D \in \mathcal{C} \qquad (12.3.10)\end{aligned}$$

such that with probability 1

$$\begin{aligned}&\sup_{C\in\mathcal{C}}\Big|\sum_{j\leq n}(I(x_j \in C) - P(C) - Y_j(C))\Big|\\&\quad\quad \leq V_n = O(n^{1/2}(\log n)^{-\lambda}) \qquad (12.3.11)\end{aligned}$$

for some measurable V_n and some $\lambda > 0$.

The proof of Theorem 12.3.2 will need the following lemmas.

Let $\{\xi_n\}$ be a sequence of strictly stationary α-mixing real random variables with $E\xi_1 = 0$, $|\xi_1| \le 1$ and $\alpha(n) \le cn^{-\rho}$, where $c > 0$, $\rho > 2$. We need an exponential bound for the partial sums of $\{\xi_i, i \ge 1\}$. Put $p = [n^{1/2-\kappa}]$, $q = [n/(2p_n)] + 1$ for some $0 < \kappa < 1/2$. Define blocks of consecutive integers H_j and I_j, $1 \le j < q$ of length p each and H_q consisting of $n - 2p(q-1)$ integers. Then $\mathrm{Card} H_q \le 2p$. The order of the blocks is $H_1, I_1, \cdots, H_{q-1}, I_{q-1}, H_q$. We leave no gaps between blocks. Put

$$y_j = \sum_{i \in H_j} \xi_i, \quad z_j = \sum_{i \in I_j} \xi_i.$$

It is easy to see that

$$\sigma^2 := E\xi_1^2 + 2 \sum_{n \ge 2} E\xi_1 \xi_n < \infty.$$

Note that for sufficiently large n

$$Ey_j^2 < \begin{cases} 2\sigma^2 p, & 1 \le j < q, \\ 5\sigma^2 p, & j = q. \end{cases} \tag{12.3.12}$$

Denote $\mathcal{L}_j = \sigma(y_1, \cdots, y_j)$. We write

$$y_j = Y_j + v_j, \quad 1 \le j \le q,$$

where $v_j = E(y_j|\mathcal{L}_{j-1}), Y_j = y_j - E(y_j|\mathcal{L}_{j-1})$. It is clear that $(Y_j, \mathcal{L}_j, 1 \le j \le q)$ is a martingale difference sequence and from Lemma 1.2.1, it follows that

$$\|v_j\|_2 \le 2\|y_j\|_\infty \alpha^{1/2}(p) \le c\, p^{1-\rho/2}. \tag{12.3.13}$$

Lemma 12.3.1. *Let $A \ge \sigma^2$. We have*

$$P\Big\{\sum_{j \le q} E(Y_j^2|\mathcal{L}_{j-1}) \ge 4An\Big\} \le cA^{-2}p^{2-\rho}.$$

Proof. By the Hölder inequality we have

$$E(Y_j^2|\mathcal{L}_{j-1}) \le E(y_j^2|\mathcal{L}_{j-1}). \tag{12.3.14}$$

By Lemma 1.2.1, we have

$$\|E(y_j^2|\mathcal{L}_{j-1}) - Ey_j^2\|_2 \le 4\|y_j\|_\infty^2 \alpha^{1/2}(p) \le 4p^{2-\rho/2}.$$

Thus by the Minkowski and Chebyshev inequalities we have

$$P\Big\{\sum_{j\le q}(E(y_j^2|\mathcal{L}_{j-1}) - Ey_j^2) \ge An\Big\}$$
$$\le 16n^{-2}A^{-2}(qp^{2-\rho/2})^2 \le cA^{-2}p^{2-\rho}.$$

Hence by (12.3.12) and (12.3.14), the probability in question does not exceed

$$P\Big\{\sum_{j\le q} E(y_j^2|\mathcal{L}_{j-1}) \ge \sum_{j\le q} Ey_j^2 + An\Big\} \le cA^{-2}p^{2-\rho}.$$

Lemma 12.3.1 is proved.

Lemma 12.3.2. *For all $R > 0$ we have*

$$P\Big\{\Big|\sum_{j\le q} Y_j\Big| > 5Rn^{1/2}\Big\} \le c\{\exp(-R^2/A) + A^{-2}p^{2-\rho}\}.$$

Proof. Let M be the index j of H_j or I_j containing n. Define

$$U_k = \begin{cases} \sum_{j\le k} Y_j, & \text{if } k \le M,\\ U_M, & \text{if } k > M;\end{cases}$$
$$s_k^2 = \begin{cases} \sum_{j\le k} E(Y_j^2|\mathcal{L}_{j-1}), & \text{if } k \le M,\\ s_M^2, & \text{if } k > M.\end{cases}$$

Obviously $U_j - U_{j-1} = Y_j \le 4p =: C$. If $R \le An^{1/2}/p$, we set $\lambda = R/(4An^{1/2})$, $K = 20An$, so that $\lambda C \le 1$. Put

$$T_k = \exp(\lambda U_k - \frac{1}{2}\lambda^2(1 + \frac{1}{2}\lambda C)s_k^2).$$

By Lemma 5.4.1 and Corollary 5.4.1 of Stout (1974) and Lemma 12.3.1 we have

$$P\Big\{\sup_{k\ge 0} U_k > 5Rn^{1/2}\Big\}$$
$$\le P\Big\{\sup_{k\ge 0} T_k > \exp\Big(\lambda^2 K - \frac{1}{2}\lambda^2\Big(1 + \frac{1}{2}\lambda C\Big)s_q^2\Big)\Big\}$$
$$\le c\Big\{\exp(-R^2/A) + A^{-2}p^{2-\rho}\Big\}.$$

But if $R > An^{1/2}/p$ we set $\lambda = 1/(4p)$ and $K = 20Rpn^{1/2}$. Then $\lambda C = 1$ and by the same calculation we obtain

$$\begin{aligned} P\Big\{ \sup_{k\geq 0} U_k > 5Rn^{1/2} \Big\} &\leq c\exp\Big(\frac{-20Rpn^{1/2} + 3An}{16p^2}\Big) + A^{-2}p^{2-\rho} \\ &\leq cA^{-2}p^{2-\rho}. \end{aligned}$$

Lemma 12.3.2 is proved.

Lemma 12.3.3. *We have*

$$P\Big\{\Big|\sum_{j\leq q} y_j\Big| \geq 7Rn^{1/2}\Big\} \leq c\Big(\exp(-R^2/A) + n^{1+\rho\kappa-\rho/2}(A^{-2} + R^{-2})\Big).$$

Proof. We have

$$\Big|\sum_{j\leq q} y_j\Big| \leq \Big|\sum_{j\leq q} Y_j\Big| + \Big|\sum_{j\leq q} v_j\Big|.$$

By the Chebyshev and Minkowski inequalities we obtain from (12.3.13) and expressing p and q in terms of n

$$P\Big\{\Big|\sum_{j\leq q} v_j\Big| \geq Rn^{1/2}\Big\} \leq cR^{-2}n^{-1}(qp^{1-\rho/2})^2 \leq cR^{-2}np^{-\rho}.$$

Combining it with Lemma 12.3.2 implies Lemma 12.3.3.

Lemma 12.3.3 remains valid if we replace y_j by $z_j, 1 \leq j < q$ and set $z_q = 0$. Hence we have an exponential bound for the partial sums of $\{\xi_i, i \geq 1\}$ as follows

$$\begin{aligned} P\Big\{\Big|\sum_{k\leq n} \xi_k\Big| > 14Rn^{1/2}\Big\} \\ \leq c\Big(\exp(-R^2/A) + n^{1+\rho\kappa-\rho/2}(A^{-2} + R^{-2})\Big). \end{aligned} \tag{12.3.15}$$

Lemma 12.3.4. *If the hypotheses of Theorem 12.3.2 are satisfied, then for any given $\varepsilon > 0$, there exists a $\delta > 0$, $\delta \leq c\varepsilon^6$ and $n_0 \leq c\exp\big(-1/(4\varepsilon)\big)$ such that for all $n \geq n_0$*

$$P^*\{\sup(|\nu_n(C) - \nu_n(D)| : C, D \in \mathcal{C}, P_x(C \triangle D) < \delta) > \varepsilon\} \leq c\exp\Big(-\frac{1}{8\varepsilon}\Big),$$

where

$$\nu_n = n^{1/2}(P_n - P_x),$$
$$P_n(B) = n^{-1}\sum_{j=1}^{n} I(x_j \in B),$$
$$P_x(B) = P(x_1 \in B),\ B \in \mathcal{A}.$$

Proof. Let r be so large that

$$2^r > \varepsilon^{-6}. \tag{12.3.16}$$

Put $\delta_k = 2^{-r-k}(k = 0,1,2,\cdots)$, $m_k = N(\delta_k, \mathcal{C}, P_x)$, $d_i = (i+1)^{-2}\varepsilon/32$. Take sets $A_{k1},\cdots,A_{km(k)}$ as in the definition of $N(\delta_k,\mathcal{C},P_x)$, so that for every $C \in \mathcal{C}$ and $k = 0,1,\cdots$ there exist $r(k) = r(k,C)$ and $s(k) = s(k,C)$ such that $A_{kr(k)} \subset C \subset A_{ks(k)}$ and $P_x(A_{ks(k)} \backslash A_{kr(k)}) < \delta_k$. Denote

$$B_k := B_k(C) = A_{ks(k)} \backslash A_{k+1,s(k+1)}, \quad D_k := D_k(C) = A_{k+1,s(k+1)} \backslash A_{ks(k)}.$$

Then $P_x(B_k) < \delta_k$ and $P_x(D_k) < \delta_{k+1} < \delta_k$.

Put $n_0 := n_0(\varepsilon) = \varepsilon^2/(256\delta_0^2)$. For every $n > n_0$ there is a unique $k = k(n)$ such that

$$1/2 < 8\delta_k n^{1/2}/\varepsilon \le 1. \tag{12.3.17}$$

Then for $n > n_0$, $k = k(n)$, $\delta = \delta_k$ and $C \in \mathcal{C}$, $r = r(k,C)$, $s = s(k,C)$ we have

$$\nu_n(A_{kr}) - \varepsilon/8 \le \nu_n(A_{kr}) - \delta n^{1/2} \le \nu_n(C) \le \nu_n(A_{ks}) + \varepsilon/8. \tag{12.3.18}$$

From this we obtain

$$\begin{aligned} &|\nu_n(A_{ks(k)}) - \nu_n(A_{0s(0)})| \\ &\le \sum_{i=0}^{k-1} |\nu_n(A_{is(i)}) - \nu_n(A_{i+1,s(i+1)})| \\ &\le \sum_{i=0}^{k-1} (|\nu_n(B_i)| + |\nu_n(D_i)|). \end{aligned} \tag{12.3.19}$$

Let $\mathcal{B}_i$ be a collection of sets $B = A_{is}\backslash A_{i+1,t}$ or $B = A_{i+1,t}\backslash A_{i,s}$ with $P_x(B) < \delta_i$. Then for every $C \in \mathcal{C}$, $B_i(C)$ and $D_i(C) \in \mathcal{B}_i$. The number of sets in $\mathcal{B}_i$

$$\mathrm{Card}(\mathcal{B}_i) \le 2m(i)m(i+1).$$

Now let us estimate $P\{|\nu_n(B)| > d_i\}$, $B \in \mathcal{B}_i$. Put $\xi_n = I(x_n \in B) - P_x(B)$. Since

$$|E\xi_1\xi_n| \le \|\xi_1\|_4\|\xi_n\|_4(\alpha(n-1))^{1/2} \le P_x^{1/2}(B)n^{-2-\tau},$$

we have

$$\sigma^2 = E\xi_1^2 + 2\sum_{n=2}^{\infty} E\xi_1\xi_n \le cP_x^{1/2}(B).$$

Hence we can take $\delta_i^{1/2}$ such that $\sigma^2 \le \delta_i^{1/2}$, it follows from (12.3.15) that

$$P\{|\nu_n(B)| > d_i\} \le c\Big(\exp\Big\{-\frac{d_i^2}{196}\delta_i^{-1/2}\Big\} + n^{-\tau-1+\kappa(2\tau+4)}(\delta_i^{-1} + i^2)\Big).$$

Then from (12.3.8) and taking $\kappa = 1/(8\tau + 16)$ we have

$$\begin{aligned} p_i &:= P\{|\nu_n(B)| > d_i \text{ for some } B \in \mathcal{B}_i\} \\ &\le c\exp\Big((-\varepsilon^2/(\delta_i^{1/2}196 \cdot 32^2(i+1)^4)) \cdot \delta_i^{-2\tau} + n^{-\tau-3/4}\delta_i^{-1-2\tau}\Big). \end{aligned}$$

Therefore by (12.3.16) and (12.3.17) and taking r large enough, we obtain

$$\begin{aligned} \sum_{i=0}^{k} p_i &\le c\Big(\exp\Big(-\frac{1}{\varepsilon}\Big) + n^{-1/4}\varepsilon^{-(1+2\tau)}\Big) \\ &\le c\exp\Big(-\frac{1}{8\varepsilon}\Big) \end{aligned}$$

for $n \ge n_0$. Put $Q_n = P(V_n > \varepsilon/8)$, where

$$\begin{aligned} V_n = \sup\{|\nu_n(A_{kr}) - \nu_n(A_{ks})| &: A_{kr} \subset A_{ks}, \\ &P(A_{ks}\backslash A_{kr}) < \delta_k,\ r, s = 1, \cdots, m_k\}. \end{aligned}$$

Then from (12.3.15) and (12.3.17) it follows that for $n \ge n_0$

$$\begin{aligned} Q_n &\le c\delta_k^{-2\tau}\exp\Big(\frac{-\varepsilon^2}{64}\cdot\delta_k^{-1/2}\Big) + n^{-\tau-3/4}\delta_k^{-1-2\tau} \\ &\le c\exp\Big(-\frac{1}{8\varepsilon}\Big). \end{aligned}$$

By using (12.3.16) again we have

$$\begin{aligned} p_0 &:= P\{\sup(|\nu_n(A_{0i}) - \nu_n(A_{0j})| : P(A_{0i}\triangle A_{0j}) < 3\delta_0) > \varepsilon/4\} \\ &\le c\delta_0^{-2\tau}\exp\Big(-\frac{\varepsilon^2}{16}\delta_0^{-1/2}\Big) + n^{-1-\tau+\rho\kappa}\delta_0^{-1-2\tau} \\ &\le c\exp\Big(-\frac{1}{8\varepsilon}\Big). \end{aligned}$$

Lemma 12.3.4 is proved.

Proof of Theorem 12.3.2.

Let S be the space of all bounded real-valued functions on $\mathcal{C}$. If $f \in S$ we set $\|f\| = \sup_{C\in\mathcal{C}} |f(C)|$. If $x \in A$ we set $h(x) = I(x \in C) - P(C)$. Thus $h : A \to S$. Let $m \geq 1$ and put $\varepsilon = m^{-1/(6\tau)}$. Find δ and n_0 according to Lemma 12.3.4. Then $\delta \leq c\, m^{-1/\tau}$ and $n_0 \leq c\exp\left(m^{1/(6\tau)}/4\right)$. Let $A_1, \cdots, A_d$, $d = N(\delta)$, be sets such that for all $C \in \mathcal{C}$ there exists A_r with $P(A_r \triangle C) < \delta$ and r minimal. Now $N(\delta) \leq c\delta^{-\tau} \leq cm$. We define $\Lambda_m : S \to S$ by $\Lambda_m h(x) = I(x \in A_r) - P(A_r)$ if $h(x) = I(x \in C) - P(C)$ with $P(A_r \triangle C) < \delta$. Then $\dim \Lambda_m S \leq cm$. From Lemma 12.3.4 (with $D = A_r$) we conclude that (12.3.4) is satisfied with $g(m) = m^{-1/(6\tau)}$ and $\theta = 1 - 1/(6\tau)$. The result follows now from Theorem 12.3.1 choosing β close to 1, but subject to $1 - \theta/4 \leq \beta < 1$. Theorem 12.3.2 is proved.

From Theorem 12.3.2 we have the following theorem immediately.

Theorem 12.3.3. *Let $\{X_j, j \geq 1\}$ be a sequence of strictly stationary α-mixing d-dimensional random vectors with a distribution $F(x)$ and*

$$\alpha(n) = O(n^{-4-2d}). \tag{12.3.20}$$

Denote $g_n(s) = I(X_n \leq s) - F(s)$, for $s \in R^d$. So the series below defining the covariance function

$$\Gamma(s, s') = Eg_1(s)g_1(s') + \sum_{n=2}^{\infty} E(g_1(s)g_n(s') + g_n(s)g_1(s'))$$

converges absolutely for $s, s' \in R^d$. Then there exists a sequence $\{Y_j, j \geq 1\}$ of i.i.d. Gaussian processes, defined on the same probability space (Ω, Σ, P), indexed by $s \in R^d$ with

$$EY_1(s) = 0, \quad EY_1(s)Y_1(s') = \Gamma(s, s'), \quad s, s' \in R^d$$

and a positive constant λ depending on d only such that with probability 1

$$\sup_{s\in R^d}\Big|\sum_{j=1}^{n}(I(X_j \leq s) - F(s) - Y_j(s))\Big| = O(n^{1/2}(\log n)^{-\lambda}).$$

Proof. Let P be the probability measure induced by F. Let $F_i(s_i)$, $1 \leq i \leq d$ be the i-th marginal of $F(s)$, $s = (s_1, \cdots, s_d)$. Let $r \geq 1$ be given. We define

$$s_{ij} = \mathrm{inv}F_i(j2^{-r}), \quad 1 \leq i \leq d, \quad 0 \leq j \leq 2^r.$$

Let $\mathcal{C}$ be the collection of all intervals $C = (-\infty, s]$, $s \in R^d$. For any $C \in \mathcal{C}$ there exist A_p and A_q both of the form $(-\infty, (s_{1j_1}, \cdots, s_{dj_d})]$ for some $(s_{1j_1}, \cdots, s_{dj_d})$ such that

$$A_p \subset C \subset A_q \quad \text{and} \quad P(A_q \backslash A_p) \le d\,2^{-r}.$$

The collection of all such sets A_p has cardinality $\le 2^{dr}$, i.e. $N(d2^{-r}, \mathcal{C}, P) \le 2^{dr}$. Hence by interpolation (12.3.8) is satisfied with $\tau = d$ and $c = (2d)^d$ and (12.3.10) holds because of (12.3.20).

Remark 12.3.1. Philipp and Pinzur (1980) gave an almost sure approximation of the multivariate empirical process by a Kiefer process. Let $\{X_n, n \ge 1\}$ be a strictly stationary α-mixing sequence of random vectors in R^q with continuous distribution F and

$$\alpha(n) = O(n^{-4-q(1+\varepsilon)})$$

for some $0 \le \varepsilon \le 1/4$. The empirical process of $\{X_n, n \ge 1\}$ is defined as

$$R(s,t) = [t](F_{[t]}(s) - F(s)), \quad t \ge 0,\ s \in R^q.$$

Let

$$\Gamma(s,s') = E\{g_1(s)g_1(s')\} + \sum_{n=2}^{\infty} E\{g_1(s)g_n(s') + g_n(s)g_1(s')\}, \qquad s,\ s' \in R^q$$

where $g_n(s) = I(X_n \le s) - F(s)$. Then without changing its distribution we can redefine the empirical process $\{R(s,t), s \in R^q, t \ge 0\}$ on a richer probability space on which there exists a Kiefer process $\{K(s,t), s \in R^q, t \ge 0\}$ with covariance function $(t \wedge t')\Gamma(s,s')$ and a constant $\lambda = \lambda(q,\varepsilon)$ such that

$$\sup_{t \le T} \sup_{s \in R^q} |R(s,t) - K(s,t)| = O(T^{1/2}(\log T)^{-\lambda}) \quad \text{a.s.} \tag{12.3.21}$$

When $q = 1$ and $\{X_n, n \ge 1\}$ are uniformly distributed over $[0,1]$, (12.3.21) coincides with the result by Yoshihara (1979), but with less mixing rate. By Theorem 1.15.1 in Csörgő and Révész (1981), (12.3.21) implies that the Strassen-type law of iterated logarithm holds true for

$$\{\eta_s(t) = R(s,t)/\sqrt{2t \log\log t}, 0 \le s \le 1\}.$$

12.4 Moduli of continuity of empirical processes

Let $\{X_n, n \geq 1\}$ be a sequence of random variables with a common distribution $F(x)$, $\{\beta_n(t), -\infty < t < \infty\}, n = 1, 2, \cdots$, a sequence of its empirical processes. The moduli of continuity of empirical processes are defined as follows:

$$w_n(a_n) = \sup_{\substack{|t-s|\leq a_n \\ -\infty<s<t<\infty}} |\beta_n(t) - \beta_n(s)|.$$

Stute (1982) proved for the i.i.d. case the following theorem.

Theorem 12.4.1. *Suppose that*
(i) $0 < a_n < 1$, $a_n \downarrow$ *and* $na_n \uparrow \infty$,
(ii) $na_n/\log n \to \infty$,
(iii) $\log a_n^{-1}/\log\log n \longrightarrow 0$.
Then

$$\lim_{n\to\infty} (a_n \log a_n^{-1})^{-1/2} w_n(a_n) = 1 \quad a.s. \tag{12.4.1}$$

Definition 12.4.1. Let $0 < \lambda \leq 1$, $A \subset R$. The function $g(x)$ is said to satisfy the uniformly local λ-order Lipschtz (λ-ulL) condition on A, if there exist $\delta > 0$, $M < \infty$ such that

$$\sup_{x\in A} |g(x+z) - g(x)| \leq M|z|^{\lambda}, \quad |z| \leq \delta. \tag{12.4.2}$$

Zhou (1994) discussed the moduli of continuity of empirical processes when the sample is φ-mixing or α-mixing, and proved the following theorems.

Theorem 12.4.2. *Let* $\{X_n, n \geq 1\}$ *be a sequence of strictly stationary* φ*-mixing random variables with a common distribution* $F(x)$*. Suppose that* $F(x)$ *satisfies the* 1*-ulL condition and* $\sum_{i=1}^{\infty} \varphi^{1/2}(2^i) < \infty$*. If there is a sequence of positive integers* $\{m_n\}$ *such that*

$$1 \leq m_n \leq n, \quad \frac{n}{m_n}\varphi(m_n) \leq A \quad \text{and} \quad \left(\frac{\log n}{na_n}\right)^{1/2} m_n \leq C,$$

where A *and* C *are two constants, then*

$$w_n(a_n) = O((a_n \log n)^{1/2}) \quad a.s. \tag{12.4.3}$$

Theorem 12.4.3. *Let $\{X_n, n \geq 1\}$ be a sequence of strictly stationary α-mixing random variables with a common distribution $F(x)$. Suppose that $F(x)$ satisfies the 1-ulL condition and $\alpha(n) = O(\rho^n)$ for some $0 < \rho < 1$ and $a_n \to 0 (n \to \infty)$. Then for any $0 < \theta < 1$ we have*

$$w_n(a_n) = O\Big(a_n^{\frac{1-\theta}{2}} \log^2 n\Big) \quad a.s. \tag{12.4.4}$$

Remark 12.4.1. If $F(x)$ satisfies the λ-ulL condition $(0 < \lambda < 1)$ on $A(\subset R)$, then (12.4.3) and (12.4.4) are rewritten as

$$w_n(a_n, A) := \sup_{t,s\in A, |t-s|\leq a_n} |\beta_n(t) - \beta_n(s)| = O((a_n^\lambda \log n)^{1/2}) \quad \text{a.s.} \tag{12.4.5}$$

and

$$w_n(a_n, A) = O\Big(a_n^{\frac{1-\theta}{2}\lambda} \log^2 n\Big) \quad \text{a.s.} \tag{12.4.6}$$

Remark 12.4.2. Because of Lemma 12.4.1 below, the condition $\sum_{n=1}^{\infty} \varphi(n) < \infty$, which is required in Zhou (1994), is weakened to $\sum_{i=1}^{\infty} \cdot\varphi^{1/2}(2^i) < \infty$ in Theorem 12.4.2.

The proof of Theorems will need the following Bernstein type inequalities.

Lemma 12.4.1. *Let $\{X_n, n \geq 1\}$ be a φ-mixing sequence with $EX_n = 0$, $|X_n| \leq d$, $EX_n^2 \leq D$ and $\sum_{i=1}^{\infty} \varphi^{1/2}(2^i) < \infty$. Then there is $C_1 = C_1(\varphi(\cdot)) > 0$ such that*

$$P\Big\{\Big|\sum_{i=1}^{n} X_i\Big| > \varepsilon\Big\}$$
$$\leq \exp\Big\{3\sqrt{e}n\frac{\varphi(m)}{m} - \alpha\varepsilon + C_1\alpha^2 Dn\Big\}$$

where α is a real number, m is a positive integer satisfying $m \leq n, \alpha \leq n$ and $\alpha m d \leq 1/4$.

Lemma 12.4.1 is an improvement of Lemma 1 of Collomb (1984). Its proof follows Collomb's lines provided one uses Lemma 2.2.2 instead of Lemma 1.2.10 which is employed by Collomb (cf. Lemma 11.1.1).

Lemma 12.4.2. (Doukhan, Leon and Portal 1984) *Let $\{X_i, i \geq 1\}$ be an α-mixing sequence with $EX_i = 0, |X_i| \leq 1$ and $\alpha(n) \leq C\rho^n$. Denote*

$\sigma = \sup \|X_i\|_\gamma$, *where* $\gamma = 2/(1-\theta), 0 < \theta < 1$. *Then there exist* C_1, C_2 *which depend only on* $\alpha(\cdot)$ *such that*

$$P\Big\{\Big|\sum_{i=1}^{n} X_i\Big| > \varepsilon\Big\} \le C_1\theta^{-1}\exp\{-C_{2,n}\varepsilon^{1/2}/(n^{1/4}\sigma^{1/2})\}$$

where

$$C_{2,n} = \begin{cases} C_2 & \text{if } n^{1/2}\sigma \le 1, \\ C_2 n^{1/4}\sigma^{1/2} & \text{if } n^{1/2}\sigma > 1. \end{cases}$$

Proof of Theorem 12.4.2.

As we have mentioned in the beginning of this chapter, we need only to consider uniform empirical processes. And by the 1-ulL condition, it is enough to show

$$\sup_{0\le t,s\le 1}\ \sup_{|t-s|\le Ma_n} |\alpha_n(t) - \alpha_n(s)| = O((a_n \log n)^{1/2}) \quad \text{a.s.} \tag{12.4.7}$$

where $\alpha_n(\cdot)$ is a uniform empirical process. Without loss of generality, we can assume that $M = 1$ in (12.4.2).

Divide the interval $[0,1]$ into K_n subintervals by points $t_0 = 0, t_j = j/K_n$, $j = 1, \cdots, K_n$, where $K_n = [a_n^{-1}\log n]$. Denote

$$V_n(t) = \sup_{|t-s|\le a_n} |\alpha_n(t) - \alpha_n(s)|.$$

For any fixed $t, s \in [0,1]$ with $|t-s| \le a_n$, there are two cases:

(i) s and t both fall into the same subinterval, i.e. there is a j, $0 \le j \le K_n - 1$ such that $s, t \in [t_j, t_{j+1}]$. Then

$$\begin{aligned} |\alpha_n(t) - \alpha_n(s)| &\le |\alpha_n(t) - \alpha_n(t_j)| + |\alpha_n(t_j) - \alpha_n(s)| \\ &\le 2\max_{0\le j\le K_n} V_n(t_j). \end{aligned}$$

(ii) s and t fall into the different subintervals, i.e. there exist j and $r, 1 \le j+1 \le r \le K_n$ such that $s \in [t_j, t_{j+1}]$, $t \in [t_r, t_{r+1}]$. Since $|s-t| \le a_n$, $t_r - t_{j+1} \le a_n$. Then

$$\begin{aligned} &|\alpha_n(t) - \alpha_n(s)| \\ &\quad\le |\alpha_n(t) - \alpha_n(t_r)| + |\alpha_n(t_r) - \alpha_n(t_{j+1})| + |\alpha_n(t_{j+1}) - \alpha_n(s)| \\ &\quad\le 3\max_{0\le j\le K_n} V_n(t_j). \end{aligned}$$

Hence, in any case, we have

$$\sup_{|t-s|\le a_n} |\alpha_n(t) - \alpha_n(s)| \le 3\max_{1\le j\le K_n} V_n(t_j). \tag{12.4.8}$$

For any fixed j, $1 \le j \le K_n$, divide the interval $[t_j - a_n, t_j + a_n]$ by points

$$\eta_{jr} = t_j + r\frac{a_n}{b_n}, \quad r = -b_n, -b_n + 1, \cdots, b_n - 1, b_n,$$

where $b_n = B[(na_n/\log n)^{1/2}]$, constant B will be specified later on. Denote $\phi_{jr} = n^{-1/2}|\alpha_n(t_j) - \alpha_n(\eta_{jr})|$. For any given $s \in [t_j - a_n, t_j + a_n]$, there is an r, $-b_n \le r \le b_n$ such that $s \in [\eta_{jr}, \eta_{j,r+1}]$. By monotoncity of the empirical distribution $E_n(t)$, we have

$$\begin{aligned} n^{-1/2}V_n(t_j) &= \sup_{|t_j - s| \le a_n} |(E_n(t_j) - t_j) - (E_n(s) - s)| \\ &\le \max_{-b_n \le r \le b_n} \max_{\eta_{jr} \le s \le \eta_{j,r+1}} |(E_n(t_j) - t_j) - (E_n(s) - s)| \\ &\le \max_{-b_n \le r \le b_n} \{|E_n(t_j) - t_j - (E_n(\eta_{jr}) - \eta_{j,r+1})|, \\ &\qquad |E_n(t_j) - t_j - (E_n(\eta_{j,r+1}) - \eta_{j,r+1})|\} \\ &\le \max_{-b_n \le r \le b_n} \{\phi_{jr}, \phi_{j,r+1}\} + |\eta_{j,r+1} - \eta_{jr}| \\ &\le \max_{-b_n \le r \le b_n} \{\phi_{jr}\} + a_n/b_n. \end{aligned} \tag{12.4.9}$$

Write

$$\phi_{jr} = \Big|\frac{1}{n}\sum_{i=1}^{n}[I(\eta_{jr} < U_i \le t_j) - (t_j - \eta_{jr})]\Big| =: \Big|\sum_{i=1}^{n} Z_i\Big|. \tag{12.4.10}$$

Obviously $|Z_i| \le 2/n$, $EZ_i = 0$ and $EZ_i^2 \le a_n/n^2$. Take $\varepsilon = B(a_n \log n/n)^{1/2}$ and $\alpha = (B^{-1}na_n^{-1}\log n)^{1/2}$. By the assumption of the theorem, for large B we have

$$\alpha m_n d = \Big(\frac{n\log n}{Ba_n}\Big)^{1/2} m_n \frac{2}{n} = 2\Big(\frac{\log n}{Bna_n}\Big)^{1/2} m_n \le \frac{1}{4}.$$

By Lemma 12.4.1 we have

$$\begin{aligned} &P\Big\{\Big|\sum_{i=1}^{n} Z_i\Big| > \varepsilon\Big\} \\ &\le C_1 \exp\Big\{-\alpha\varepsilon + C_1\alpha^2 a_n/n\Big\} \\ &\le C_1 \exp\Big\{-B^{1/2}\log n(1 - C_1/B^{3/2})\Big\}. \end{aligned} \tag{12.4.11}$$

where $C_1 = \exp(3\sqrt{e}A)$. Choosing B large enough, we have

$$\begin{aligned} &P\Big\{\max_{0 \le j \le K_n} \max_{-b_n \le r \le b_n} |\phi_{jr}| > B(a_n \log n/n)^{1/2}\Big\} \\ &\le cK_n b_n n^{-B^{1/2}/2} \le cn^{-2}. \end{aligned}$$

Therefore, from the Borel-Cantelli lemma it follows that

$$\max_{0\le j\le K_n}\max_{-b_n\le r\le b_n}|\phi_{jr}|\le B(a_n\log n/n)^{1/2}\quad \text{a.s.} \tag{12.4.12}$$

Combining it with (12.4.8) and (12.4.9) yields (12.4.7). Theorem 12.4.2 is proved.

Proof of Theorem 12.4.3.

The proof is along the same lines of that of Theorem 12.4.2 with Lemma 12.4.2 instead of Lemma 12.4.1. Let $0<\theta<1$, $\gamma=2/(1-\theta)$. It is clear that $E|Z_i|^\gamma\le 2n^{-\gamma}a_n$, $i=1,\cdots,n$. Hence $\sigma:=\sup\{\|Z_i\|_\gamma : i=1,\cdots,n\}\le 2^{1/\gamma}n^{-1}a_n^{1/\gamma}$ and $n^{1/2}\sigma\le 2^{1/\gamma}n^{-1/2}a_n^{1/\gamma}\le 1$. By Lemma 12.4.2 with $\varepsilon=\varepsilon_n=B(n^{-1/2}a_n^{(1-\theta)/2}\log^2 n)$, we have

$$\begin{aligned}P\{\phi_{jr}\ge\varepsilon_n\}&\le C_1\exp\{-C_2B(n^{-1}a_n^{1-\theta}\log^4 n)^{1/4}/n^{-1/4}a_n^{1/(2\gamma)}\}\\&\le C_1\exp\{-C_2B\log n\}.\end{aligned}$$

Therefore for large B, we have

$$P\Big\{\max_{0\le j\le K_n}\max_{-b_n\le r\le b_n}\phi_{jr}>\varepsilon_n\Big\}\le cn^{-2}.$$

Chapter 13 Convergence of Some Statistics with a Mixing Sample

Large sample theory in statistics is an important subject. In general, the sample is assumed to be independent. But, in some practical cases, the observations are dependent. In this part, we shall give some large sample properties for several interesting and useful statistics, such as U-statistics, error variance estimations in linear models, density function estimations with a mixing sample.

13.1 U-statistics

Let $\{X_n, n \geq 1\}$ be a strictly stationary sequence with a common distribution $F(\cdot)$, $h : R^m \to R$ be a symmetric function in its m arguments. A U-**statistic** is given by

$$U_n = \binom{n}{m}^{-1} \sum_{1 \leq t_1 < \cdots < t_m \leq n} h(X_{t_1}, \cdots, X_{t_m}) \quad n \geq m.$$

Here h is called a *kernel function* of U_n. This class of statistics was introduced by Hoeffding (1948) as a generalization of the sample mean. Many statistics of interest fall within this class or may be approximated by a member of this class.

Remark 13.1.1. U-statistics are closely connected with another class of statistics, the so-called *von-Mises statistics* (von Mises 1947) defined by

$$V_n = n^{-m} \sum_{t_1=1}^{n} \cdots \sum_{t_m=1}^{n} h(X_{t_1}, \cdots, X_{t_m}) \qquad n \geq 1.$$

These two kinds of statistics have similar limit behavior. So we discuss only U-statistic as a representative.

A kernel h is called degenerate (for the distribution F) if for all choices of $a_i, 1 \leq i \leq m$ and every $j \in \{1, \cdots, m\}$,

$$Eh(a_1, \cdots, a_{j-1}, X_j, a_{j+1}, \cdots, a_m) = 0;$$

a U-statistic will be called degenerate if the corresponding kernel has this property.

First of all, we introduce an important tool in deriving the asymptotic theory of U-statistics, Hoeffding's projection method. Put

$$\theta = \int \cdots \int h(x_1, \cdots, x_m) \prod_{k=1}^{m} dF(x_k), \tag{13.1.1}$$

$$h_r(x_1, \cdots, x_r) = \int \cdots \int h(x_1, \cdots, x_m) \prod_{i=r+1}^{m} dF(x_i), \quad r = 1, \cdots, m-1,$$

and

$$\widetilde{h}_r(x_1, \cdots, x_r) = h_r(x_1, \cdots, x_r) - \theta, \quad r = 1, \cdots, m-1.$$

The projection of U_n is defined as

$$\widehat{U}_n = \frac{m}{n} \sum_{i=1}^{n} \widetilde{h}_1(X_i) + \theta.$$

$U_n - \widehat{U}_n$ may itself be expressed as a U-statistic

$$\begin{aligned} U_n - \widehat{U}_n &= \binom{n}{m}^{-1} \sum_{1 \leq t_1 < \cdots < t_m \leq n} H(X_{t_1}, \cdots, X_{t_m}) \\ &=: R_n, \end{aligned} \tag{13.1.2}$$

where

$$H(x_1, \cdots, x_m) = h(x_1, \cdots, x_m) - \widetilde{h}_1(x_1) - \cdots - \widetilde{h}_1(x_m) - \theta$$

is a degenerate kernel. We call R_n the remainder of U_n.

At first, modifying definition of a φ-mixing sequence, we call a sequence $\{X_n, n \geq 1\}$ φ^*-mixing or φ-mixing in both directions of time if the sequence itself and the time reversed sequence are φ-mixing, that is

$$\varphi^*(n) := \sup_{k \in \mathbb{N}} \sup_{\substack{A \in \mathcal{F}_1^k, \\ B \in \mathcal{F}_{k+n}^{\infty}}} \max\{|P(B|A) - P(A)|, |P(B|A) - P(B)|\} \to 0 \quad \text{as} \quad n \to \infty.$$

Obviously, φ^*-mixing implies φ-mixing.

In this section, we shall establish weak and strong convergence for a φ^*-mixing sequence.

Denker and Keller (1983) proved the CLT and their rate of convergence, functional CLT and a.s. approximation by a Wiener process. Combining the results of weak convergence and strong approximation for a φ-mixing sequence in Chapters 5 and 9, we can weaken the conditions on moments and/or $\varphi^*(n)$.

13.1.1 Bounds for the remainder R_n

Let $\{X_n, n \geq 1\}$ be a strictly stationary φ^*-mixing sequence. Assume

$$s^2 := \sup_{1 \leq t_1 < \cdots < t_m} E(h(X_{t_1}, \cdots, X_{t_m}))^2 < \infty. \tag{13.1.3}$$

First, we cite two lemmas given by Denker and Keller (1983). Lemma 13.1.1 is a conditional version of Lemma 1.2.8 with $p = q = 2$. Let $\mathcal{A}, \mathcal{B}, \mathcal{B}_1, \mathcal{B}_2$ be sub-σ-fields of $\mathcal{F}$. For probabilities P and Q on $\mathcal{F}$ define the distance of P and Q over $\mathcal{A}$ given $\mathcal{B}$ by

$$d(P, Q : \mathcal{A}|\mathcal{B}) = \sup_{A \in \mathcal{A}, B \in \mathcal{B}} |P(A|B) - Q(A|B)|.$$

Moreover put

$$d(P : \mathcal{A}|\mathcal{B}) = \sup_{A \in \mathcal{A}, B \in \mathcal{B}} |P(A|B) - P(A)|.$$

Lemma 13.1.1. *Let f_1 and f_2 be an $\mathcal{A} \vee \mathcal{B}_1$- and $\mathcal{A} \vee \mathcal{B}_2$-measurable function, and let P, Q_1 and Q_2 be probability measures coinciding on $\mathcal{A}$. Then*

$$\begin{aligned}
&|E_P(f_1 f_2) - E_P[E_{Q_1}(f_1|\mathcal{A}) \cdot E_{Q_2}(f_2|\mathcal{A})]| \\
&\quad \leq (4 + 2\sqrt{2}) \\
&\quad\quad \cdot \max\{d^{1/2}(P : \mathcal{B}_1|\mathcal{A} \vee \mathcal{B}_2), d^{1/2}(P, Q_1 : \mathcal{B}_1|\mathcal{A}), d^{1/2}(P, Q_2 : \mathcal{B}_2|\mathcal{A})\} \\
&\quad\quad \cdot \{E_P(f_1^2)^{1/2}, E_{Q_1}(f_1^2)^{1/2}\} \cdot \max\{E_P(f_2^2)^{1/2}, E_{Q_2}(f_2^2)^{1/2}\}.
\end{aligned}$$

Lemma 13.1.2. *Let f be an $\mathcal{A} \vee \mathcal{B}$-measurable function, and let $P_n (n \geq 1)$ and Q be probability measures coinciding on $\mathcal{A}$. If*

$$\lim_{n \to \infty} d(P_n, Q : \mathcal{B}|\mathcal{A}) = 0,$$

then

$$E_Q|f| \le \liminf_{n\to\infty} E_{P_n}|f|.$$

Lemma 13.1.3. *For given $\varepsilon > 0$, there exists a $C = C_\varepsilon > 0$ such that*

$$ER_n^2 \le Cn^{-2+\varepsilon}s^2 \quad (n \ge m).$$

Proof. By (13.1.2), it suffices to estimate the variance of a degenerate U-statistics. We shall show: if h is degenerate, then

$$\binom{n}{m}^2 EU_n^2 \le cn^{2(m-1)+\varepsilon}s^2. \tag{13.1.4}$$

For $\mathbf{a} = (a_1, \cdots, a_m), \mathbf{b} = (b_1, \cdots, b_m) \in \mathbb{N}^m$, we put

$$W(\mathbf{a}, \mathbf{b}) = \sum h(X_{t_1}, \cdots, X_{t_m}), \tag{13.1.5}$$

where the summation extends over all indices $t_1, \cdots, t_m$ satisfying $a_i \le t_i \le b_i$ and $t_i \ne t_j$ for $1 \le i \ne j \le m$. Putting $\mathbf{1} = (1, \cdots, 1) \in \mathbb{N}^m$, we thus have

$$\binom{n}{m} U_n = \frac{1}{m!} W(\mathbf{1}, n\mathbf{1}). \tag{13.1.6}$$

For the estimation of $E(W(\mathbf{1}, n\mathbf{1}))^2$ we proceed recursively, decomposing $W(\mathbf{1}, n\mathbf{1})$ into sums over smaller index-blocks (This is inspired by the proof of Theorem 2.1.2).

We need some preparations: let $(X_n^{(1)})_{n\ge1}, \cdots, (X_n^{(m)})_{n\ge1}$ be m independent copies of the sequence $(X_n)_{n\ge1}$, and put for $q \in \{1, \cdots, m\}^m$

$$W(\mathbf{a}, \mathbf{b}; q) = \sum h(X_{t_1}^{(q_1)}, \cdots, X_{t_m}^{(q_m)}),$$

where the summation extends over the same index-set as in (13.1.5). Let

$$\mathrm{I}_n = \{(\mathbf{a}, \mathbf{b}) \in \mathbb{N}^{2m} : b_i = a_i + n - 1, a_i = a_j \text{ or } |a_i - a_j| \ge n \text{ for all } i, j\}$$

and define

$$\tau(n) = \sup\{E(W(\mathbf{a}, \mathbf{b}; q))^2 : (\mathbf{a}, \mathbf{b}) \in \mathrm{I}_n, q \in \{1, \cdots, m\}^m\}.$$

Consider fixed non-negative integers k, l, p, n with $n = kl + p$. For each $(\mathbf{a}, \mathbf{b}) \in \mathrm{I}_n$ and $q \in \{1, \cdots, m\}^m$ we have by the Hölder inequality and the triangle inequality:

$$\begin{aligned} &|E(W(\mathbf{a}, \mathbf{b}; q))^2 - E(W(\mathbf{a}, \mathbf{b} - p\mathbf{1}; q))^2| \\ &\qquad \le (\tau(n)^{1/2} + k^m \tau(l)^{1/2}) mpn^{m-1}s, \end{aligned} \tag{13.1.7}$$

where we have tacitly assumed that

$$\sup_{1 \leq t_1 < \cdots < t_m} E(h(X_{t_1}^{(q_1)}, \cdots, X_{t_m}^{(q_m)}))^2 \leq s^2 \tag{13.1.8}$$

for all $q \in \{1, \cdots, m\}^m$, what can easily be proved using repeatedly Lemma 13.1.2.

Now decompose $W(\mathbf{a}, \mathbf{b} - p\mathbf{1}; q)^2$ as

$$W(\mathbf{a}, \mathbf{b} - p\mathbf{1}; q)^2 = \sum_{u,v \in \{0, \cdots, k-1\}^m} W(\mathbf{a} + l\mathbf{u}, \mathbf{a} + l(\mathbf{u} + \mathbf{1}) - \mathbf{1}; q) \cdot W(\mathbf{a} + l\mathbf{v}, \mathbf{a} + l(\mathbf{v} + \mathbf{1}) - \mathbf{1}; q).$$

Each $\Gamma(\mathbf{u}) := W(\mathbf{a} + l\mathbf{u}, \mathbf{a} + l(\mathbf{u} + \mathbf{1}) - \mathbf{1}; q)$ is determined by m blocks of time coordinates of length l (possibly counting a block several times). Denote for a fixed pair $(\mathbf{u}, \mathbf{v})$ these $2m$ blocks by $B_1, \cdots, B_{2m}$ and assume $\inf B_i \leq \inf B_{i+1} (i = 1, \cdots, 2m - 1)$. With this convention it is not hard to see that

$$\begin{aligned} \operatorname{Card}\{(\mathbf{u}, \mathbf{v}) : \sup B_1 + l \geq \inf B_2 \text{ and } \inf B_{2m} - l \leq \sup B_{2m-1}\} \\ \leq C_m k^{2m-2}, \end{aligned} \tag{13.1.9}$$

where C_m is a combinatorial constant depending only on m. For all pairs $(\mathbf{u}, \mathbf{v})$ not belonging to the set described in (13.1.9) we can apply Lemma 13.1.1 in the following way: assume that $\sup B_1 + l < \inf B_2$ and that B_1 is a block stemming from $\Gamma(\mathbf{u})$. (The remaining three cases are treated exactly in the same way.)

Choose $f_1 = \Gamma(\mathbf{u}), f_2 = \Gamma(\mathbf{v}), \mathcal{B}_1 = \sigma(X_t, t \in B_1), \mathcal{B}_2$ trivial and

$$\mathcal{A} = \sigma(X_t, t \in B_1 \cup \cdots \cup B_{2m}).$$

Observing (13.1.8) we then get

$$|E\Gamma(\mathbf{u})\Gamma(\mathbf{v})| \leq (4 + 2\sqrt{2})\varphi^*(l)^{1/2}\tau(l),$$

because h is degenerate. It is for this application of Lemma 13.1.1 that we have to introduce independent copies of the original process, and since in two of above mentioned four cases, we have to single out the block B_{2m} (instead of B_1), we need φ-mixing in both directions of time. Combining the last estimate with (13.1.7) and (13.1.9) we obtain

$$\begin{aligned} W(\mathbf{a}, \mathbf{b}; q)^2 \leq\ & C_m k^{2m-2}\tau(h) + k^{2m}(4 + 2\sqrt{2})\varphi^*(h)^{1/2}\tau(l) \\ & + (\tau(n)^{1/2} + k^m \tau(l)^{1/2}) mpn^{m-1}s \end{aligned}$$

and taking the supremum over $(\mathbf{a}, \mathbf{b}) \in \mathbf{I}_n$ and $q \in \{1, \cdots, m\}^m$

$$\tau(n) \le k^{2m-2}\tau(l)(C_m + (4+2\sqrt{2})k^2\varphi^*(l)^{1/2}) \\ + (\tau(n)^{1/2} + k^m\tau(l)^{1/2})mpn^{m-1}s. \qquad (13.1.10)$$

Take k to be large enough such that $(C_m + (4+2\sqrt{2}))k^{-\varepsilon} \le 1/4$. Then choose $n_0 = \min\{s : k^2\varphi^*(s)^{1/2} \le 1\}$. For $l \ge n_0$ and $p \le k$, (13.1.10) implies

$$(\tau(n)^{1/2} - \frac{1}{2}mkn^{m-1}s)^2 \le (\frac{1}{2}k^{m-1+\varepsilon/2}\tau(l)^{1/2} + k^{1-\varepsilon/2}mkn^{m-1}s)^2,$$

and hence

$$\tau(n)^{1/2} \le \frac{1}{2}k^{m-1+\varepsilon/2}\tau(l)^{1/2} + (\frac{1}{2} + k^{1-\varepsilon/2})mkn^{m-1}s. \qquad (13.1.11)$$

Given n choose $l_0, l_1, \cdots, l_r$ such that $l_0 = n, l_{i-1} = kl_i + p_i$ for some $0 \le p_i < k (i = 1, \cdots, r)$ and $n_0 \le l_r < kn_0$. Apply (13.1.11) to each pair (l_{i-1}, l_i) to obtain

$$\tau(l_{i-1})^{1/2} \le \frac{1}{2}k^{m-1+\varepsilon/2}\tau(l_i)^{1/2} + Al_{i-1}^{m-1}s,$$

where $A = (\frac{1}{2} + k^{1-\varepsilon/2})mk$. By induction this leads to

$$\begin{aligned} \tau(n)^{1/2} &\le 2^{-r}k^{r(m-1+\varepsilon/2)}\tau(l_r)^{1/2} + An^{m-1+\varepsilon/2}s\sum_{j=0}^{r-1}2^{-j} \\ &\le n^{m-1+\varepsilon/2}\Big(\frac{1}{2^r}\frac{l_r^m s}{l_r^{m-1+\varepsilon/2}} + 2As\Big) \\ &\le n^{m-1+\varepsilon/2}((kn_0)^{1-\varepsilon/2} + 2A)s, \end{aligned}$$

since $n \ge k^r l_r$ and $l_r \le kn_0$. If we put $C = ((kn_0)^{1-\varepsilon/2} + 2A)^2(m!)^{-2}$, we get

$$\tau(n) \le Cs^2n^{2m-2+\varepsilon}(m!)^2 \qquad (13.1.12)$$

and (13.1.4) follows from (13.1.6).

Lemma 13.1.4. *Assume that condition (13.1.3) is satisfied. Then for any $\varepsilon > 0$ and $c_N > 0$ we have*

$$R_n = O(n^{-3/4+\varepsilon}) \quad a.s.$$

and

$$P\Big\{\max_{1\le n\le N} n|R_n| \ge c_N\Big\} = O(N^{1/2+\varepsilon}c_N^{-2}).$$

Proof. By (13.1.2) it clearly suffices to prove the lemma for a degenerate U-statistic with kernel h. Put

$$Z(p,q) = W(\mathbf{1},(p+q)\mathbf{1}) - W(\mathbf{1},p\mathbf{1}) \quad (p,q \in \mathbb{N}).$$

If $2^{r-1} \le n < 2^r$ and $n = \sum_{i=1}^r d_i 2^{r-i}$ denotes the dyadic expansion of n,

$$W(\mathbf{1},n\mathbf{1}) = Z(0,n) = \sum_{k=1}^{r} Z\Big(\sum_{i=1}^{k-1} d_i 2^{r-i}, d_k 2^{r-k}\Big).$$

For $r,u \in \mathbb{N}, l = 1,\cdots,r$ and $j = 1,\cdots,2^l$ consider the sets

$$E_{j,l}^{r,u} = \{|Z((j-1)2^{r-l}, 2^{r-l})| \ge \alpha_{r,u}\},$$

where $\alpha_{r,u}$ are constants to be chosen later. We shall show below that

$$E(Z(p,q))^2 = O(q(p+q)^{b-1}), \quad b = 2m - 3/2 + \varepsilon. \tag{13.1.13}$$

By the Chebyshev inequality $P(E_{j,l}^{r,u}) = O(\alpha_{r,u}^{-2} 2^{b(r-l)} j^{b-1})$.

The a.s. bound for R_n follows now from the Borel-Cantelli lemma, because for $\alpha_{r,1} = 2^{b(r-1)/2}(r-1)^3/r$ we have

$$\begin{aligned}
&P\Big\{\max_{2^{r-1}\le n\le 2^r} |Z(0,n)| \ge n^{b/2}(\log n)^3\Big\} \\
&\le \sum_{l=1}^{r}\sum_{j=1}^{2^l} P(E_{j,l}^{r,1}) + O(r^{-3}).
\end{aligned}$$

To prove the maximal inequality, let R, N be given such that $2^{R-1} \le N < 2^R$. Putting $\alpha_{r,N} = 2^{(r-1)(m-1)r^{-1}} c_N$ for $r \le R$, it follows that

$$\begin{aligned}
&P\Big\{\max_{1\le n\le N} n^{-m+1}|Z(0,n)| \ge c_N\Big\} \\
&\le P\Big\{\bigcup_{r=1}^{R}\bigcup_{n=2^{r-1}}^{2^r-1}\{|Z(0,n)| \ge n^{m-1}c_N\}\Big\} \\
&\le \sum_{r=1}^{R}\sum_{l=1}^{r}\sum_{j=1}^{2^l} P(E_{j,l}^{r,N}) = O(N^{1/2+\varepsilon}(\log N)^3 c_N^{-2}).
\end{aligned}$$

We still have to prove (13.1.13). If $q > p$, it follows from (13.1.12) that

$$\begin{aligned}
E(Z(p,q))^2 &\le \{(EZ(0,p)^2)^{1/2} + (EZ(0,p+q)^2)^{1/2}\}^2 \\
&= O((p+q)^{2m-2+\varepsilon}) = O(q(p+q)^{b-3/2}).
\end{aligned}$$

If $p \geq q \geq p^{1/2}$ we obtain similarly

$$E(Z(p+q))^2 = O((p+q)^{2m-2+\varepsilon}) = O(q(p+q)^{b-1}).$$

Now consider the case of $q < p^{1/2}$. Put $\widetilde{I}_n = \{(\mathbf{a}, \mathbf{b}) \in \mathbb{N}^{2m} : b_i = a_i+n-1$ for exact $m-1$ coordinates and $a_i = b_i$ else; $a_i = a_j$; or $|a_i - a_j| \geq n$ for all $i, j\}$ and define

$$\widetilde{\tau}(n) = \sup\{E(W(\mathbf{a}, \mathbf{b}; q))^2 : (\mathbf{a}, \mathbf{b}) \in \widetilde{I}_n, q \in \{1, \cdots, m\}^m\}.$$

Then similarly to (13.1.12), we can show

$$\widetilde{\tau}(n) \leq Cn^{2m-3+\varepsilon} s^2$$

for some $C > 0$. Hence we obtain

$$\begin{aligned} E(Z(p,q))^2 &= E\Big(\sum_{k=0}^{q-1} Z(p+k,1)\Big)^2 \\ &\leq \Big(\sum_{k=0}^{q-1} (E(Z(p+k,1))^2)^{1/2}\Big)^2 \\ &= O(q^2(p+q)^{2m-3+\varepsilon}) = O(q(p+q)^{b-1}). \end{aligned}$$

This finishes the proof of Lemma 13.1.4.

13.1.2 WIP for U_n

Let

$$\sigma_n^2 = E\Big(\sum_{i=1}^{n} \widetilde{h}_1(X_i)\Big)^2,$$

$$W_n(t) = \frac{nt}{m\sigma_n}(U_{[nt]} - \theta), \quad 0 \leq t \leq 1.$$

Theorem 13.1.1. *Let h be a non-degenerate kernel. Assume that condition (13.1.3) is satisfied and $\sigma_n^2 \to \infty$. Moreover, assume that for any $\varepsilon > 0$*

$$\lim_{n\to\infty} \frac{n}{\sigma_n^2} E\widetilde{h}_1(X_1)^2 I(|\widetilde{h}_1(X_1)| > \varepsilon\sigma_n) = 0.$$

Then $W_n \Longrightarrow W$ as $n \to \infty$.

Proof. Put

$$\widehat{W}_n(t) = \frac{nt}{m\sigma_n}(\widehat{U}_{[nt]} - \theta), \quad 0 \leq t \leq 1.$$

By Corollary 5.1.4, $\widehat{W}_n \Longrightarrow W$ as $n \to \infty$. Theorem 2.1.2 implies that $\sigma_n^2/n^{1-\varepsilon} \to \infty$ as $n \to \infty$ for any $\varepsilon > 0$. Hence by the maximal inequality of Lemma 13.1.4 we obtain

$$\begin{aligned} &P\Big\{\sup_{0\le t\le 1} \frac{nt}{m\sigma_n}|R_{[nt]}| \ge \varepsilon\Big\} \\ &\le P\Big\{\sup_{0\le t\le 1} nt|R_{[nt]}| \ge \varepsilon m n^{(1-\varepsilon)/2}\Big\} \\ &= O(n^{-1/2+2\varepsilon}). \end{aligned}$$

Hence the result follows from (13.1.2).

Remark 13.1.1. Denker and Keller (1983) and Zhang (1989) discussed the Berry-Esseen inequality for U-statistics with a φ^*-mixing sample. We write the Zhang's result without any proof:

Let $\{X_n, n \ge 1\}$ be a strictly stationary φ^*-mixing sequence, U_n be a U-statistic with kernel $h(x_1, x_2)$. Denote

$$\sigma^2 = Eh_1^2(X_1) - \theta^2 + 2\sum_{k=1}^{\infty}\{Eh_1(X_1)h_1(X_{k+1}) - \theta^2\}.$$

Suppose that there exist constants $C > 0, \beta > 0$ such that

$$\varphi^*(n) \le Ce^{-\beta n}.$$

If $\sigma^2 > 0$ and

$$\sup_{1\le i<j\le n} E|h(X_i, X_j)|^3 < \infty,$$

then for any $\varepsilon > 0$ we have

$$\sup_x \Big|P\Big\{\frac{U_n - \theta}{\sqrt{VarU_n}} \le x\Big\} - \Phi(x)\Big| \le cn^{-1/2+\varepsilon}.$$

13.1.3 Strong approximation by a Wiener process for U_n

Theorem 13.1.2. *Let h be a non-degenerate kernel. Assume that*

$$\sup_{1\le t_1<\cdots<t_m} E|h(X_{t_1}, \cdots, X_{t_m})|^{2+\delta} < \infty \quad \textit{for some} \quad \delta > 0. \qquad (13.1.14)$$

Moreover, assume that

(i) $\sigma_n^2 \ge c_1 n$ *for some* $c_1 > 0$,

(ii) $\varphi^*(n) \le c_2 n^{-\alpha}$ *for some* $c_2 > 0$ *and* $\alpha > 0$.

Then one can redefine $\{X_n, n \geq 1\}$ *without changing its distribution on a richer probability space together with a Wiener process* $W(\cdot)$ *such that*

$$\frac{n}{m}(U_n - \theta) - W(\sigma_n^2) = O(\sigma_n^{2/(2+\delta)}(\log \sigma_n)^{1+\varepsilon+(1+\lambda)/(2+\delta)}) \quad a.s.$$

for any $\varepsilon > 0$, *where* $\lambda = 2(\log 3)/\log \tau^{-1}$ *and* $\tau = 1 - 2(\alpha - 1)/\alpha(2+\delta)$.

Proof. By Lemma 13.1.2, condition (13.1.14) implies that $E|\widetilde{h}_1(X_1)|^{2+\delta} < \infty$. Therefore from Remark 9.1.1, we conclude that we can redefine the sequence $\{\widetilde{h}_1(X_n), n \geq 1\}$ on a new probability space together with a Wiener process $W(\cdot)$ such that

$$\sum_{i=1}^{n} \widetilde{h}_1(X_i) - W(\sigma_n^2) = O(\sigma_n^{2/(2+\delta)}(\log \sigma_n)^{1+\varepsilon+(1+\lambda)/(2+\delta)}) \quad \text{a.s.}$$

In fact, it is not hard to see that on the new probabilty space we also can redefine $\{X_n\}$ itself, for example by considering strong invariance principles for R^2-valued random vectors. By Lemma 13.1.4, $nR_n = O(n^{1/4+\varepsilon})$ a.s., hence the theorem is proved.

13.1.4 SLLN for U_n

Wang (1994) proved a SLLN for U-statistics with a φ^*-mixing sample. We consider only the case of $m = 2$. In order to prove the theorem, We need the following lemma, which was proved in the proof of Theorem 3 in Babbel (1989):

Lemma 13.1.5. *Let* h *be a degenerate kernel. Suppose that condition (13.1.3) is satisfied and*

$$\varphi^*(n) = O(n^{-(4+\delta)}) \quad \textit{for some } \delta > 0. \tag{13.1.15}$$

Then

$$E\Big(\max_{1\leq i\leq n} U_i\Big)^2 \leq cn^{-2}s^2.$$

Theorem 13.1.3. *Assume that condition (13.1.15) is satisfied and*

$$\sup_{n\geq 2} E|h(X_1, X_n)| < \infty. \tag{13.1.16}$$

Then we have

$$U_n \longrightarrow \theta \quad a.s. \quad as \quad n \to \infty.$$

where $\theta = \int\int h(x_1, x_2)dF(x_1)dF(x_2)$.

Proof. For $k \in \mathbb{N}$, put

$$\begin{aligned}
h^{(k)}(x_1, x_2) &= h(x_1, x_2) I(|h(x_1, x_2)| \le 2^{2k}), \\
\theta^{(k)} &= \int\int h^{(k)}(x_1, x_2) dF(x_1) dF(x_2), \\
\widetilde{h}_1^{(k)}(x) &= \int h^{(k)}(x, y) dF(y), \\
H^{(k)}(x_1, x_2) &= h^{(k)}(x_1, x_2) - \widetilde{h}_1^{(k)}(x_1) - \widetilde{h}_1^{(k)}(x_2) + \theta^{(k)}
\end{aligned}$$

and

$$\begin{aligned}
U_n^{(k)} &= \binom{n}{2}^{-1} \sum_{1 \le i < j \le n} h^{(k)}(X_i, X_j), \\
\widetilde{U}_n^{(k)} &= \frac{2}{n} \sum_{j=1}^{n} \widetilde{h}_n^{(k)}(X_j), \\
\Delta_n^{(k)} &= \binom{n}{2}^{-1} \sum_{1 \le i < j \le n} H^{(k)}(X_i, X_j).
\end{aligned}$$

The last one is a U-statistic with a degenerate kernel $H^{(k)}$. It is easy to see that

$$U_n^{(k)} = \widetilde{U}_n^{(k)} + \Delta_n^{(k)} - \theta^{(k)}$$

and hence we can write

$$\begin{aligned}
&\limsup_{n \to \infty} |U_n - \theta| \\
&\qquad \le \limsup_{k \to \infty} \max_{2^k \le n < 2^{k+1}} \{ |\widetilde{U}_n^{(k)} - 2\theta^{(k)}| + |\Delta_n^{(k)}| \\
&\qquad + |U_n - U_n^{(k)}| + |\theta - \theta^{(k)}| \}. \qquad (13.1.17)
\end{aligned}$$

Obviously, condition (13.1.16) implies that

$$\theta - \theta^{(k)} \to 0 \quad \text{as } k \to \infty. \qquad (13.1.18)$$

Applying Lemma 13.1.5 and condition (13.1.16), we have for any $\varepsilon > 0$

$$\begin{aligned}
&\sum_{k=1}^{\infty} P\Big\{\max_{2^k \le n < 2^{k+1}} |\Delta_n^{(k)}| \ge \varepsilon\Big\} \\
&\le c\varepsilon^{-2} \sum_{k=1}^{\infty} 2^{-2k} \sup_{n\ge 2} Eh^2(X_1, X_n) I(|h(X_1, X_n)| \le 2^{2k}) \\
&\le c\varepsilon^{-2} \sup_{n\ge 2} \sum_{k=1}^{\infty} 2^{-2k} \sum_{j=1}^{k} Eh^2(X_1, X_n) I(2^{2(j-1)} < |h(X_1, X_n)| \le 2^{2j}) \\
&\le c\varepsilon^{-2} \sup_{n\ge 2} \sum_{k=1}^{\infty} Eh^2(X_1, X_n) I(2^{2(j-1)} < |h(X_1, X_n)| \le 2^{2j}) \sum_{k=j}^{\infty} 2^{-2k} \\
&\le c\varepsilon^{-2} \sup_{n\ge 2} E|h(X_1, X_n)| < \infty,
\end{aligned}$$

which implies

$$\limsup_{k\to\infty} \max_{2^k \le n < 2^{k+1}} |\Delta_n^{(k)}| = 0 \quad \text{a.s.} \tag{13.1.19}$$

Moreover

$$\begin{aligned}
&\sum_{k=1}^{\infty} P\Big\{\bigcup_{2^k \le n < 2^{k+1}} \{U_n \ne U_n^{(k)}\}\Big\} \\
&\le \sum_{k=1}^{\infty} P\Big\{\bigcup_{2^k \le n < 2^{k+1}} \Big(\bigcup_{1\le i<j\le n} \{|h(X_i, X_j)| \ge 2^{2k}\}\Big)\Big\} \\
&\le \sum_{k=1}^{\infty} 2^{2(k+1)} \sup_{n\ge 2} P\{|h(X_1, X_n)| \ge 2^{2k}\} \\
&\le 4 \sup_{n\ge 2} \sum_{k=1}^{\infty} 2^{2k} \sum_{j=k}^{\infty} P\{2^{2j} \le |h(X_1, X_n)| < 2^{2(j+1)}\} \\
&\le c \sup_{n\ge 2} \sum_{j=1}^{\infty} E|h(X_1, X_n)| I(2^{2j} \le |h(X_1, X_n)| < 2^{2(j+1)}) \\
&\le c \sup_{n\ge 2} E|h(X_1, X_n)| < \infty,
\end{aligned}$$

which implies

$$\limsup_{k\to\infty} \max_{2^k \le n < 2^{k+1}} |U_n - U_n^{(k)}| = 0 \quad \text{a.s.} \tag{13.1.20}$$

Estimate the first term of the right hand side of (13.1.17). Put

$$\widehat{h}_1^{(k)}(x) = \widetilde{h}_1^{(k)}(x) I(|\widetilde{h}_1^{(k)}(x)| \le 2^k)$$

and write

$$
\begin{aligned}
|\widetilde{U}_n^{(k)} - 2\theta^{(k)}| \le & \Big|\frac{2}{n}\sum_{i=1}^{n}(\widehat{h}_1^{(k)}(X_i) - E\widehat{h}_1^{(k)}(X_i))\Big| \\
& + \Big|\frac{2}{n}\sum_{i=1}^{n}(\widetilde{h}_1^{(k)}(X_i) - \widehat{h}_1^{(k)}(X_i))\Big| \\
& + 2|\theta^{(k)} - E\widehat{h}_1^{(k)}(X_1)|. \qquad (13.1.21)
\end{aligned}
$$

Consider $|\theta^{(k)} - E\widehat{h}_1^{(k)}(X_1)|$ first. We have

$$
|\theta^{(k)} - E\widehat{h}_1^{(k)}(X_1)| = |E\widetilde{h}_1^{(k)}(X_1)I(|\widetilde{h}_1^{(k)}(X_1)| > 2^k)| \to 0 \qquad (13.1.22)
$$

as $k \to \infty$. Similarly to (13.1.20), we have

$$
\sum_{k=1}^{\infty} P\Big\{\bigcup_{2^k \le n < 2^{k+1}} \Big\{\sum_{i=1}^{n} \widetilde{h}_1^{(k)}(X_i) \ne \sum_{i=1}^{n} \widehat{h}_1^{(k)}(X_i)\Big\}\Big\} < \infty,
$$

which implies

$$
\limsup_{k\to\infty} \max_{2^k \le n < 2^{k+1}} \Big|\frac{2}{n}\sum_{i=1}^{n}(\widetilde{h}_1^{(k)}(X_i) - \widehat{h}_1^{(k)}(X_i))\Big| = 0 \quad \text{a.s.} \qquad (13.1.23)
$$

Furthermore, $\{\widehat{h}_1^{(k)}(X_n) - E\widehat{h}_1^{(k)}(X_n), n \ge 1\}$ is a strictly stationary and φ^*-mixing sequence with mixing coefficients $\varphi^*(n) = O(n^{-(4+\delta)})$. Hence using Lemma 2.2.10 we obtain

$$
\begin{aligned}
\sum_{k=1}^{\infty} & P\Big\{\max_{2^k \le n < 2^{k+1}} \Big|\frac{2}{n}\sum_{i=1}^{n}(\widehat{h}_1^{(k)}(X_i) - E\widehat{h}_1^{(k)}(X_i))\Big| \ge \varepsilon\Big\} \\
& \le c\sum_{k=1}^{\infty} 2^{-k} E(\widehat{h}_1^{(k)}(X_1))^2 \\
& \le c\sum_{k=1}^{\infty} E|\widetilde{h}_1^{(k)}(X_1)| < \infty,
\end{aligned}
$$

which implies

$$
\limsup_{k\to\infty} \max_{2^k \le n < 2^{k+1}} \frac{2}{n}\sum_{i=1}^{n}(\widehat{h}_1^{(k)}(X_i) - E\widehat{h}_1^{(k)}(X_i)) = 0 \quad \text{a.s.} \qquad (13.1.24)
$$

Combining (13.1.22)-(13.1.24) with (13.1.21) yields

$$
\limsup_{k\to\infty} \max_{2^k \le n < 2^{k+1}} |\widetilde{U}_n^{(k)} - 2\theta^{(k)}| = 0 \quad \text{a.s.} \qquad (13.1.25)
$$

(13.1.17)-(13.1.20) and (13.1.25) together imply the conclusion of Theorem 13.1.3.

13.2 Error variance estimations in linear models

Consider a linear regression model

$$Y_i = x_i'\beta + e_i, \quad i = 1, 2, \cdots,$$

where $\{x_i\}$ is a known p-demensional design sequence, $\{Y_i\}$ is a sequence of observed response, β is an unknown p-dimensional vector, and $\{e_i\}$ is a random error sequence which is strictly stationary, with

$$Ee_1 = 0, \quad \sigma^2 := Ee_1^2 > 0, \quad \nu := Ee_1^4 < \infty. \tag{13.2.1}$$

On the basis of the residual sum of squares, estimation of σ^2 is

$$\widehat{\sigma}^2(n) = \frac{1}{n - r_n}\Big\{\sum_{j=1}^{n} e_j^2 - \sum_{i=1}^{r_n}\Big(\sum_{j=1}^{n} a_{ij}^{(n)} e_j\Big)^2\Big\},$$

where r_n is the rank of matrix $X_n = (x_1, \cdots, x_n)'$ and stable to $r_n \le p$ when n is large enough, and $(a_{ij}^{(n)})$ is an n-th real orthogonal matrix decided by design matrix X_n. Put

$$\tau = \nu - \sigma^4 + 2\sum_{j=2}^{\infty} E(e_1^2 - \sigma^2)(e_j^2 - \sigma^2).$$

If $0 < \tau < \infty$, define random functions $Z_n(\cdot)$ of $C[0,1]$ as follows:

$$Z_n(0) = 0, \quad Z_n(i/n) = (i - r_i)(\widehat{\sigma}_i^2 - \sigma^2)/\sqrt{n\tau}, \quad i = 1, \cdots, n,$$

$$Z_n(\cdot) \text{ is linear in } \Big[\frac{i-1}{n}, \frac{i}{n}\Big].$$

Moreover define the random function $Z(\cdot)$ of $C[0,\infty]$ by

$$Z(t) = ([t] - r_{[t]})(\widehat{\sigma}^2([t]) - \sigma^2).$$

Lin (1984) and Lu (1986) studied weak invariance principle and strong approximation for $Z_n(\cdot)$ and $Z(\cdot)$ respectively. We shall concentrate our attention on the sequence $\{e_i\}$ which is φ-mixing. Similar methods can be used to study other kinds of mixing error sequences.

13.2.1 Weak invariance principle

For any $q > 0$, define

$$a(q) = \inf\Big\{a : \sup_{A:P(A)\leq \nu/q^4} \sup_i Ee_i^4 I_A \leq a\Big\}. \tag{13.2.2}$$

It is clear that $a(q) \downarrow 0$ as $q \uparrow \infty$. Denote the inverse function of $a = a(q)$ by $q = q(a)$. Obviously, it is non-increasing. Let $a = d(t)$ be the solution of the equation $a/q(a)^5 = t$. Putting $g(b) = d(2^{1/4}/b)$, we have $g(b) \downarrow 0$ as $b \to \infty$. Let $t_n(\uparrow \infty)$ be maximum integers satisfying $t_n^2 g(t_n^{-1/2} n^{1/4}) = o(1)$. It is easy to see the existence of such t_n. Lin (1984) showed:

Theorem 13.2.1. *Let the random error sequence $\{e_i\}$ with (13.2.1) be strictly stationary and φ-mixing. Suppose that mixing coefficients $\varphi(n)$ satisfy*

(i) $\sum_{n=1}^{\infty} \varphi^{1/2}(n) < \infty$.

Then $\tau < \infty$. If, in addition, $\tau > 0$ and

(ii) $nt_n^{-1}\varphi(t_n) = o(1)$,

then

$$Z_n \Longrightarrow W \quad \text{as } n \to \infty.$$

In order to prove the WIP, we need some lemmas.

Lemma 13.2.1. *Let $\{\alpha_i^{(n)}, i = 1, \cdots, n\}$ be a sequence of series of random variables, which are independent within each series, satisfying*

$$E{\alpha_i^{(n)}}^4 I(|\alpha_i^{(n)}| \geq q) \leq a(q), \quad i = 1, \cdots, n, \tag{13.2.3}$$

where $a(q)$ is defined in (13.2.2). Then we have

$$b^4 P(|X| \geq b) = O(g(b)) \quad \text{as} \quad b \to \infty \tag{13.2.4}$$

uniformly in the class $\mathcal{F}$ of the random variables with the form of $X = \sum_{i=1}^{n} \cdot a_i \alpha_i^{(n)}$ where a_k satisfy $\sum_{k=1}^{n} a_k^2 \leq 1$.

Proof. Let $f_i := f_{in}$ and $F_i := F_{in}$ be the characteristic function and the distribution function of $\alpha_i^{(n)}$ respectively, $f_i^{(k)}$ k-th derivative of f_i. We first show that

$$f_i(t) = \sum_{k=0}^{4} \frac{f_i^{(k)}(0)}{k!} t^k + o(t^4) \quad \text{as } t \to 0 \tag{13.2.5}$$

uniformly in $i = 1, \cdots, n$. To this end, write

$$f_i(t) - \sum_{k=0}^{4} \frac{f_i^{(k)}(0)}{k!} t^4 = -\frac{t^4}{4!} \int_{-\infty}^{\infty} (1 - e^{it\theta x}) x^4 dF_i(x), \tag{13.2.6}$$

where $|\theta| \leq 1$. Put $t = \pm a(q)/q^5$. For $|x| \leq q$,

$$|1 - e^{it\theta x}| \leq |t\theta x| \leq a(q)/q^4.$$

Hence (13.2.6) and (13.2.3) imply that

$$\Big| f_i(t) - \sum_{k=0}^{4} \frac{f_i^{(k)}(0)}{k!} t^4 \Big| \leq \frac{t^4}{24} \Big\{ \int_{|x|<q} |1 - e^{it\theta x}| x^4 dF_i(x) + \int_{|x|\geq q} 2x^4 dF_i(x) \Big\} \leq t^4 a(q)/8$$

for $i = 1, \cdots, n$. Note that $a(q)$ is independent of i and $a(q) \downarrow 0$ as $q \to \infty$. Uniformity of (13.2.5) is proved. (13.2.5) can be rewritten as

$$\log f_i(t) = \sum_{j=1}^{4} b_{ij} t^j + g_i(t), \quad i = 1, \cdots, n, \tag{13.2.7}$$

where $g_i(t)/t^4 = O(a(q))$ as $t \to 0 (q \to \infty)$ uniformly in $i = 1, \cdots, n$. Let f and F be the characteristic function and the distribution function of $X = \sum_{i=1}^{n} a_i \alpha_i^{(n)}$ respectively. Then (13.2.7) implies

$$\log f(t) = \sum_{i=1}^{n} \log f_i(a_i t) = \sum_{j=1}^{4} \Big(\sum_{i=1}^{n} b_{ij} a_i^j \Big) t^j + \sum_{i=1}^{n} g_i(a_i t)$$

where

$$\sum_{i=1}^{n} g_i(a_i t) \leq c a(q) \sum_{i=1}^{n} (a_i t)^4 \leq c d(t) t^4.$$

And further

$$\begin{aligned} f(t) &= \sum_{k=0}^{4} \frac{f^{(k)}(0)}{k!} t^k + O(t^5 + d(t) t^4) \\ &= \sum_{k=0}^{4} \frac{f^{(k)}(0)}{k!} t^k + O(d(t) t^4), \end{aligned}$$

since $t/d(t) \to 0$ as $t \to 0$ by the definition of $d(\cdot)$. Then, uniformly in f

$$\begin{aligned}
O(d(t)) &= f^{(4)}(0) - \{f(4t) - 4f(2t) + 6f(0) \\
&\quad - 4f(-2t) + f(-4t)\}/(2t)^4 \\
&= \int x^4 dF(x) - \int \Big(\frac{e^{itx} - e^{-itx}}{2t}\Big)^4 dF(x) \\
&= \int \Big(1 - \Big(\frac{\sin tx}{tx}\Big)^4\Big) x^4 dF(x) \\
&\geq \int_{|x|\geq b} \Big(1 - \Big(\frac{\sin tx}{tx}\Big)^4\Big) x^4 dF(x) \\
&\geq \frac{1}{2} \int_{|x|\geq b} x^4 dF(x) \geq \frac{1}{2} b^4 P(|X| \geq b)
\end{aligned}$$

provided $(bt)^4 \geq 2$, which implies (13.2.4).

Remark 13.2.1. Similarly, if we assume that for some integer $m > 0$ and δ, $0 \leq \delta < 1$,

$$\max_{1\leq i\leq n} E|\alpha_i^{(n)}|^{m+\delta} I(|\alpha_i^{(n)}| > q) \longrightarrow 0 \quad \text{as } q \to \infty$$

instead of condition (13.2.3), we have

$$P\{|X| \geq b\} = o(b^{-m-\delta}) \quad \text{as } b \to \infty$$

uniformly in $\mathcal{F}$ defined in Lemma 13.2.1.

Lemma 13.2.2. *Under the conditions (i) and (ii) in Theorem 13.2.1, for any* $\varepsilon > 0$

$$nP(|X| \geq n^{1/4}\varepsilon) = o(1)$$

uniformly in random variables with form $X = \sum_{k=1}^n a_k e_k$ *where* $\sum_{k=1}^n a_k^2 \leq 1$.

Proof. Put

$$h_n = [nt_n^{-1}/2], \quad \xi_j = \sum_{k=2jt_n+1}^{(2j+1)t_n} a_k e_k,$$

$$\eta_j = \sum_{k=(2j+1)t_n+1}^{2(j+1)t_n} a_k e_k, \quad j = 0, 1, \cdots, h_n - 1,$$

$$\xi_{h_n} = \sum_{k=2h_nt_n+1}^{n} a_k e_k.$$

For any $q > 0$, we are going to estimate

$$E\xi_j^4 I\Big(|\xi_j| > 5^{1/4} t_n^{1/2}\Big(\sum_{k=2jt_n+1}^{(2j+1)t_n} a_k^2\Big)^{1/2} q\Big).$$

For brevity, we only consider the case when $j = 0$, and put the event

$$B = \Big\{|\xi_0| > 5^{1/4} t_n^{1/2}\Big(\sum_{k=1}^{t_n} a_k^2\Big)^{1/2} q\Big\}.$$

We have

$$\begin{aligned} E\xi_0^4 I_B = & \sum_{k=1}^{t_n} a_k^4 E e_k^4 I_B + \sum_{p\neq q} a_p^2 a_q^2 E e_p^2 e_q^2 I_B \\ & + \sum_{p\neq q} a_p^3 a_q E e_p^3 e_q I_B + \sum_{p\neq q\neq r} a_p^2 a_q a_r E e_p^2 e_q e_r I_B \\ & + \sum_{p\neq q\neq r\neq s} a_p a_q a_r a_s E e_p e_q e_r e_s I_B. \end{aligned} \tag{13.2.8}$$

Obviously, none of the first two sums in the right hand side of (13.2.8) is exceeded by

$$\Big(\sum_{k=1}^{t_n} a_k^2\Big)^2 \max_{1\le k\le t_n} E e_k^4 I_B.$$

Put $M = \max_{1\le k\le t_n} E e_k^4 I_B$. For the third sum, there is

$$\Big|\sum_{p\neq q} a_p^3 a_q E e_p^3 e_q I_B\Big| \le M \sum_{p=1}^{t_n} |a_p^3| \sum_{q=1}^{t_n} |a_q| \le M t_n^{1/2}\Big(\sum_{k=1}^{t_n} a_k^2\Big)^2.$$

Similarly the absolute value of the fourth sum has the upper bound $M t_n (\sum_{k=1}^{t_n} a_k^2)^2$. For the fifth sum, its absolute value has the upper bound $M t_n^2 (\sum_{k=1}^{t_n} a_k^2)^2$. Therefore we obtain

$$E\xi_0^4 I_B \le 5 M t_n^2 \Big(\sum_{k=1}^{t_n} a_k^2\Big)^2. \tag{13.2.9}$$

Referring to the estimation of $E\xi_0^4 I_B$, it is easy to see that $E\xi_0^4$ has upper bound $5\nu t_n^2 \left(\sum_{k=1}^{t_n} a_k^2\right)^2$, which implies $P(B) \le \nu/q^4$. From (13.2.2), we have $M \le a(q)$. Inserting it into (13.2.9) yields

$$\frac{1}{5} t_n^{-2} \Big(\sum_{k=1}^{t_n} a_k^2\Big)^{-2} E\xi_0^4 I_B \le a(q). \tag{13.2.10}$$

For $\xi_j, j = 1, \cdots, h_n$, we have almost the same conclusions (for ξ_{h_n}, there may be a difference of a constant). Let $\{\xi'_j, j = 0, 1, \cdots, h_n\}$ be independent random variables such that ξ'_j obeys the same distribution as ξ_j. By (13.2.10),

$$\left\{5^{-1/4} t_n^{-1/2} \Big(\sum_{k=2jt_n+1}^{(2j+1)t_n} a_k^2 \Big)^{-1/2} \xi'_j, j = 0, 1, \cdots, h_n \right\}$$

satisfies the conditions given in Lemma 13.2.1. Choosing $\left(\sum_{k=2jt_n+1}^{(2j+1)t_n} a_k^2\right)^{1/2}$ as a_k and $t_n^{-1/2} n^{1/4} \varepsilon$ as b in Lemma 13.2.1 ($\varepsilon > 0$ is given arbitrarily), we obtain

$$t_n^{-2} n P\left\{ t_n^{-1/2} \Big| \sum_{j=0}^{h_n} \xi'_j \Big| \geq t_n^{-1/2} n^{1/4} \varepsilon \right\} = O(g(t_n^{-1/2} n^{1/4} \varepsilon)).$$

Because of the choice of t_n, we have

$$nP\left\{ \Big| \sum_{j=0}^{h_n} \xi'_j \Big| \geq n^{1/4} \varepsilon \right\} = O(t_n^2 g(t_n^{-1/2} n^{1/4})) = o(1). \tag{13.2.11}$$

Furthermore by Lemma 1.2.9

$$\begin{aligned} &\Big| E \exp\Big(itn^{-1/4} \sum_{j=0}^{h_n} \xi'_j \Big) - E \exp\Big(itn^{-1/4} \sum_{j=0}^{h_n} \xi_j \Big) \Big| \\ &\quad \leq (h_n + 1) \varphi(t_n) \leq \frac{1}{2} n t_n^{-1} \varphi(t_n). \end{aligned}$$

By condition (ii) in Theorem 13.2.1, $n^{-1/4} \sum_{j=0}^{h_n} \xi'_j$ has the same limit distribution as $n^{-1/4} \sum_{j=0}^{h_n} \xi_j$. Thus from (13.2.11)

$$nP\left\{ \Big| \sum_{j=0}^{h_n} \xi_j \Big| \geq n^{1/4} \varepsilon \right\} = o(1).$$

For η_j, we have the same relation. Combining these two results implies the conclusion of the lemma.

Proof of Theorem 13.2.1.

Obviously it is enough to prove the WIP. Define $U_n(\cdot)$ and $V_n(\cdot)$ by $U_n(0) = 0, V_n(0) = 0$,

$$U_n(i/n) = \sum_{j=1}^{n}\Big(e_j^2 - \frac{i-r_i}{i}\sigma^2\Big)/\sqrt{n\tau},$$

$$V_n(i/n) = \sum_{j=1}^{r_i}\Big(\sum_{k=1}^{i} a_{jk}^{(i)} e_k\Big)^2/\sqrt{n\tau}, \quad i = 1,\cdots,n,$$

both U_n and V_n are linear in $\Big[\frac{i-1}{n}, \frac{i}{n}\Big]$.

We have

$$Z_n = U_n - V_n.$$

By Theorem 5.1.1

$$U_n \Longrightarrow W \quad \text{as } n \to \infty.$$

Hence in order to prove the theorem, it suffices to show that for any $\varepsilon > 0$

$$P\Big\{\sup_{0\le t\le 1} |V_n(t)| > \varepsilon\Big\} \to 0, \quad \text{as} \quad n \to \infty. \tag{13.2.12}$$

Since $r_i \le p$, (13.2.12) is equivalent to

$$P\Big\{\max_{1\le i\le n}\Big|\sum_{k=1}^{i} a_k^{(i)} e_k\Big| \ge (n\tau)^{1/4}\varepsilon^{1/2}\Big\} \to 0 \quad \text{as} \quad n \to \infty$$

for any $\{a_k^{(n)}\}$ satisfying $\sum_{k=1}^{n} a_k^{(n)^2} \le 1$. By Lemma 13.2.2, we have

$$P\Big\{\max_{1\le i\le n}\Big|\sum_{k=1}^{i} a_k^{(i)} e_k\Big| \ge (n\tau)^{1/4}\varepsilon^{1/2}\Big\}$$

$$\le n \max_{1\le i\le n} P\Big\{\Big|\sum_{k=1}^{i} a_k^{(i)} e_k\Big| \ge (n\tau)^{1/4}\varepsilon^{1/2}\Big\} \to 0 \quad \text{as} \quad n \to \infty.$$

The proof of the theorem is complete.

Remark 13.2.2. When $\{e_i\}$ is a strictly stationary m-dependent sequence, one can take t_n to be a constant, and hence condition (ii) is satisfied. When $\{e_i\}$ is a bounded sequence, we can take $a(q) \equiv 0$ and t_n, for example, to be $[n^{2/3}]$. Hence condition (ii) is also satisfied.

Remark 13.2.3. Lin (1984) also gave the WIP when $\{e_i\}$ is α-mixing.

13.2.2 Strong approximation

Using the strong appximation result for a φ-mixing sequence (cf. Theorem 9.1.1), Lu (1986) showed the following theorem.

Theorem 13.2.2. *Let the random error sequence $\{e_i\}$ with (13.2.1) be strictly stationary and φ-mixing. Suppose that $E|e_1|^{8+\delta} < \infty$ for some $0 < \delta \leq 1$ and*

$$\varphi(n) = o(n^{1-4(2+\varepsilon)(10-\delta)/\delta\theta}) \quad as \quad n \to \infty \tag{13.2.13}$$

where $0 < \theta < 1$ and $\varepsilon > 0$. Then

$$Z(t) - W(t) = O(t^{1/4}(\log t)^{9/4+\varepsilon}) \quad a.s.$$

In order to prove the theorem, we need a lemma:

Lemma 13.2.3. *Under the conditions of Theorem 13.2.2,*

$$X = o(n^{1/8}) \quad a.s.$$

where $X = \sum_{k=1}^n a_k e_k$ with $\sum_{k=1}^n a_k^2 \leq 1$.

Proof. Put

$$d_n = [n^{\delta\theta/4(10-\delta)}], \qquad h_n = [n/2d_n],$$

$$\xi_j = \sum_{k=2jd_n+1}^{(2j+1)d_n} a_k e_k, \qquad \eta_j = \sum_{k=(2j+1)d_n+1}^{2(j+1)d_n} a_k e_k, \quad j = 0, 1, \cdots, h_n - 1,$$

$$\xi_{h_n} = \sum_{k=2h_nd_n+1}^{n} a_k e_k.$$

Similarly to (13.2.9),

$$\begin{aligned}
E|\xi_0|^{8+\delta} &\leq E\Big(\xi_0^8 \sum_{k=1}^{d_n} |a_k e_k|^\delta\Big) \\
&\leq \sum_k |a_k|^{8+\delta} E|e_k|^{8+\delta} + \sum_{k\neq j}\Big(\sum_{i=1}^{8} |a_k|^i |a_j|^{8-i+\delta} E|e_k|^i |e_j|^{8-i+\delta}|\Big) \\
&\quad + \cdots + \sum_{k_i\neq k_j, i\neq j} |a_{k_1}\cdots a_{k_8}||a_{k_9}|^\delta |E|e_{k_1}\cdots e_{k_8}||e_{k_9}|^\delta| \\
&\leq c d_n^{(10-\delta)/2} \Big(\sum_{k=1}^{d_n} a_k^2\Big)^{(8+\delta)/2}.
\end{aligned}$$

Let $\{\xi_j', j = 0, 1, \cdots, h_n\}$ be such independent random variables that ξ_j' obeys the same distribution as ξ_j. Put

$$\xi_j'' = d_n^{-\frac{10-\delta}{2(8+\delta)}} \Big(\sum_{k=2jd_n+1}^{(2j+1)d_n} a_k^2 \Big)^{-1/2} \xi_j'$$

and

$$a_j' = \Big(\sum_{k=2jd_n+1}^{(2j+1)d_n} a_k^2 \Big)^{1/2}.$$

Then

$$\sum_{j=0}^{h_n} {a_j'}^2 \le \sum_{k=1}^{n} a_k^2 \le 1.$$

Applying Remark 13.2.1 to $X = \sum_{j=0}^{h_n} a_j' \xi_j''$ and $b = \varepsilon d_n^{-(10-\delta)/2(8+\delta)} n^{1/8}$, we obtain

$$\begin{aligned} & P\Big\{ \Big| \sum_{j=0}^{h_n} \xi_j' \Big| \ge \varepsilon n^{1/8} \Big\} \\ &= P\Big\{ d_n^{-\frac{10-\delta}{2(8+\delta)}} \Big| \sum_{j=0}^{h_n} \xi_j' \Big| \ge \varepsilon d_n^{-\frac{10-\delta}{2(8+\delta)}} n^{\frac{1}{8}} \Big\} \\ &= o(d_n^{5-\delta/2} n^{-1-\delta/8}) = o(n^{-1-\delta(1-\theta)/8}). \end{aligned} \tag{13.2.14}$$

By φ-mixing property,

$$\begin{aligned} & |P\{\xi_1 + \xi_2 \le x\} - P\{\xi_1' + \xi_2' \le x\}| \\ & \quad \le \int |P\{\xi_1 \le x - u | \xi_2 = u\} - P\{\xi_1 \le x - u\}| dP\{\xi_2 \le x\} \\ & \quad \le \varphi(d_n), \end{aligned}$$

and hence

$$\begin{aligned} & \Big| P\Big\{ \sum_{j=0}^{h_n} \xi_j \le x \Big\} - P\Big\{ \sum_{j=0}^{h_n} \xi_j' \le x \Big\} \Big| \\ & \qquad \le O(h_n \varphi(d_n)) = O(n^{-1-\varepsilon}). \end{aligned} \tag{13.2.15}$$

Combining (13.2.14) with (13.2.15) implies

$$P\Big\{ \Big| \sum_{j=1}^{h_n} \xi_j \Big| \ge \varepsilon n^{1/8} \Big\} = O(n^{-1-\varepsilon}).$$

By the Borel-Cantelli lemma

$$\Big|\sum_{j=0}^{h_n} \xi_j\Big| = o(n^{1/8}) \quad \text{a.s.}$$

Similarly

$$\Big|\sum_{j=0}^{h_n-1} \eta_j\Big| = o(n^{1/8}) \quad \text{a.s.}$$

The lemma is proved.

Proof of Theorem 13.2.2.
Write

$$\begin{aligned} Z(t) &= \sum_{k=1}^{[t]} (e_k^2 - \sigma^2) + r_{[t]}\sigma^2 - \sum_{i=1}^{r_{[t]}} \Big(\sum_{j=1}^{[t]} a_{ij}^{([t])} e_j\Big)^2 \\ &=: X(t) + r_{[t]}\sigma^2 - \sum_{i=1}^{r_{[t]}} \Big(\sum_{j=1}^{[t]} a_{ij}^{([t])} e_j\Big)^2. \end{aligned}$$

From Lemma 13.2.3, we have

$$\Big(\sum_{j=1}^{[t]} a_{ij}^{([t])} e_j\Big)^2 = o(t^{1/4}) \quad \text{a.s.} \quad \text{as } t \to \infty. \tag{13.2.16}$$

By Theorem 9.1.1, there exists a Wiener process $\{W(t), t \geq 0\}$ such that for any $\varepsilon > 0$

$$X(t) - W(t) = O(t^{1/4} (\log t)^{9/4+\varepsilon}) \quad \text{a.s.} \quad \text{as } t \to \infty. \tag{13.2.17}$$

Combining (13.2.16) with (13.2.17) implies the conclusion of the theorem.

13.3. Density estimations

Let $\{X_n, n \geq 1\}$ be a sequence of R^d-valued random variables with a common density function $f(x)$. In general, there are two kinds of estimations for $f(x)$. The first is the so-called kernel estimation, which is defined by

$$f_n(x) = (nh_n^d)^{-1} \sum_{i=1}^{n} K\Big(\frac{x - X_i}{h_n}\Big), \tag{13.3.1}$$

where the window width $h_n \downarrow 0$ as $n \to \infty$. Another kind is the so-called nearest neighbor estimation, which is defined by

$$\widehat{f}_n(x) = k_n/\{n|S(x, a_n(x))|\}, \tag{13.3.2}$$

where $k_n, 1 \le k_n \le n$, is the given integer, $a_n(x)$ is the distance from x to the k_n-th closest X_i (in $X_1, \cdots, X_n$), $S(x,a)$ is the hypersphere of center x having the radius a and $|S(x,a)| = L(S(x,a))$, where L denotes the Lebesgue measure in R^d.

In this section we always assume $\{X_n\}$ is φ-mixing. Other kinds of mixing sequences can be studied similarly.

13.3.1 *Kernel estimation*

Many authors, such as Lin(1983), Masry and Györfi (1987), Shao (1990), Cai (1991), Peligrad (1992), Fan and Xue (1993) have studied limit behavior of the kernel estimation (KE) of the density function for a mixing sequence.

Let $\{X_n, n \ge 1\}$ be a R^d-valued φ-mixing sequence with a common unknown density function $f(x) = f(x_1, \cdots, x_d)$. Consider the KN $f_n(x)$ defined by (13.3.1). Peligrad (1992) showed the following result:

Theorem 13.3.1. *Suppose that D is a compact subset of R^d and f is continuous on an ε-neighborhood of D. Suppose that K satisfies the following conditions:*

(1) $K(\cdot)$ *is a density on* R^d,
(2) $K(x) \le K_1 < \infty$ *for any* $x \in R^d$,
(3) $\|x\|^{d+1}K(x) \to 0$ *as* $x \to \infty$,
(4) $\int \|x\|K(x)\,dx = K_2 < \infty$,
(5) $K(\cdot)$ *is Lipschitz of order* γ *on* R^d.

Then

$$\sup_{x\in D} |f_n(x) - f(x)| = O(h_n + d_n^{1/2} \log n/(nh_n^d)^{1/2}) \quad a.s. \tag{13.3.3}$$

where $d_n = \exp\Big(2\sum_{i=1}^{[\log n]} \varphi^{1/2}(2^i)\Big)$. *If, in addition, the condition*

$$\sum_{n=1}^{\infty} \varphi^{1/2}(2^n) < \infty \tag{13.3.4}$$

is satisfied, then for $h_n = O\Big((\log^2 n/n)^{1/(d+2)}\Big)$

$$\sup_{x\in D} |f_n(x) - f(x)| = O((\log^2 n/n)^{1/(d+2)}) \quad a.s. \tag{13.3.5}$$

Remark 13.3.1. If condition 3) is weakened into
(3)′ $\|x\|^d K(x) \to 0$ as $x \to \infty$,
condition (4) is dropped, and condition (13.3.4) is replaced by

$$\lim_{n\to\infty} \varphi(n) < 1/2$$

and

$$nh_n^d/(d_n \log^2 n) \to \infty \quad \text{as } n \to \infty,$$

we have

$$\sup_{x\in D} |f_n(x) - f(x)| \to 0 \quad \text{a.s. as } n \to \infty.$$

Remark 13.3.2. Under independence assumption, the rate of the maximal deviation of the estimation from the true density is due to Kuelbs (1976) and its size is $O((\log\log n)^{1/2}/n^{1/(2(d+1))})$. (13.3.5) improves the speed of this convergence to $O((\log^2 n/n)^{1/(d+2)})$.

Remark 13.3.3. Shao (1990) further improved the speed to $O((\log n/n)^{1/(d+2)})$, but a stronger mixing condition, i.e. $\varphi(n) = O(n^{-(2+d)})$, is required.

Proof of Theorem 13.3.1.

Obviously by Lemma 2.2.2, under condition (13.3.4), (13.3.3) with $h_n = O((\log^2 n/n)^{1/(d+2)})$ implies (13.3.5). We prove (13.3.3). Write

$$\begin{aligned}
&\sup_{x\in D}|f_n(x) - f(x)| \\
&\quad \le \sup_{x\in D} |f_n(x) - Ef_n(x)| + \sup_{x\in D} |Ef_n(x) - f(x)| \\
&\quad =: \Delta_1 + \Delta_2.
\end{aligned} \tag{13.3.6}$$

Estimate Δ_1 first. Because D is compact, we can choose a covering of D with $l(n)$ balls $B_1, \cdots, B_{l(n)}$ of centers $t_1, \cdots, t_{l(n)}$ having the radius

$$R_n = h_n(n^{-1}h_n^d \log^2 n)^{1/2\gamma};$$

the number $l(n)$ can be chosen less than $O(h_n^{-d}(n/(h_n^d \log^2 n))^{d/(2\gamma)})$. Without loss of generality, we can assume $h_n > n^{-2/d}$. Hence

$$l(n) = O(n^{2+3d/(2\gamma)}). \tag{13.3.7}$$

Let x be in D and define

$$S_j(x) = h_n^{-d/2} \sum_{i=1}^{j} \Big(K\Big(\frac{x - X_i}{h_n}\Big) - EK\Big(\frac{x - X_i}{h_n}\Big)\Big).$$

So

$$\Delta_1 = \sup_{x\in D}(nh_n^{d/2})^{-1}S_n(x).$$

By condition 5), for every $x \in B_k$, we have

$$|S_n(x) - S_n(t_k)| \le CnR_n^\gamma h_n^{-d/2-\gamma} \le C(n\log^2 n)^{1/2} \tag{13.3.8}$$

for some $C > 0$. And by Lemma 2.2.2, there exists $G > 0$ such that

$$ES_n^2(x) \le Gnd_n. \tag{13.3.9}$$

Moreover by condition 2)

$$\begin{aligned} h_n^{-d/2}\Big|K\Big(\frac{x-X_i}{h_n}\Big) - EK\Big(\frac{x-X_i}{h_n}\Big)\Big|/(Gnd_n)^{1/2} \\ \le 2K_1(Gnd_nh_n^d)^{-1/2} =: C_n. \end{aligned}$$

Now let $A(> C)$ be a positive number specified later on. By (13.3.8) we have

$$\begin{aligned} &P\Big\{\sup_{x\in D}|S_n(x)| \ge 2A(n\log^2 n)^{1/2}\Big\} \\ &\le \sum_{k=1}^{l(n)}\Big(P\Big\{\sup_{x\in B_k}|S_n(x) - S_n(t_k)| \ge A(n\log^2 n)^{1/2}\Big\} \\ &\quad + P\{|S_n(t_k)| \ge A(n\log^2 n)^{1/2}\}\Big) \\ &\le l(n)\max_{1\le k\le l(n)} P\{|S_n(t_k)| \ge A(n\log^2 n)^{1/2}\}. \end{aligned} \tag{13.3.10}$$

Estimate the probability of the right hand side of (13.3.10). It is clear that for any $\eta > 0$, $0 < \eta < 1/2$, there are $p \ge 1$ and $A > 0$ such that for $n \ge p$

$$\varphi(p) + \max_{1\le i\le n} P\{|S_n(t_k) - S_i(t_k)|/(Gnd_n)^{1/2} \ge A\} \le \eta.$$

By a combination of (2.2.18) and (2.2.19) in Lemma 2.2.7, we have

$$\begin{aligned} &P\Big\{\max_{1\le j\le n}|S_j(t_k)|/(Gnd_n)^{1/2} \ge x + 2A + 2pC_n\Big\} \\ &\le \frac{\eta}{1-\eta}P\Big\{\max_{1\le j\le n}|S_j(t_k)|/(Gnd_n)^{1/2} \ge x\Big\}. \end{aligned} \tag{13.3.11}$$

Let $B_n = 2A + 2pC_n$ and $M_n = \max_{1\le j\le n}|S_j(t_k)|/(Gnd_n)^{1/2}$. Obviously, for any $\alpha_n > 0$

$$E\exp(\alpha_n M_n) \le \exp(\alpha_n B_n) + \alpha_n\int_{B_n}^\infty \exp(\alpha_n x)P(M_n > x)\,dx.$$

After changing x to $x + B_n$, by (13.3.11) we get

$$E\exp(\alpha_n M_n) \le \exp(\alpha_n B_n) + \frac{\eta}{1-\eta} E\exp\{\alpha_n(M_n + B_n)\}.$$

Letting $\alpha_n = (2B_n)^{-1}\log(\eta^{-1} - 1)$ yields

$$E\exp(\alpha_n M_n) \le ((\eta^{-1} - 1)^{-1/2} - (\eta^{-1} - 1)^{-1})^{-1} =: g(\eta).$$

Put $\alpha = \inf_n \alpha_n = (4A)^{-1}\log(\eta^{-1} - 1)$. Then

$$\begin{aligned} P\{\alpha(\log n)^{-1}(Gnd_n)^{-1/2}|S_n(t_k)| \ge 3d/(2\gamma) + 4\} \\ \le g(\eta)\exp\{-(3d/(2\gamma) + 4)\log n\} \\ = O(n^{-(3d/(2\gamma)+4)}) \end{aligned} \tag{13.3.12}$$

uniformly in k as $n \to \infty$. Putting $A = 2((3d/2\gamma + 4)G^{1/2}/\log(\eta^{-1} - 1))^{1/2}$, which implies $A = (3d/2\gamma + 4)G^{1/2}/\alpha$. From (13.3.7), (13.3.10) and (13.3.12) we obtain

$$\sum_{n=1}^{\infty} P\Big\{\sup_{x\in D}|S_n(x)| \ge 2A(nd_n\log^2 n)^{1/2}\Big\} < \infty. \tag{13.3.13}$$

Therefore

$$\Delta_1 = O\Big(\Big(\frac{d_n\log^2 n}{nh_n^d}\Big)^{1/2}\Big) \quad \text{a.s.} \quad \text{as } n \to \infty.$$

As for Δ_2, by the well-known Bochner-Parzen theorem (cf. Parzen 1962), under the conditions of the theorem, we have

$$\Delta_2 = O(h_n) \quad \text{as } n \to \infty.$$

The proof of the theorem is completed.

13.3.2 Nearest neighbor estimation

Chai (1984) studied strong consistency of nearest neighbor estimation (NNE) of the density function for a mixing sequence. Using the improved Bernstein inequality, Lemma 12.4.1, Chai's theorems hold under the weaker mixing condition.

Theorem 13.3.2. *Suppose that condition (13.3.4) is satisfied and k_n in (13.3.2) satisfy*

$$k_n \to \infty, \quad k_n/n \to 0 \quad \textit{as } n \to \infty. \tag{13.3.14}$$

Then $k_n/\sqrt{n} \to \infty$ *implies*

$$\widehat{f}_n(x) \xrightarrow{P} f(x) \quad a.s.\ x \in R^d(L); \tag{13.3.15}$$

and $\sum_{n=1}^{\infty} \exp(-ck_n^2/n) < \infty$ *for any* $c > 0$ *implies*

$$\widehat{f}_n(x) \to f(x) \quad a.s., \quad a.s.\ x \in R^d(L). \tag{13.3.16}$$

Proof. Denote V_d the volume of an unit ball in R^d, μ and μ_n the distribution of X_1 and the empirical distribution of $X_1, \cdots, X_n$ respectively. For any given $\varepsilon > 0$ put

$$\begin{aligned}
b_n(x) &= (f(x)+\varepsilon)V_d n/k_n, \\
b'_n(x) &= (f(x)-\varepsilon)V_d n/k_n, \\
S_n(x,b) &= S(x, b_n^{-1/d}(x)), \\
S_n(x,b') &= S(x, b'^{-1/d}_n(x)).
\end{aligned}$$

Then

$$\begin{aligned}
&P\{|\widehat{f}_n(x) - f(x)| > \varepsilon\} \\
&\quad \le P\{\widehat{f}_n(x) - f(x) > \varepsilon\} + P\{\widehat{f}_n(x) - f(x) < -\varepsilon\} \\
&\quad \le P\{\mu_n(S_n(x,b)) - \mu(S_n(x,b)) \ge \frac{k_n}{n} \\
&\qquad - \mu(S_n(x,b))\} + P\{\mu_n(S_n(x,b')) \\
&\qquad - \mu(S_n(x,b')) \ge \frac{k_n}{n} - \mu(S_n(x,b'))\}
\end{aligned} \tag{13.3.17}$$

(If $f(x) \le \varepsilon$, the second term of the right hand side of the first and the second inequality sign disappears). By the well-known Lebesgue density theorem we have that as $n \to \infty$

$$\begin{aligned}
&\mu(S_n(x,b))/|S_n(x,b)| \to f(x) \quad \text{a.s. } x \in R^d(L), \\
&\mu(S_n(x,b'))/|S_n(x,b')| \to f(x) \quad \text{a.s. } x \in R^d(L).
\end{aligned}$$

Let the exceptional sets be denoted by D and D' respectively. Put $E = D^c \bigcap D'^c$. Then for any $x \in E$ and n large enough,

$$\begin{aligned}
\mu(S_n(x,b)) &\le \frac{k_n}{n}(f(x) + \frac{\varepsilon}{2})/(f(x)+\varepsilon), \\
\mu(S_n(x,b')) &\ge \frac{k_n}{n}(f(x) - \frac{\varepsilon}{2})/(f(x)-\varepsilon).
\end{aligned}$$

Put $a(x) = \varepsilon/(2(f(x)+\varepsilon))$ and $a'(x) = \varepsilon/(2(f(x)-\varepsilon))$. For any $x \in E$

$$\begin{aligned}
&P\{|\widehat{f}_n(x) - f(x)| > \varepsilon\} \\
&\quad \le P\{|\mu_n(S_n(x,b)) - \mu(S_n(x,b))| \ge \frac{k_n}{n} a(x)\} \\
&\qquad + P\{|\mu_n(S_n(x,b')) - \mu(S_n(x,b'))| \ge \frac{k_n}{n} a'(x)\} \\
&\quad =: I_{n1} + I_{n2}.
\end{aligned}$$

Let $\xi_i = I(X_i \in S_n(x,b)) - \mu(S_n(x,b))$, $i = 1, \cdots, n$. Then using Lemma 12.4.1 we have

$$I_{n1} \le 2\exp\Big\{-c\frac{k_n^2}{n} a^2(x)\Big\},$$

and similarly

$$I_{n2} \le 2\exp\Big\{-c\frac{k_n^2}{n} a'^2(x)\Big\}.$$

They imply (13.3.15) if $k_n/\sqrt{n} \to \infty$ and (13.3.16) if $\sum_{n=1}^{\infty} \exp(-ck_n^2/n) < \infty$ for any $c > 0$.

Remark 13.3.4. By a finer analysis, we can obtain a rate of a.s. consistency, i.e.

$$\widehat{f}_n(x) - f(x) = o(r_n^{-1}) \quad \text{a.s.}$$

where $r_n = n^{\beta}, 0 < \beta < 1/(2(d+1))$, if $f(\cdot)$ satisfies the local Lipschitz condition on x and $f(x) > 0$ and conditions (13.3.4) and (13.3.14) are satisfied.

Next we consider uniform strong consistency in the case of $d = 1$. At this time, $\widehat{f}_n(x) = k_n/(2na_n(x))$, $x \in R$. We need some lemmas. Let F and F_n be the distribution function of X_1 and the empirical distribution function of $X_1, \cdots, X_n$ respectively. Define the empirical process of $\{X_n, n \ge 1\}$:

$$R(s,t) = [t](F_{[t]}(s) - F(s)), \quad s \in R, t \ge 0.$$

Lemma 13.3.1. (Berkes and Philipp 1977) *Suppose that*

$$\varphi(n) = O(n^{-5-\delta}) \quad \textit{for some } \delta \in (0, 1/4). \tag{13.3.18}$$

Then there exists a version $K(s,t)$ of a Kiefer process such that

$$\sup_{t \le T} \sup_{s \in R} |R(s,t) - K(F(s),t)| = O(T^{1/2}(\log T)^{-\lambda}) \quad \textit{a.s.}$$

for some $\lambda > 0$.

Lemma 13.3.2.(Csörgő and Révész 1981) *For a Kiefer process $K(s,t)$,*

$$\limsup_{t\to\infty} \sup_{0\le s\le 1} |K(s,t)|/(t\log\log t)^{1/2} = 1/\sqrt{2} \quad a.s.$$

Theorem 13.3.3. *Suppose that (13.3.18) is satisfied and $\{k_n\}$ satisfies*

$$k_n/n \to 0 \quad \text{and} \quad k_n/(n\log\log n)^{1/2} \to \infty. \tag{13.3.19}$$

And suppose that f is uniformly continuous in R. Then

$$\sup_x |\widehat{f}_n(x) - f(x)| \to 0 \quad a.s.$$

proof. By Lemma 13.3.1, for any $\varepsilon > 0$ there is a constant $a > 0$ such that

$$P(A_n, i.o.) \le \varepsilon, \tag{13.3.20}$$

where $A_n = \left\{\sup_x |R(x,n) - K(F(x),n)| \ge an^{1/2}(\log n)^{-\lambda}\right\}$. Similarly, by Lemma 13.3.2 for any $\varepsilon > 0$ there is a constant $b > 0$ such that

$$P(B_n, i.o.) \le \varepsilon, \tag{13.3.21}$$

where $B_n = \left\{\sup_{0\le s\le 1} |K(s,n)| \ge b(n\log\log n)^{1/2}\right\}$. Put

$$d_n(x) = \frac{k_n}{2n(f(x)+\varepsilon)}, \quad d_n'(x) = \frac{k_n}{2n(f(x)-\varepsilon)}$$

and

$$S_n(x,d_n) = (x - d_n(x), x + d_n(x)),$$
$$S_n(x,d_n') = (x - d_n'(x), x + d_n'(x)).$$

Then similarly to (13.3.17)

$$\begin{aligned} P\Big\{\bigcup_{n\ge m}\Big\{\sup_x|\widehat{f}_n(x) - f(x)| > \varepsilon\Big\}\Big\} \\ &\le P\Big\{\bigcup_{n\ge m}\bigcup_x\{\mu_n(S_n(x,d_n)) \\ &\quad - \mu(S_n(x,d_n)) \ge \frac{k_n}{n} - \mu(S_n(x,d_n))\}\Big\} \\ &\quad + P\Big\{\bigcup_{n\ge m}\bigcup_{x:f(x)>\varepsilon}\{\mu_n(S_n(x,d_n')) \\ &\quad - \mu(S_n(x,d_n')) \ge \frac{k_n}{n} - \mu(S_n(x,d_n'))\}\Big\} \\ &=: J_{m1} + J_{m2}. \end{aligned} \tag{13.3.22}$$

By uniform continuity of f, $M := \sup_x f(x) < \infty$, and further for any x and large n,

$$
\begin{aligned}
\mu(S_n(x,d_n)) &\le \frac{k_n}{n}(f(x)+\frac{\varepsilon}{2})/(f(x)+\varepsilon),\\
\frac{k_n}{n} - \mu(S_n(x,d_n)) &\ge \frac{k_n}{n}\cdot\frac{\varepsilon}{2(f(x)+\varepsilon)} \ge \frac{k_n}{n}\cdot\frac{\varepsilon}{2(M+\varepsilon)} =: p_n,\\
\mu(S_n(x,d_n')) &\ge \frac{k_n}{n}(f(x)-\frac{\varepsilon}{2})/(f(x)-\varepsilon),\\
\mu(S_n(x,d_n')) - \frac{k_n}{n} &\ge \frac{k_n}{n}\cdot\frac{\varepsilon}{2(f(x)-\varepsilon)}\\
&\ge \frac{k_n}{n}\cdot\frac{\varepsilon}{2(M-\varepsilon)} =: q_n, \quad \text{as } f(x)>\varepsilon.
\end{aligned}
$$

Therefore

$$
\begin{aligned}
J_{m1} &\le P\Big\{\bigcup_{n\ge m}\Big\{\sup_x |\mu_n(S_n(x,d_n)) - \mu(S_n(x,d_n))| \ge p_n\Big\}\Big\}\\
&\le 2P\Big\{\bigcup_{n\ge m}\Big\{\sup_x |F_n(x)-F(x)| \ge \frac{p_n}{2}\Big\}\Big\}\\
&\le 2P\Big\{\bigcup_{n\ge m}\Big\{\sup_x \Big|\frac{R(x,n)-K(F(x),n)}{n}\Big| \ge \frac{p_n}{4}\Big\}\Big\}\\
&\quad + 2P\Big\{\bigcup_{n\ge m}\Big\{\sup_x \frac{|K(F(x),n\ )|}{n} \ge \frac{p_n}{4}\Big\}\Big\}\\
&=: 2J_{m1}^{(1)} + 2J_{m1}^{(2)}.
\end{aligned}
$$

By condition (13.3.19) we have

$$\sqrt{n}p_n(\log n)^{\lambda} \to \infty \quad \text{as } n\to\infty.$$

Hence it follows from (13.3.20) that

$$
\begin{aligned}
J_{m1}^{(1)} &\le P\Big\{\bigcup_{n\ge m}\Big\{\sup_x \Big|\frac{R(x,n)-K(F(x),n)}{n^{1/2}(\log n)^{-\lambda}}\Big| \ge \frac{n^{1/2}p_n(\log n)^{\lambda}}{4}\Big\}\Big\}\\
&\le P\Big\{\bigcup_{n\ge m} A_n\Big\} \to 0 \quad \text{as } m\to\infty.
\end{aligned}
$$

Similarly (13.3.21) implies

$$J_{m1}^{(2)} \le P\Big\{\bigcup_{n\ge m} B_n\Big\} \to 0 \quad \text{as } m\to\infty.$$

Thus $J_{m1} \to 0$ as $m \to \infty$. Moreover

$$J_{m2} \le P\Big\{ \bigcup_{n \ge m} \Big\{ \sup_x |\mu_n(S_n(x, d_n')) - \mu(S_n(x, d_n'))| \ge q_n \Big\}\Big\}.$$

In the same way as for J_{m1}, we have $J_{m2} \to 0$ as $m \to \infty$. Then it follows from (13.3.22) that for any $\varepsilon > 0$

$$P\Big\{ \bigcup_{n \ge m} \Big\{ \sup_x |\widehat{f}_n(x) - f(x)| > \varepsilon \Big\}\Big\} \to 0 \quad \text{as } m \to \infty.$$

This completes the proof of the theorem.

Remark 13.3.5. If f possesses a bounded second derivative then for $k_n = [n^{7/10}]$ and any $C_n \to \infty$ we have

$$\sup_x |\widehat{f}_n(x) - f(x)| = o(n^{-1/5}(\log\log n)^{1/2} C_n) \quad \text{a.s.}$$

Remark 13.3.6. In Yu (1993) a simple and useful nonparametric estimator of a density $f(x)$ based on a sample $X_1, \cdots, X_n$ has been defined. If $m = m_n$ is a positive integer, the nonparametric estimator $f_n(x)$ of $f(x)$ is defined by dividing $2m_n$ by n times the length of the smallest interval containing x which consists of $2m_n$ of the n observations in which the half of them lie on the left side of x and the other half on the right side. Formally,

$$f_n(x) = \begin{cases} \dfrac{2m_n}{n(X_{(2m_n+j)} - X_{(j+1)})}, & x \in [X_{(m_n+j)}, X_{(m_n+j+1)}), \\ & \quad \text{for } j = 0, 1, \cdots, , n - 2m_n; \\ 0 & x < X_{(m_n)} \text{ or } x \ge X_{(n-m_n+1)}, \end{cases}$$

where $(X_{(1)}, \cdots, X_{(n)})$ is the order statistic of $(X_1, \cdots, X_n)$. Yu (1993) showed that in the case of i.i.d. observations $f_n(x)$ converges to $f(x)$ (in probability and a.s.) under some conditions. Yu (1995) studied that the rate of strong uniform convergence for the estimator defined as above when the observations satisfy a φ-mixing or an α-mixing condition.

Chapter 14 Strong Approximations for Other Kinds of Dependent Random Variables

The almost sure invariance principle for sums of weakly dependent random variables has been given by Philipp and Stout (1975), such as lacunary trigonometric series, φ-mixing or α-mixing sequence of random variables, Gaussian sequence and additive functional of Markov chains . The strong approximations of sums of φ-mixing and α-mixing sequences have been studied in Chapter 9. In this chapter we shall investigate the strong approximations of sums of other kinds of dependent random variables, including lacunary trigonometric series with weights, a class of Gaussian sequences and additive functional of Markov processes. All of these improve essentially and comprehensively the results in Philipp and Stout's monograph (1975).

14.1 Lacunary trigonometric series with weights

Let $\{n_k, k \geq 1\}$ be a sequence of positive real numbers with

$$n_{k+1}/n_k \geq 1 + q/k^r \tag{14.1.1}$$

for all k and some $q > 0$, $0 \leq r < 1/2$. It is said to be a *lacunary sequence* when $r = 0$. Let $\{a_k, k \geq 1\}$ be a sequence of non-zero real numbers. Put

$$A_n^2 = \frac{1}{2}\sum_{k=1}^{n} a_k^2, \qquad B_n^2 = \frac{1}{2}\sum_{k=1}^{n} a_k^4. \tag{14.1.2}$$

Suppose that $A_n \to \infty$ and that there exist constants δ, β with $0 < \delta \leq$

$1, \beta > 0$ such that

$$a_k = O(A_k^{1-\delta}), \tag{14.1.3}$$

$$A_{2k} = O(A_k), \tag{14.1.4}$$

$$k^\beta = O(A_k). \tag{14.1.5}$$

In this section, $([0,1), \mathcal{B}, P)$ denotes a probability space, where $\mathcal{B}$ consists of the Lebesgue measurable sets of [0,1), and P is the Lebesgue measure on $\mathcal{B}$. We consider the trigonometric series

$$S(A_n^2, \omega) = \sum_{k=1}^{n} a_k \cos 2\pi n_k \omega, \quad \omega \in [0,1). \tag{14.1.6}$$

For $t \geq 0$, put

$$S(t, \omega) = S(A_n^2, \omega), \quad \text{if } A_n^2 \leq t < A_{n+1}^2, \tag{14.1.7}$$

where $A_0 = 0$.

The strong approximations of $S(t)$ by a Wiener process $W(t)$ were first discussed by Gaposhkin (1966). Later on, in the case of $r = 0$, Philipp and Stout (1975) proved the almost sure invariance principle for lacunary trigonometric series with weights under condition (14.1.3) and obtained the approximation order of $1/2 - \lambda$ for each $\lambda < \delta/32$. In a special case o unweighted summands, they obtained the approximation order of $5/12 + \lambda$ for each $\lambda > 0$ and conjectured that the constant 5/12 would be replaced by 1/3. Sun (1984) showed this fact. Shao (1987) improved all these results comprehensively. He studied the general case of lacunary trigonometric series with weights and pointed out that Sun's order is not the best possible, and obtained, when the case of $r = 0$, the order of 1/4 which is the best if one were only to use the Skorohod embedding scheme. Furthermore, in some particular cases, the logarithmic order is obtained

Theorem 14.1.1. *Suppose that conditions (14.1.3)–(14.1.5) are satisfied and that $\delta\beta > r$. Then without changing the distribution of $\{S(t), t \geq 0\}$, we can redefine the process $\{S(t), t \geq 0\}$ on a richer probability space together with a Wiener process $\{W(t), t \geq 0\}$ such that*

$$S(t) - W(t) = O\left(t^{(2-\delta)/4 + r/(4\beta)} \log^2 t\right) \quad a.s. \tag{14.1.8}$$

Theorem 14.1.2. *Under the conditions of Theorem 14.1.1 and the assumption of $|a_k|$ non-increasing, we have*

$$S(A_n^2) - W(A_n^2) = O\Big(A_n^{r/(2\beta)}(B_n^{1/2} \vee A_n^{r/(2\beta)}) \log^2 A_n\Big) \qquad a.s. \quad (14.1.9)$$

We immediately obtain the approximation order for lacunary trigonometric series with unweighted summands from the above general theorems.

Corollary 14.1.1. *If $a_k \equiv 1$, then we have*

$$S(t) - W(t) = O\Big(t^{(1+2r)/4} \log^2 t\Big) \qquad a.s.$$

In the case of $r = 0$ the result (14.1.10) is an essential improvement of Theorem 3.1 in Philipp and Stout (1975), the order of $1/4$ is the best possible provided $S(t)$ is constructed via the Skorohod embedding method.

Corollary 14.1.2. *Suppose that the condition of Theorem 14.1.2 and condition (14.1.5) are satisfied (for some $\beta > 0$), and $B_N = O(1)$, $|a_k| \downarrow 0$. Then when $r = 0$ we have*

$$S(t) - W(t) = O(\log^2 t) \qquad a.s.$$

The proofs of Theorems need the following lemmas.

Lemma 14.1.1. *Let $\xi_1, \cdots, \xi_n$ be a sequence of random variables. Put*

$$S_k = \sum_{j=1}^{k} \xi_j, \quad M_k = \max_{1 \le j \le k} |S_j|.$$

Suppose that there exists a sequence $\{c_k\}$ such that for all $0 \le i < j \le n$

$$E|S_j - S_i|^2 \le \sum_{i<k\le j} c_k,$$

then

$$EM_k^2 \le \Big(\sum_{i=1}^{k} c_i\Big)\Big(\frac{\log 2k}{\log 2}\Big)^2$$

for each $k \le n$.

The proof refers to Theorem 2.4.1 in Stout (1974).

Lemma 14.1.2. *For any* $v \geq 0$,

$$\sum_{j\geq k} \frac{A_j^v}{n_j} = O\Big(\frac{A_k^v}{n_k} k^r\Big), \tag{14.1.10}$$

$$\sum_{j>k} \frac{A_j^v}{n_j - n_k} = O\Big(\frac{A_k^v}{n_k} k^{2r}\Big). \tag{14.1.11}$$

Proof. (14.1.4) implies that there exists a constant $C > 0$ such that

$$A_{k+j} = O\Big(\Big(\frac{k+j}{j}\Big)^C A_j\Big)$$

for all $k, j \geq 1$. And by $n_{k+1}/n_k \geq 1 + q/k^r$, we have

$$\begin{aligned}\frac{n_{k+j}}{n_k} &\geq \prod_{i=1}^{j}\Big(1 + \frac{q}{(k+i)^r}\Big)\\ &\geq c\exp\Big(\frac{q((k+j)^{1-r} - k^{1-r})}{2(1-r)}\Big),\end{aligned} \tag{14.1.12}$$

where we have used the well-known formula

$$\sum_{k=1}^{n} k^{-r} = \frac{n^{1-r}}{1-r} + u + O(n^{-r}), \quad 0 \leq r < 1/2,$$

where $u = u(r)$ is a constant. Hence

$$\begin{aligned}\sum_{j\geq k} \frac{A_j^v}{n_j} &= O\Big(\frac{A_k^v}{n_k k^c}\Big(\sum_{j\geq k} j^c \exp\Big(-\frac{q\, j^{1-r}}{2(1-r)}\Big)\Big)\exp\Big(\frac{q\, k^{1-r}}{2(1-r)}\Big)\Big)\\ &= O\Big(\frac{A_k^v}{n_k} k^r\Big).\end{aligned}$$

Thus (14.1.10) is proved. The proof of (14.1.11) is similar.

Lemma 14.1.3. *Let* $W(t)$ *be a Wiener process and* $\{t_n\}$ *be a sequence of random variables. Suppose that there exists a sequence* $\{b_n\}$ *of real numbers with* $b_n = o(n)$, *such that*

$$t_n - n = O(b_n) \quad a.s., \tag{14.1.13}$$

then

$$W(t_n) - W(n) = O(b_n^{1/2} \log^{1/2} n) \quad a.s. \tag{14.1.14}$$

Proof. (14.1.13) implies that there exists a constant C such that

$$|t_n - n| \le Cb_n \quad \text{a.s.},$$

hence

$$|W(t_n) - W(n)| \le \sup_{0\le t\le n-Cb_n} \sup_{0\le s\le 2Cb_n} |W(t+s) - W(t)| \quad \text{a.s.}$$

In terms of a well-known result (cf. Theorem 3.2B in Hanson and Russo 1983), we have

$$\sup_{0\le t\le n-Cb_n} \sup_{0\le t\le 2Cb_n} |W(t+s) - W(t)|$$
$$= O\Big(\Big(b_n\Big(\log \frac{n+Cb_n}{2Cb_n} + \log\log(n+Cb_n)\Big)\Big)^{1/2}\Big) \quad a.s.$$

(14.1.14) follows from the above relations.

We now define an increasing sequence $\mathcal{F}_k$ of σ-fields as follows. Put $p = 2/\beta + 4$. For each integer k, let r_k be the largest integer i such that

$$2^i \le A_k^p n_k. \tag{14.1.15}$$

We define $\mathcal{F}_k$ to be the σ-field generated by the intervals of the form

$$U_{vk} = [v2^{-r_k}, (v+1)2^{-r_k}), \quad 0 \le v < 2^{r_k}.$$

Putting $\xi_k(\omega) = a_k cos 2\pi n_k \omega, \quad X_k(\omega) = E(\xi_k|\mathcal{F}_k)$, we obtain by (14.1.15), for each $k,\ j \ge 1$

$$E(\xi_{k+j}|\mathcal{F}_j) = O(a_{k+j}(1 \wedge A_j^p n_j/n_{k+j})). \tag{14.1.16}$$

Lemma 14.1.4. *We have*

$$\sum_{k=1}^{\infty} |\xi_k - X_k| = O(1). \tag{14.1.17}$$

Proof. We have from (14.1.15)

$$|\xi_k - X_k| = O(A_k^{-p} a_k) = O(A_k^{-p+1}).$$

Hence (14.1.17) follows from (14.1.5).

Lemma 14.1.5. *We have*

$$\sum_{1\le j\le N} EX_j^2 - A_N^2 = O(1). \tag{14.1.18}$$

Proof. Noting that

$$E\xi_j^2 - a_j^2/2 = a_j^2 \sin 2\pi n_j/2\pi n_j,$$
$$\sum_{j\le N} E\xi_j^2 - A_N^2 = O(\sum_{j\le N} a_j^2/n_j) = O(1)$$

and

$$\sum_{j\le N}(EX_j^2 - E\xi_j^2) = O(\sum_{j\le N} a_j^2 A_j^{-p}) = O(1),$$

we get (14.1.18).

We can now represent X_j as

$$X_j = Y_j + u_j - u_{j+1}, \tag{14.1.19}$$

where $\{Y_j, \mathcal{F}_j\}$ is a martingale difference sequence and $u_1 = 0$

$$u_j = \sum_{k=0}^{\infty} E(X_{j+k}|\mathcal{F}_{j-1}), \quad j \ge 2. \tag{14.1.20}$$

Lemma 14.1.6. *We have for all j, $1 \le j < n-1$,*

$$\sum_{j<i<k\le n} |E(X_iX_k|\mathcal{F}_j)| = Cn^{2r}A_n^{2-2\delta}\log^2 A_n, \tag{14.1.21}$$

where constant C does not depend on j and n.

Proof. By the definition of X_k, we admit

$$E(X_iX_k|\mathcal{F}_j) = E(X_i\xi_k|\mathcal{F}_j).$$

Write

$$E(\xi_kX_i|\mathcal{F}_j) = \sum_{v=0}^{2^{r_j}-1} I(U_{vj})b_v.$$

We have

$$\begin{aligned}\frac{2^{-r_j}}{a_k a_i} b_v = &\sum_{l=0}^{2^{r_i-r_j}-1} 2^{r_i} \int_{v2^{-r_j}+l2^{-r_i}}^{v2^{-r_j}+(l+1)2^{-r_i}} \cos 2\pi n_k t\,dt \\ &\cdot \int_{v2^{-r_j}+l2^{-r_i}}^{v2^{-r_j}+(l+1)2^{-r_i}} \cos 2\pi n_i t\,dt \\ = &\, 2^{r_i+1} \frac{\sin 2\pi n_i 2^{-r_i-1} \sin 2\pi n_k 2^{-r_i-1}}{(2\pi)^2 n_i n_k} \\ &\cdot \sum_{l=0}^{2^{r_i-r_j}-1} \Big(\cos 2\pi(n_k - n_i)\Big(v2^{-r_j} + \Big(l+\frac{1}{2}\Big)2^{-r_i}\Big) \\ &+ \cos 2\pi(n_k+n_i)\Big(v2^{-r_j} + \Big(l+\frac{1}{2}\Big)2^{-r_i}\Big)\Big).\end{aligned}$$

Using the equality

$$\sum_{v=0}^{n-1} \cos(av+b) = \frac{\sin an/2}{\sin a/2} \cos(b + a(n-1)/2),$$

where a and b are any real numbers with $\sin(a/2) \neq 0$, we obtain

$$\begin{aligned}\frac{2^{-r_j} b_v}{a_k a_i} = &\frac{2^{r_i+1} \sin 2\pi n_i 2^{-r_i-1} \sin 2\pi n_k 2^{-r_i-1}}{(2\pi)^2 n_i n_k} \\ &\cdot \Big(\frac{\sin 2\pi(n_k-n_i)2^{-r_j-1}}{\sin 2\pi(n_k-n_i)2^{-r_i-1}} \cos 2\pi(n_k-n_i)\Big(v+\frac{1}{2}\Big)2^{-r_j} \\ &+ \frac{\sin 2\pi(n_k+n_i)2^{-r_j-1}}{\sin 2\pi(n_k+n_i)2^{-r_i-1}} \cos 2\pi(n_k+n_i)\Big(v+\frac{1}{2}\Big)2^{-r_j}\Big).\end{aligned}$$

Hence

$$b_v = O(a_k a_i 2^{r_j}(1 + 2^{-r_j} n_i)/n_k)$$

for all i, k. And we also have for i, k with $n_k + n_i \leq 2^{r_i-1}$,

$$b_v = O(a_k a_i (1 \wedge 2^{r_j}/(n_k - n_i)).$$

For each $i : j < i < n$, take $k_0(i) = \max\{k : (n_k + n_i) \leq 2^{r_i-1}\} \wedge (n-1)$. (14.1.12) and (14.1.15) imply that $k_0(i) - i = O(i^r \log A_i)$, hence

$$\begin{aligned}\sum_{j<i<k\leq n} &E(X_i X_k | \mathcal{F}_j) \\ &= O\Big(\sum_{i=j+1}^{n-1} \sum_{k>k_0(i)}^{n} |a_k a_i| \Big(\frac{2^{r_j}}{n_k} + \frac{n_i}{n_k}\Big) \\ &\quad + \sum_{i=j+1}^{n-1} \sum_{k=i+1}^{k_0(i)} |a_k a_i| \Big(1 \wedge \frac{2^{r_j}}{n_k - n_i}\Big)\Big). \qquad (14.1.22)\end{aligned}$$

From Lemma 14.1.2 and (14.1.15), the first part of the right hand side of (14.1.22) is bounded by

$$A_n^{2-2\delta}\sum_{i=j+1}^{n-1}\frac{(2^{r_j}+n_i)}{n_{k_0(i)}}n^r = O\Big(A_n^{2-2\delta}\sum_{i=j+1}^{n-1}\frac{(2^{r_j}+n_i)}{n_iA_i^p}n^r\Big)$$
$$= O(A_n^{2-2\delta}n^{2r}).$$

Take $i(j)=\max\{i:\ 2^{r_j}\ge n_i i^{-3r}\}\wedge(n-1)$. Then (14.1.12) and (14.1.15) imply that $i(j)-j=O(j^r\log A_j)$. Again from Lemma 14.1.2, the second part of (14.1.22) is bounded by

$$A_n^{2-2\delta}\Big(\sum_{i=j}^{i(j)}\sum_{i<k\le k_0(i)}1+\sum_{i>i(j)}^{n-1}\sum_{k>i}^{k_0(i)}\frac{2^{r_j}}{n_k-n_i}\Big)$$
$$=O\Big(A_n^{2-2\delta}\Big(n^{2r}\log^2 A_n+\sum_{i>i(j)}^{n-1}\frac{2^{r_j}}{n_{i+1}}i^{2r}\Big)\Big)$$
$$=O\Big(A_n^{2-2\delta}\Big(n^{2r}\log^2 A_n+\frac{2^{r_j}(i(j))^{3r}}{n_{i(j)+1}}\Big)\Big)$$
$$=O(n^{2r}A_n^{2-2\delta}\log^2 A_n).$$

The lemma is proved.

Lemma 14.1.7. *We have*

$$u_j=O(j^rA_j^{1-\delta}\log A_j). \tag{14.1.23}$$

Proof. Using (14.1.16), (14.1.20) and Lemma 14.1.2, we find

$$u_j=\sum_{k=0}^{\infty}E(\xi_{k+j}|\mathcal{F}_{j-1})$$
$$=O\Big(\sum_{k=0}^{\infty}A_{k+j}^{1-\delta}(1\wedge A_j^pn_j/n_{k+j})\Big)$$
$$=O(j^rA_j^{1-\delta}\log A_j).$$

Lemma 14.1.8. *For each $k,n(k<n)$,*

$$\sum_{k<i\le n}E(X_iu_{n+1}|\mathcal{F}_i)=O(A_n^{2-2\delta}n^{2r}\log^2 A_n); \tag{14.1.24}$$

$$\sum_{k<i\le n}E(X_iu_k|\mathcal{F}_k)=O(A_n^{2-2\delta}n^r\log A_n). \tag{14.1.25}$$

Proof. By (14.1.20) we have

$$E(u_{n+1}|\mathcal{F}_i) = \sum_{j=0}^{\infty} E(\xi_{j+n+1}|\mathcal{F}_i)$$
$$= O\Big(\sum_{j=0}^{\infty} A_{j+n+1}^{1-\delta}(1 \wedge A_i^p n_i/n_{j+n+1})\Big).$$

Hence

$$\sum_{i=k+1}^{n} E(X_i u_{n+1}|\mathcal{F}_i) = O\Big(\sum_{i=k+1}^{n}\sum_{j=0}^{\infty}(1 \wedge A_i^p n_i/n_{j+n+1})A_{j+n+1}^{1-\delta}A_n^{1-\delta}\Big).$$

Taking $k_0 = [c_0 n^r \log A_n]$, where c_0 will be specified later, we have

$$\sum_{i=k+1}^{n}\sum_{j=0}^{\infty}(1 \wedge A_i^p n_i/n_{j+n+1})A_{j+n+1}^{1-\delta}$$
$$\le \sum_{i=n-k_0+1}^{n}\sum_{j=0}^{k_0} A_{j+n+1}^{1-\delta} + \sum_{i=k+1}^{n-k_0}\sum_{j=0}^{\infty}\frac{A_i^p n_i}{n_{j+n+1}}A_{j+n+1}^{1-\delta}$$
$$+ \sum_{i=n-k_0+1}^{n}\sum_{j=k_0+1}^{\infty}\frac{A_i^p n_i}{n_{j+n+1}}A_{j+n+1}^{1-\delta}$$
$$= O\Big(\sum_{i=k+1}^{n-k_0}\frac{A_i^p n^r n_i}{n_{n+1}}A_{n+1}^{1-\delta} + k_0^2 A_n^{1-\delta}$$
$$+ \sum_{i=n-k_0+1}^{n}\frac{A_i^p n_i}{n_{k_0+n+1}}(k_0+n+1)^r A_{k_0+n+1}^{1-\delta}\Big)$$
$$= O\Big(n^{3r}A_n^{p+1-\delta}\exp\Big(\frac{(n-k_0)^{1-r}-n^{1-r}}{2(1-r)}q\Big) + k_0^2 A_n^{1-\delta}$$
$$+ A_n^p n^r k_0 \exp\Big(\frac{n^{1-r}-(n+k_0)^{1-r}}{2(1-r)}q\Big)\Big)$$
$$= O(A_n^{1-\delta}n^{2r}\log^2 A_n).$$

The last equality holds so long as we take c_0 sufficiently large. This proves (14.1.24); the proof of (14.1.25) is similar.

Lemma 14.1.9. *We have*

$$\sum_{j=1}^{n} EY_j^2 - A_n^2 = O(n^{2r}A_n^{2-2\delta}\log^2 A_n). \tag{14.1.26}$$

Proof. Since $\{Y_j\}$ is a martingale difference sequence, we have

$$\sum_{j=1}^{n} EY_j^2 = E\Big(\sum_{j=1}^{n} Y_j\Big)^2$$
$$= E\Big(\sum_{j=1}^{n} X_j\Big)^2 + 2Eu_{n+1}\sum_{j=1}^{n} X_j + Eu_{n+1}^2.$$

Hence (14.1.26) follows from Lemma 14.1.5 - Lemma 14.1.8.

Lemma 14.1.10. *Under the assumption $\delta\beta > r$, we have*

$$\sum_{j=1}^{n} Y_j^2 - A_n^2 = O(A_n^{2-\delta} n^r \log^3 A_n) \quad a.s. \tag{14.1.27}$$

Proof. Note that (14.1.5) and $\delta\beta > r$ imply $n^r = O(A_n^\delta)$. Applying Lemma 14.1.1, the Borel-Cantelli lemma and the subsequence method, we need only to show that for any $0 \le m \le n$

$$E\Big(\sum_{m<j\le n} (Y_j^2 - EY_j^2)\Big)^2 = O\Big(\Big(\sum_{m<j\le n} EY_j^2\Big) A_n^{2-2\delta} n^{2r} \log^2 A_n\Big), \tag{14.1.28}$$

where the constant implied by O does not depend on m, n. Observe that

$$E\Big(\sum_{m<j\le n} (Y_j^2 - EY_j^2)\Big)^2$$
$$= \sum_{m<j\le n} E(Y_j^2 - EY_j^2)^2 + 2\sum_{m<k\le n} E(Y_k^2 - EY_k^2)\Big(\sum_{k<i\le n} Y_i^2\Big),$$

where

$$\sum_{m<k\le n} E(Y_k^2 - EY_k^2)\Big(\sum_{k<i\le n} Y_i^2\Big)$$
$$= \sum_{m<k\le n} E(Y_k^2 - EY_k^2)\Big(\sum_{k<i\le n} Y_i\Big)^2$$
$$= \sum_{m<k\le n} E(Y_k^2 - EY_k^2)\Big(\sum_{k<i\le n} X_i^2 + 2\sum_{k<i<j\le n} X_i X_j$$
$$+ 2\sum_{k<i\le n} X_i(u_{n+1} - u_{k+1}) + (u_{n+1} - u_{k+1})^2\Big)$$
$$=: \sum_{m<k\le n} (I_1(k) + I_2(k) + I_3(k) + I_4(k)).$$

By Lemma 14.1.6 – Lemma 14.1.8, it follows that

$$\max_{2\le i\le 4} I_i(k) = O(n^{2r} A_n^{2-2\delta} EY_k^2 \log^2 A_n). \tag{14.1.29}$$

We now show that (14.1.29) holds for $I_1(k)$ as well. Since

$$\Big|\sum_{k<i\le n} X_i^2 - \sum_{k<i\le n} \xi_i^2\Big| = O(1),$$

it suffices to show that (14.1.29) holds for $E(Y_k^2 - EY_k^2)\left(\sum_{k<i\le n} \xi_i^2\right)$. Noting that Y_k is $\mathcal{F}_k$-measurable and writing

$$Y_k = \sum_{i=0}^{2^{r_k}-1} d_i I(U_{ik}),$$

we have by the definition of $\mathcal{F}_k$

$$\begin{aligned}
&E(Y_k^2 - EY_k^2)\Big(\sum_{j=k+1}^{n} \xi_j^2\Big) \\
&= \sum_{i=0}^{2^{r_k}-1} d_i^2 \sum_{j=k+1}^{n} \int_{i2^{-r_k}}^{(i+1)2^{-r_k}} a_j^2 \cos^2 2\pi n_j t\, dt \\
&\quad - 2^{-r_k}\Big(\sum_{i=0}^{2^{r_k}-1} d_i^2\Big) \sum_{j=k+1}^{n} \int_0^1 a_j^2 \cos^2 2\pi n_j t\, dt \\
&= \sum_{i=0}^{2^{r_k}-1} d_i^2 \sum_{j=k+1}^{n} a_j^2 \frac{\sin 2\pi n_j 2^{-r_k}}{\pi n_j} \cos 4\pi n_j \Big(i+\frac{1}{2}\Big)2^{-r_k} \\
&\quad - 2^{-r_k}\Big(\sum_{i=0}^{2^{r_k}-1} d_i^2\Big) \sum_{j=k+1}^{n} a_j^2 \frac{\sin 4\pi n_j}{2\pi n_j} \\
&= O\Big(2^{-r_k}\Big(\sum_{i=0}^{2^{r_k}-1} d_i^2\Big)\Big(\sum_{j=k+1}^{n} \frac{a_j^2}{n_j} + \sum_{j=k+1}^{n} a_j^2 (1 \wedge A_k^p n_k/n_j)\Big)\Big) \\
&= O(A_n^{2-2\delta} n^r EY_k^2 \log A_n).
\end{aligned}$$

This proves that (14.1.29) holds for $I_1(k)$. For EY_k^4, we have

$$EY_k^4 = O(EY_k^2 A_k^{2-2\delta} k^{2r} \log^2 A_k).$$

This proves (14.1.28), and the lemma follows.

Lemma 14.1.11. *We have*

$$\sum_{j=1}^{n}(E(Y_j^2|\mathcal{F}_{j-1}) - Y_j^2) = O(n^r A_n^{2-\delta}\log^3 A_n) \quad a.s. \tag{14.1.30}$$

Proof. Put $R_j = Y_j^2 - E(Y_j^2|\mathcal{F}_{j-1})$. Then $\{R_j, \mathcal{F}_j\}$ is a martingale defference sequence, and we have

$$ER_k^2 = O(EY_k^4) = O(EY_k^2 A_k^{2-2\delta}k^{2r}\log^2 A_k).$$

The proof is verbatim as that of Lemma 14.1.10.

By a martingale version of the Skorohod representation theorem, there exists a probability space, on which a Wiener process and a sequence of non-negative random variables T_i are defined such that

$$\Big\{W\Big(\Big(\sum_{j\le m} T_j\Big), m \ge 1\Big\} \quad \text{and} \quad \Big\{\sum_{j\le m} Y_j, m \ge 1\Big\}$$

have the same distribution. Hence on the new probability space, without loss of generality we can redefine $\{Y_j\}$ by

$$Y_j = W\Big(\sum_{i\le j} T_i\Big) - W\Big(\sum_{i<j} T_i\Big)$$

and can keep the same notation. Write

$$\mathcal{L}_m = \sigma\Big\{W\Big(\sum_{j=1}^{k} T_i\Big), k \le m\Big\},$$

$$\mathcal{A}_m = \sigma\Big\{W(t), 0 \le t \le \sum_{j=1}^{m} T_j\Big\}.$$

It is clear that $\mathcal{L}_m \subset \mathcal{A}_m, m \ge 1$, and T_j is $\mathcal{A}_j$- measurable. By the embedding theorem, for every $j \ge 1$, $ET_j = EY_j^2$, and

$$E(T_j|\mathcal{A}_{j-1}) = E(Y_j^2|\mathcal{A}_{j-1}) = E(Y_j^2|\mathcal{L}_{j-1}) \quad \text{a.s.} \tag{14.1.31}$$

Moreover, for any $v > 1$ there is a C_v such that

$$E|T_j|^v = C_v E|Y_j|^{2v}. \tag{14.1.32}$$

Lemma 14.1.12. *We have under the conditions of Theorem 14.1.1*

$$\sum_{j\le n} T_j - A_n^2 = O\Big(A_n^{2-\delta+r/\beta}\log^3 A_n\Big). \tag{14.1.33}$$

Proof. Write

$$\begin{aligned}\sum_{j\le n} T_j - A_n^2 &= \sum_{j\le n}(T_j - E(T_j|\mathcal{F}_{j-1})) + \sum_{j\le n}(E(Y_j^2|\mathcal{F}_{j-1}) - Y_j^2)\\ &\quad + \sum_{j\le n} Y_j^2 - A_n^2 =: I_1 + I_2 + I_3.\end{aligned}$$

Put $Z_j = T_j - E(T_j|\mathcal{A}_{j-1})$. Then $\{(Z_j, \mathcal{A}_j)\}$ is a martingale difference sequence and $EZ_j^2 = O(EY_j^4)$. By the proof analogous to that of Lemma 14.1.11 we obtain

$$I_1 = O(A_n^{2-\delta} n^r \log^3 A_n) \quad \text{a.s.} \tag{14.1.34}$$

The lemma follows from (14.1.5), (14.1.27) (14.1.30) and (14.1.34).

Proof of Theorem 14.1.1.
Note that

$$\sum_{j\le n} Y_j - \sum_{j\le n} X_j = u_{n+1} = O(A_n^{1-\delta} n^r \log A_n).$$

The theorem follows from Lemmas 14.1.3 and 14.1.12.

Proof of Theorem 14.1.2.

Note that $\delta = 1$ now. By the proof analogous to that of all the above lemmas under the conditions of Theorem 14.1.2, we also have

$$\sum_{j\le n} T_j - A_n^2 = O(n^r(n^r \vee \overline{B}_n) \log^3 A_n) \quad \text{a.s.,} \tag{14.1.35}$$

where $\overline{B}_n = \sum_{k\le n} a_k^2 EY_k^2$.

Since

$$\sum_{j\le k} EY_j^2 = O\Big(\sum_{j\le k} a_j^2\Big),$$

and a_j^2 is non-increasing, we have

$$\overline{B}_n = O\Big(\sum_{k\le n} a_k^4\Big)$$

by using the Abel transformation. Hence

$$\sum_{j\le n} T_j - A_n^2 = O\Big(n^r(n^r \vee \overline{B}_n) \log^3 A_n\Big) \quad \text{a.s.} \tag{14.1.36}$$

The theorem follows from (14.1.36) and Lemma 14.1.3.

Remark 14.1.1. If $k^\beta = O(A_k)$ for some $\beta > 0$, $|a_k| \downarrow$, and $|a_k| = O(k^{-\theta})$ for some $r \le \theta < 1/2$, then by the proof similar to that of Theorem 14.1.1, we have

$$S(A_n^2) - W(A_n^2) = O(C_n^{1/4} \log^2 A_n) \quad \text{a.s.}$$

where $C_n = \sum_{k \le n} |a_k|^{4-2r/\theta}$.

Let $\overline{S}_n(\omega) = \sum_{k \le n} \sqrt{2} \cos 2\pi n_k \omega$, $\omega \in [0,1)$. Combining Corollary 14.1.1 with the results of the increments of a Wiener process, we can obtain a.s. limiting behavior of increments of the sums $\overline{S}_n$.

14.2 A class of Gaussian sequences

Let $\{X_n, n \ge 1\}$ be a centered sequence of Gaussian random variables. Under some conditions, including $E\Big(\sum_{k=m+1}^{m+n} X_k\Big)^2 = n\sigma^2 + O(n^{1-\varepsilon})$ for some $\varepsilon > 0$, $\sigma^2 > 0$ and $EX_m X_{m+n} = O(n^{-2})$, Philipp and Stout (1975) established strong approximations of partial sums

$$S(t) = \sum_{k \le t} X_k \qquad t \ge 0$$

by a Wiener process with order $O(t^{1/2-\lambda})$, where $0 < \lambda < \frac{1}{60} \wedge \frac{4\varepsilon}{15}$. By applying the property of a symmetric matrix and the circle-plate theorem about eigenvalues of a matrix, Shao (1985) proved an ideal result as follows:

Theorem 14.2.1. *Suppose that there exist $C_i > 0, i = 1, 2, 3,$ such that for every n*

$$E\Big(\sum_{k=m+1}^{m+n} X_k\Big)^2 \ge C_1 n \quad \textit{uniformly in } m \tag{14.2.1}$$

$$EX_n^2 \le C_2 \tag{14.2.2}$$

and

$$\gamma(n) := \sup_m |EX_m X_{m+n}| \le C_3 n^{-3/2-\lambda} \tag{14.2.3}$$

for some $\lambda > 0$. Then without changing the distribution of $\{S(t), t \ge 0\}$, we can redefine the process $\{S(t), t \ge 0\}$ on a richer probability space together with a Wiener process $\{W(t), t \ge 0\}$ such that

$$S(t) - W(b_t) = O(\log^{1/2} t) \qquad a.s. \tag{14.2.4}$$

where

$$b_t = b_{[t]} = a_t + r_t, \quad a_t = ES^2(t), \quad r_t = O\Big(\sum_{k=1}^{[t]} k^{-1/2-\lambda}\Big). \tag{14.2.5}$$

Corollary 14.2.1. *If the condition (14.2.3) is strengthened by*

$$\gamma(n) \le Cn^{-2}(\log n)^{-1-\varepsilon} \tag{14.2.3'}$$

for some $\varepsilon > 0$, then we have

$$S(t) - W(a_t) = O(\log^{1/2} t) \qquad a.s. \tag{14.2.4'}$$

Corollary 14.2.2. *Suppose that $\{X_n, n \ge 1\}$ is a centered stationary Gaussian sequence and (14.2.3′) is satisfied, then*

$$\sigma^2 := EX_1^2 + 2\sum_{k=2}^{\infty} EX_1X_k$$

converges absolutely. If $\sigma^2 > 0$, then, assuming $\sigma^2 = 1$, we have

$$S(t) - W(t) = O(\log^{1/2} t) \qquad a.s.$$

The proof of the theorem needs the following lemmas.

Lemma 14.2.1. *Let A be a real symmetric matrix of order n with eigenvalues $\lambda_1, \cdots, \lambda_n$. Denote $\lambda = \max_{1\le i\le n} |\lambda_i|$. Then for any row vector $\mathbf{C}$ we have*

$$|\mathbf{CAC'}| \le \lambda \mathbf{CC'} \tag{14.2.6}$$

where $\mathbf{C'}$ is the transposed vector of $\mathbf{C}$.

Proof. We need only to prove that matrices $\lambda I - A$ and $\lambda I + A$ are non-negative definite. By the well-known property of a matrix, there exists a real orthogonal matrix U such that $U'AU$ is a diagonal matrix Λ with the diagonal elements which is just equal to the eigenvalues of the matrix A. Therefore we have

$$\lambda I - A = U(\lambda I - \Lambda)U'.$$

It is clear that the eigenvalues of $\lambda I - \Lambda$ are all non-negative, so that the eigenvalues of the real symmetric matrix $\lambda I - A$ are also non-negative.

Thus the matrix $\lambda I - A$ is non-negative definite, similarly, so is the matrix $\lambda I + A$. The lemma is proved.

Lemma 14.2.2. *At least one of the following inequalities is satisfied for eigenvalues of any matrix* $A = (a_{ij})_{n\times n}$*:*

$$|\lambda - a_{ii}| \le \sum_{j=1, j\neq i}^{n} |a_{ij}| \quad i = 1, \cdots, n. \tag{14.2.7}$$

This result, the so-called circle-plate theorem, is due to Gerschgorin (cf. Franklin 1968, p.161 Theorem 1).

Denote $\mathcal{F}_n = \sigma\{X_k, 1 \le k \le n\}$. Write

$$\begin{aligned} Y_n &= \sum_{k=0}^{\infty}(E(X_{n+k}|\mathcal{F}_n) - E(X_{n+k}|\mathcal{F}_{n-1})) \\ &= X_n + u_{n+1} - u_n \end{aligned} \tag{14.2.8}$$

where $u_1 = 0$ and

$$u_n = \sum_{k=0}^{\infty} E(X_{n+k}|\mathcal{F}_{n-1}), \quad n = 2, 3, \cdots. \tag{14.2.9}$$

It is clear that $\{Y_n, \mathcal{F}_n\}$ is a martingale difference sequence. We shall prove below that the series in (14.2.9) is convergent under the assumptions of Theorem 14.2.1.

Lemma 14.2.3. *If (14.2.1), (14.2.2) and (14.2.3) are satisfied, and for any* $k \ge 1$

$$EX_k^2 \ge \sum_{j=1, j\neq k}^{\infty} |EX_k X_j| + 1, \tag{14.2.10}$$

then

$$\|u_n\|_2 = O(1).$$

Proof. Let A be the covariance matrix of $(X_1, \cdots, X_j)$ and $\mathbf{C} = (EX_1X_{j+k}, \cdots, EX_jX_{j+k})$. Then, by (5.22) in Philipp and Stout (1975), we have

$$E(E^2(X_{j+k}|\mathcal{F}_j)) = \mathbf{C}A^{-1}\mathbf{C}', \tag{14.2.11}$$

for any $j, k \ge 1$. By condition (14.2.10) and Lemmas 14.2.1 and 14.2.2, we obtain

$$\mathbf{C}A^{-1}\mathbf{C}' \le \mathbf{C}\mathbf{C}'. \tag{14.2.12}$$

Denote

$$u_{nk} = E(X_{n+k}|\mathcal{F}_{n-1}), \quad k = 0, 1, \cdots; n = 1, 2, \cdots. \tag{14.2.13}$$

It follows from (14.2.11), (14.2.12) and (14.2.3) that

$$\begin{aligned} Eu_{nk}^2 &\leq \sum_{i=1}^{n-1} (EX_{n+k}X_i)^2 \\ &\leq C_3^2 \sum_{i=1}^{n-1} (n+k-i)^{-3-2\lambda} \leq C_3^2 (k+1)^{-2-2\lambda}. \end{aligned} \tag{14.2.14}$$

Thus we have

$$\|u_n\|_2 \leq \sum_{k=0}^{\infty} \|u_{nk}\|_2 = O(1).$$

Proof of Theorem 14.2.1.

1) We first prove Theorem 14.2.1 under condition (14.2.10). By a well-known property of a Gaussian sequence (cf. Ibragimov and Rozanov 1978, p.14), Y_n, which is defined by (14.2.8), is a linear combination of $X_1, \cdots, X_n$, so that $\{Y_n, n \geq 1\}$ is also a Gaussian sequence. By (14.2.8) $\{Y_n, n \geq 1\}$ is a martingale difference sequence. Therefore $Y_n, n = 1, 2, \cdots$ are independent. Then we can construct a Wiener process $\{W(t), t \geq 0\}$ from $\{Y_n, n \geq 1\}$. Let $b_t = \sum_{k \leq t} EY_k^2$. We redefine

$$Y_n = W(b_n) - W(b_{n-1}), \quad n = 1, 2, \cdots$$

and from (14.2.8) we have

$$\sum_{k \leq n} X_k - \sum_{k \leq n} Y_k = -u_{n+1}.$$

Note that $u_n, n = 1, 2, \cdots$, are normal with uniformly bounded variances, then we have

$$u_n = O(\log^{1/2} n) \qquad \text{a.s.}$$

i.e.

$$\sum_{k \leq n} X_k - \sum_{k \leq n} Y_k = O(\log^{1/2} n) \quad \text{a.s.}$$

or

$$S(t) - W(b_t) = O(\log^{1/2} t) \quad \text{a.s.} \tag{14.2.15}$$

Furthermore

$$
\begin{aligned}
b_n - a_n &= 2E\Big(\sum_{k=1}^{n} X_k u_{n+1}\Big) + Eu_{n+1}^2 \\
&= 2E\Big(\sum_{k=1}^{n}\sum_{j=1}^{\infty} X_k X_{n+j}\Big) + Eu_{n+1}^2 \\
&= O\Big(\sum_{k=1}^{n}\sum_{j=1}^{\infty}\gamma(n+j-k)\Big) = O\Big(\sum_{k=1}^{n} k^{-1/2-\lambda}\Big).
\end{aligned}
$$

Then Theorem 14.2.1 holds true under (14.2.10).

2) Now we prove Theorem 14.2.1 in general case. Put $l = [312C_3^2/C_1^2]$, define

$$
X_m^* = \sum_{(m-1)l<k\le ml} X_k \quad m = 1, 2, \cdots, \quad S_t^* = \sum_{k\le t} X_k^*.
$$

Then $\{X_n^*\}$ is also a Gaussian sequence satisfying (14.2.1), (14.2.2) and (14.2.3). It follows from (14.2.1) that

$$
EX_n^{*2} \ge C_1 l. \tag{14.2.16}
$$

We prove that $\{X_n^*\}$ satisfies (14.2.10). In fact

$$
\begin{aligned}
&\sum_{\substack{j=1\\ j\ne n}}^{\infty} |EX_n^* X_j^*| \\
&\quad = \sum_{\substack{j=1\\ j\ne n}}^{\infty} \Big|E\Big(\sum_{(n-1)l<k\le nl} X_k\Big)\Big(\sum_{(j-1)l<i\le jl} X_i\Big)\Big| \\
&\quad \le C_3 \sum_{\substack{j=1\\ j\ne n}}^{\infty} \sum_{(n-1)l<k\le nl}\ \sum_{(j-1)l<i\le jl} |i-k|^{-3/2} \\
&\quad \le 2C_3 \sum_{k=1}^{l}\sum_{m=0}^{\infty} (m+k)^{-3/2} \\
&\quad \le 6C_3 \sum_{k=1}^{l} k^{-1/2} \le 6C_3(1 + 2l^{1/2}).
\end{aligned} \tag{14.2.17}
$$

By the definition of l, it is easy to verify that the right hand side of (14.2.17) does not exceed $C_1 l - 1$, that is to say, $\{X_n^*\}$ satisfies (14.2.10).

For any fixed n, there exists an m such that $(m-1)l \le n < ml$ and hence

$$
|S_n - S_m^*| \le \max_{(m-1)l\le k<ml} |X_k|. \tag{14.2.18}
$$

Note that $\{X_n, n \geq 1\}$ is a Gaussian sequence with uniformly bounded variances. Then we have

$$\max_{(m-1)l \leq k < ml} |X_k| = O(\log^{1/2} m) \quad \text{a.s.} \tag{14.2.19}$$

and

$$\begin{aligned} a_n - a_m^* &= E\Big(\sum_{k=1}^{n} X_k\Big)^2 - E\Big(\sum_{k=1}^{ml} X_k\Big)^2 \\ &= -2E\Big(\sum_{k=1}^{n} \sum_{j=n+1}^{ml} X_k X_j\Big) - E\Big(\sum_{j=n+1}^{ml} X_j\Big)^2 \\ &= O\Big(\sum_{k=1}^{\infty} \gamma(k)\Big) = O(1), \end{aligned} \tag{14.2.20}$$

where $a_m^* = E(S_m^*)^2$. It follows from 1), (14.2.18)–(14.2.20) that the proof of Theorem 14.2.1 is completed.

14.3 The non-negative additive functional of a Markov process

Let $\mathbf{X} = \{X_t, t \geq 0\}$ be a homogeneous, right continuous, strong-Feller Markov process, which is defined on a probability space $(\Omega, \mathcal{F}, P)$ with values in a complete, σ-compact measurable metric space $(E, \rho, \mathbf{B})$, satisfying the following conditions:

(i) for any $x \in E$, $t > 0$ and open set $U \in \mathbf{B}$, the stationary transition function of the Markov process $\mathbf{X}$

$$p(t, x, U) > 0; \tag{14.3.1}$$

(ii) for every $a \in E$ there exists a compact set K such that

$$P_a\{X_t(\omega) \in K \text{ for some } t \geq 0\} = 1, \tag{14.3.2}$$

where $P_a(\cdot) = P(\cdot | X_0 = a)$.

From (i) and (ii) it follows that $\mathbf{X}$ is a recurrent strong Markov process.

For every $U \in \mathbf{B}$, put

$$\tau_U(\omega) = \begin{cases} \inf\{t, t \geq 0, X_t(\omega) \in U\}, & \text{if the set is non-empty,} \\ \infty & \text{otherwise.} \end{cases}$$

Define the exit distribution

$$h^U(a, S) = P_a\{X_{\tau_U}(\omega) \in S, \tau_U < \infty\} \quad \text{for } a \in E,\ S \in \mathbf{B}.$$

Denote

$$h^U f(x) = \int_{\overline{U}} h^U(x, dy) f(y) \qquad \text{for } f(x) \in B(\overline{U}),$$

where $B(\overline{U})$ is the set of all bounded measurable function on $\overline{U}$.

Let $\mathbf{H}= \{K, L\}$ be the collection of all set-pairs satisfying the following conditions:

(H_1) K and L are closed subsets of E and have an interior point at least,

(H_2) K and/or L are compact,

(H_3) (i) $K \subset E \backslash L$, $E\backslash L$ is a connected open set, or

(ii) $L \subset E\backslash K$, $E\backslash K$ is a connected open set.

For any given set-pair $(K, L) \in \mathbf{H}$, $x \in K$, put

$$T^K(x, U) = \int_E h^L(x, dy) h^K(y, U), \tag{14.3.3}$$

where $U \in \mathbf{B}$, $U \subset K$. By recurrence $T^K(x, U)$ is a one-step transition function. Define the transformation

$$T^K f(x) = \int_E T^K(x, dy) f(y) \tag{14.3.4}$$

for $x \in K$, $f \in B(K)$.

Without loss of generality, assume that K is compact, and (i) of (H_3) is satisfied. Then for T^K there exists a unique invariant probability measure μ on K. Moreover, suppose that $\mathbf{X}$ satisfies

(iii) $T^K\{B(K)\} \subset C(K) := \{\text{all continuous function on } K\}$.

Let τ be a stopping time of the process $\mathbf{X}$, put $\Omega_\tau = \{\tau(\omega) < \infty\}$. Then

$$\mathcal{N}_\tau := \{A : A \subset \Omega_\tau, \forall t \geq 0, A \cap (\tau \leq t) \in \mathcal{N}_t\}$$

is a σ-field of Ω_τ, where $\mathcal{N}_t = \sigma\{X_s, 0 \leq s \leq t\}$. Let $\mathcal{N}^*$ be a σ-field of Ω which was introduced in Dynkin (1963 (3, 3.5)). Define the shift operators θ_t from $\mathcal{N}^*$ to $\mathcal{N}^*$ and the shift operator θ_τ from $\mathcal{N}^*$ to $\Omega_\tau \mathcal{N}^*$ satisfying

$$\theta_\tau A = \bigcup_{t \geq 0} \{\theta_t A,\ \tau(\omega) = t\} \subset \Omega_\tau \qquad A \in \mathcal{N}^*, \tag{14.3.5}$$

where the operations of union, intersection and complement are keeped by the operator θ_τ, and we have

$$\theta_\tau\{X_t \in \Gamma\} = \{X_{t+\tau} \in \Gamma\} \qquad \Gamma \in \mathbf{B}. \tag{14.3.6}$$

For given $(K, L) \in \mathbf{H}$, we define random functions on Ω as follows:

$$\begin{aligned} \tau_1 &= \tau_1(K, L, \omega) \\ &= \begin{cases} \inf\{t, t \geq 0, X_t(\omega) \in K\} & \text{if the set is non-empty,} \\ \infty & \text{otherwise;} \end{cases} \\ \sigma_n &= \sigma_n(K, L, \omega) \\ &= \begin{cases} \inf\{t, t > \tau_n, X_t(\omega) \in L\} & \text{if the set is non-empty,} \\ \infty & \text{otherwise;} \end{cases} \\ \tau_{n+1} &= \tau_{n+1}(K, L, \omega) \\ &= \begin{cases} \inf\{t, t > \sigma_n, X_t(\omega) \in K\} & \text{if the set is non-empty,} \\ \infty & \text{otherwise,} \end{cases} \end{aligned}$$

for $n \geq 1$. By Doob (1953), τ_n, σ_n $(n \geq 1)$ are the almost sure finite stopping times of the process $\mathbf{X}$. Denote $\xi_n(\omega) = X_{\tau_n}(\omega)$. Its transition function $T^K(x, U)$ satisfies the Doeblin condition and there exists a unique ergodic set, non-periodic. Moreover $T^K(x, U)$ satisfies

$$|(T^K)^n(x, U) - \mu(U)| < C\delta^n, \tag{14.3.7}$$

where $0 < \delta < 1$, C is a constant. Put

$$P_\mu(B) = \int_E P_a(B)\mu(da) \qquad B \in \mathcal{F}. \tag{14.3.8}$$

Then P_μ is a probability measure on $(\Omega, \mathcal{F})$, and $P_a(B) = 1$ implies $P_\mu(B) = 1$. Put

$$E_\mu f(\cdot) = \int_\Omega f(\cdot) P_\mu(d\omega).$$

At last, let $\phi = \{\phi_t^s(\omega)\}$ be a non-negative, strongly measurable, homogeneous additive functional on Ω, i.e. $\{\phi_t^s, 0 \leq s \leq t\}$ is a family of real valued functions satisfying the following condtions:

(ϕ_1) For any $s \leq t$, $\phi_t^s(\omega)$ is non-negative $\overline{\mathcal{N}}_t$-measurable, where $\overline{\mathcal{N}}_t$ is the completion of $\mathcal{N}_t$ in probability space $(\Omega, \mathcal{F}, P)$,

(ϕ_2) for any $\omega \in \Omega$ and $s \leq t \leq u$, $\phi_u^s = \phi_t^s + \phi_u^t$,

(ϕ_3) for any $\omega \in \Omega$ and $h \geq 0, s \leq t$, $\theta_h \phi_t^s = \phi_{t+h}^{s+h}$,

(ϕ_4) for any $0 \leq u \leq v$, the bivariate function $\phi_t^0(\omega)$ of point (t, ω) is $\mathbf{B}_{[u,v]} \times \overline{\mathcal{N}_v^u}$-measurable on $[u, v] \times \Omega$, where $\mathbf{B}_{[u,v]}$ is the σ-field of interval $[u, v]$, $\overline{\mathcal{N}_v^u}$ is the completion of $\mathcal{N}_v^u = \sigma(X_t, u \leq t \leq v)$.

It is easy to check that

$$\theta_{\sigma_1} \tau_n = \tau_{n+1} - \sigma_1. \tag{14.3.9}$$

Denote

$$y_n = \phi_{\tau_{n+1}}^{\tau_n}, \quad z_n = \tau_{n+1} - \tau_n.$$

Then

$$\theta_{\sigma_1} y_n = y_{n+1}, \qquad \theta_{\tau_1} z_n = z_{n+1}. \tag{14.3.10}$$

Lemma 14.3.1. *$\{y_n, n \geq 1\}$ is a strictly stationary φ-mixing sequence of random variables with $\sum_{n=1}^{\infty} \varphi^{1/2}(n) < \infty$.*

Particularly, $\{z_n, n \geq 1\}$ and $\{w_n = y_n + cz_n, n \geq 1\}$ are also the strictly stationary φ-mixing sequences.

Proof. In order to prove $\{y_n\}$ is a strictly stationary sequence, it needs only to show that θ_{σ_1} on P_μ is an operator preserving measure. Indeed, from (14.3.8) and Dynkin (1963 Theorem 3.11), for any $B \in \mathcal{N}_0 := \sigma\{X_t, t \geq 0\}$ we have

$$P_\mu(\theta_{\sigma_1} B) = \int_{\Omega_{\sigma_1}} P_{X_{\sigma_1}}(B) P_\mu(dx) = P_\mu(B).$$

In order to prove $\{y_n\}$ is φ-mixing, we first show the following two facts:

(i) $\sigma\{y_1, \cdots, y_{m-1}\} \subset \sigma\{\xi_1, \cdots, \xi_m\} \subset \mathcal{N}_{\tau_m}$,

(ii) $\sigma\{y_{m+k-1}, y_{m+k}, \cdots\} \subset \sigma\{\xi_{m+k}, \xi_{m+k+1}, \cdots\}$.

Since $\mathcal{N}_{\tau_{m-1}} \subset \mathcal{N}_{\tau_m}$ and $\xi_m = X_{\tau_m}$ is $\mathcal{N}_{\tau_m}$-measurable, ξ_k, $1 \leq k \leq m$ are also $\mathcal{N}_{\tau_m}$-measurable. Therefore

$$\sigma\{\xi_1, \cdots, \xi_m\} \subset \mathcal{N}_{\tau_m}.$$

Thus (i) holds true, if we can prove

$$\sigma\{y_n \in A, \ A \in \mathcal{B}_R\} \subset \sigma\{\xi_{n+1} \in \Gamma, \ \Gamma \in \mathbf{B}\} \tag{14.3.11}$$

where $\mathcal{B}_R$ is Borel σ-field of R_+. Put

$$\Lambda = \{A : \ (y_n \in A) \in \sigma\{\xi_{n+1} \in \Gamma, \Gamma \in \mathbf{B}\}, A \in \mathcal{B}_R\},$$

it is easy to verify that Λ is a λ-system and contains a π-system

$$\Pi = \{[0, t] : \ (y_n \in [0, t)) \in \sigma\{\xi_{n+1} \in \Gamma, \Gamma \in \mathbf{B}\}\}.$$

Then $\sigma(\Pi) \subset \Lambda$ that implies (14.3.11). The proof of (ii) is similar.

Thus for any $A \in \sigma\{y_k, 1 \leq k \leq m-1\}$, $B \in \sigma\{y_k, k \geq m+n-1\}$, by the strong Markov property and (14.3.7) we have

$$\begin{aligned} P_\mu(AB) &= E_\mu(E_\mu(I(A)I(B)|\mathcal{N}_{\tau_m})) \\ &= E_\mu(I(A)E_\mu(I(B)|\mathcal{N}_{\tau_m})) = E_\mu(I(A)E_\mu(I(B)|\xi_m)) \end{aligned}$$

and

$$\begin{aligned}
&|P_\mu(AB) - P_\mu(A)P_\mu(B)| \\
&\quad = |E_\mu(I(A)E_\mu(I(B)|\xi_m)) - E_\mu I(A)E_\mu I(B)| \\
&\quad = \Big|E_\mu\Big\{I(A)\int_E E_\mu(I(B)|\xi_{m+n}=\eta)\Big[(T^K)^n(\xi_m,d\eta)-\mu(d\eta)\Big]\Big\}\Big| \\
&\quad \le E_\mu\Big\{I(A)\int_E E_\mu(I(B)|\xi_{m+n}=\eta)V_n(\xi_m,d\eta)\Big\} \\
&\quad \le E_\mu I(A)V_n(\xi_m,E) \le P_\mu(A)C\delta^n, \qquad (14.3.12)
\end{aligned}$$

where $V_n(\xi_m, A) = |(T^K)^n(\xi_m, A) - \mu(A)|$. Taking $\varphi(n) = C\delta^n(0 < \delta < 1)$ we obtain $\{y_n\}$ is a φ-mixing sequence with $\sum_{n=1}^{\infty}\varphi^{1/2}(n) < \infty$.

Particularly, letting $\phi_t^s = t - s$, we have $y_n = z_n$, so that the result of $\{z_n\}$ holds true. Moreover, from (i) and (ii)

$$\sigma\{w_k, 1 \le k \le m-1\} \subset \sigma\{\xi_k, 1 \le k \le m\} \subset \mathcal{N}_{\tau_m},$$
$$\sigma\{w_k, k \ge m+n-1\} \subset \sigma\{\xi_k, k \ge m+n\}.$$

$\{w_n\}$ is also a strictly stationary φ-mixing sequence. Lemma 14.3.1 is proved.

Lemma 14.3.2. *Let ϕ_t^s and ψ_t^s be the non-negative, strongly measurable, homogeneous additive functionals of $\mathbf{X}$ with the finite $\alpha_\phi = E_\mu y_1$, $\alpha_\psi = E_\mu\psi_{\tau_2}^{\tau_1} \neq 0$. Then*

(a)

$$P_a\Big\{\lim_{t\to\infty}\phi_t^0/\psi_t^0 = \alpha_\phi/\alpha_\psi\Big\} = P\Big\{\lim_{t\to\infty}\phi_t^0/\psi_t^0 = \alpha_\phi/\alpha_\psi\Big\} = 1 \qquad (14.3.13)$$

where

$$P(A) = \int_E P_a(A)P_0(da), \qquad (14.3.14)$$

P_0 is an initial distribution.

(b) Particularly, letting $l(t)$ to be a positive integer-valued random variable such that

$$\tau_{l(t)} \le t < \tau_{l(t)+1} \qquad (14.3.15)$$

and $\alpha_z = E_\mu z_n$, we have

$$P_a\Big\{\lim_{t\to\infty}\frac{t}{l(t)} = \alpha_z\Big\} = P\Big\{\lim_{t\to\infty}\frac{t}{l(t)} = \alpha_z\Big\} = 1, \qquad (14.3.16)$$

$$P_a\Big\{\lim_{t\to\infty}\frac{\tau_{l(t)}}{t} = 1\Big\} = P\Big\{\lim_{t\to\infty}\frac{\tau_{l(t)+1}}{t} = 1\Big\} = 1. \qquad (14.3.17)$$

Proof. By the strict stationarity and φ-mixing property of $\{y_n\}$, $\{y_n\}$ is an ergodic sequence and its invariant σ-field $\mathcal{U}$ is trivial. By (14.3.8) for any $A \in \mathcal{U}$

$$P_a(A) = P_\mu(A) = 0 \text{ or } 1.$$

Hence we have

$$P_\mu\Big\{\lim_{n\to\infty} \frac{1}{n}\sum_{k=1}^{n} y_k = \alpha_\phi\Big\} = 1.$$

And since $\limsup \frac{1}{n}\sum_{k=1}^{n} y_k$ and $\liminf \frac{1}{n}\sum_{k=1}^{n} y_k$ are $\mathcal{U}$-measurable, we have

$$P_a\Big\{\lim_{n\to\infty} \frac{1}{n}\sum_{k=1}^{n} y_k = \alpha_\phi\Big\} = 1. \tag{14.3.18}$$

Obviously $l(t) \to \infty (t \to \infty)$. Write

$$\phi_t^0 = \phi_{\tau_1}^0 + \sum_{k=1}^{l(t)-1} y_k + \phi_t^{\tau_{l(t)}}. \tag{14.3.19}$$

Since $\tau_1 < \infty$ a.s., for any given $\varepsilon > 0$ and large t we have

$$P_a\{\phi_{\tau_1}^0 / l(t) \geq \varepsilon\} = 0. \tag{14.3.20}$$

By the non-negativity of ϕ_t^s, for large t we also have

$$P_a\{\phi_t^{\tau_{l(t)}} / l(t) \geq \varepsilon\} = P_a\{\phi_{\tau_{l(t)+1}}^{\tau_{l(t)}} / l(t) \geq \varepsilon\} = 0. \tag{14.3.21}$$

Combining (14.3.18)–(14.3.21), we get

$$P_a\Big\{\lim_{t\to\infty} \phi_t^0 / l(t) = \alpha_\phi\Big\} = 1. \tag{14.3.22}$$

Similarly we have

$$P_a\Big\{\lim_{t\to\infty} \psi_t^0 / l(t) = \alpha_\psi\Big\} = 1. \tag{14.3.23}$$

Putting (14.3.22), (14.3.23) and (14.3.14) together yields (14.3.13).

Note that (14.3.16) is a special case of (14.3.22) with $\phi_t^s = t - s$. Moreover

$$\frac{\tau_{l(t)}}{t} = \frac{\tau_1}{t} + \frac{l(t)-1}{t} \cdot \frac{1}{l(t)-1} \sum_{k=1}^{l(t)-1} z_k.$$

From $\tau_1 < \infty$ a.s., (14.3.18) and (14.3.16), (14.3.17) holds true.

Now let us consider the process $S(t)$ generated by the additive functional ϕ_t^0 as follows:

$$S(t) = \phi_t^0 - Mt.$$

Put

$$\psi_t^s = t - s, \quad M = \alpha_\phi / \alpha_\psi.$$

Lu (1986) proved the following theorem

Theorem 14.3.1. *Let ϕ_t^s be the non-negative, strongly measurable, homogeneous additive functional of $\mathbf{X}$, $w_n = y_n - M z_n$. Suppose that for some $0 < \delta \leq 2$,*

$$E_\mu y_n^{2+\delta} < \infty, \qquad E_\mu z_n^{2+\delta} < \infty.$$

Then $E_\mu w_n^{2+\delta} < \infty$ and

$$\sigma_w^2 = E_\mu w_1^2 + 2\sum_{k=2}^{\infty} E_\mu w_1 w_k$$

converges absolutely. Without loss of generality assume that $\sigma_w^2/\alpha_\psi = 1$, then without changing the distribution of $S(t)$, we can redefine the process $\{S(t), t \geq 0\}$ on a richer probability space together with a Wiener process $\{W(t), t \geq 0\}$ such that for any given $\varepsilon > 0$ a.s.

$$S(t) - W(t) = \begin{cases} O\left(t^{\frac{1}{2+\delta}+\varepsilon}\right) & \text{if } 0 < \delta < 2, \\ O\left(t^{\frac{1}{4}}(\log t)^{\frac{9}{4}+\varepsilon}\right) & \text{if } \delta = 2. \end{cases}$$

Proof. We can write

$$S(t) = \phi_t^0 - Mt = \phi_{\tau_1}^0 + \sum_{k=1}^{l(t)-1} w_k + \phi_t^{\tau_{l(t)}} - M(t - \tau_{l(t)} + \tau_1).$$

Denote $Z(t) = \sum_{k=1}^{l(t)-1} w_k$, we have

$$S(t) - Z(t) = \phi_{\tau_1}^0 + \phi_t^{\tau_{l(t)}} - M(t - \tau_{l(t)} + \tau_1). \tag{14.3.24}$$

We prove that the right hand side of (14.3.24) a.s. equals to $O((t \log\log t)^{1/(2+\delta)})$. By Lemma 14.3.1 $\{y_n\}$ is a strictly stationary φ-mixing sequence with $\varphi(n) = Ce^{-\theta n}$, $\theta > 0$. Moreover $\{y_n^{(2+\delta)/2}, n \geq 1\}$ is also a strictly stationary φ-mixing sequence with

$$E(y_n^{(2+\delta)/2})^2 = Ey_n^{2+\delta} < \infty.$$

Therefore by the law of iterated logarithm we have

$$\sum_{k=1}^{n}(y_k^{(2+\delta)/2} - Ey_k^{(2+\delta)/2}) = O((n \log\log n)^{1/2}) \quad \text{a.s.}$$

From the non-negativity of additive functional ϕ_t^s and Lemma 14.3.2 we obtain

$$\phi_t^{\tau_{l(t)}} \le y_{l(t)} = O((l(t)\log\log l(t))^{1/(2+\delta)}) = O((t\log\log t)^{1/(2+\delta)}) \quad \text{a.s.}$$

Similarly we also have

$$t - \tau_{l(t)} \le z_{l(t)} = O((t\log\log t)^{1/(2+\delta)}) \quad \text{a.s.} \tag{14.3.25}$$

Obviously $\phi_{\tau_0}^0 = O(1)$, $M\tau_1 = O(1)$. Thus we obtain

$$S(t) - Z(t) = O((t\log\log t)^{1/(2+\delta)}) \quad \text{a.s.}$$

For the strictly stationary φ-mixing sequence $\{w_n\}$, from Theorems 9.1.1 and 9.1.2 we have a.s.

$$\sum_{k=1}^{n} w_k - W(n\sigma_w^2) = \begin{cases} O\Big(t^{\frac{1}{2+\delta}+\varepsilon}\Big) & \text{if } 0<\delta<2 \\ O\Big(t^{\frac{1}{4}}(\log t)^{\frac{9}{4}+\varepsilon}\Big) & \text{if } \delta = 2. \end{cases} \tag{14.3.26}$$

Now we write

$$\begin{aligned} &Z(t) - W(t) \\ &= \Big(\sum_{k=1}^{l(t)-1} w_k - W((l(t)-1)\sigma_w^2)\Big) + \Big(W((l(t)-1)\sigma_w^2) - W(t)\Big) \\ &=: I_1 + I_2. \end{aligned} \tag{14.3.27}$$

From Lemma 14.3.2 and (14.3.26) we have a.s.

$$I_1 = \begin{cases} O\Big(l(t)^{\frac{1}{2+\delta}+\varepsilon}\Big) = O(t^{\frac{1}{2+\delta}+\varepsilon}) & \text{if } 0<\delta<2 \\ O\Big(l(t)^{\frac{1}{4}}(\log l(t))^{\frac{9}{4}+\varepsilon}\Big) = O(t^{\frac{1}{4}}(\log t)^{\frac{9}{4}+\varepsilon}) & \text{if } \delta = 2. \end{cases} \tag{14.3.28}$$

On the other hand, by the law of iterated logarithm for the strictly stationary φ-mixing sequence $\{z_n\}$ we have

$$\sum_{k=1}^{n} z_k - n\alpha_\psi = O((n\log\log n)^{1/2}) \quad \text{a.s.}$$

Hence

$$\tau_{l(t)} - l(t)\alpha_\psi = O((l(t)\log\log l(t))^{1/2}) = O((t\log\log t)^{1/2}) \quad \text{a.s.,}$$

so that

$$t - l(t)\alpha_\psi = t - l(t)\sigma_w^2 = O((t\log\log t)^{1/2}) \quad \text{a.s.}$$

By theorem 1.2.2 of Csörgő and Révész (1981) we have

$$I_2 = W((l(t)-1)\alpha_\psi) - W(t) = O((t \log\log t)^{1/4}) \quad \text{a.s.}$$

Combining it with (14.3.25)–(14.3.27) we have a.s.

$$S(t) - W(t) = \begin{cases} O\left(t^{\frac{1}{2+\delta}+\varepsilon}\right) & \text{if } 0 < \delta < 2 \\ O\left(t^{\frac{1}{4}}(\log t)^{\frac{9}{4}+\varepsilon}\right) & \text{if } \delta = 2. \end{cases}$$

Theorem 14.3.1 is proved.

Remark 14.3.1. From Theorem 14.3.1 we can give a weak invariance principle and a law of iterated logarithm for the additive functional of a Markov process.

Appendix Slowly Varying Function

Definition A1. A positive and measurable function $R(x)$ on $[A, \infty)$ for some $A \geq 0$ is called *regularly varying* at infinite point with an exponent α, if for any $a > 0$

$$\lim_{x \to \infty} R(ax)/R(x) = a^{\alpha}. \tag{A1}$$

Rewrite a regularly varying function $R(x)$ with an exponent α as

$$R(x) = x^{\alpha} L(x). \tag{A2}$$

Then, by (A1), we have

$$\lim_{x \to \infty} L(ax)/L(x) = 1.$$

Definition A2. A regularly varying function $L(x)$ with the exponent $\alpha = 0$ is called a *slowly varying function.*

We shall list some main propositions on a slowly varying function. For their proof, we refer to Seneta (1976) and Ibragimov and Linnik (1971).

There are Karamata's representation theorem for a slowly varying function.

Theorem A1. *Let $L(x)$ be a slowly varying function defined on $[A, \infty)$, $A \geq 0$. Then there exists a positive $B \geq A$ such that for any $x \geq B$*

$$L(x) = \exp\Big\{\eta(x) + \int_B^x \frac{\varepsilon(t)}{t}\, dt\Big\},$$

where $\eta(x)$ is a bounded measurable function on $[B, \infty)$ with $\eta(x) \to c(|c| < \infty)$ as $x \to \infty$ and $\varepsilon(x)$ is a continuous function on $[B, \infty)$ with $\varepsilon(x) \to 0$ as $x \to \infty$.

Using this representation theorem, we can derive many useful properties. In the sequel, we always assume that $L(x)$, $L_1(x)$, $L_2(x)$ are slowly varying functions.

Property A1. For any $a \geq 0$.

$$\lim_{x\to\infty} L(x+a)/L(x) = 1.$$

Property A2. For any $\varepsilon > 0$,

$$\lim_{x\to\infty} x^{\varepsilon} L(x) = \infty, \quad \lim_{x\to\infty} x^{-\varepsilon} L(x) = 0.$$

Property A3. $\sup_{2^k \leq t \leq 2^{k+1}} L(t)/L(2^k) \to 1$, as $k \to \infty$.

Property A4. Let $a = a(x) \to 0$ as $x \to \infty$. Then for any $\varepsilon > 0$,

$$\lim_{x\to\infty} a^{\varepsilon} \frac{L(ax)}{L(x)} = \lim_{x\to\infty} a^{\varepsilon} \frac{L(x)}{L(ax)} = 0.$$

Property A5. $(\log L(x))/\log x \to 0$ as $x \to \infty$.

Property A6. For any real number α, $L^{\alpha}(x)$, $L_1(x)L_2(x)$ and $L_1(x)+L_2(x)$ all are slowly varying. Moreover, if $L_2(x) \to \infty$ as $x \to \infty$, then $L_1(L_2(x))$ also is slowly varying.

Property A7. Define $\overline{L}(x)$ and $\underline{L}(x)$ by

$$x^{\gamma}\overline{L}(x) = \sup_{B \leq t \leq x} t^{\gamma} L(t),$$
$$x^{\gamma}\underline{L}(x) = \inf_{x \leq t} t^{\gamma} L(t),$$

where $\gamma > 0$ is an arbitrary constant. Then $\overline{L} \sim L$ and $\underline{L} \sim L$.

Remark A1. As a consequence of Property A7, $x^{\gamma}L(x)$ is equal asymptotically to a non-decreasing regularly varying function with the same exponent γ.

Property A8. For $R_1(x) = x^{\gamma}L_1(x)$, $\gamma > 0$, there exists a regularly varying function $R_2(x) = x^{1/\gamma}L_2(x)$ such that

$$R_1(R_2(x)) \sim x, \; R_2(R_1(x)) \sim x \quad \text{as } x \to \infty.$$

$R_2(x)$ here is defined asymptotically uniquely, i.e. if the above relations hold true with R_3 instead of R_2 and $R_3(x) \to \infty$ as $x \to \infty$, then $R_3(x) \sim x^{1/\gamma}L_2(x)$.

Property A9. Let $L(x)$ is a positive slowly varying function on $[A, \infty)$. Assume that $R(x) = x^\gamma L(x)$ is non-decreasing on $[A, \infty]$ for some $\gamma > 0$. For $x \geq R(A)$ let

$$R^*(x) = \inf\{y : y \in [A, \infty),\ R(y) \geq x\}.$$

Then $R^*(x) = x^{1/\gamma} L^*(x)$, where $L^*(x)$ is a slowly varying function and $R^*(x)$ is an inverse function of $R(x)$ with the meaning in Property A8.

References

ALDOUS, D. J.
(1978) Weak convergence of randomly indexed sequences of random variables. Math. Proc. Camb. Phil. Soc. **83**, 117-126.

ALEXANDER, K. S., PYKE, R.
(1986) A uniform central limit theorem for set-indexed partial-sum processes with finite variance. Ann. Probab. **14**, 582-597.

ANDREWS, D.
(1984) Nonstrong mixing autoregressive processes. J. Appl. Probab. **21**, 930- 934.

BABBEL, B.
(1989) Invariance principle for U-statistics and von-Mises functionals. J. Sta- tist. Plan. Infer. **22**, 337-354.

BAI, Z. D., SU, C.
(1985) The complete convergence for partial sums of iid random variables. Scientia Sinica, **28**, 1261-1277

BASS, R. F.
(1985) Law of the iterated logarithm for set-indexed partial-sum processes with finite variance. Z. Wahrsch. verw Gebiete **70**, 591-608.

BASS, R. F., PYKE, R.
(1984) Functional law of the iterated logarithm and uniform central limit theorem for partial-sum processes indexed by sets. Ann. Probab. **12**, 13-34.
(1985) The space $D(A)$ and weak convergence for set-indexed processes. Ann. Probab. **14**, 582-597.

BAUM, L. E., KATZ, M.
(1965) Convergence rate in the law of large numbers. Trans. Amer. Math. Soc. **120**, 108-123.

BENNETT, G.

(1962) Probability inequalities for the sum of independent random variables. J. Amer. Statist. Assoc. **59**, 33-45.

BERBEE, H. C. P.
(1987) Convergence rates in the strong law for bounded mixing sequences. Probab. Theory Rel. Fields, **74**, 255-270.

BERKES, I., MORROW, G. J.
(1981) Strong invariance principle for α-mixing random fields. Z. Wahrsch. Verw. Gebiete, **57**, 15-37.

BERKES, I., PHILIPP, W.
(1979) Approximation theorems for independent and weakly dependent random variables. Ann. Probab. **7**, 29-54.

BERNSTEIN, S. N.
(1926) Sur l'extension du théorème limite du calcul des probabilités aux sommes de quantités dépendantes. Math. Ann. **97**, 1-59.

BILLINGSLEY, P.
(1968) Convergence of Probability Measures. J. Wiley, New York.

BLUM, J. R., HANSON, D. L., KOOPMANS, L.
(1963) On the strong law of large numbers for a class of stochastic processes. Z. Wahrsch. Verw Gebiete, **2**, 1-11.

BOLTHAUSEN, E.
(1982) On the central limit theorem for stationary mixing random fields. Ann. Probab. **10**, 1047-1050.

BOROVKOV, A. A.
(1973) On the rate of convergence for the invariance principle. Theory Probab. Appl. **18**, 207-225. (By Russian).

BOSQ, D.
(1975) Inegalite de Bernstein pour Les Processus Stationnaires et melangeants. C. R. Acad. Sci. Paris Ser. A **281**, 1095-1098.

BRADLEY, R. C.
(1983) Approximation theorem for strong mixing random variables. Mich. Math. J. **30**, 69-81.
(1984) Counterexamples to the central limit theorem under strong mixing conditions. Colloq. Math. Soc. János Bolyai 36. Limit Theorems in Probab. and Statist. Veszprém (Hungary) 1982. 153-172.

(1984) The central limit question under ρ-mixing. Rocky Mountain. J. Math. **17**, 95-114.

(1988) A central limit theorem for stationary ρ-mixing sequences with infinite variance. Ann. Probab. **16**, 313-332.

(1992) On the spectral density and asymptotic normality of weakly dependent random fields. J. Theoretical Probab. **5**, 355-373.

BRADLEY, R. C., BRYC, W.

(1985) Multilinear forms and measures of dependence between random variables, J. Multivariate Anal. **16**, 335-367.

BROWN, B. M.

(1971) Martingale central limit theorems. Ann. Math. Statist. **42**, 59-66.

BULINSKII, A. V., ZHURBENKO, I. G.

(1981) Limit Theorem for Stochastic Processes. M. Moskov, Gos. Uh-T. (by Russian).

BURKHOLDER, D. L.

(1973) Distribution function inequalities for martingales. Ann. Probab. **1**, 19-42.

CAI, Z. W.

(1991) Strong consistency and rates for recursive non-parameter conditional probability density estimates under (α, β)-mixing conditions. Stochastic Processes Their Appl. **38**, 323-333.

CHAI, G. X.

(1989) Consistency for nearest neighbor density estimators of a stationary sence. Acta Math. Sinica, **32**, 423-432. (By Chinese).

CHEN, D. Q.

(1991) A uniform central limit theorem for non-uniform φ-mixing random fields. Ann. Probab. **19**, 636-649.

CHEN, X. R., WU, Y. H.

(1989) Strong law for a mixing sequence. Acta Math. Appl. Sinica, **5**, 367-371.

CHOW, Y. S.,

(1965) Local convergence of martingales and the law of large numbers. Ann. Math. Statist. **36**, 552-558.

CHOW, Y. S., TEICHER, H.
(1978) Probability Theory, Springer-Verlag. New York.

COLLOMB, G.
(1984) Propriétés de convergence presque complète du prédicteur à noyau. Z. Wahrsch. verw. Gebiete, **66**, 441-460.

CRAMER, H.
(1946) Mathemetical Mathods of Statistics. Princeton Univ. Press. Princeton, N. J.

CSÖRGŐ, M., CSÖRGŐ, S., HORVÁTH, L.,
(1986) An Asymptotic Theory for Empirical Reliability and Concentration Processes. Lecture Notes in Statistics Vol.33, Springer-Verlag, New York.

CSÖRGŐ, M., CSÖRGŐ, S., HORVÁTH, L., MASON, D. M.
(1986a) Weighted empirical and quantile processes. Ann. Probab. **14**, 31-85.
(1986b) Normal and stable convergence of integral functions of the empirical distribution function. Ann. Probab. **14**, 86-118.

CSÖRGŐ, M., HORVÁTH, L.
(1993) Weighted Approximations in Probability and Statistics. Wiley, Chichester.

CSÖRGŐ, M., RÉVÉSZ, P.
(1981) Strong Approximations in Probability and Statistics. Academic Press, New York.

DABROWSKI, K. I.
(1982) A note on a theorem of Berkes and Philipp for dependent sequences. Statist. Probab. Lett. **1**, 53-55.

DAVYDOV, YU. A.
(1968) On convergence of distributions generated by stationary stochastic processes. Probab. Theory Appl. **13**, 730-737. (by Russian).
(1973) Mixing condition for Markov chains. Probab. Theory Appl. **18**, 312- 328. (by Russian).

DEHLING, H.
(1983) A note on a theorem of Berkes and Philipp. Z. Wahrsch. verw Gebiete, **62**, 39-42.

DEHLING, H., DENKER, M. M., PHILIPP, W.
(1986) Central limit theorems for mixing sequences of random variables under minimal conditions. Ann. Probab. **14**, 1359-1370.

DENKER, M.
(1985) Uniform integrability and the central limit theorem for strongly mixing processes. Progress in Probab. and Statist. Dependence in Probab. and Statist. 269-274.

DENKER, M., KELLER, G.
(1983) On U-statistics and Von-Mises statistics for weakly dependent processes. Z. Wahrsch. verw Gebiete **64**, 505-522.

DOBRUSHIN, R. L.
(1956) The central limit theorem for non-stationary Markov chain. Probab. Theory Appl. **1**, 72-88, 365-425. (by Russian).
(1968) The description of the random fields by its conditional distributions. Probab. Theory Appl. **13**, 201-229. (by Russian).

DOBRUSHIN, R. L., NAHAPETIAN, B. S.
(1974) Strong convexity of pressure for a lattice system of classical statistical physics. Teoret. Mat. Fiz. **20**, 223-234. (By Russian).

DOOB, J. L.
(1953) Stochastic Processes. Wiley, New York.

DOUKHAN, P.
(1994) Mixing: Properties and Examples. Lecture Notes in Statist. No. 85, Springer-Verlag.

DOUKHAN, P., GUYON, X.
(1991) Mélange pour des processes linéaires spatiaur. C. R. Acad. Sci. Paris Sér. 1, **313**, 465-470.

DOUKHAN, P., LEON, J., PORTAL, F.
(1984) Vitesse de convergence dans le théoréme central limite pour des variables aléatoires mélangeantes à valeurs dans un espace de Hilbert. C. R. Acad. Sci. Paris Sér. 1, **289**, 305-308.

DOUKHAN, P., MASSART, P., RIO, E.
(1994) The functional central limit theorem for strong mixing processes. Ann. Inst. H. Poincaré Probab. Statist. **30**, 63-82.

DU, W. C.

(1993) The law of large numbers for a φ-mixing sequence. J. of Suzhou Univ. **9**, 1-6. (By Chinese).

DUDLEY, R. M.
(1973) Sample functions of the Gaussian process. Ann. Probab. **1**, 66-103.
(1974) Metric entropy of some classes of sets with differentiable boundaries. J. Appr. Theory **10**, 207-336.
(1979) Central limit theorems for empirical measures. Ann. Probab. **6**, 899- -929. Correction, ibid. **7**, 909-911.

DYNKIN, E. B.
(1963) Markov Processes. M. Fismatgiz.

ERDÖS, P., RÉNYI, A.
(1959) On Cantor's series with convergent $\sum 1/q_n$. Ann. Univ. Sci. Budapest, Sect. Math. **2**, 93-109.

ERICKSON, R. V.
(1981) Lipschitz smoothness and convergence with applications to the central limit theorem for summation processes. Ann. Probab. **9**, 831-851.

FAN, J. K., XUE, L. G.
(1993) Strong uniform consistency for kernel estimates of derivatives of a density function under φ-mixing samples. Appl. Math. J. Chin. Univ. **8**, 273-278. (By Chinese).

FRANKLIN, J. N.
(1968) Matrix Theory. Prentice-Hall, Inc. Englewood Cliffs, New Jersey.

GAEUSSLER, P.
(1983) Limit theorems for empirical processes indexed by classes of sets allowing a finite-dimensional parametrization. Probab. Math. Statist. **4**, 1-12.

GAPOSHKIN, V. F.
(1966) Lacunary series and independent functions. Russian Math. Survey **21**, 3-82.

GEBELEIN, H.
(1941) Das Statistische Problem das Karelationals Variation und eigenwert-Problem und sein Zusammenhang mit Ausgleischungrechnung. Z. Angew. Math. Mech. **21**, 364-379.

GNEGENKO, B. V., KOLMOGOROV, A. M.
(1949) Limit Distributions for Sums of Independent Random Variables. Nauka, Moscow (Russian), 1949. English Edition: Reading, Mass.: Addison-wesley 1954.

GOLDIE, C. M., GREENWOOD, P. E.
(1986a) Characterisations of set-indexed Brownian motion and associated conditions for finite-dimensional convergence. Ann. Probab. **14**, 802-816.
(l986b) Variance of set-indexed sums of mixing random variables and weak convergence of set-indexed processes. Ann. Probab. **14**, 817-839.

GORDIN, M. I.
(1969) On the central limit theorem for stationary process, Soviet Math. Dokl. **10**, 1174-1176.

GORODEZKII, V. V.
(1982) The invariance principle for stationary random fields satisfying the strong mixing condition. Probab. Theory Appl. **27**, 358-363. (by Russian).
(1983) On the rate of convergence in the invariance principle for strong mixing sequences. Probab. Theory Appl. **28**, 780-785. (by Russian).
(1984) The central limit theorem and an invariance principle for weakly dependent random fields. Soviet Math. Dokl. **29**, 529-532. (by Russian).

GRIN, A. G.
(1991) Norming sequences to limit theorems for weakly dependent variables. Probab. Theory Appl. **36**, 285-300. (by Russian).

GUYON, X.
(1992) Modèles stationnaires sur $\mathbb{Z}^d$ et propriétés du second order. (Preprint.)

HALL, P., HEYDE, C. C.
(1980) Martingale Limit Theory and Its Application. Acad. Press, New York.

HANSON, D. L., RUSSO, R. P.
(1983) Some results on increments of the Wiener process with applications to lag sums of i.i.d. random variables. Ann. Probab. **11**, 609-623.

HEINRICH, L.
(1985) Non-uniform estimates, moderate and large deviations in the central limit theorem for m-dependent random variables. Nath. Nachr.

121, 107-121.

HERRNDORF, N.
(1983a) The invariance principle for φ-mixing sequences. Z. Wahrsch. verw Gebiete, **63**, 97-108.
(1983b) Stationary strongly mixing sequence not satisfying the central limit theorem. Ann. Probab. **11**, 809-813.
(1984) A functional central limit theorem for weakly dependent sequence of random variables. Ann. Probab. **12**, 141-153.
(1985) A functional central limit theorem for strongly mixing sequence of random variables. Z. Wahrsch. verw Gebiete, **69**, 541-550.

HEYDE, C. C., SCOTT, D. J.
(1973) Invariance principles for the law of the iterated logarthm for martingale and processes with stationary increments. Ann. Probab. **1**, 428-437.

HIPP, C.
(1979) Convergence rates of the strong law for stationary mixing sequences. Z. Wahrsch. verw. Gegiete, **49**, 49-62.

HIRSCHFELD, O.
(1935) A connect between correlation and contengency. Math. Proc. Camb. Phil. Soc. **31**, 520-524.

HOEFFDING, W.
(1948) A class of statistics with asymptotically normal distribution. Ann. Math. Statist. **19**, 293-325.

HOEFFDING, W., ROBBINS, H.
(1948) The central limit theorem for dependent random variables. Duke Math. J., **15**, 773-780.

HOPF, E.
(1937) Ergodentheorie. Ergebnisse Math., 5, No.2. Springer, Berlin (Reprinted by Chelsea, New York, 1948).

HSU, P. L., ROBBINS, H.
(1947) Complete convergence and the law of large numbers. Proc. Nat. Acad. Sci. U.S.A. **33**, 25-31.

IBRAGIMOV, I. A.
(1962) Some limit theorems for stationary processes. Probab. Theory Appl. **7**, 349-382. (1959) Dakl. AN USSR, **125**, 711-714. (By Russian).

(1975) A note on the central limit theorem for dependent random variables. Probab. Theory Appl. **20**, 134-139. (By Russian).

IBRAGIMOV, I. A., LINNIK, YU. V.,
(1971) Independent and Stationary Dependent Sequences of Random Variables. Nauka, M. (Russian), 1965. English Edition: Woltors-Noordhoff, Groningen 1971.

IBRAGIMOV, I. A., SOLEV, V. N.
(1969) A condition for regularity of a Gaussian stationary process. Soviet Math. Dokl. **10**, 371-375.

IBRAGIMOV, I. A., ROZANOV, U. A.
(1978) Gaussian Random Processes. Springer-Verlag New York.

IOSIFESCU, J.
(1977) Limit theorem for φ-mixing sequences. A survey. Proc. 5th Conf. on Probab. Theory, Sept. 1-6, 1974, Brasov, Romania (Editura Acad. R. S. R., Bucuresti) 51-57.

IOSIFESCU, M., THEODORESCU, R.
(1969) Random Processes and Learning. Springer-Verlag, New York.

JAIN, N. C., PRUITT, W. E.
(1975) The other law of the iterated logarithm. Ann. Probab. **3**, 1046-1049.

KOLMOGOROV, A. N., ROZANOV, U. A.
(1960) On the strong mixing conditions of a stationary Gaussian process. Probab. Theory Appl. **2**, 222-227. (By Russian).

KOMLÓS, J., MÁJOR, P., TUSNÁDY, G.
(1975) An approximation of partial sums of independent R.V.'s and the sample DF. I. Z. Wahrsch. verw Gebiete, **32**, 111-131.

KONG, F. C., ZHANG, M. J.
(1994) A note on the complete convergence for ρ-mixing sequence. Chin. Sci. Bull. **39**, 778-781.

KUELBS, J.
(1976) Estimation of the multidimensional probability density function. MRC Technical Report No. 1646, Univ. of Wisconsin, Madison, Wisconsin.

LAI, T. L.

(1977) Convergence rates and r-quick versions of the strong law for stationary mixing sequences. Ann. Probab. **5**, 693-706.

LIN, Z. Y.

(1981) Limit theorems for a class of weakly dependent random variables. Chin. Ann. Math. **2**, 181-185. (By Chinese).

(1982) The weak convergence for dependent random variables with infinite variance. Chin. Ann. Math. **3**, 35-44. (By Chinese).

(1983) The kernel estimators of density function for a φ-mixing sequence. Science Bull. **28**, 709-713.

(1984) Invariane principle for the error variance estimators when the sample is weakly dependent. Science Bull. **29**, 584-587.

(1987) On Csörgő-Révész 's increments of sums of non-i.i.d. random variables, Scientia Sinica, **30**, 921-931.

(1988) A note on C-R increments of partial sums, Appl. Math. J. Chinese Univ. **3**, 449-453. (By Chinese).

(1989) The increments of partial sums of a dependent sequence when the moment generating functions do not exist. Acta Math. Sinica, New Series **5**, 289- 296.

(1991) On the increments of partial sums of a φ-mixing sequence. Probab. Theory Appl. **36**, 344-354.

LIN, Z. Y., LU, C. R.

(1992) Strong Limit Theorems. Kluwer Academic Publishers and Science Press, Dordrecht-Beijing.

LU, C. R.

(1984) Weak convergence of empirical processes of sequence of stationary random variables. Chin. Ann. Math. **5B**, 731-735.

(1985) Weak convergence for a ψ-mixing sequence. J. Hangzhou Univ. **12**, 7- 11. (By Chinese).

(1986 a) Strong approximations for error variance estimators when the sample is weakly dependent. Science Bull. **31**, 505-509.

(1986 b) Strong approximations for additive functional of Markov processes. J. Engineering Math. **3**, 113-116. (By Chinese).

(1992) A uniform central limit theorem for set-indexed partial sum processes when the variance is infinite. Appl. Math. J. Chinese Univ. **7B**, 157-162.

(1993) A note of the convergence rate of the invariance principle for dependent random variables. J. Hangzhou Univ. **20**, 397-401. (By Chinese).

(1995) Weak convergence for nonuniform φ-mixing random fields. Chin. Ann. Math. (to appear).

(1995) Weak convergence for α-mixing random fields. (to appear).

MASRY, E., GYÖRFI, L.
(1987) Strong consistency and rates for recursive probability density estimators of stationary processes. J, Multivariate Anal. **22**, 79-83.

MCLEISH, D. L.
(1975) Invariance principle for dependent variables. Z. Wahrsch. verw Gebiete, **57**, 495-507.

MÓRICZ, F.
(1982) A general moment inequality for the maximum of partial sums of singale series. Acta Sci. Math. Hung. **44**, 67-75.

NAHAPETIAN, B. S.
(1980) The central limit theorems for random fields with mixing conditions. In Multicomponent System (R.L. Dobrushin and Ya.G. Sinai. eds) Advances in Probab. **6**, 531-548. Dekker, New York.
(1987) An approach to limit theorems for dependent random variables. Probab. Theory Appl. **32**, 589-594. (By Russian).

OODAIRA, H., YOSHIHARA, K. I.
(1972) Functional central limit theorem for strictly stationary processes satis- fying the strong mixing condition. Kodai Math. Sem. Rep. **24**, 259-269.

PARZEN, E.
(1962) On estimation of a probability density function and mode. Ann. Math. Statist. **33**, 1065-1076.

PELIGRAD, M.
(1982) Invariance principles for mixing sequences of random variables. Ann. Probab. **10**, 968-981.
(1983) A note on two measures of dependent and mixing sequence. Adv. in Appl. Probab. **15**, 461-464.
(1985) An invariance principle for φ-mixing sequences. Ann. Probab. **13**, 1304-1313.
(1986) Recent advances in the central limit theorem and its weak invariance principle for mixing sequences of random variables. Dependence in Probab. and Statist. Eberlin, E and Taqqu, M. S. (eds) Pregress in Prob. and Stat. Birckhauser. **11**, 193-223.
(1987) On the central limit theorem for ρ-mixing sequences of random variables. Ann. Probab. **15**, 1387-1394.

(1990) On Ibragimov-Iosifescu conjecture for φ-mixing sequences. Stochastic Processes Their Appl. **35**, 293-308.

(1992) Properties of uniform consistency of the kernel estimatiors of density and of regression functions under dependence assumptions. Stoch. Stoch. Reports, **40**, 147-168.

(1994) Convergence of stopped sums of weakly dependent random variables. (Preprint).

PELIGRAD, M., SHAO, Q. M.

(1995) Estimation of the variance of partial sums for ρ-mixing random variables. J. Multivariate Anal. **52**, 140-157.

PETROV, V. V.

(1975) Sums of Independent Random Variables. Springer, Berlin.

PHILIPP, W.

(1984) Invariance principle for sums of mixing random elements and the multivariate empirical process. In: Révész, P.(ed.) Colloq. Math. Soc. János Bolyai **36**, Limit theorems in probab. and statist. Veszprém (Hungary) 1982. 843-873.

PHILIPP, W., PINZUR, L.

(1980) Almost sure approximation theorems for the multivariate empirical process. Z. Wahrsch. verw Gebiete, **54**, 1-13.

PHILIPP, W., STOUT, W.

(1975) Almost Sure Invariance Principle for Partial Sums of Weakly Dependent Random Variables. Mem. Amer. Math. Soc., No.161.

PROHOROV, YU. V.

(1956) The convergence of random processes and limit theorems in probability theory. Probab. Theory Appl. **1**, 177-238. (By Russian).

RÉNYI, A.

(1958) On mixing sequences of sets. Acta Math. Acad. Sci. Hung. **9**, 215-228.

(1960) On the central limit theorem for the sum of a random number of independent random variables. Acta Math. Acad. Sci. Hung. **11**, 97-102.

RIO, E.

(1993) Covariance inequalities for strongly mixing processes. Ann. Inst. Henri. Poincaré, **29**, 587-597.

ROSENBLATT, M.

(1956) A central limit theorem and a strong mixing condition. Proc. Nat. Acad. Sci. USA. **42**, 43-47.
(1971) Markov Processes. Structure and Asymptotic Behaviour. Springer Verlag.

ROZANOV, Y. A., VOLCONSKI, V. A.
(1959, 1961) Some limit theorems for random function I, II. Probab. Theory. Appl. **4**, 186-207; **6**, 202-215. (By Russian).

SAKHANENKO, A. I.
(1981) On the rate of convergence of the invariance principle. The Third Vilnius Conference on Probab. Theory and Math. Statist. (Abstract), 135- 136. V.2 Vilnius.
(1984) Rate of convergence in the invariance principle for variables with exponential moments that are not identically distributed, Limit Theorems for Sums of Random Variables, Trudy Inst. Math., **3**, "Nauka"" Sibirsk. Otdel., Novosibirsk, 4-49.

SEN, P. K.
(1971) A note on weak convergence of empirical processes for sequences of φ- mixing random variables. Ann. Math. Statist. **42**, 2131-2133.
(1974) Weak convergence of multidimensional empirical processes for stationary φ-mixing processes. Ann. Probab. **2**, 147-154.

SENETA, E.
(1976) Functions of Regular Variation. Lect. Notes in Math. No.508, Springer-Verlag, New York.

SHAO, Q. M.
(1985) Almost sure invariance principle for Gaussian processes. Chinese J. Appl. Probab. Statist. **1**, 43-46. (By Chinese).
(1986) Weak convergence of multidimensional empirical processes in the strong mixing case. Chin. Ann. Math. **7A**, 547-552. (By Chinese).
(1987) Strong approximations on lacunary trigonometric series with weights. Scientia Sinica, **30**, 796-806.
(1988a) An inequality of moments and its application. Acta Math. Sinica. **31**, 736-747. (By Chinese).
(1988b) A remark on the invariance principle for a ρ-mixing sequence. Chin. Ann. Math. **9A**, 409-412. (By Chinese).
(1989a) Limit theorems for independent and dependent random variables. Univ. of Chinese Science and Techneque Ph.D.Thesis.
(1989b) On the invariance principle for ρ-mixing sequences. Chin. Ann. Math. **10B**, 427-433.

(1989c) On complete convergence for a ρ-mixing sequence. Acta Math. Sinica **32**, 377-393. (By Chinese).

(1990) Exponential inequalities and density estimation under dependent assumptions. manuscript.

(1991) A further discussion on the complete convergence for independent random variables. Chinese J. Appl. Probab. Statist. **7**, 174-188. (By Chinese).

(1993a) On the invariance principle for stationary ρ-mixing sequence with in- finite variance. Chin. Ann. Math. **14B**, 27-42.

(1993b) Almost sure invariance principles for mixing sequences of random variables. Stochastic Processes Appl. **48**, 319-334.

(1993c) On complete convergence for α-mixing sequence. Statist. Probab. Lett. **16**, 279-287.

(1995) Maximal inequality for partial sums of ρ-mixing sequences. Ann. Prob. **23**, 948-965.

SHAO, Q. M., LU, C. R.

(1987) Strong approximations for partial sums of weakly dependent random variables. Scientia Sinica. **30**, 575-587.

SHAO, Q. M., YU, H.

(1995) Weighted weak convergence for empirical processes of dependent sequence. (to appear).

SHORACK, G. R., WELLNER, J. A.

(1986) Empirical Processes with Applications to Statistics. Wiley, New York.

SIRAZGIMOV, S. H., GAFROV, M. U.

(1987) Method of Series in the Boundary Problem of Random Walk. Nauka, Taschkent. (By Russian).

STOUT, W.

(1974) Almost Sure Convergence. New York, Academic Press.

STRITTMATTER, W.

(1990) Strong approximations for set indexed partial sum processes and empirical processes of mixing random fields. Probab. in Banach Space. Vol.**7** 201-268.

STUTE, W.

(1982) The oscillation behavior of empirical process. Ann. Probab. **10**, 86-107.

SU, C.
(1989) On convergence of tail probability series in law of large numbers. Science in China, **32**, 1423-1436.

SU, Z. G.
(1992) Strong approximations for set-indexed partial-sum processes of φ-mixing random fields. Acta Math. Sinica. Ser.A **35**, 101-111. (By Chinese).

SUN, Z. G.
(1984) An approximation of lacunary trigonometric series by Brownian motion. Scientia Sinica. **27**, 905-914.

SUNKLODAS, J.
(1984) Rate of convergence in the central limit theorem for random variables with strong mixing. Lithuanian Math. J. **24**, 182-190.

TIKHOMIROV, A. N.
(1980) On the rate of convergence in the central limit theorem for weakly dependent random variables. Probab. Theory Appl. **25**, 800-818. (By Russian).

UTEV, S. S.
(1981) Remark on the rate of convergence in the invariance principle. Sibirsk. Math. J. **22**, 206-209. (By Russian).
(1984) Inequalities for sums of weakly dependent random variables and estimates of the comvergence rate in the invariance principle. Advance in Probab. Theory. 73-114.
(1990) On the CLT for the series scheme of random variables with the φ-mixing. Probab. Theory Appl., **35**, 110-117. (By Russian).

VON-MISES, R. V.
(1947) On the asymptotic distribution of differentiable statistical functions. Ann. Math. Statist. **18**, 309-348.

WANG, Q. Y.
(1994) The strong law of large numbers for U-statistics with a φ-mixing sample. (To appear, by Chinese).

WANG, Y. B.
(1993) The 0-1 law for the mixing sequence and its application in the complete convergence. J. Sys. Sci. and Math. Sci. **13**, 42-52. (By Chinese).

XUE, L. G.
(1994) Convergence rates of the strong law of large numbers for a mixing sequence. J. Sys. Sci. and Math. Sci. **14**, 213-221. (By Chinese).

YOKOYAMA, R.
(1980) Moment bounds for stationary mixing sequences. Z. Wahrsch. verw Gebiete, **52**, 45-57.

YOSHIHARA, K. I.
(1975) Weak convergence of multidimensional empirical processes for strong mixing sequences of stochastic vectors. Z.Wahrsch. verw Gebiete, **33**, 133-137.
(1978) Note on multidimensional empirical processes for stationary φ-mixing random vectors. J. Multivariate Anal. **8**, 584-588.
(1979) Convergence rates of the invariance principle for absolutely regular sequences. Yokohama Math. J. **27**, 49-55.

YU, J.
(1993) Consistency of an alternative nearest neighbor probability density function estimator. Statistical Research Report 1993-6, Umeå University. Revised version, originally published in Acta Math. Scie. 1986.
(1995) Uniform convergence rates for a nearest neighbor density estimator under dependence assumptions. Statistical Research Report 1995-3, Umeå University. (To appear).

ZHANG, Y.
(1989) On Berry-Esseen bounds of U-statistics for dependent sample. J. of Hangzhou Univ. **16**, 392-398. (By Chinese).

ZHOU, Y.
(1995) Convergence of oscillation moduli of empirical process with dependent sample and its application. Chin. Ann. Math. (To appear, by Chinese).

ZHURBENKO, I. G.
(1984) On mixing conditions for random process with values in Hilbert space. Soviet Math. Dokl. **30**, 465-467. (By Russian).

ZUPAROV, T. M.
(1991) On the rate of convergence in the central limit theorem for weakly dependent variables. Probab. Theory Appl. **36**, 635-643. (By Russian).

Index

Other Mathematics and Its Applications titles of interest:

P.M. Alberti and A. Uhlmann: *Stochasticity and Partial Order. Doubly Stochastic Maps and Unitary Mixing.* 1982, 128 pp. ISBN 90-277-1350-2

A.V. Skorohod: *Random Linear Operators.* 1983, 216 pp. ISBN 90-277-1669-2

L.M. Stancu-Minasian: *Stochastic Programming with Multiple Objective Functions.* 1985, 352 pp. ISBN 90-277-1714-1

L. Arnold and P. Kotelenez (eds.): *Stochastic Space-Time Models and Limit Theorems.* 1985, 280 pp. ISBN 90-277-2038-X

Y. Ben-Haim: *The Assay of Spatially Random Material.* 1985, 336 pp. ISBN 90-277-2066-5

A. Pazman: *Foundations of Optimum Experimental Design.* 1986, 248 pp. ISBN 90-277-1865-2

P. Kree and C. Soize: *Mathematics of Random Phenomena. Random Vibrations of Mechanical Structures.* 1986, 456 pp. ISBN 90-277-2355-9

Y. Sakamoto, M. Ishiguro and G. Kitagawa: *Akaike Information Criterion Statistics.* 1986, 312 pp. ISBN 90-277-2253-6

G.J. Szekely: *Paradoxes in Probability Theory and Mathematical Statistics.* 1987, 264 pp. ISBN 90-277-1899-7

O.I. Aven, E.G. Coffman (Jr.) and Y.A. Kogan: *Stochastic Analysis of Computer Storage.* 1987, 264 pp. ISBN 90-277-2515-2

N.N. Vakhania, V.I. Tarieladze and S.A. Chobanyan: *Probability Distributions on Banach Spaces.* 1987, 512 pp. ISBN 90-277-2496-2

A.V. Skorohod: *Stochastic Equations for Complex Systems.* 1987, 196 pp. ISBN 90-277-2408-3

S. Albeverio, Ph. Blanchard, M. Hazewinkel and L. Streit (eds.): *Stochastic Processes in Physics and Engineering.* 1988, 430 pp. ISBN 90-277-2659-0

A. Liemant, K. Matthes and A. Wakolbinger: *Equilibrium Distributions of Branching Processes.* 1988, 240 pp. ISBN 90-277-2774-0

G. Adomian: *Nonlinear Stochastic Systems Theory and Applications to Physics.* 1988, 244 pp. ISBN 90-277-2525-X

J. Stoyanov, O. Mirazchiiski, Z. Ignatov and M. Tanushev: *Exercise Manual in Probability Theory.* 1988, 368 pp. ISBN 90-277-2687-6

Other Mathematics and Its Applications titles of interest:

E.A. Nadaraya: *Nonparametric Estimation of Probability Densities and Regression Curves.* 1988, 224 pp. ISBN 90-277-2757-0

H. Akaike and T.Nakagawa: *Statistical Analysis and Control of Dynamic Systems.* 1998, 224 pp. ISBN 90-277-2786-4

A.V. Ivanov and N.N. Leonenko: *Statistical Analysis of Random Fields.* 1989, 256 pp. ISBN 90-277-2800-3

V.Paulauskas and A. Rackauskas: *Approximation Theory in the Central Limit Theorem. Exact Results in Banach Spaces.* 1989, 176 pp. ISBN 90-277-2825-9

R.Sh. Liptser and A.N. Shiryayev: *Theory of Martingales.* 1989, 808 pp. ISBN 0-7923-0395-4

S.M. Ermakov, V.V. Nekrutkin and A.S. Sipin: *Random Processes for Classical Equations of Mathematical Physics.* 1989, 304 pp. ISBN 0-7923-0036-X

G. Constantin and I. Istratescu: *Elements of Probabilistic Analysis and Applications.* 1989, 488 pp. ISBN 90-277-2838-0

S.Albeverio, Ph. Blanchard and D. Testard (eds.): *Stochastics, Algebra and Analysis in Classical and Quantum Dynamics.* 1990, 264 pp. ISBN 0-7923-0637-6

Ya.I. Belopolskaya and Yu.L. Dalecky: *Stochastic Equations and Differential Geometry.* 1990, 288 pp. ISBN 90-277-2807-0

A.V. Gheorghe: *Decision Processes in Dynamic Probabilistic Systems.* 1990, 372 pp. ISBN 0-7923-0544-2

V.L. Girko: *Theory of Random Determinants.* 1990, 702 pp. ISBN 0-7923-0233-8

S.Albeverio, PH. Blanchard and L. Streit: *Stochastic Processes and their Applications in Mathematics and Physics.* 1990, 416 pp. ISBN 0-9023-0894-8

B.L. Rozovskii: *Stochastic Evolution Systems. Linear Theory and Applications to Non-linear Filtering.* 1990, 330 pp. ISBN 0-7923-0037-8

A.D. Wentzell: *Limit Theorems on Large Deviations for Markov Stochastic Process.* 1990, 192 pp. ISBN 0-7923-0143-9

K. Sobczyk: *Stochastic Differential Equations. Applications in Physics, Engineering and Mechanics.* 1991, 410 pp. ISBN 0-7923-0339-3

G. Dallaglio, S. Kotz and G. Salinetti: *Distributions with Given Marginals.* 1991, 300 pp. ISBN 0-7923-1156-6

A.V. Skorohod: *Random Processes with Independent Increments.* 1991, 280 pp. ISBN 0-7923-0340-7

L. Saulis and V.A. Statulevicius: *Limit Theorems for Large Deviations.* 1991, 232 pp. ISBN 0-7923-1475-1

A.N. Shiryaev (ed.): *Selected Works of A.N. Kolmogorov*, Vol. 2: *Probability Theory and Mathematical Statistics.* 1992, 598 pp. ISBN 90-277-2795-X

Yu.I. Neimark and P.S. Landa: *Stochastic and Chaotic Oscillations.* 1992, 502 pp. ISBN 0-7923-1530-8

Y. Sakamoto: *Categorical Data Analysis by AIC.* 1992, 260 pp. ISBN 0-7923-1429-8

Lin Zhengyan and Lu Zhuarong: *Strong Limit Theorems.* 1992, 200 pp. ISBN 0-7923-1798-0

J. Galambos and I. Katai (eds.): *Probability Theory and Applications.* 1992, 350 pp. ISBN 0-7923-1992-2

N. Bellomo, Z. Brzezniak and L.M. de Socio: *Nonlinear Stochastic Evolution Problems in Applied Sciences.* 1992, 220 pp. ISBN 0-7923-2042-5

A.K. Gupta and T. Varga: *Elliptically Contoured Models in Statistics.* 1993, 328 pp. ISBN 0-7923-2115-4

B.E. Brodsky and B.S. Darkhovsky: *Nonparametric Methods in Change-Point Problems.* 1993, 210 pp. ISBN 0-7923-2122-7

V.G. Voinov and M.S. Nikulin: *Unbiased Estimators and Their Applications.* Volume 1: *Univariate Case.* 1993, 522 pp. ISBN 0-7923-2382-3

V.S. Koroljuk and Yu.V. Borovskich: *Theory of* U-*Statistics.* 1993, 552 pp. ISBN 0-7923-2608-3

A.P. Godbole and S.G. Papastavridis (eds.): *Runs and Patterns in Probability: Selected Papers.* 1994, 358 pp. ISBN 0-7923-2834-5

Yu. Kutoyants: *Identification of Dynamical Systems with Small Noise.* 1994, 298 pp. ISBN 0-7923-3053-6

M.A. Lifshits: *Gaussian Random Functions.* 1995, 346 pp. ISBN 0-7923-3385-3

M.M. Rao: *Stochastic Processes: General Theory.* 1995, 635 pp. ISBN 0-7923-3725-5

Yu.A. Rozanov: *Probability Theory, Random Processes and Mathematical Statistics.* 1995, 267 pp. ISBN 0-7923-3764-6

L. Zhengyan and L. Chuanrong: *Limit Theory for Mixing Dependent Random Variables.* 1996, 426 pp. ISBN 0-7923-4219-4

GPSR Compliance

The European Union's (EU) General Product Safety Regulation (GPSR) is a set of rules that requires consumer products to be safe and our obligations to ensure this.

If you have any concerns about our products, you can contact us on ProductSafety@springernature.com

In case Publisher is established outside the EU, the EU authorized representative is:

Springer Nature Customer Service Center GmbH
Europaplatz 3
69115 Heidelberg, Germany

Batch number: 10583998

Printed by Printforce, the Netherlands